(Frontispiece.)

A striking instance of adaptive behavior. The photographs are scenes from moving pictures taken of a rat while transferring a dish of food over a wall to an enclosed space it had been trained to leave in reaching the food place. Since the rat was reluctant about feeding in the exposed area, it overcame the difficulty in the above manner. (*From Hamilton and Ballachey*, 1934. *Jour. Genet. Psychol.*, 45, *p.* 261.)

PRINCIPLES

OF

ANIMAL PSYCHOLOGY

BY

N. R. F. MAIER, Ph.D.

Professor of Psychology,
University of Michigan

AND

T. C. SCHNEIRLA, Sc.D.

Curator, Department of Animal Behavior,
The American Museum of Natural History

DOVER PUBLICATIONS, INC.
New York

Published in the United Kingdom by Constable and Company Limited, 10 Orange Street, London, W.C.2.

This new Dover edition, first published in 1964, is an enlarged and corrected republication of the work first published by the McGraw-Hill Book Company, Inc., in 1935. A new introduction and list of supplementary references have been prepared by the authors especially for this Dover edition. Also included are five supplementary articles never before collected in book form.

Standard Book Number: 486-61120-5
Library of Congress Catalog Card Number: 64-15514

Manufactured in the United States of America

Dover Publications, Inc.
180 Varick Street
New York 14, N.Y.

To

L. W. S. AND A. A. M.

PREFACE TO DOVER EDITION

Since the early 1930's, interest in problems in the psychology and behavior of animals has increased and broadened greatly. The continued demand for this book warrants reissuing it in a form adapted to a wider group of readers. The chapters in the Supplement are added for a contemporary discussion and for current references on problems raised in the book itself.

Organization of the book along phyletic lines emphasizes the importance of an evolutionary approach to the psychological and behavioral study of animals. No linear arrangement of behavior types was advanced; rather, discussions of the evidence led to the recognition of qualitative as well as quantitative differences in the organization and underlying abilities of major adaptive patterns. The doctrine of psychological levels represents a concept basic to comparative psychology, often neglected by those who would apply a single general formula to the adaptive behavior of animals. We need to carry out comparisons of phyletic types as well as of more closely related groups. Also, thoroughgoing comparisons are needed of the abilities and organization underlying behavior, based upon analyses of the interrelationships among these properties, as against the mere listing of data and behavioral descriptions that often passes as comparative psychology.

We favor theory as a tool and not as an immutable end in itself. The classical distinction between the innate and the acquired, the unlearned and the learned, has become more questionable than ever in the face of mounting evidence on behavioral development. In Articles I and II of the Supplement, therefore, the concepts of "maturation" and "experience" are extended into new applications for the ontogeny of behavior. Ethological terms such as "imprinting," IRM, and "releasers," however convenient they often seem to be, are still essentially hypothetical, do not emphasize developmental comparisons among the major animal types, and are phenomenalistic rather than calling attention to the quantitative aspects and range of early behavioral ontogeny. The last point is the aim of the concepts of "approach-fixation" and "withdrawal-fixation" advanced by us in Article I of the Supplement.

A survey of our 1935 Subject Index brings out important qualitative contrasts as suggested by terms such as "sensory integration," "perception," "variability," and "plasticity." These ways of emphasizing qualitative differences among the psychological resources of different animal types we find supported by subsequent developments, as regards learning theory, for example. With respect to learning, our 1935 terms "contiguity" and "selection," extended in our 1942 article (Article III in the Supplement) and elsewhere, find strong present-day support. The fact that items such as food and shock can have dual roles in learning, serving according to conditions either as reinforcing or as selective agents, deserves increased emphasis. We would extend our differentiation of *ability* and of *motivational factors* in behavior study, to distinguish between what is learned and what motivationally selected, against the questionable practice of merging motivational factors into learning as presumed stimulus-excitatory agents.

Our emphasis upon the need for analytically studying situational differences is borne out, for example, by a contrast of *contiguity*, or "classical" conditioning methods in which a specific unconditioned stimulus is used to control the critical response, with methods employing the maze or problem box in which no such means is used to control response. Or, as another example, the methods of Köhler and of Adams permit the influence of insight whereas the methods of Guthrie and Harlow exclude opportunities for insight. In Articles III and V of the Supplement we distinguish the importance of such differences both for results and for theory in the study of learning. In these articles we also suggest a resolution of the problem of S-S *vs.* S-R bonds, which from Hull's notions of conditioning and Tolman's cognitive views might be considered an issue in pure conditioning.

Unnecessary conflicts in learning theory are propagated when one specific type of test situation is emphasized to the exclusion of others. Properly to stress differences that arise in learning according to the prevalent test situation, we suggest functional terms such as "contiguity" and "selectivity" rather than the conventional terms "classical" and "instrumental" conditioning, as the former is historical only and the latter carries insufficient emphasis on functional differentiation. Emphasis upon operations, although convenient for reporting results, often obscures what effects the particular test situation has upon the animal. Thus the tendency of Skinner operationists to overlook the effects of their planful controls over the subject's behavior during an elicited sequence of responses in lever-pressing or teaching-machine situations turns attention from processes of selectivity and of organization in learning in favor of an emphasis upon automatization.

Biologists find existing animals to be an incomplete series of numerous ramifications in which transitional ancestral types have been widely eliminated through natural selection. The psychologist, who must therefore study animal types as he finds them and must rank them from his evidence, must also expect to find differing complexities and patterns of variables intervening between structure and behavior on each level. From experience, we know what difficulties may be anticipated in relating behavioral organization and capacities to morphological turning points, for at progressively higher psychological levels the relationship of structure to behavior becomes increasingly devious, especially through the intervention of new variables.

The comparative approach to behavior study in man and lower animals can become more useful and significant through specific techniques such as those employing drugs and biochemical agents. As another example, the use of implanted electrodes to apply intraneural stimulative effects at different levels has widening potentialities for the investigation of developmental and social phenomena. It is to be hoped that such potentialities can be realized through adequate emphasis upon training the researchers in the principles and disciplines of comparative psychology. Through use of the comparative methodology, psychological differences as well as similarities between man and lower animals may become known.

From the stage of behavior modification through simple tissue adaptation in the lowest animals, great advances are found in modifying behavior, first through contiguity and then also through selective learning. Our qualitative distinction between learning and reasoning, the latter a capacity to solve problems by combining previously disparate experiences, now finds additional support through research on animal problem reactions and on creativity in man. In man, this last capacity advances beyond the plasticity evidenced by certain other mammals in their selective learning and perhaps also in their reasoning to a stage of increasing novelty of solution gained through an ability to accomplish the *fragmentation* of entire learned patterns and thereby to reorganize and recombine the results of experience in a creative manner.

Comparative investigation and study of behavioral development and capacities on different animal levels have increasingly stimulated investigations on human capacities and personality. One important reason for encouraging such a discipline is to understand each animal in its own psychological terms; another is to know what similarities and what differences exist among animals including man. As a further ideal, the effective use of the comparative method in basic studies of animal, child, and social psychology is certain to advance the conceptual development of all these fields and thus benefit zoology, psychology, and science generally.

N. R. F. Maier
T. C. Schneirla

Ann Arbor, Mich.
New York, N.Y.
December, 1963

PREFACE TO FIRST EDITION

This work is designed to serve as a systematic textbook of animal behavior for courses in psychology and biology. Part I covers the behavior of animals below the mammals, a comparative treatment in which we have endeavored not only to characterize each important animal type but also to work out certain fundamental principles of animal adjustment. In Parts II and III these principles are developed further as the major problems of animal psychology are attacked in connection with the behavior of mammals.

This organization of the parts results from an attempt to represent the manner in which the subject has developed experimentally, and also to suggest a solution of the frequent conflict between the consideration of animal behavior and the treatment of psychological problems.

It is hoped that a systematic treatment of animal behavior will stimulate interest in comparative psychology by serving as the foundation upon which human psychology must rest. Many problems which cannot be adequately analyzed in human subjects may be successfully attacked when lower animals are used as subjects. The volume of research in animal psychology has been increasing rapidly in recent years as science has come to recognize the fundamental importance of applying concepts to the entire series of living organisms. Introductory texts in general psychology express this advance by incorporating more evidence from studies on infra-human animals. It is probable that in the near future a training in psychology which does not include a study of lower animals will be regarded as inadequate.

At the end of each chapter a short list of suggested readings is appended, to indicate some general treatments of important topics from sources available in most libraries. In order to keep the general bibliography within bounds, only those titles are included which are given reference in the textbook.

Grateful acknowledgment is due to Professors B. D. Thuma and G. B. Vetter, for their valuable criticisms of the whole manuscript. Drs. Q. F. Curtis, O. C. Ingebritsen, K. C. Pratt, and others, have been very helpful in their criticisms of parts of the work. The dedication expresses only in a small way our appreciation for the valuable assistance which our wives have given.

N. R. F. MAIER.
T. C. SCHNEIRLA.

ANN ARBOR, MICH.,
NEW YORK, N. Y.,
October, 1935.

CONTENTS

PAGE

SUPPLEMENT TO DOVER EDITION

PRINCIPLES OF ANIMAL PSYCHOLOGY

GENERAL INTRODUCTION

The Nature of Organization in the Living Individual.—The purpose of this volume is to arrive at an understanding of the principles which underlie the adjustment of animals, as living organisms, to their environments. Our approach to the subject is based upon the following interpretation of the fundamental concepts "life" and "organism." *Life* may be defined as that condition in which a body demonstrates the functions *irritability*, *motility*, and *reproductivity*. The *organism* or living individual is that aggregation of elements which displays the functions of life in a self-consistent manner. The physiologist finds that the parts of an organism are capable of functioning together in a well-integrated manner; and the psychologist finds this organization expressed in behavior, in the activities of the organism as it adjusts to its environment.

The unity and organization in the organism are the key to its behavior as an individual. Our first general task is to consider, for each outstanding type of animal, the evidence which is essential to an understanding of the manner in which adaptation to environmental conditions is effected. The "psychological standing" of a given animal is determined by the degree to which its total organization of activities has progressed and by the general nature of this organization in connection with the overcoming of difficulties which are encountered in its adjustment to surrounding conditions.

The "interesting things which animals do" are fascinating material for study but have been greatly misinterpreted by some who have been content merely to describe the acts and to draw from this superficial treatment certain general conclusions as to their nature. This applies particularly to stories of the performances of domestic animals, which the proud owners usually are loath to evaluate in the light of all the evidence and in comparison with the abilities of other animals. A psychological understanding of such abilities is to be gained only through a careful examination of the evidence in the light of a critical and sound judgment gained in a study of behavior principles. If the path seems to be difficult at first, the student may assure himself that the game is well worth the candle.

Major Problems and Order of Treatment.—In a study of the adaptation of animals to their environments, it is necessary (1) to analyze the behavior of characteristic animal types in relation to their structural equipment; (2) to compare and to contrast the behavior of the various types of animal; (3) to trace the development of complexity in adaptation through various animal groups; (4) to analyze certain problems systematically, without special consideration to animal forms; and (5) to study individual variations in behavior.

Part I represents a survey of the major types of animal which stand below the mammals in their psychological position, and that part of the book attacks problem (1) in particular, as well as problems (2) and (3). Part II prepares the way for Part III, in which we find it possible to treat problem (4) and problem (5) in studying the behavior of the highest animals, the mammals. This procedure is made the most profitable one, and the necessary one, by the fact that psychologists in their experimental investigations of mammals have directed their efforts primarily toward the solution of certain problems and not toward the understanding of a given animal as such. This fact is probably attributable to a natural tendency to view results in terms of their eventual application to an understanding of the psychological propensities of man. When man compares himself with other animals, he makes the basis of evaluation one which throws into relief his own admirable superiorities and advantages.

The order in which the major animal groups are treated here is dependent upon their respective psychological positions as indicated by the available evidence. For instance, the organization of activities in the behavior of the worm shows certain important advances over that of the starfish, and accordingly worms follow echinoderms in our treatment, although often placed before them in zoological classification. The mammals are treated last of all because it is in their behavior that the problems of animal psychology reach their most complex form. The study of lower animals gives us a keen appreciation of the basic principles upon which complex behavior rests in higher groups. This prepares us for a more fundamental understanding of the most involved psychological organization, that of the mammal, which otherwise defies adequate analysis.

The Selection of Material for Study.—In a survey such as the present one, it would not be possible to treat directly in a comprehensive manner the great mass of experimental literature which is available in this branch of psychology. Therefore we have selected representative work and have dealt with the results which are most essential to a systematic study of animal adjustment. Typical experiments have been given fairly detailed treatment, so that the student may form a critical attitude while he is gaining a knowledge of essential methods and types of evidence. Because of limited space, many important experiments and much valuable evidence could not be included in the present survey, and at times the selection of experimental material for consideration has been forced to a somewhat arbitrary basis. We believe, however, that the inclusion of additional material would verify in further detail the principles arrived at from the evidence presented.

Since a constructive treatment is desired, evidence has been critically dealt with in a negative way only when such procedure contributes to an understanding of method, or when the evidence concerns an important controversial point. The lessons taught by inadequate experiments are better learned from the study of other contributions which compensate for the mistakes.

The specialists will note that certain topics have been wholly left out or briefly treated. Some matters have seemed either unessential for a treatment of basic principles, or too little known to contribute substantially to a study of psychological development. However, problems such as these usually will be found dealt with in other ways. For instance, although the phenomenon of bird song is not given special attention, similar issues are taken up in our study of the development of the pecking response in the chick.

Similarly, the book does not include separate chapters on the structures of animals, on methodology, and on history. Each of these topics is related in some specific way to each major problem in behavior, and is so treated. Structures are considered in connection with the behavior phenomena which they underlie, and new structures are described only when their importance for behavior can be demonstrated as essential to the psychology of the animal concerned.

Although methodology is extremely important, it is questionable whether a general method can be outlined for the entire subject, since the method employed depends upon the animal under investigation and upon the nature of the theory which is being tested. "Scientific method" as a general subject constitutes a logical method for arriving at knowledge. Qualitatively it is not different from any other kind of fact gathering. A method that leads to new and dependable information should be used in the investigation of a given problem, but what method this is can be determined only from experience and thought. "Training in science" teaches one where pitfalls are likely to be found, but the nature of the pitfalls and the special means of avoiding them depend upon the problem and the animal under investigation. New methods grow out of new concepts, and no general framework for these can be outlined *a priori*.

As for a special study of history, the lessons which it teaches are largely if not entirely embodied in a study of the contemporary condition of the science. Accordingly, special reference is made to historical events only when they have some special significance for the understanding of a problem, and reference is then made in connection with the problem for which the event is of particular importance. Those who are interested in history as a special subject are referred to separate treatments such as the excellent one offered by Warden, Jenkins, and Warner (1934).

The Logical Treatment of Evidence.—Each phenomenon in animal behavior represents a problem. The comparative psychologist first attempts to understand the causal factors which contribute to a given phenomenon. To take a simple example, an animal may work for its food at night mainly because of the nature of its sensitivity to temperature and to light, and because it is able to learn that its slow locomotion makes food-capture impossible in the daytime. The experimenter may determine the nature of the causal factors by observing the phenomenon in the animal's typical environment (*i.e.*, its "habitat") and by investigating it under laboratory conditions. It is to be hoped that the future will bring more studies of animals under field conditions, as a guide to better laboratory investigations. It must be kept in mind that an experiment is also an observation, but one in which the use of apparatus and of more adequate control of conditions than is possible in the field permits a thorough investigation of factors which underlie a phenomenon. By observation and by experiment the scientist obtains evidence, which must then be interpreted to improve understanding of the phenomenon and to guide further investigation.

The reader should have the opportunity to evaluate the logical basis on which the authors have studied and organized the evidence.

The investigation of animal behavior is of the greatest value to the science of psychology because it yields an abundance of facts concerning behavior adjustments and permits the thorough testing of logical interpretations of these facts. The growth of any science is dependent upon the maturity and quality of its theories. Hence our selection of evidence for treatment has been largely determined by a consideration of its importance to systematic interpretation. This makes it necessary to state our attitude toward theory.

Once the psychologist has studied the variety of phenomena which are presented in the behavior of different types of animal, and understands the nature of the causal factors in the respective cases, he is equipped to formulate a general interpretation which constitutes his theory of animal behavior. Since a theory is an interpretation of facts, it follows that when facts are patterned to form a unified account of a phenomenon a theory results. When items of evidence (*i.e.*, facts) are organized into patterns, the items take on meaning. It is this meaning which makes the evidence interesting to students of the subject. Because the "meaning" of facts depends upon the theory which is developed from them, different theories impart different meanings to the same facts. The student then must choose that one among the alternatives which teaches him most about the phenomenon.

One criterion is of the greatest importance. A good theory must stand the test of consistency. For this reason, in case each of several alternative theories appears to satisfy this test in explaining a given phenomenon, the value of all of the theories is necessarily decreased. When such cases arise scientists work with greater energy to obtain more facts, since further evidence is certain to disclose inconsistencies among the alternative lines of explanation and hence to show which is the most adequate.

The importance of theorizing to science is evident. The theory produces a number of *hypotheses*, *i.e.*, assumptions which necessarily follow from it. These must be subjected to experimental test. In creating hypotheses, the theory presents suggestions for research, and thus in one sense may be evaluated in terms of the amount of further investigation which is inspired by it. The scientist, motivated by interest in the way his hypotheses lead to the discovery of facts, will endeavor to make his theory adequate and contributive. As the scientific study of a given phenomenon progresses the verification of certain hypotheses broadens knowledge and enhances the value of some one among the alternative theoretical explanations. In this manner the science ascertains the most satisfactory theory and eliminates the unsatisfactory alternatives. A theory which cannot be tested naturally fails in this respect and is unsatisfactory for pragmatic reasons.[1]

When a science has matured in its theories intelligent prediction of further advances is possible. It is then that the science may make a distinct social contribution. Immature sciences may contribute to practical problems, but their contributions are necessarily limited and specific in nature. The emphasis of practical application at the expense of theoretical contributions is the long route to scientific progress and therefore to social progress. It is of course true that an experiment designed for direct and "practical" purposes may be of some immediate value, but unless the investigation has been planned upon the broad basis of theoretical understanding the results are almost certain to have a very limited scope of application.

These statements appear justified by the manner in which the subject of animal psychology has developed. Judging from the number of highly motivated scientists who are working in this field, and the quality of work which is being performed, the future of our subject is a very promising one indeed.

[1] This would appear to be the reason why contemporary psychology has largely discarded the concept "mind" as the basis for a theoretical explanation of the phenomenon of "awareness." Since psychology has not yet reached the stage at which a satisfactory theory of awareness can be formulated (Lashley, 1923), there is nothing to be gained from a discussion of whether given animals are "conscious." To illustrate the point in another way, science has outgrown mystical theories such as "vitalism," because they subordinate the critical treatment of facts to the emotionally biased attitude of the observer, which is all too frequently traceable to an early training in the doctrines of superstition. For similar reasons, an exaggerated and almost religious attention to "viewpoints" in contemporary psychology has actually delayed the advance of the science.

PART I

PRINCIPLES DISCERNIBLE IN THE BEHAVIOR OF INFRAMAMMALIAN ANIMALS

The "organism" is the living individual, and the activities of the organism constitute its behavior. The environment of an individual being is that totality of energy changes which may stimulate it and influence its behavior.

In the lowest organisms behavior is dependent upon the nature of the structural equipment possessed by an individual. In consequence the lower animal may be regarded as essentially a creature of the environment in which it is able to live. Survival is possible only if that environment elicits responses which tend to perpetuate the animal. The appearance of new structures which make possible different lines of behavior increases the chances that the animal type will become better adapted to its environment, unless the appearance of unfavorable characters leads to its extinction.

The reason for the possession of structural equipment which makes adaptation (fitness for the environment) possible lies in the history of the race, *i.e.*, in the evolution of the given type of animal, and is not to be found in a study of the individual member of the species. We should not be tempted to take the easy-way-out of explaining behavior in terms of the end it attains, a special and hybrid form of teleology, unless the facts warrant such an interpretation. For instance, we cannot say that the "scallop flees from the starfish because it recognizes an enemy which will devour it." Rather, we know only that the starfish presents stimuli which act upon the scallop in a given manner and produce "flight" as a response. Actually, it is only among the highest animals that the individual is able to anticipate the results of its actions and to regulate its behavior accordingly. In any case teleological explanations cannot be adopted in advance, but each animal form must be described as the evidence warrants and dictates.

The environment of the lowest animals plays upon their equipment in a manner which is directly responsible for the nature of their behavior. Since the "environment" of a given animal depends upon the nature of the stimuli to which it is sensitive, animals will have different environments according to differences in their equipment of receptors. In our study of successively higher animal groups, we shall find that advance in the scope of sensitivity makes possible other developments of great importance for the adequacy of adjustment and hence for the psychological standing of various animal types.

The environment furnishes energy which acts upon a *receptor* in the animal, and sets into effect an excitation which is conducted by a *nervous system* or some other transmissive mechanism to parts capable of action, the *effectors*. We must first work out the fundamental properties of behavior for animals in which these three agencies are not differentiated or specialized to a great extent. In the lowest animals the nervous system serves as a simple bridge which connects receptors with action structures. The nature of behavior in these animals therefore expresses the stimulus in a fairly direct manner. Higher in the series, however, the receptors become more specialized and the environment is correspondingly widened. Then it is the increase in the structural complexity of the nervous system which makes possible an advance in the complexity of the relationship between the animal and its environment.

In Part I we shall trace the manner in which the basic behavior mechanisms expand and become supplemented by new abilities. For the inframammalian animals the chief task of the comparative psychologist is to characterize each principal type in terms of the properties which are essential to its adjustment. If clearly distinguished in animals that stand low in the psychological scale, the discovered principles will greatly assist the student in grasping more difficult phases of the subject in which specialization introduces greater complexity into behavior.

CHAPTER I

FUNDAMENTAL CHARACTERISTICS OF BEHAVIOR: PLANTS AND PROTISTA[1]

INTRODUCTION: THE SEED PLANT AS A BEHAVING ORGANISM

Growth Responses.—The Spermatophyta or seed plants are distinguished from animals by their lack of locomotion (and by their

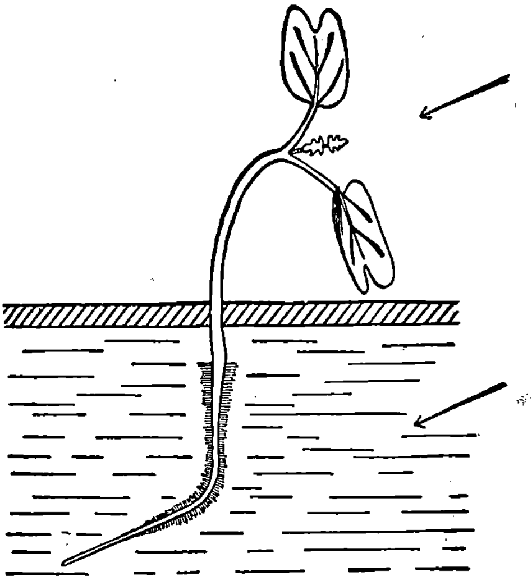

FIG. 1.—Growth responses of a plant to light. The stem has bent toward the light, while the root has bent away from it and the leaves are turned at right angles to the direction of its rays. (*From Haupt.*)

possession of chlorophyll), but they are nevertheless behaving organisms, since they demonstrate irritability and respond to energy

[1] Among the *Metaphyta* (multicellular plants) the phylum *Spermatophyta* (seed plants) is most important for the present subject. The superphylum *Protista* includes the *Protophyta* (single-celled plants, such as bacteria, desmids, algae) and the phylum *Protozoa* (single-celled animals). Phylum *Protozoa* includes the following classes: *Rhizopoda* (ameboid form, pseudopods characteristically present, Fig. 2); *Flagellata* (commonly with flagella, Fig. 5); and *Ciliata* (ciliated at some life stage, Fig. 7).

changes in their environment. "Growth responses" are observable in all seed plants, and are brought about by effects which alter the physiology of the tissues. The rooted plant typically bends toward light which shines upon it from one side. This response depends upon a reduction of growth rate on the more intensely illuminated side, and an increased growth of tissues on the darkened side of the stalk (Fig. 1). The position which is assumed by the stalk, leaves,

TABLE 1.—PLANTS AND PROTISTA

RECEPTOR EQUIPMENT	FEATURES OF SENSITIVITY
Chemical. No specialized receptors. In protozoa, temporary or permanent anterior end most sensitive.	*Metaphyta:* Leaves, growing parts affected by gases in air; roots by chemicals in soil. *Protista:* Withdrawal from strong anterior stimulation, approach to weak stimulation.
Mechanical. *Metaphyta:* Ectoderm generally sensitive to contact; papillae (protuberances) heighten effect in many plants, as do special hairs. *Protista:* General protoplasmic iritability. Cilia (ciliates) and flagella (flagellates) may be considered specialized receptors.	Tendrils of plants coil around roots; roots push way through soil under influence of contact. Special responses (*e.g.*, in insectivorous plants) depend upon touching of irritable parts. *Protista:* contact closely related to chemical stimulation in feeding; effects probably the same. Pseudopod formation of rhizopod controlled by contact, as is the "backing reaction" of ciliates and flagellates.
Light. *Metaphyta:* epidermal layers light-sensitive; in some, rounded epidermal cells have lenslike function. Chlorophyll (green substance) acted upon by light. *Protista:* Protoplasm directly acted upon by light. Spots of light-sensitive pigment in some flagellates.	Photosynthesis (action of light on chlorophyll) important for metabolism, the basis of nutrition and growth. Light controls turgor responses such as "day-night" movements, and the direction of growth. In *Protista* light usually controls movement: controls protoplasmic flow in *Amoeba*, whipping of flagellum in flagellates.

Other Effects. Starch grains in roots of certain plants press differently according to position of root, thus influencing movements. Pressure of fluids in roots and in stems acts in similar manner to control position. Temperature exerts a general effect on protoplasmic activities (p. 23), controlling rate of reproduction, locomotion, and physiological processes. Optimum differs among groups, *e.g.*, near 22°C. for *Amoeba proteus*.

CONDUCTION

Metaphyta: Passage of fluids from one region to another; or sap stream may carry chemical from stimulated zone to another locality, thus influencing response.

Protista: Energy conducted through open protoplasm from the stimulated point, as difference in electrical potential at interfacial surfaces of protoplasmic globules. In ciliates a "neuromotor" apparatus described: protoplasmic fibrils leading from ciliated regions to specially contractile protoplasm (*e.g.*, retractor strands, membranelles).

MOVEMENT AND ACTION

Metaphyta: slow position changes through growth (stalk-bending), or more rapid turgor movements, *e.g.*, closing of leaves at night.

Protista: Pseudopodia formed through open protoplasm in some rhizopods, through opening in a shell in others. Permanent and specially contractile pseudopodia in some. Rhythmically coordinated beating of cilia in the ciliate movement; local ciliary action carries particles into gullet in feeding. Whipping of flagellum pulls the flagellate forward, carries particles into gullet.

and other growing parts of the plant depends upon the manner in which stimulating agents such as light, moisture and gases in the air, and temperature changes, act upon the tissues. Mechanical stimuli (contact with objects) and internal changes (fluid movements attributable to "gravity" and to growth) cause many types of movement. A tendril coils around a rod because growth is reduced on the side which touches the surface; a horizontally held root turns downward because of pressure and growth changes within it.

Turgor Responses.—Some rooted plants are capable of local movements which occur more rapidly than is possible through growth changes. Such activities, attributable to rapid changes in the fluid content of cells, are called *turgor movements*. In the "Venus's-fly-trap" *Dionaea*, for instance, the closure of the two halves of a barb-edged leaf when the bristles upon it are touched[1] is attributable to a rapid increase in the fluid content of cells on the under side of the leaf. In this way the plant entraps insects which alight upon its leaves, the prey being digested by secretions within the closed leaf before the leaf is slowly opened by further growth. The leaves of the "sensitive plant" *Mimosa* suddenly droop when the plant is touched, or when illumination is decreased in intensity. Normally a given leaf group is held erect by the pulvinus, a little cushion of cells situated beneath the base of the main stem; but many stimuli cause the pulvinus cells to lose fluid to surrounding cells, so that the leaf cluster is not supported. Such responses, as well as slower movements, may be elicited by means of the rapid intercellular passage of fluids, a means by which excitation is conducted from the region first affected to other parts of the organism. Devices capable of more rapid conduction of excitation have been described for certain plants (Bose, 1926).

Plants Exhibit the Chief Properties of Behavior.—Seed plants are therefore to be considered behaving organisms, since they display varying degrees and types of sensitivity, conduction of excitation from one part of the individual to another, and the ability to respond in various ways to energy. Understanding of phenomena such as those cited increases with knowledge of the organic processes set up in the living substance of the plant by external energy. Certain fundamental problems have been raised here which may be examined to greater advantage in the behavior of animals.

[1] To produce closure of the leaf, the tactual stimulus must be presented by an object which contains nitrogen in some form.

A Basic Principle of Adjustment

The "Optimum" Illustrated in the Behavior of Bacteria.—Organisms in general make most readily those responses which tend to promote life activities; that is to say, behavior is fundamentally adaptive. The *principle of the optimum* is basic to all adaptation. The optimum is that environmental condition (*e.g.*, a given temperature) which best promotes the representative physiological processes (*i.e.*, the metabolism) of the organism. In conditions above or below this value the normal metabolism of the organism is interfered with, and general behavior is consequently changed. The manner in which a representative protophyte, the bacterium, adjusts to its environment, illustrates this principle nicely. We may quote the following typical observation from Jennings (1906*a*).

> If we place water containing many Spirilla on a slide, allowing some small air bubbles to remain beneath the cover-glass, we find after a time that the bacteria are collecting about the bubbles. The course of events in forming the collections is seen to be as follows: At first the Spirilla are scattered uniformly, swimming in all directions. They pass close to the air bubble without change in the movements. But gradually the oxygen throughout the preparation becomes used up, while from the air bubble oxygen diffuses into the water. After a time therefore the bubble must be conceived as surrounded by a zone of water impregnated with oxygen. Now the bacteria begin to collect about the bubble. They do not change their direction of movement and swim straight toward the center of diffusion of the oxygen. On the contrary the movement continues in all directions as before. A Spirillum swimming close to the bubble into the oxygenated zone does not at first change its movement in the least. It swims across the zone until it reaches the other side, where it would again pass out into the water containing no oxygen. Here the reaction occurs; the organism reverses its movement and swims in the opposite direction. . . . This is continued, the direction of movement being reversed as often as the organism comes to the outer boundary of the zone of oxygen within which it is swimming. . . . As a result of this way of acting the bacterium of course remains in the oxygenated area. . . . The finding of the oxygen then depends upon the usual movements of the bacteria—not upon movements specially set in operation or directed by the oxygen (p. 29).

In this situation, the circular zone of water around the bubble represents the optimum for Spirillum, since it contains oxygen, an essential ingredient in the characteristic protoplasmic processes (metabolism) of these bacteria. Hence the movement of the bacteria is not interrupted as it carries them into the zone. When a given individual within the oxygenated area swims partly into the surrounding water, which contains less oxygen, there occurs a disturbance of protoplasmic activities which alters swimming activity so that the direction of progress is changed. Hence a collection of bacteria forms within the oxygen-saturated zone. This has been called the *trap phenomenon.*

The basis of this phenomenon is the fact that the individual bacterium of this species inherits a type of protoplasm[1] the balanced activity of which is interfered with in deoxygenated water, but not in an oxygenated medium. Other types of bacteria (anaerobic) are killed by oxygen, and such bacteria collect in unoxygenated water, since their forward locomotion is interfered with *only* when oxygenated water is entered. *It is therefore the protoplasmic constitution of an organism which determines its optimum, and hence primarily governs the nature of its adaptive responses.*

Principles of Behavior in a Representative Protozoan

Amoeba Best Known among the Protozoa.—Nearly 9,000 species of the phylum *Protozoa,* living in a great variety of environments (*e.g.*, fresh and salt water; in the bodies of animals), present a multiplicity of behavior types. Most *Protozoa* are microscopic in size, but for animal behavior, size is often a feature of slight importance in comparison with the manner in which the parts of the organism cooperate in the adjustment of the whole to its environment. The *Amoeba*[2] serves as an excellent representative of the phylum, particularly because of the extent to which it has been investigated.

The protoplasm of *Amoeba* is differentiated into a *pellicle* (the outermost layer, thin and tough); the *plasmagel layer* (a thickened wall of fairly solid but elastic protoplasm) lying next within; and an inner mass of *plasmasol* (relatively fluid protoplasm) which contains the nucleus and other "inclusions" (Fig. 2). The form of the body is subject to constant change. *Amoeba* is an organism, and its parts

[1] Protoplasm is the viscous substance of which living organisms are essentially composed.

[2] *Amoeba* is a member of the class *Rhizopoda* (order *Lobosa*), one of the *Protozoa* characterized by the production of blunt fingerlike protuberances of protoplasm called *pseudopodia.*

function in a well-integrated manner as the animal responds to stimuli. It will be profitable to study the conditions which govern this organization in behavior, since the principles involved in the adjustments of this protozoan are fundamentally the same as those which account for less directly approachable behavior in higher animals.

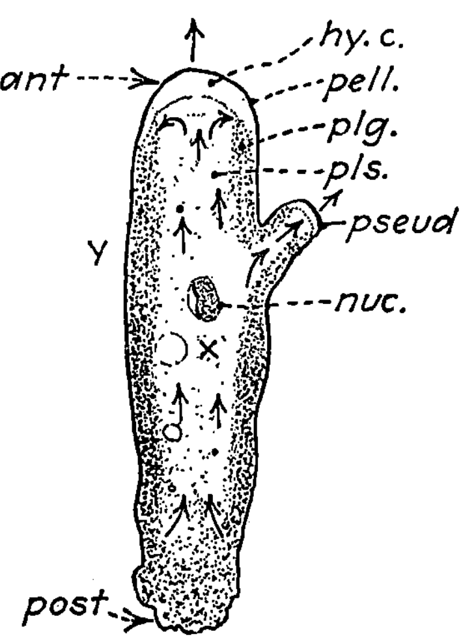

FIG. 2.—Schematic view of a moving *Amoeba* (from above). *ant*, anterior pole at which gelation begins; *post*, posterior pole at which solation begins and plasmasol stream starts to flow forward. The direction of plasmasol streaming is indicated by arrows. In a few moments the portion of the plasmasol stream now at *x* in the interior will have flowed forward to a point opposite *y* in the plasmagel layer, which will then be farther from the anterior end. *Hy. c.*, hyalin cap; *nuc.*, nucleus; *pell.*, pellicle; *plg.*, plasmagel layer; *pls.*, plasmasol layer; *pseud*, newly formed pseudopod. (*Modified from Mast, Pantin.*)

The behavior of *Amoeba* at a given time depends upon three factors: (1) the strength and localization of externally presented stimuli; (2) the internal condition of the organism; and (3) the general nature of the surrounding medium (*e.g.*, its temperature). An adequate understanding of amoeboid movement should be gained first if we are to study the manner in which the above three factors bring about variations in behavior.

An Analysis of Amoeboid Movement.—When an *Amoeba* is lightly touched in one place the wall of the organism bulges somewhat in the stimulated region, and a fingerlike protuberance, a *pseudopodium*, is formed there. Subsequently the entire animal may move in the direction of the stimulated point, with the newly formed pseudopodium in the lead (Fig. 2). The most adequate theory (Hyman, 1917; Pantin, 1924; Mast, 1926) postulates a reversible chemical reaction set into effect by the stimulus, which basically involves changes in the state of protoplasm between the *gel* (relatively solid) and the *sol* (relatively fluid) condition.[1] A weak stimulus sets into effect a

[1] Protoplasm is a colloid, a suspension of particles of one density in a fluid of different density.

for the different effects exerted upon protoplasmic activity by stimuli of different intensities.

Localization Most Important when Intense Stimuli Are Involved. We have established the point that the more intense the stimulus, the more likely it is to change the general direction of locomotion. When strong stimuli are used, therefore, the *locality of application* is a very important factor. An experiment by Hyman (1917) nicely demonstrated this. A needle was applied on each test with sufficient force to pierce the pellicle of the *Amoeba* at a given point, either the "anterior" end, one "side," or the "posterior" end (Fig. 3), with respect to the direction of locomotion. In all cases (1) the stimulated region contracted more or less (depending upon its previous activity), while the "advance end" contracted powerfully so that its surface became very wrinkled. In consequence, the protoplasm flowed toward the middle of the body, which bulged out, and (2) after a delay, new pseudopodia formed opposite the stimulated point and the *Amoeba* moved rapidly away toward this side. Thus a strong stimulus, wherever applied, typically forces movement in which the stimulated locality becomes the "posterior" end. On the other hand, when weaker intensities are used to stimulate, the location of the stimulus often is less important than is the nature of the previous pattern of movement. The second of our three general factors is responsible for this difference.

2. *The Internal Condition of the Organism: The Temporarily Dominant Activity Pattern, a Physiological Gradient.*—Observation of amoeboid locomotion encourages the working conclusion that a given pattern of activity remains in force until it is interfered with by outside stimulation. When an *Amoeba* is moving in a given direction, the temporary "anterior end" may be considered dominant over other portions of the body, since it is evidently from this region that the excitation is conducted which first puts other parts into their subordinate roles. Physiologically, the dominant anterior portion is the most active part of the *Amoeba*, and Hyman (1917) has demonstrated a decrease in the rate of physiological activity from this part to the posterior end. At a given time, then, there exists in *Amoeba* a "physiological gradient" which corresponds in nature to the apparent dominance-subordination relationships of the parts in activity.

Child's (1924) experiments demonstrated for other organisms that killing agents such as dyes or strong acids first affect the parts which are physiologically

change in the colloidal state of the protoplasm, a change which is conducted throughout the organism and which makes possible a movement toward the stimulated side. For an *Amoeba* which is weakly stimulated on one side, the phenomenon may be theoretically described as follows:

1. *Formation of a Pseudopodium in the Stimulated Locality.*—Energy furnished by the stimulus causes the protoplasm in that locality to become more *sol* (*i.e.*, more fluid). An increase in acidity occurs in the vicinity, and there is a local increase in the water content of the protoplasm. This forces a bulge in the outer wall and marks the beginning of a pseudopodium. If the change continues and eventually dominates the entire organism, the process of pseudopod formation becomes true movement.

2. *The Amoeba Becomes a Contractile "Plasmagel Sac."*—The *Amoeba* now may be described as an elongated sac, a *plasmagel sac* filled with fluid, the "open" end of the sac (Fig. 2) having been made very *sol* by the stimulus. As fluid protoplasm reaches the anterior clear region, carried by a central current which has formed, it is deflected to the walls of the pseudopod. There it becomes *gel* (more solid), and builds onto the "rim" of the plasmagel sac. As the anterior end advances, more protoplasm is constantly added to the plasmagel rim in this way. A given amount of substance thus added to the wall in time comes to lie closer and closer to the "posterior" end of the *Amoeba*, not because this substance actually moves, but because plasmasol flows forward through the tube of which it forms a part. The closer this plasmagel substance lies to the posterior end the more water it loses, and hence the more *gel* (solid) it becomes. For this reason the plasmagel sac has the power of contractility, especially near the posterior end, and squeezes its fluid contents forward in the direction of locomotion.

3. *Features Which Complete the Continuous Gel-sol Cycle.*—From the central region near the posterior end a *stream of plasmasol* flows through the central part of the moving *Amoeba* toward the anterior end. This fluid protoplasm apparently is "squeezed" forward by contraction of the plasmagel, possibly is "drawn" forward as well, as the anterior end takes up water from protoplasm in other parts. The strongly contracted protoplasm at the posterior end begins to "melt" in the central part and is then forced forward in the plasmasol stream. Upon reaching the anterior end, this protoplasm gels as it is added to the rim of the plasmagel tube.

The reversible reaction thus set up by the stimulus continues in force until its pattern (*i.e.*, its operation in different parts of the *Amoeba*) is altered by the effect of further stimulation.

Sometimes an *Amoeba* moves for more than a few minutes without undergoing major changes in direction, but typically the form and direction of movement fluctuate constantly (Schwitalla, 1924). As one activity pattern becomes altered, a side of the organism which has served as "anterior" may next become "posterior," and the leading pseudopod may start to shrink and lose its dominance as a new pseudopodium forms. We are now in a position to treat the principal factors which govern the behavior of *Amoeba* at a given time and which are responsible for variations in the movement of this organism.

The Principal Determining Factors in Amoeboid Behavior. 1. *Intensity and Localization of the Stimulus.*—Other things equal, the energy value of the stimulus determines whether the *Amoeba* moves toward or away from the source. Stimuli of *weak intensity* characteristically elicit movement toward the stimulated locality, the so-called "positive response." A floating *Amoeba proteus* commonly "follows up" the pseudopod which first touches the substratum, and a crawling *Amoeba* moves toward a small object which brushes against one side, a response important for food taking (pp. 23*f*.). Edwards (1923) used a capillary pipette to introduce a weakly concentrated chemical close to the *Amoeba*, and typically a pseudopod formed in the locality, followed by movement toward the stimulated side. Schaeffer (1916) reported cases in which individuals moved toward particles of glass without having touched them and judged that the response was elicited by small quantities of light reflected from the glass. As described above, weak stimulation produces movement toward the source by setting into effect a local liquefaction of protoplasm which determines a new pattern of internal activity.

Direction of Response as Dependent upon Stimulus Intensity.—The more intense the stimulus which is applied to *Amoeba* the more does the next movement of the animal diverge in direction from the stimulated locality. Thus far there have been no experiments in which contact or chemical stimuli of graded intensities were applied to *Amoeba* under adequate observational conditions, but Mast (1926) has performed such an experiment with light.[1] Light of a given

[1] Folger (1927) obtained results which led him to conclude that light and contact of equivalent intensities exert qualitatively identical effects upon the protoplasm of *Amoeba*.

intensity was directed momentarily onto the advance end o ing *Amoeba*, and the observer noted any change in the plasmasol or any interruptions or changes that might occur i movement. Five degrees of intensity were employed, rangi I, the weakest, to V, the strongest.

The results were clear-cut. The greater the intensity of t which was used to stimulate the advancing end the more wid was its interruption of the forward flow of plasmasol and the more likely was this flow to reverse in its direction or the reversal to remain in force. The four lowest stimulus intensities interfered with the previous movement in relation to their respective strengths, but even intensity IV was not sufficient to force a general change in locomotion. A still more intense stimulus, degree V, was effective not only in completely stopping forward locomotion, but also in setting up a reversed plasmasol current which continued and made possible general movement in a direction opposite to that taken when the stimulus was applied.

FIG. 3.—Reaction of *Am* to strong tactual stimula In case 1 the stimulus is app to the posterior end, in 2 to side, in 3 to the advancing Stage *b* shows the resulting g eral contraction in each case, c shows the direction of s sequent locomotion. (*Redra from Hyman*, 1917.)

The four weakest intensities of stimulation used by Mast did not interfere with the resumption of forward movement in the original direction, although they did alter the protoplasmic flow to some extent. The response in each of these cases might therefore called a "positive" one, in contrast with the "negative" response the strongest intensity, V. But one doubts the value of suc arbitrary expressions as "positive," since they poorly express th important fact that the effect of a stimulus depends upon its intensit (other things equal). The "positive" and the "negative" i amoeboid behavior do not represent a mysterious ability on the par of the organism to diagnose the "beneficiality" of the stimulus but are only the crude and arbitrary designations of some observers

the most active, next affect less active parts, and so on. Hyman (1917) introduced such toxic agents as potassium cyanide into the medium of a moving *Amoeba*. The advancing end was first broken down, and the disintegration of protoplasm progressed regularly from that part toward the posterior end of the organism. If several pseudopodia existed at the time, the most recent and active one of them was first affected, and others were attacked in the order of their original formation.

The anterior or dominant end of the advancing *Amoeba* may be considered a temporarily specialized receptor, inasmuch as stimuli which are too weak in strength to be effective elsewhere on the body bring observable changes in movement when applied to this portion. For the same reason the tip of a pseudopod is more sensitive than is its base (Verworn, 1889). In the same sense the plasmagel sac may be considered a temporarily specialized contractile agent, which is dominated by protoplasmic changes conducted from the anterior end.

To repeat, the nature of the existing physiological gradient (the pattern of activity which prevails at a given time) is of importance in determining the effect of new stimuli. Weak stimuli may change the direction of movement somewhat when presented to the sensitive anterior end, but prove much less effective (or are ineffective) when presented to other parts of the organism which are subordinate at the time. Stronger stimuli are relatively more effective in altering the prevailing pattern of activity.

Weak Stimuli May Become Effective through Summation.—A stimulus may be too weak to alter general behavior, but may nevertheless set up protoplasmic changes which summate with the energy effects of stimuli presented shortly afterward. For instance, a swimming diatom (a protophyte) may brush against the right side of an *Amoeba's* principal pseudopod, without detectably changing the movement of the animal. But when immediately afterward the *Amoeba* comes into direct contact with the side of a filament, it changes direction somewhat toward the right. The observer may conclude that the weak contact which just preceded the broad frontal contact had left an effect which increased the susceptibility of the right side of the forward end to stimuli presented shortly thereafter. An experiment by Folger (1927) suggests that such occurrences account for much of the variability in the normal behavior of *Amoeba*. A mechanical stimulus (the dropping of a 300-mg. wire through a glass tube to strike one end of the microscope slide) was presented first in a typical trial. Then three seconds later a light was flashed momentarily onto the *Amoeba* (by reflection from the microscope mirror). Table 2 gives representative results for two orders of presentation, and for control tests in which the stimuli were presented at well-separated intervals.

TABLE 2

Manner of presentation	Number of trials	No definite responses
I. Stimuli presented at well-separated intervals:		
Mechanical shock alone	21	2
Sudden illumination alone	5	0
II. One stimulus closely following the other:		
Mechanical shock, followed by sudden illumination.	6	4
Sudden illumination, followed by mechanical shock.	26	16

Neither stimulus used by Folger was sufficiently intense to produce an observable change in behavior when presented alone. However, the fact that such weak stimuli do affect *Amoeba* was shown by the ability of the second stimulus, when it closely followed the first (series II), to produce a response which otherwise could not have been elicited by it. Hence slight protoplasmic changes induced by very weak stimuli may be summated with the effect of stimuli which follow. This fact shows the necessity of taking into account the condition of the organism when diagnosing its behavior.

Adaptation to Continued Stimulation.—The *Amoeba* is subject to adaptation in that a steadily maintained stimulus may come to have a decreased effect. When the protozoan is brought from the dark into bright light (as when the student begins his observation by reflecting strong light onto a microscope slide) there occurs a general contraction of the body which may persist for some time. During this quiescent period (*cf.* p. 18) the *Amoeba* appears incapable of movement, but after a few minutes the protoplasm begins to flow again and soon locomotion is resumed. At this time the light, which at first elicited a definite response, evidently has ceased to stimulate the *Amoeba.* The continued action of the stimulus apparently has set up a protoplasmic change which diminishes the effect of light for the time. Once the internal change (the adaptation) has been brought about, the organism gives no further indication of response to the stimulus until some abrupt change occurs in its intensity. The same is true of chemicals, temperature, and other types of stimuli. This accounts for Jennings' (1906*a*) observation that "change" is the most effective means of stimulation. Change *is stimulation.*

It is noteworthy that if the intensity of the stimulus is not increased abruptly but by degrees, the adaptation of the organism may be induced so gradually that the observer finds it difficult to detect a behavior change at any time. This important difference in the effect of gradual and of sudden increases in intensity suggests that adaptation depends upon relatively slow alterations in the balance of the opposite phases of a reversible chemical reaction in the protoplasm.

A suggestive experiment was performed by Mast and Pusch (1924). A dark-adapted *Amoeba* was permitted to enter a beam of bright light which stopped its forward movement. On successive trials a count was made of the number of small pseudopodia put out toward the light (and stopped by it) before the general direction of locomotion was reversed. When trials were made at 3-minute intervals, the number of projected pseudopodia decreased to zero within 10 or 20 repetitions of the test; and an interval of 24 hours did not remove the protoplasmic change.

It is apparent that *the readiness with which Amoeba responds to external stimulation is not constant.* Frequently the animal appears very sluggish and fails to react readily (if at all) to stimuli of moderate or of weak intensities. At other times it is very active in locomotion and very weak stimuli are effective. Schaeffer (1916, 1920) observed that an *Amoeba* which had ingested a considerable quantity of assimilable material might remain quiescent for as long as one hour. The mere intake of substances is not responsible for such changes, since the ingestion of amounts of carbon or glass particles did not, in Schaeffer's observations, greatly alter the condition of reactivity. However, assimilable (digestible) substances did induce the sluggish condition. Digestion of the material apparently changes the protoplasm so that it is not affected by slight external energy changes. It is evident that this phenomenon is closely related to "adaptation."

The Inherited Constitution of the Protoplasm.—This factor is basic to all of the influences upon internal condition, since it determines the characteristic level of response. Under equivalent conditions stable and predictable differences appear in the behavior of individuals representing various species of *Amoeba*. For three closely related species Schaeffer (1920) has described the characteristic differences in form and in behavior given in Table 3. Such differences, which

TABLE 3

Characteristic	Species of *Amoeba*		
	Amoeba proteus	*Amoeba discoides*	*Amoeba dubia*
Average size during locomotion, microns	600	450	400
Usual number of pseudopods during locomotion	5	3	12
General resistance to external conditions	Very great	Slight	Greater than *A. discoides*
Effect of mechanical stimulation	Responsive	Slightly responsive	Very responsive
Reaction to carmine particles	Readily ingested, but rejected in few minutes	About same as *Amoeba proteus*	Eaten only occasionally; often retained for hours

become much more striking when widely separated species are compared, depend upon the possession by each individual of a protoplasm which is characteristic of its species, a protoplasm which is directly transmitted to each of the two daughter individuals which arise from the division of a parent *Amoeba*.[1] The problem of the manner in which the characteristic species protoplasm is transmitted from one individual to its descendants has been treated suggestively by Lillie (1918).

3. *The General Nature of the Surrounding Medium.*—The properties of the fluid medium which surrounds a protozoan have much to do with its "level of activity" (readiness to respond) at a given time. A change in surrounding conditions may so alter the internal condition of the organism that the vigor of movement and of reaction to stimulation is considerably changed. Individuals taken from different cultures, the waters of which differ chemically, may show wide variations in sensitivity to equivalent stimuli. Pantin (1924)

[1] In binary fission, which is the usual method of reproduction in *Amoeba*, the nucleus of the parent individual divides, after which the animal elongates and finally splits into two new Amoebae. This is a type of "asexual reproduction."

reported that in a medium of lower salt content than its normal one a marine *Amoeba* became more fluid and formed pseudopodia which were broader and more responsive to stimulation than the usual ones; but when salt content was supra-optimal the animal put out pseudopodia less readily, and these appendages proved subnormal in sensitivity. Mast (1931) found that the higher the acidity of the surrounding medium the more susceptible was the individual *Amoeba* to the action of light. Schwitalla (1924) varied the temperature of a water bath which surrounded the observation "well" across which an *Amoeba* moved, and found that the rate of locomotion differed correspondingly. At 0°C. the *Amoeba* was practically immobile; at 10°C. the range of speeds was 0.5 to 4.5 mm. per second; at 20°C. it was 1.2 to 10.7 mm. per second; and at 30°C. it was 5.3 to 6.5 mm. per second. It should be added that under conditions which increase its general rate of movement the *Amoeba* is also more responsive to stimulation.

Feeding Activities and Amoeba's Adjustment to Its Environment. We have studied the nature of the factors which determine the behavior of *Amoeba*. This study also constitutes a survey of the adaptiveness of behavior, *i.e.*, its suitability to or its conformity with the conditions of the environment. This organism, like others which have long survived, is so constituted that its characteristic activities promote its life. But this does not mean that the relative "beneficiality" of an act determines whether or not an *Amoeba* shall commit it. The actual reasons for the adaptiveness of behavior are indirect ones, so far as the contemporary *Amoeba* is concerned. Such matters lie beyond the scope of our study, since they necessitate examination of the colloid chemistry of different species protoplasms, in relation to the probable nature of the agencies which have influenced the evolution of such protoplasms in the race.

The Three Phases of Food Taking.—Food taking, a strikingly adaptive feature of normal amoeboid behavior, is subject to investigation along the lines of our preceding analysis. In this activity three stages may be arbitrarily distinguished: (1) movement toward the source of a weak stimulus; (2) ingestion, taking the stimulating substance through the body wall; and (3) assimilation, internally breaking down the substance (or releasing it through the body wall after a time).

1. The intensity of stimulation governs whether an *Amoeba* approaches a stimulus source or moves in another direction, as we

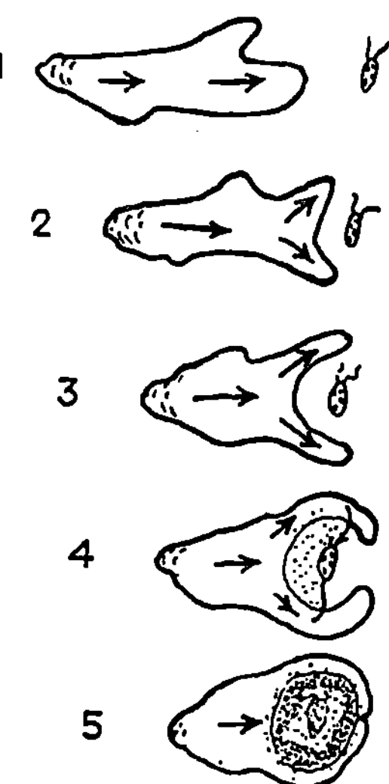

FIG. 4.—The "food-cup" reaction of *Amoeba*. (1) A small organism rests fairly quietly against the bottom. An *Amoeba* happens to be moving toward it. (2) When the *Amoeba* is fairly close, its main or leading pseudopod begins to divide. (3) The pseudopods flow somewhat to the sides as the *Amoeba* continues to move forward, and they partly surround the organism without actually touching it. (4) The hollowing of the central part of the forward end is now marked by the appearance of a thin upper sheet of plasmagel which flows domelike over the prey as the pseudopods continue to surround it. (5) The pseudopods have met and fused with each other; the upper sheet of protoplasm has completely covered the space encircled by the pseudopods and has fused with them, as has the floor protoplasm of the food cup. Now the small organism is digested during a quiescent period of several minutes. (*Redrawn from Schaeffer*, 1920.)

have seen. Schaeffer (1916) observed *Amoeba* approaching a grain of soluble material (tyrosin, an amino acid) at a distance of 60 to 100μ. The animal commonly moves toward or extends pseudopodia toward objects which have moved past it, apparently responding to weak contact stimulation furnished by water currents. A well-fed *Amoeba* usually fails to approach an object which does not furnish a chemical effect as well as a tactual one, *i.e.*, summation of stimuli is necessary when stimuli are very weak.

2. Ingestion of the object is essentially a continuation of the approach, and the strength of stimulation is an important determinant of whether the response continues into this phase. As Schaeffer (1916) observed, an active *Amoeba* will ingest a great variety of substances, including some which afford only a mechanical stimulus (*e.g.*, carbon particles, glass). Such substances, when moving, provide a summative effect. Chemically effective substances, such as inert materials which are soluble, are normally more effectively ingested than are objects which present merely a weak tactual stimulus. Minute swimming organisms are most readily ingested of all objects, apparently owing to the fact that they furnish weak contact and a weak chemical effect in summation.

The extent of body surface acted upon by the object, in relation to the strength of the stimulus, would appear to account for many of the differences in the mode of ingestion. The "food cup" (Fig. 4) apparently is produced by objects which stimulate relatively intensely a limited amount of body surface, so that the outer protoplasm "hollows out" in the locality as the captor continues to move forward. In contrast, *Amoeba* usually "flows around" a large object with its body surface constantly touching the substance during ingestion.

3. Whether the object is assimilated or is eventually ejected depends upon its solubility in the *Amoeba's* protoplasm. Non-digestible materials (*e.g.*, soot, sand grains) are ejected sooner or later in contractile vacuoles, but chemically soluble substances are soon broken down and assimilated. It happens that most of the substances which stimulate *Amoeba* effectively, and thus are ingested readily, also have the chemical properties which make for ready assimilation. However, as Schaeffer (1920) reported, there are many assimilable substances (*e.g.*, gelatin particles) which are digestible but usually are not ingested.

This brief survey suggests that the problem of food taking should not be approached from the misleading standpoint of "food value," but that the proper procedure is to ascertain those properties of stimulation and of protoplasmic activity which produce the observed results.

Special Problems in Protozoan Behavior

Although the various species of *Ciliata* and *Flagellata* (see footnote, p. 9) vary greatly in form and in behavior, adaptation in these protozoan classes is subject to the controlling factors which have been demonstrated for *Amoeba* as a representative of the *Rhizopoda*. A consideration of certain special problems will illustrate this point.

The Significance of Permanently Differentiated Protoplasms. *Amoeba Compared with More Specialized Protozoa.*—In the representative *Amoeba*, activity patterns (physiological gradients) are temporary in nature, and under appropriate conditions external factors may make any portion of the organism the dominant and controlling locality. The most sensitive part of the *Amoeba* is the momentarily dominant or leading part. In other protozoans, permanent sensitivity gradients exist; that is, the permanently specialized anterior end is typically most sensitive and there is a regular decrease in sensitivity toward the posterior end. One of the crucial advances in animal adaptation is the appearance of *permanent physiological gradients* in single-celled organisms. Some important consequences of this fact for sensitivity and for action may be briefly outlined.

Permanently Differentiated Sensitivity.—The stable differentiation of highly sensitive protoplasms is well illustrated by the specialization of sensitivity to light in the flagellate, *Euglena.* This special sensitivity is mediated by the "eyespot," a reddish spot of oily pigment which lies near the base of the flagellum, close to the anterior end (Fig. 5). *Euglenae* which are swimming about in a dish pass into a lighted spot without change in movement, but gradually collect in this area, since each individual is whipped around by its flagellum whenever entrance into the shaded border zone suddenly decreases the intensity of light to which the eyespot has become adapted (Engelmann, 1882). Similarly, a *Euglena* remains within an area of moderately intense light after having entered by chance from an intensely illuminated area which surrounds it. In this case, when the organism reaches the brightly illuminated border zone, the sudden increase in illumination sets up an abrupt change in movement through its action on the specialized photoreceptor.

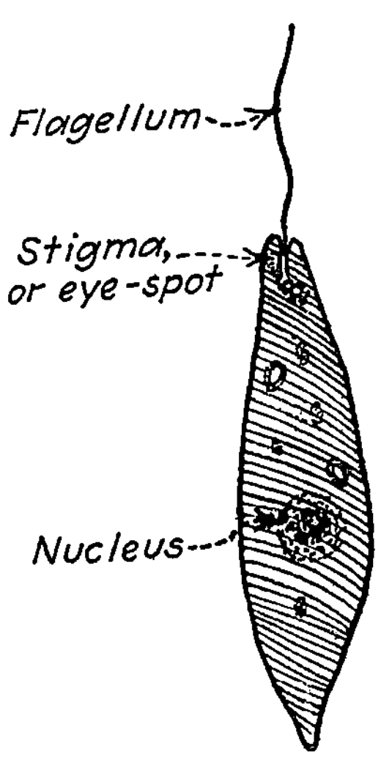

FIG. 5.—The flagellate protozoan, *Euglena.* (*Redrawn from Parker and Haswell.*)

Permanently Specialized Contractility.—The permanent differentiation of protoplasms similarly makes possible *specialized motor functions.* The flagellum of *Euglena* is a specially contractile "Organelle," the whipping stroke of which pulls the animal forward with sensitive end in advance. The fine hairlike cilia which cover the body of *Paramecium,* a representative ciliate, beat together in wavelike fashion so that the organism is rotated forward about its longitudinal axis. Owing to the asymmetrical shape of this organism, movement is in the form of a double spiral (Fig. 6, 1–6), *i.e.,* the anterior end describes larger spirals than does the posterior end. This fact is of great importance for the manner in which movement is modified under certain stimulating conditions, as we shall find.

The possession of permanently differentiated action elements makes it possible for these organisms to behave in a highly varied and versatile manner under the influence of changing external and internal conditions, and thereby the precision of adaptation is increased. Among the ciliates, in particular, the characteristic movement (forward rotation) attributable to the action of cilia becomes the basis for a diversity of behavior types among groups that vary greatly in general body form.

In some of the *Protozoa*, especially the stalked forms (*e.g.*, *Stentor*, Fig. 7), there are specially contractile protoplasmic strands which extend from near the mouth opening toward the posterior end. When the organism is stimulated effectively, these "myonemes" contract suddenly, thereby shortening the body. In normal feeding, a stream of water carrying suspended particles is drawn into the mouth funnel by the cilia lining it, but if a large object bumps against the anterior end, the backward jerk is elicited, and the substance is whisked off in water currents without being ingested. This permits a certain amount of "selection" in the feeding of *Stentor* and similar organisms.

Specialized Conductile Protoplasms.—In the *Amoeba*, conduction of excitation occurs through "open protoplasm," although the direction of conduction differs according to the pattern of activity in force at the time. In more specialized *Protozoa*, the lines of conduction through the protoplasm (which in *Amoeba* depend upon the flux of conditions) are fixed in terms of protoplasmic differences between the specialized anterior end and other parts. The existence of specialized conductile protoplasms is attested by a large body of results. For instance, Yocom (1918) cut certain protoplasmic fibrils in a parasitic ciliate and reported interference with irritability and with movement.

In summary, it may be said that specially irritable protoplasms increase the range of energy to which the organism may respond, while specialization in conductile and action protoplasms increases the rapidity as well as the variety of response. The differentiation of a stable physiological gradient, with a specialized dominant anterior, is the foundation for all of these factors which make for advances in behavior. We turn to an examination of the increased variability in behavior which is a consequence of these structural improvements.

The Significance of Variability in the Normal Mode of Locomotion. *Typical Variability in the Forward Swimming of Paramecia.*—A number of observers, particularly Jennings (1906*a*) and Alverdes (1922), have pointed out the great variability to which the normal movement of the ciliate *Paramecium* is subject. The width of the spiral is constantly changing and its form varies constantly as the animal slightly swerves or widely swings in its course. The speed of swimming frequently changes, as well. *Paramecium* may glide slowly for a distance with one side in contact with a surface, and the

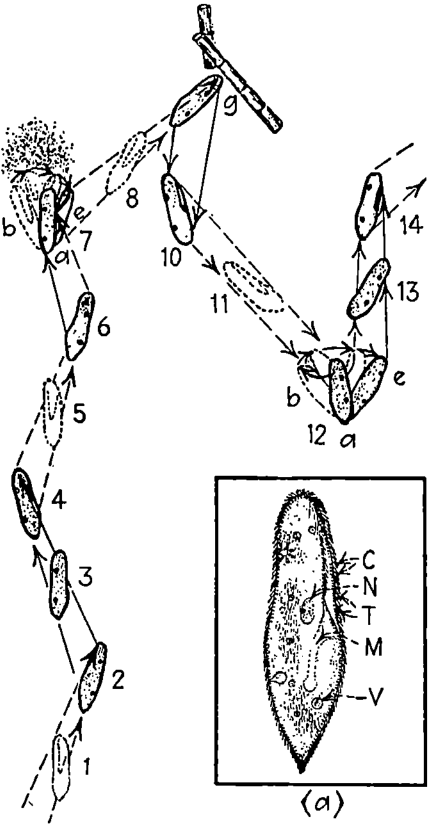

FIG. 6.—The normal forward spiraling (1–7*a*) and the reversing reaction (9–12*a*) of *Paramecium.* At 7*a* a chemical stimulus causes the organism to swivel (7*a* to 7*e*) and next to move forward in a new direction (7*e*–9). At 9, contact with a filament causes the animal to spiral backward. At the end of its backward excursion it swivels (12*a*–12*e*) and again spirals forward. (*Modified from Jennings,* 1906*a, and from Alverdes,* 1922.) (*a*) *Paramecium. C,* cilia; *M,* mouth; *N,* nucleus; *T,* tentaculocysts; *V,* vacuole. (*Modified from Parker and Haswell.*)

animal often floats gently against an object and comes to rest with most of its cilia motionless.

Many of these behavior variations may be traced to the intensity changes of stimulation, or to the manner in which stimuli are encountered, as the following instances suggest. If a surface is struck at an obtuse angle, the *Paramecium* may *ricochet* from it without an observable interruption of swimming except an enforced change in its direction. *Circling or veering* has also been observed. As the organism nears a drop of chemical there may occur a slight increase in the width of the spiral, accounting for an arching in the course which may carry it around the drop. Sometimes, upon gently bumping an object in its course or upon entering a zone of chemical stimulation, the *Paramecium* slows in swimming so that forward progress ceases for a short time (Fig. 6, 7 *a–e*). However, during this moment the animal continues to rotate, so that when the general ciliary activity is resumed, the body is driven in a new direction by the beating of its cilia.

The "Backing Reaction" and Its Variability.—When a contact or chemical stimulus is fairly intense (as when a rapidly swimming *Paramecium* bumps against a surface), a more or less vigorous backing response is given (Fig. 6, 9–12*e*). This reaction may be described as follows:

1. The *Paramecium* spirals backward from the point of contact, carrying through a distance which appears to vary with the strength of stimulation, but which is usually not much greater than the body length. Often, as noted above, the stimulus is so weak that the animal does not give its backing response at all, but swimming is merely arrested for a moment.

2. Then the animal pauses, but the body continues to rotate, so that the anterior pole circles widely (Fig. 6, 12*a–e*).

3. Forward movement is promptly resumed, and the Paramecium moves in a direction which depends upon the position of the front end in its spiral at the time forward locomotion began (Fig. 6, 12*e*–13).

Jennings (1906*a*) observed that, when very strong stimuli were used, the animal usually backed through a greater distance before forward locomotion reasserted itself. Alverdes (1922) elicited the backing reaction by stimulating with various salts, and found that the distance through which the animal backed was greater when the stimulating chemicals were more concentrated. In vigorous responses the eventual change in direction of progress from the original course is much greater, since the front end of the body swings through a wider circle during the backward pause. In contrast, when Jennings used a weak stimulus such as water 3 to 4°C. higher than the optimal temperature, the *Paramecium* would merely pause in place, its anterior end swinging in a narrow circle owing to the continued rotation, and then would proceed with but a slight change in direction. As Jennings wrote,

> Between this delicate reaction and the violent one first described there exists every intermediate gradation, depending on the intensity of stimulation.

The Adaptive Significance of Variability in Locomotion.—This raises the question of the adaptive significance of such variations in behavior. The close attunement of the *Paramecium* to its environment is made possible by an extensive specialization in its protoplasmic structures, and it is a fact of great importance for variability in behavior that the animal is capable of movement in three dimensions. This ciliate is very successful, ordinarily, in escaping the locality of a strong stimulus. Two reasons are apparent: (1) The backing reaction is then stronger, and carries through a greater distance than in responses to weaker stimuli; and (2) the more strongly irritated animal rotates more widely at the end of its backward movement, and is thus more likely to set off in a very different direction. *Withdrawal from the stimulus source thus depends upon whether the stimulus is adequate to alter the chemistry of the animal's protoplasm in a manner that interrupts normal swimming.*

It is significant that chemical compounds such as acetic acid or zinc chloride normally elicit the backing response when in strengths much less than that capable of destroying the protoplasm. According to Jennings (1906*a*), potassium chloride produces the backing reaction at a minimal strength of 1/20 per cent, although it does not destroy protoplasm below a strength of 1 per cent. In contrast, chromic acid does not elicit the backing reaction below 1/150 per cent, a concentration at which this chemical destroys protoplasm. Further, Jennings found that *Paramecia* collect in solutions of acid salts (*e.g.*, copper sulphate) after having entered readily from the surrounding medium, although the "entrapping" substances are actually toxic and soon kill the organisms. In many cases the backing reaction appears under these conditions, but only when it is too late. Such behavior is to be accounted for only in terms of the action of different values of energy upon the chemistry of protoplasm, thus affecting swimming, and not by vaguely assuming that the *Paramecium* possesses an ability to distinguish "injurious" substances from others.

The responses we have described adjust the organism to its environment, *i.e.*, they are "adaptive." Repeated backing reactions are likely to take the ciliate out of the zone of strong (or supraoptimal) stimulation sooner or later, unless death comes first, since each repetition of the stimulus interrupts forward swimming and changes the direction of locomotion. A *Paramecium* thus escapes from a close tangle of plant filaments, since *one of the many* backing reactions given in response to contact with filaments will so change its direction that the next forward progress carries the organism through some small opening. Expressions such as "trial and error" are out of place here, since they encourage the misleading and fallacious teleological assumption that the animal is "trying to escape injury."

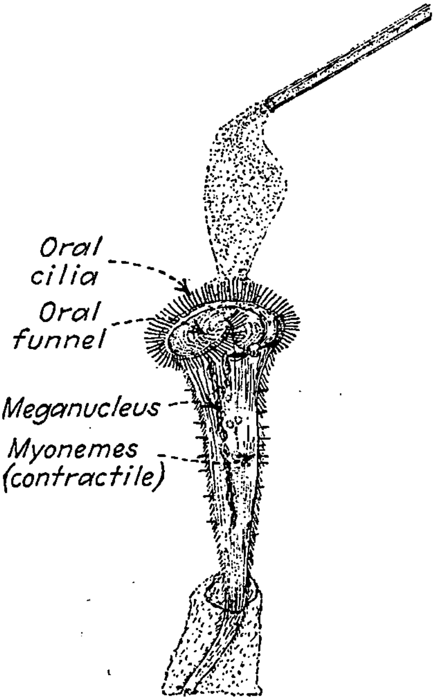

"Adaptation," and Modification of Behavior under Continued Stimulation.—In the more specialized *Protozoa*, not only may behavior vary considerably under changing external stimulation, but a sequence of identical stimuli may produce successive changes in behavior. This fact greatly extends the possibilities of adaptation, as the following case will show.

FIG. 7.—*Stentor roeselii* as in Jennings' carmine experiment. (*Redrawn from Jennings*, 1906*a*.)

Jennings' "Carmine Experiment" with the Ciliate, Stentor.—*Stentor* is a little trumpet-shaped ciliate (Fig. 7) which is usually attached to the leaf of a water plant by the base of its stalk, about which a cup of mucus has been secreted by the protoplasm. The free funnel-shaped end is equipped with large border cilia, the beating of which forms a vortex of water in which small particles are carried into the interior of the animal. Jennings (1906*a*) directed a constant stream of fine carmine particles from a capillary pipette onto the funnel of a *Stentor*. At first there was no observable change in activity and the ciliary action continued; the stimulus apparently was below the excitation threshold. But as carmine continued to drop, the following behavior changes appeared in succession:

1. The anterior part of the stalk contracted, so that the funnel twisted around as it was jerked a short distance toward the base. The reaction was repeated as the particles continued to fall upon the disc and occurred more promptly and vigorously each time.

2. After this response, the direction of the ciliary stroke was changed for an instant, accounting for a reversal in the direction of the water current at the funnel. This response was repeated a few times in rapid succession.

3. The animal then contracted more completely (than in 1) upon its stalk, the body twisting in one direction as it was pulled back into the mucus tube. Each time *Stentor* re-extended, the following contraction came more promptly, and was more vigorous.

4. As the carmine continued to fall, the animal at last contracted repeatedly and violently while still in its tube. In the end this usually broke it free from its attachment to the substratum; whereupon it whirled off through the water, carried in the typical ciliate spiral by the beating of its body cilia.

Factors Producing a Sequence of Reactions under Weak Stimulation. Let us analyze this phenomenon, to ascertain the reasons for the occurrence of successively different responses. First of all, increasingly stronger responses may be elicited if stimuli are presented at sufficiently short intervals, so that their effects may be *summated.* As Jennings (1906*a*) found, a weak tactual stimulus fails to elicit a response from *Stentor* after five or six repetitions, if it is repeated at intervals of 1 min. Hence the time interval between stimuli must be short, or summation does not occur.

The work of Danisch (1921) on *Vorticella*, a related ciliate, is significant here. Mechanical stimuli of controlled and graded intensities were presented to the normally active organism. If successive stimuli were separated by a sufficient time interval (25 sec.), the effects of each application were found to "wear off" so that the next stimulus produced virtually the same response. In another test, however, the summation effect was obtained with stimuli of the same strength as those used in the first test (*i.e.*, 1,500 ergs energy value), because of the short time interval of 10 to 15 sec. between stimuli. This result held even when the stimulus intensity was reduced to 1,000 ergs. Hence if the time interval is short, or the stimulus sufficiently intense, the effects of each application remain in the protoplasm and are summated with those of the following stimulus to produce a stronger response. A strong stimulus produces changes which survive longer, and are therefore more effectively summated. A single stimulus of sufficient strength, applied to a stalked ciliate, produces a response closely similar to the one which is eventually elicited by the repetition of weak stimulation.

But when a weak stimulus is repeated, why is a series of different responses obtained, instead of an increasingly vigorous repetition of the same reaction? In the stalked ciliate there is valid evidence for the existence of an excitation—conduction gradient (pp. 18*f.*) between funnel end and attached end.[1] The highly irritable funnel margin is the first part of the organism to respond to the repeated weak stimulus. As the effects of stimulation are summated the excitation is next conducted to the highly contractile fibrils arranged along the stalk. Thereby the less irritable parts of the organism further toward the base are successively brought into play, until finally the entire organism responds vigorously. It is not difficult to understand the fact that if the carmine dropping ceases for a time, the first response to be elicited on its resumption is less like the last one of the preceding series the longer the "rest period" has been.

Jennings' results are thus accounted for by the phenomenon of *summation* of protoplasmic changes, and by the existence of an *excitation-conduction gradient* in *Stentor*.

The Capacity of Protozoa for Altered Behavior

Temporary Alteration of the Backing Reaction in Paramecium.— Experiments already reported (Folger, p. 20; Mast and Pusch, p. 21; Jennings, Danisch, above) show that protoplasmic changes induced by stimulation may alter behavior during their relatively short survival time. In the case of *Paramecium*, attempts have been made to study the time limit and the nature of such changes. Smith (1908) found that a *Paramecium* drawn with water into a capillary tube too small in diameter to permit the animal to swing around would initially give repeated backing reactions to successive contacts with the surface film. After a number of such responses, however, the animal somehow doubled up and turned around when the film was touched. Day and Bentley (1911) confirmed him in finding a decrease in the number of backing reactions that preceded

[1] We owe much to Herrick (1924) and to Child (1924) for having emphasized the significance of the axial gradient principle for animal psychology.

the successful reversal on consecutive tests. Within a few trials (average 8.6) many subjects were able to double and reverse direction upon the first contact with the film. The experimenters applied the term "learning" to the phenomenon.

Buytendijk (1919) repeated the work, and attributed the change in behavior to an increased flexibility of the body wall due to the accumulation of acid in the protoplasm during the activity.[1] *Paramecia* which he made more flexible by immersion in chloroform doubled and reversed direction promptly upon contacting the surface film in a capillary tube. Such a protoplasmic change evidently survives for some time, since Day and Bentley found that certain of their *Paramecia*, after swimming about in an open medium for 10 to 20 min., still showed the effects of the previous activity upon being returned to the capillary tube.

Temporary Alteration of the Feeding Reaction in Paramecium.— A related phenomenon is of interest. Metalnikow (1912) discovered that *Paramecium* takes carmine less readily after having been in a carmine-saturated medium for a time. He put specimens into a carmine medium for 24 hr., and then introduced them into a carmine-free medium for 30 to 60 min. When the vacuoles were emptied of carmine the animals were again placed in the carmine medium for 1 hr. and a record was made of the amount ingested by them. Using the same method, Losina-Losinsky (1931) has shown that the tendency to take carmine less readily lasts for three days in *Paramecia* which have gotten rid of their supply after having been surfeited with the substance. More surprising than that, the behavior change appears to some extent in the next three generations of daughter individuals. Overfeeding with carmine apparently so alters the protoplasm that while the effect persists, carmine fails to exert its normal stimulative action upon the organism.

"Primitive Learning" in Protozoa.—Such changes, together with others we have described (adaptation to light, summation of stimuli) resemble those involved in the "learning" of higher animals in that they are attributable to the individual's own experience and that while they persist the normal activities of the organism are altered. However, the behavior changes they permit are very temporary in

[1] The accumulation of salts in the water, as products of the animal's own activity, together with a rise in concentration due to evaporation, might partially hydrolize the protoplasm and thus increase for a time the flexibility of the body wall.

nature, and the organism reverts each time to its original condition, with no trace of the experience being evidenced in its subsequent activities. Apparently the element that is lacking here is a specialized conduction system which may be more permanently altered by the effects of experience. Nevertheless, these phenomena in protozoan behavior show certain characteristics which are common to all learning, and which are therefore of importance for any study of the origin of the learning capacity in the animal series.

Suggested Readings

Calkins, G. 1926. *Biology of the Protozoa.* Philadelphia: Lea & Febiger. (1933, 2d ed.) Chap. IV.

Cowdry, E. (Ed.) 1924. *General cytology.* University of Chicago Press. Secs. IV, V, VI.

Jennings, H. S. 1906. *Behavior of the lower organisms.* (1923 ed.) Columbia University Press. Chaps. III and IV.

Mast, S. O. 1909. The reactions of *Didinium nasutum* (Stein) with special reference to the feeding habits and the functions of trichocysts. *Biol. Bull.*, **16,** 91–118.

Schaeffer, A. 1920. *Ameboid movement.* Princeton University Press. Pp. 8–40.

Warden, C. J., Jenkins, T. N., and Warner, L. 1934. *Introduction to comparative psychology.* New York: Ronald. Chap. 9.

CHAPTER II

FIRST PROBLEMS IN THE BEHAVIOR OF MULTICELLULAR ANIMALS

INTRODUCTION

A multicellular or metazoan animal is composed of numerous units or "cells," which are arranged in layers. Each unit has its limiting wall, and in this respect (often in others, as well) is similar to a protozoan. The cells cooperate in the activities of the whole organism, but in different ways in correspondence with structural specializations among them. Our study concerns not the individual units as such but the properties which determine the behavior of the aggregate.

The principles that hold in protozoan behavior are also fundamental to the behavior of multicellular animals. However, because of the peculiar structural properties of the multicellular organism, its potentialities for complex activities and for modifiable behavior are not limited as are those of the protozoan. The specialization of groups of structurally distinct units will be found a much more promising basis for the evolution of higher forms of behavior than the specialization of protoplasm within a single limiting wall.

THE RELATIVITY OF THE TERM "ORGANISM": PHYLUM PORIFERA[1]

The Free-swimming Larval Sponge.—The immature sponge, the larva, grows by virtue of the division of a fertilized ovum (egg), and escapes from the body of the adult as a sac-shaped two-layered mass of cells. The flagella of its outer cells drive it through the water, with its closed end ahead. At this early stage the larva swims near the surface, and moves toward light. Finally it remains near

[1] The phylum *Porifera* (sponges, Fig. 8) includes animals with two body layers and with numerous pores in the body wall. Most sponges live in salt water, but there are some fresh-water forms.

TABLE 4.—PORIFERA AND COELENTERATA

Porifera: Cells of osculum (Fig. 8) and pore cells in body wall directly acted upon by mechanical (current) and chemical stimuli; adult sponge virtually insensitive to light. No specialized receptors for types of energy. No response to changes in position. Adult sponge sessile; osculum closes in quiet water. Water drawn in pores, out osculum, by beating of flagellated interior cells. Neuroid conduction (pp. 38*f.*); no specialized conduction elements.

COELENTERATES

RECEPTOR EQUIPMENT	FEATURES OF SENSITIVITY
Chemical. Pits containing cells located at intervals around bell margin in many medusae. Hairs project to outside from "primary-sense cells" in the epithelium of medusa and hydroid.	These considered organs of contact-chemical sensitivity. Tentacles and stomodaeum (lips and gullet)' in the hydroid; tentacles, bell margin and manubrium (mouth tube) in the medusa, especially sensitive both to chemical and to tactual stimulation.
Mechanical. Probably same epithelial cells sensitive to both contact and chemical. Statocysts, clapperlike structures, in cavities around margin of bell in some medusae.	Sea anemone's disc expands in response to tactual effect of water currents; contracts in still water. In some medusae, statocyst held to have static (equilibrium) function; in most regulates strength of muscular action in swimming.
Light. Symbiotic algae in green *Hydra;* brown *Hydra* lacks these. Sense clubs or tentaculocysts at intervals around bell margin in *Scyphomedusae*, contain statocysts, also ocelli, pigmented pits usually with lens structure above. Others have ocelli alone.	Green *Hydrae* collect in weak or moderately strong light; brown *Hydrae* unresponsive to light. Medusae collect in shaded regions of their medium. Many polyps contract in bright light, expand tentacles and feed in dim light, a day-night rhythm.

CONDUCTION

In *Hydra* and lower coelenterates, sensory cells in body wall have branches which extend to muscle cells below (Fig. 11*A*), these branches serve as conductile elements. "Nerve net," a network of interwinding cell processes between body wall layers, is typical of higher coelenterates (Fig. 11*C*). In some medusae nerve net concentrated near marginal sense organs, in others it forms "nerve ring" around bell margin. Nerve net conducts diffusely, and with a decrement (p. 50).

ACTION SYSTEM

Contractile cells specialized, arranged in sheets. Tentacles contractile in hydroids, body bends about, or moves about on base of stalk as in sea anemone. Some medusae move by means of circular muscle alone, the bell closing with each contraction; others have both circular (closing) and radial (opening) muscles; which work reciprocally. Nematocysts (barbed thread ejected when capsule explodes) plentiful in tentacles and body wall of most coelenterates.

the bottom, a change which may be due to the disappearance of a photosensitive pigment as well as to an increase in weight. The flagellated cells are turned to the inside in the course of development, and the sponge becomes an attached adult.

Status of Behavior in the Adult Sponge. *Typical Activity in the Sessile Adult.*—In strong contrast to the larva, the adult sponge is fixed in position and is incapable of moving actively as a whole. The simplest adult type resembles a vase with sievelike walls (Fig. 8*B*). Continuous beating of the flagellated cells (which now line chambers in the walls) draws a current of water through the inhalent pores of the walls, circulates it through walls and body cavity, and forces it out in a continuous stream through a large opening (the osculum) at the upper end of the animal. Fine organic material in the water is sifted out in the walls of the sponge and digested by the individual cells. This is the typical activity of the animal.

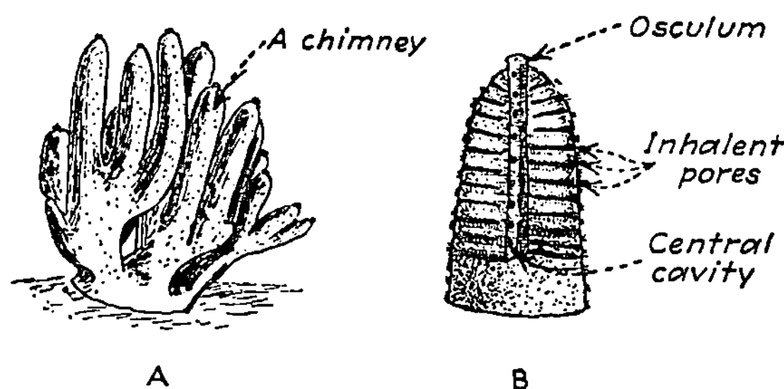

Fig. 8.—(*A*) Sketch of the marine sponge *Sycon gelatinosum*, showing chimneys. (*B*) Section of one chimney. (*Redrawn from Parker and Haswell.*)

It involves a minimum of active cooperation among the parts; or considered strictly, *no* active cooperation among the parts.

A Primitive Basis of Response.—The osculum is closed in still water; it also closes when the water contains ether, or when the temperature of the water is raised (Parker, 1919). These stimuli act directly upon a circular sheet of cells (the sphincter) which surrounds the osculum. Each cell in this sphincter[1] is independently irritable, and contracts when stimulated; hence *the individual cell combines the functions of specialized receptor and specialized effector* (Fig. 11*A*).[2]

A sluggish transmission of impulses unites the cells of the osculum in action. In contracting, a stimulated cell excites others by pulling mechanically upon them, and these cells in turn excite others in the

[1] Sphincter: A muscle that surrounds an opening which it closes in contracting.

[2] Cells that independently combine the functions of irritability and contractility are found also in the pores and canals of sponges (Wilson, 1910).

same manner. Thus stimulation of one part of the osculum may spread throughout the entire sheet of cells which comprises the sphincter. This is *neuroid transmission:* a protoplasmic transmission of impulses, non-nervous in character. The sponge lacks true nervous tissue.

Neuroid transmission is slightly more efficient in some sponges. In the marine sponge *Stylotella* (*cf. Sycon*, Fig. 8), each chimney of which has its own osculum, Parker (1919) found that the osculum of a chimney would close shortly after a cut was made in the tissue 1 cm. below it. McNair (1923) reported for a fresh-water sponge that while a needle prick caused the osculum to close and excited a contraction which passed from tip to base of the stimulated chimney, no change was seen in other chimneys as close as 5 mm. to the active one. Neuroid transmission permits a spread of activity from part to part of a given chimney, but its coordinating effect goes no further.

A Limited Functional Organization of Parts.—The parts of the adult sponge are bound together structurally, but functional relationships among them are limited. The parts cooperate only in a fortuitous manner, as when the contraction of certain cells in the oscular sphincter pulls upon other cells and causes them to contract, or when the closing of the osculum influences the beating of the flagellated cells in the wall chambers by altering the pressure of water currents within the sponge. However, Parker (1919) maintains that from such sluggish and limited neuroid transmission as we find in the sponge the true nervous conduction of higher animals has evolved.

The sponge owes its unity as an organism, feebly developed as this is, to units which differentiate during development and which remain together and multiply in the adult. These units cooperate in total behavior, not because they are functionally united, but because they are all parts of one "metabolic whole." The sponge therefore represents a very low grade organism, and one which is not so adequately fitted for behavior as a whole as is a representative protozoan. Nevertheless, *it is from this primitive "metabolic unity" that psychological unity, the product of organized functional cooperation among the parts of an animal in its behavior, develops in the animal series.*

An Improved Cooperation of Parts in Behavior: Phylum Coelenterata[1]

The Coelenterate Equipped for a Wider Adjustment.—The adult sponge does not truly "behave as a whole," since dominance-subordination relationships among its parts have not advanced beyond the metabolic level. This is animal behavior at its lowest ebb; but the protozoan ancestor which produced the sponge gave rise to other metazoa in which closer functional relationships developed among bodily parts, thus making possible an adaptation to a wider environment. In the coelenterate specialized receptor cells, when excited, release impulses through a specialized nervous tissue which lies between the two body layers and conducts diffusely to the specialized muscle tissue.[2] The result of such structure is a much better integrated organism, one capable of wider environmental adjustment.

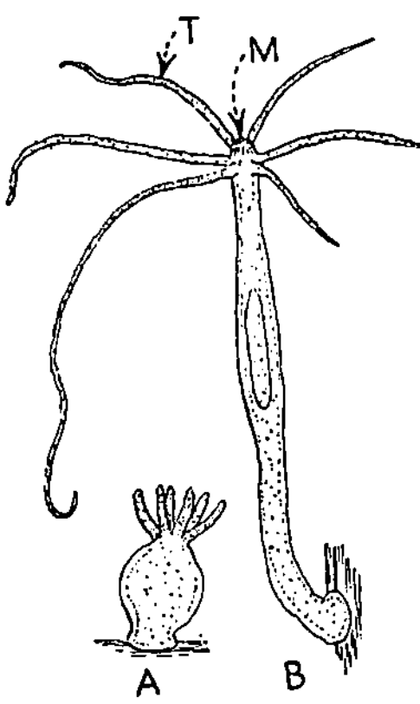

Fig. 9.—*Hydra.* (*A*) Contracted. (*B*) Extended. *M*, mouth; *T*, tentacle. (*Redrawn from Shipley and MacBride.*)

Although the free-swimming medusa and the stalked hydroid typically occur at different stages in the life history of the individual coelenterate, and are closely related biologically, their structural differences make for a very different behavior. Hence we shall treat them as separate subjects for behavior study.

Characteristics of Hydroid Behavior. *The Activities of a Representative Hydroid.*—The fresh-water *Hydra* (Fig. 9) is usually attached to an aquatic plant near the surface. At intervals of 1 to 2 min. the body elongates and the outstretched tentacles wave about. After a time the tentacles draw back, and after it has shortened, the body extends in a new direction. When a small aquatic animal

[1] The phylum *Coelenterata* includes radially symmetrical animals with a two-layered body wall, tentacles, nematocysts ("stinging cells"), and a single blind gastrovascular cavity. In class *Hydrozoa* two body forms, the hydroid (Fig. 10) and the medusoid (Fig. 12), generally occur in the life history of the same individual; class *Scyphozoa* has the medusoid form alone (large, commonly called a "jellyfish"); and class *Anthozoa* has the hydroid form alone (sea anemone).

[2] A tissue is an aggregation of cells of similar structure which typically perform in common a special function.

swims against one of the highly irritable tentacles, this member contracts about the prey and bends toward the mouth-opening, which is in the center of the crown of tentacles (Wagner, 1904). Other tentacles become excited, although they may not have touched the prey; these wave about, and if they touch the prey they also grasp it and bend toward the mouth.

Hydra moves to a new place about ten times daily; more frequently if the food supply is sparse in a given area. Its usual form of locomotion is a "looping" movement, in which tentacles and base are alternately attached to the substratum. The place in which the animal finally settles is mainly determined by the light, by the amount of oxygen in the water, and by the temperature. Except at extreme temperatures, or when the animal has been without food for some time, light is the most important factor (Wilson, 1891; Haug, 1933).

Hydra represents a great advance in sensitivity, conduction, and reaction, as compared with the sponge. Receptor cells are present in the outer body layer and are especially concentrated at the anterior end, in tentacle walls and around the mouth. These cells are excited by weaker energy than that which excites other cells of the body. However, there is no evidence that they are specialized as to the types of energy which will arouse them.[1] Thus, in *Hydra*, a given receptor cell probably is equivalently sensitive to energy in the form of chemical, mechanical, and thermal stimuli.

In the feeding of *Hydra*, the normally effective stimulus is the combined action of contact and a chemical, a summation of stimuli, but the threshold of sensitivity may be raised or lowered according to the animal's condition. Steche (1911) found that the tentacles of a *Hydra* which had not fed for some time readily responded to a wad of paper which touched them, grasped it and bent toward the mouth as in "food taking." Weak diffusing chemicals readily excite the hungry" coelenterate, because of the lowered threshold of sensitivity in the animal. In the poorly fed *Hydra*, the excitation of one tentacle promptly leads to the activity of others. This and

[1] Warden, Jenkins, and Warner (1934) favor the term "generalized receptors" for these structures, since the relative sensitivity of the body regions in coelenterates is the same for chemical, contact, and other classes of stimuli. The cell which is specifically sensitive to a given kind of energy (*e.g.*, chemical) is a later development in the animal series; the "generalized receptor" appears to be the primitive type.

other behavior such as shortening of tentacles and body contraction in response to a strong local stimulus, indicate the presence of a specialized conduction system.

The question as to how general unity is achieved in coelenterate behavior is best answered by the analysis of a complex adaptive activity.

Feeding Behavior in the Marine Hydroid (Polyp).—In the normal feeding of the expanded sea anemone (Fig. 10), as in that of *Hydra*, substances are grasped by the tentacles and passed by them to the mouth in the center of the disc. This activity has been studied by a number of investigators, and in particular by Parker (1917). First in consideration is the activity of the individual tentacle. Mucus

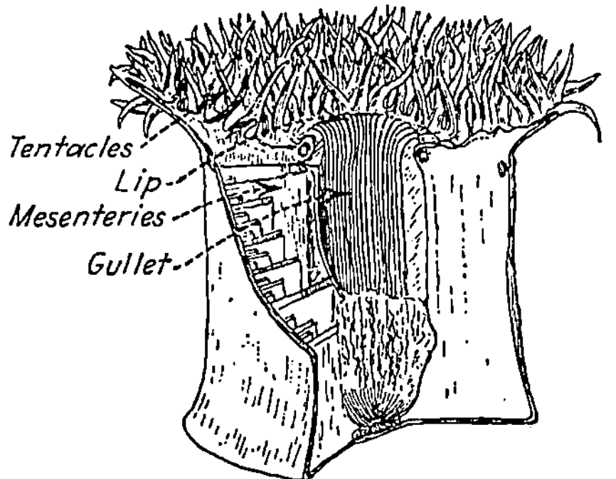

FIG. 10.—Dissection of a sea anemone, *Metridium marginatum*. (*Redrawn from Hegner.*)

secreted by certain of its cells makes the outer surface of the tentacle sticky, so that small objects readily adhere to it. Cilia which cover the tentacle beat constantly in such a manner as to sweep substances toward the tip of the member. These processes are accessory to the actual capture and ingestion of food, and as such they may be grouped with the explosion of nematocysts (barbed threads) from capsules in a local part of the tentacle which is intensely stimulated.

The adequately stimulated tentacle (p. 41) first bends outward toward the stimulating object at the margin of the disc, contracts, thereby grasping the object, and then bends toward the center of the disc. The nature of the response of the individual tentacle to given stimuli is determined by the receptive and neuromuscular equipment of the tentacle itself. When a tentacle is cut from the disc and properly mounted, it responds to stimulation just as it does when attached to the animal (Torrey, 1904). This is true in general of local structures in coelenterates.

The excitation is transmitted throughout the tentacle by a specialized conductile tissue. First, sensory cells in the outer layer (ectoderm) of the tentacles have branched threadlike prolongations, which reach the muscles beneath the surface. These cells combine the functions of sensitivity and conductivity, and according to Parker (1919) represent the second stage in the evolution of a specialized nervous system (Fig. 11*B*). But further, scattered thickly between the receptor cells and the muscle tissue there are cells with branching processes which are in simple contact with one another, the *nerve net* (Fig. 11*C*). Basically, the nerve net conducts

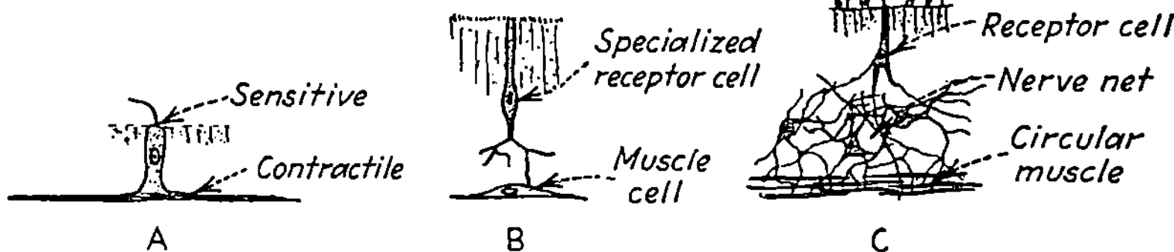

FIG. 11.—Stages in the evolution of the nervous system. (*A*) Stage at which a single cell performs the functions both of reception and of contraction (*e.g.*, cell in sponge osculum). (*B*) Stage at which a specialized receptor cell and a specialized contractile cell are directly connected (*e.g.*, in tentacle of Hydra). (*C*) Stage at which specialized conduction tissue intervenes (*e.g.*, bell of medusa). (*Modified from Parker and from Hertwig.*)

diffusely, without intrinsically determined directionalization, in contrast to the conduction of a "synaptic nervous system" (see Chaps. IV and XIV). Hence the nerve net functions as though the processes of its cells had grown together to form a continuous anastomosis. This transmission system makes the tentacle a unit in action.

Through the nerve net, excitation from the tentacle first aroused reaches others on the disc. In some polyps the excited tentacles bend at once toward the mouth, and if the food then chances to fall upon them it is carried to the center of the disc (*i.e.*, toward the tips of the tentacles) by ciliary action. *Metridium* is an example of this type. In other polyps (*e.g.*, *Aiptasia*) the secondarily aroused tentacles first bend toward the locality from which the nerve-net impulse reaches them, then bend toward the mouth when the food is touched and grasped. Conduction through the nerve net thus unites the tentacles in behavior.

The lips and stomodaeum (gullet) are opened by muscles of the complete mesenteries (Fig. 10), which contract when the nerve-net excitation reaches them from aroused tentacles. This response may be facilitated by direct stimulation of the lips, since Fleure and Walton (1907) found that the lips were capable of responding to a diffusing chemical so weak in concentration that it failed to arouse tentacles. Upon reaching the mouth, whether the substance is carried into the interior of the animal depends upon the action of *oral cilia*. The normal beating of these cilia maintains a constant flow of water *outward* from the gullet. If the nerve-net excitation is adequate (and Pantin, in 1935, has shown that it differs according to the manner in which tentacles are stimulated), the direction of this ciliary stroke is reversed, water is drawn *inward* through the mouth, and the material transported by the tentacles is ingested.

Changes in the sensitivity of the mouth region account for certain important variations in the feeding act, variations which Nagel (1894) attributed to "intelligence." Allabach (1905) discovered that, although the tentacles of a fairly well fed *Metridium* might carry a piece of filter paper to the mouth, failure of the oral cilia to reverse their stroke would then force outward bending of tentacles and transport of the object from the disc. This is in decided contrast to the fact that in a poorly fed animal the oral cilia reverse their stroke readily when the mouth region is gently touched or is stimulated with a weak chemical. In the gorged animal, even the touching of crab-meat morsels to the lips fails to reverse the action of oral cilia. Parker (1919) succeeded in bringing about all of these phenomena in pieces of ciliated membrane cut from the gullet of *Metridium*. Hence the described behavior changes are attributable to increases in the excitation threshold of local receptors (sensory adaptation), brought about by continued stimulation.

Other important changes may occur in the feeding act. Nagel (1894) found that the tentacles of the fairly well fed sea anemone contract only partially when a wad of filter paper touches them, although they contract about morsels of fish meat or about pieces of filter paper soaked in meat juice and bend toward the mouth. The tentacles of the poorly fed animal readily grasp an object such as a piece of filter paper which affords only a tactual stimulus, but after having been plied with food for a time the tentacles fail to take filter paper, and finally they may not even respond to pieces of meat. The facts suggest that while the tentacular receptors of the unfed animal have a low excitation threshold and thus are adequately stimulated by weak contact alone, continued stimulation somehow raises their threshold. Hence, in the gorged animal a summation of contact and chemical stimuli is necessary to arouse the tentacles, or even this may be ineffective. Let us see whether the change is local in nature, as this implies, or is a general one as some have maintained.

In an experiment by Jennings (1905), after tentacles on the right side of a sea anemone's disc had been continuously fed with pieces of crab meat, not only did those tentacles become unresponsive but also tentacles on the left side of the disc which had not been directly stimulated during the experiment. He attributed the result to "loss of hunger," which affected the animal as a whole. Allabach (1905) found that the tentacles of the unstimulated left side of the disc were normally responsive after continuous feeding had made those of the right side unresponsive, provided that morsels of food carried by the tentacles of the right side were removed before they had entered the mouth. Jennings' result, therefore, was produced in some manner by food which had been ingested.

Parker (1919) punctured the column wall of an anemone and injected a quantity of meat juice into the interior with a fine syringe. All of the animal's tentacles then became unresponsive. He explained that the chemical, circulating through the interior of the body, had passed into the hollow tentacles (see Fig. 10) and had diffused through their walls to the receptor cells, ultimately "adapting" these cells as if they had been continuously stimulated from the outside. In Allabach's experiment food had not reached the interior, and so the unfed tentacles were not changed in any respect. Adaptation of receptor cells in the tentacles, by gradually raising their excitation threshold, thus is responsible for the progressively decreasing responsiveness of a continuously fed sea anemone.

To recapitulate, the following factors are responsible for normal feeding behavior in the sea anemone and for variations in its form: (1) The constant beating of cilia which cover the tentacles, sweeping small objects toward the tentacle tips; the secretion of mucus on the stimulated tentacle; and the discharge of nematocysts as accessory components. (2) The contraction of a stimulated tentacle and its bending (in most polyps) first toward the stimulated side, then toward the mouth—as dependent upon the structure of the tentacle itself. (3) The condition of sensitivity of the receptor cells in the tentacles, as dependent upon preceding stimulation. (4) The nerve net, transmitting excitation from tentacles first stimulated to others and to central portions of the disc, thereby bringing the mouth region into play. (5) Reversal of the stroke of oral cilia by nerve-net impulses reaching the mouth (and by direct stimulation), causing transported substances to be swept into the mouth. (6) The susceptibility of ciliated epithelium of mouth and gullet to adaptation ("sensory fatigue").

There is no dominating center in the diffuse conduction system of the polyp; the general direction of transmission depends upon the order in which the parts are excited. For example, in feeding the tentacles are aroused first, and nerve-net impulses from them bring the mouth region into play. In the expulsion of large undigested objects from the mouth the tentacles are brought into action by nerve-net impulses originating at the center of the disc. The manner in which a given part functions (*e.g.*, the action of tentacles in feeding, or the action of the column base in locomotion[1]) essentially depends upon the structure of that part itself. However, the fact that the activity of one part may excite other parts through the nerve net unites the localities in behavior according to (1) the strength and localization of the stimulus, and (2) the condition of the animal at the time. The second factor merits further consideration.

Modification of Behavior in the Hydroid Coelenterate.—We are familiar with the manner in which the adaptation of receptor cells accounts for temporary changes in hydroid food taking (pp. 42*ff.*). There is no evidence that such changes involve the nerve net, but for other instances of temporarily altered behavior the case may be different.

Jennings (1905) observed that after an expanded *Aiptasia* had been caused to contract by the fall in succession of two or three drops of water from a height of 30 cm. onto its disc, the animal re-expanded fairly promptly but usually failed to react to similar stimuli within the next 5 min. Parker (1919) attributes the temporary change to the production of a neuromuscular state which prevents conduction for a time, and which thus resembles the "refractory period" of higher animals (Chap. XIV).

[1] In sea anemones (Parker, 1919) locomotion is effected by means of waves of muscular contraction which pass across the base from one side to the other. Any part of the periphery may lead, according to the conditions of stimulation. Parker failed to observe any change in locomotion when the upper part of an animal (with oral disc and tentacles) was cut away from the base. Pantin (1935) has demonstrated the neurophysiological basis of this condition.

The ability to maintain a given position for some time is general among sea anemones and other polyps. The column of *Aiptasia* often bends irregularly when the animal extends from a crevice into open water, and Jennings (1906) noted that these irregular postures persist for some time after the animal is removed from its place. He caused an individual to bend for some time toward the left, then stimulated it repeatedly, but this position was resumed each time until 15 general contractions had been elicited. The animal then bent to the right, and returned to that posture after an induced general contraction. The maintenance of a posture presumably is attributable to the survival of induced tonus in the neuromuscular tissues of the column. Repeated stimulation is usually required to remove a posture and induce a new one, but in certain cases a single strong stimulus (*e.g.*, an electric shock) which elicits a vigorous general response will remove the posture which is in force at the time.

Many sea anemones regularly contract at ebb tide and expand at flood tide. Bohn (1907) reported that this "tidal rhythm" persisted in *Actinia* for from 3 to 8 days after specimens had been placed in a dark laboratory tank. However, Parker (1917) failed to confirm this for *Calliactis* (*Sagartia*), and Gee (1913) failed for *Cribrina*, although both of these polyps plainly show a tidal rhythm in their normal habitat.

Under normal conditions the polyp *Metridium* shows a well-marked day-night rhythm which is controlled by periodic changes in illumination (Hargitt, 1907), but in running water in the dark specimens remained expanded for more than 36 hr. after their removal from tide pools and later behaved quite irregularly (Parker, 1917). It may be concluded that the tidal and day-night rhythms, when they occur in polyps, depend upon rhythmic changes in external conditions, and that they persist only as long as these external changes directly present periodic alterations in stimulation.

Problems in the Behavior of the Coelenterate Medusa. *The Control of Movement in Medusa.*—The mechanism of movement in the medusa (or jellyfish) best expresses the basic properties of behavior in that form of coelenterate. The medusa (Fig. 12) is moved through the water by rhythmic opening and closing movements of its bell (umbrella). Each contraction forces water from the bell cavity, and spurts the animal forward; and contraction is followed by an expansion which slows forward movement as the bell is flattened by the elasticity of its tissue. In this simplest form of medusoid movement, forward propulsion is attributable to the rhythmic contraction of a band of circular muscles which borders the bell and behaves like a sphincter (Note 1, p. 38).

The bell is contracted in response to impulses which are conducted through the nerve net spread out below the epidermis next the muscle layer (Fig. 11*C*). This excitation originates periodically in the marginal sense organs which are arranged at intervals around the periphery of the bell, and spreads symmetrically up over the bell, contracting the muscle as it passes. Romanes (1885) removed all but one of the eight marginal sense organs of *Aurelia* without

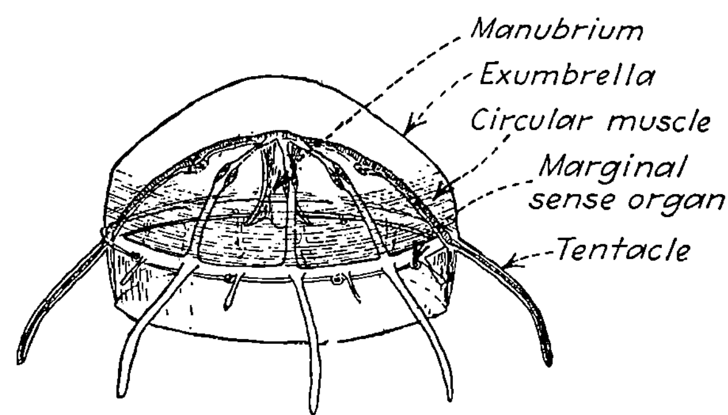

FIG. 12.—Medusa (*Rhopalonema velatum*). (*Redrawn from Hertwig's Manual.*)

interrupting the rhythmic pulsations of the bell, but contractions ceased when the remaining sense organ was extirpated.

Mayer (1908) suggested that the impulse for contraction periodically arises through the accumulation of a chemical substance (sodium oxalate) in or close to the marginal sense organs. When a contraction has exhausted one supply of this substance, more is formed during the relaxation of the bell.[1] When a sufficient quantity has accumulated the substance interacts with chemicals in the sea water, setting free powerful nerve stimulants (sodium chloride and sodium sulphate). These chemicals excite a nerve-net impulse which contracts the circular muscles and causes the bell to pulse, after which there is a relaxation of the musculature while

[1] The rate of bell contraction depends upon prevailing temperature, which indicates that the process is chemically conditioned. In *Cassiopea*, Harvey (1911) recorded 25 pulsations of the bell per minute at a temperature of 30°C., but only 8 per minute at 16°C.

the exciting chemical again accumulates. It is probable also that the stretching of nerve net in the opening of the bell plays a part in arousing the next contraction.

The manner in which the nerve impulse spreads over the bell from the margin, where it starts, is dependent upon the fact that nerve-net tissue conducts diffusely, *i.e.*, equally in all directions from the point of stimulation. Romanes (1885) and others have shown by "cutting experiments" that an impulse may pass from one part of the bell to another so long as these parts remain connected by a small isthmus of tissue.

Swimming is a more complicated process in some medusae which have a more complex neuromuscular structure. Bozler (1926) reported that in certain medusae which possess both circular muscles (which contract the umbrella) and radial muscles (which open the umbrella), the two systems of muscles work antagonistically under a somewhat involved nervous control. In *Pelagia*, for instance, the circular musculature has a lower stimulation threshold, a longer latent period, and a longer phase of contraction than has the radial musculature. In this scyphozoan the contraction of the radial muscles hastens the normal relaxation of the circular muscles, a kind of reciprocal inhibition with a nervous basis.

In certain medusae the ganglion cells and intermeshed fibers of the nerve net are more concentrated around the bell margin, forming the *nerve ring*. Such concentrations of nerve-net tissue transmit impulses more readily from the most sensitive parts (tentacles, marginal sense organs, and manubrium) to the musculature of the bell, but they do not change the manner in which bodily parts are combined in behavior.

The nerve net contains no "controlling centers," but conducts according to the manner in which stimuli act upon the receptors. Herrick (1924) had outlined its role in the following terms: (1) The nerve net brings a much larger body under the influence of stimuli, and draws parts closer together in function, than is possible with a sluggish neuroid transmission (pp. 38*f*.). However, it does not improve qualitatively upon neuroid transmission, since it does not conduct with intrinsic directionalization, as does the "synaptic nervous system" of higher animals. (2) The nerve net also permits *summation* of impulses. (3) Displacement of one activity by another may take place more efficiently through *interference* of the corresponding nerve-net impulses. Evidence of the importance of these functions will be found in the following pages.

External Stimulation and Variability in Behavior.—As we have seen, the foundation of behavior in the medusa is symmetrical movement produced by impulses which are rhythmically discharged over the bell through the nerve net, originating at the bell margin. Upon this basis is superimposed the altering effect of external stimulation, or of temporary changes in the metabolic condition of the animal. External stimuli alter the movement pattern according to the manner of their presentation. To illustrate, when a tentacle on one side of the bell is strongly stimulated, in the ensuing response, the musculature of the locality contracts most vigorously and consequently the bell is turned from the stimulated side (Yerkes, 1902*a*). The nature of responses to local stimulation is particularly dependent upon the localization of the stimulus, since nerve-net tissue conducts first and most strongly to the nearest part of the action system. That nerve-net tissue conducts with a decrement (*i.e.*, its impulses act less effectively upon muscles at greater distances from the point of stimulation) is a fact of fundamental importance for behavior in animals which are limited to this type of conduction system.[1]

Yerkes' (1904*a*) studies on reaction time in the medusa show the direct relationship that exists between stimulus strength and the promptness and nature of the response, and suggest the importance of this fact for the relative dominance of stimuli over behavior. Dark-adapted specimens of the small medusa, *Gonionemus*, reacted to weak daylight in 9.4 sec., to strong daylight in 7.0 sec., and to strong sunlight in 5.5 sec., on the average. In decided contrast, an electric shock from a four-cell battery brought a response in the average time of 0.605 sec. Strong stimuli exert a much more prompt and powerful control over behavior, and when two stimuli of different strength are presented simultaneously to different localities the nerve-net impulse excited by the stronger must occlude (*i.e.*, interfere with) the impulse of the weaker stimulus. Mayer (1906) observed that when two waves of excitation were discharged from a marginal sense organ in opposite directions around a prepared circumferential ring of bell tissue, the weaker wave would disappear

[1] In nerve-net conduction, repeated stimulation is necessary for the spread of excitation to more distant regions. Pantin (1935) has shown that a slow series of nerve-net impulses is aroused by a continued stimulus, and that succeeding impulses travel farther through the nerve net by virtue of facilitation from the effect of earlier impulses (summation). The number and frequency of impulses are greater the stronger the stimulus.

when it met the stronger wave on the opposite side of the ring. The stronger wave, however, would continue to pass around the ring in its original direction.

The Mechanism of Feeding in the Medusa.—Feeding activity varies considerably from group to group, in correspondence with differences in equipment. Its pattern is particularly dependent upon the number and length of marginal tentacles, and the length and form of the manubrium.[1] Of course, the size of the medusa is also an important factor. Medusae capture a variety of prey, and some of them are able to catch fish without suffering injury to the delicate umbrellar structures (Lebour, 1923).

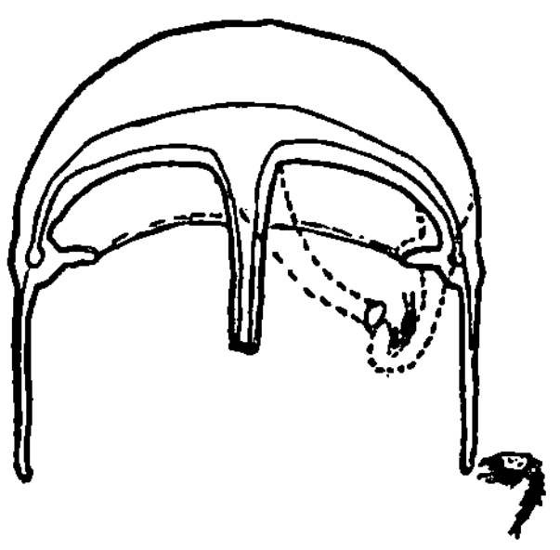

FIG. 13.—Schematic illustration of tentacle reaction and manubrium reaction in the medusa (see text).

Feeding in the medusa is a continuous activity once it is set into motion by the adequate stimulation of marginal tentacles. These contract about the object, and their further contraction pulls the object upward toward the bell. Meanwhile the manubrium has become excited to swing over toward the stimulated sector (Fig. 13); it envelops the prey with its expanding end, and shortly the process of digestion begins.

The first point of interest is the nature of the stimuli which are adequate to arouse the tentacles. Generally the tentacles are first excited to bend about by a chemical which diffuses to them from the object. If the chemical is very weak, or is inorganic, as Yerkes observed for *Gonionemus*, the tentacles shorten but slightly, and their response to weak contact is similar. But when weak contact is accompanied by a chemical stimulus, or if the touched object brushes quickly across a tentacle, the tentacular response is "contraction and shortening" as in the normal feeding act. In the last two cases summation of stimuli plays an evident part; in one case the effect of weak contact reinforces that of a weak chemical, while in the other case the effects of tactual stimuli are summated. The importance of tactual summation in accounting for the superior

[1] The manubrium projects from the center of the subumbrella, and is the external opening of the digestive tract, *i.e.*, its end is the mouth. In some medusae it is tube-shaped, in others it is divided into frill-edged arms, and there are still other types.

effectiveness of contact with a moving object has been pointed out by Kafka (1914).

This point is evidently of importance in an interesting variant from the normal mode of medusoid food taking, described by Orton (1922). The mucus-covered exumbrellar surface of *Aurelia* entraps a variety of objects (small aquatic organisms, debris) which in time are carried down to the margin of the umbrella by the constant beating of cilia. When a mass of material has accumulated at a given place on the margin of the umbrella, apparently the point of adequate stimulation of marginal receptors is reached, for the manubrium then reaches over and envelops the material.

In any species the "food response" has a typical pattern because it is a reaction elicited from an animal of given structure by stimuli which lie within a given range of intensity. The stimulus, if too weak, brings an ineffective local response; if too strong, quickly turns the animal away; but if adequate, it elicits a tentacular response which captures the object, a contraction of the bell margin which pulls tentacle and prey toward the manubrium, and an appropriate bending of the manubrium. It is perhaps unnecessary to say that the medusa does not respond to "food" but to an object which sets up certain energy changes in its tissues (*cf.* pp. 23*ff.*).

The manner in which the individual tentacle responds to stimuli, and the conditions of its adequate stimulation, depend upon the sensory and neuromuscular makeup of the tentacle itself. Yerkes (1902*b*) and others have demonstrated that tentacles cut from the bell react as do tentacles on the intact animal when equivalent stimuli are presented.

To continue our analysis of the feeding response, when the nervous excitation reaches the base of the stimulated tentacle in normal feeding, it causes a contraction of the nearby bell musculature which pulls the prey-laden tentacle toward the center of the bell. Various explanations have been offered to account for the bending of the manubrium toward the activated tentacle. The best explanation was suggested by Romanes (1885) and formulated by Loeb (1900). It may be stated as follows: From the excited marginal point x the nerve-net impulses spread diffusely over the bell, *i.e.*, radially in all directions (Fig. 14*A*). Since the wave reaches the manubrium first at the point x_1 in the figure, the manubrium contracts first and strongest on that side, and hence bends directly toward x.

In one of his experiments with *Tiaropsis indicans* (a species in which the manubrium touches stimulated marginal points with marked accuracy) Romanes made a cut concentric with the bell margin as along line *xy* in Fig. 14*B*, so that direct nerve-net conduction from any point between 1 and 3 on the margin and part 1_a–3_a of the manubrium base was prevented. When point 2 was then stimulated, the manubrium bent from one part of the bell margin to another, but did not reach 2 with normal precision. Stimulation of a point closer to 3 gave rise to successive nerve-net impulses which acted upon 3_a more effectively than upon 1_a, since the distance 3-y-3_a was shorter than the distance 3-x-1_a and since nerve net conducts with a decrement. For that reason the manubrium was more accurate in bending toward a touched marginal point closer to 3 than toward 2 Romanes (p. 113) reported

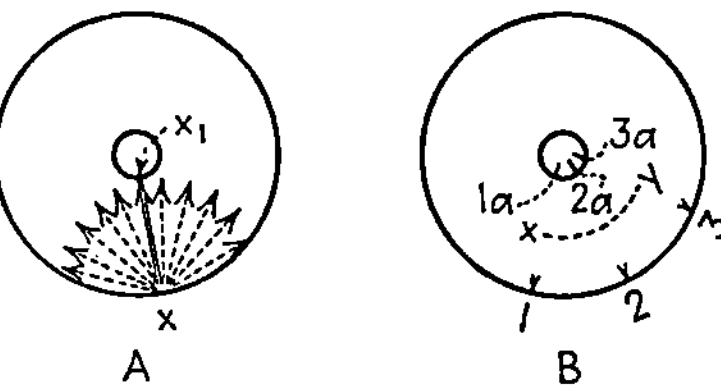

Fig. 14.—(*A*) Scheme of the manner in which nerve-net impulses, diffusing across the bell of medusa, cause the manubrium to bend toward the stimulated point *x* on the bell margin (see text). (*B*) Illustration of Romanes' and of Bozler's experiments (see text).

> . . . *gradations* between the ability of the manubrium to localize correctly and its ability to localize at all, these gradations being determined by the circumferential distance from the end of the cut and the point of stimulation.

This evidence, together with the results of Bozler's (1926) experiments, shows that the direction in which the manubrium bends during feeding is dependent upon the relative effectiveness of nerve-net impulses which reach its base on different sides.

The neuromuscular structure of the manubrium itself is responsible for the behavior of that member. A manubrium which Yerkes (1902*b*) cut from *Gonionemus* and pinned by its base in a dissecting dish bent toward whichever side of its base was stimulated, or bent and engulfed a piece of food which touched one side. Romanes (1885) proved that, although a large part of the manubrium was cut away, the remaining section of stump would still bend toward a stimulated part of the bell.

Recapitulation

We may summarize the facts for coelenterate behavior in general by reviewing those for the behavior of medusa. In the medusa, the manner in which tentacles and manubrium behave is locally determined, since the parts respond in their characteristic manners even when isolated, if adequately stimulated. The responses of a tentacle depend upon the specific effect which energy exerts upon its tissues. This means that any variation in the stimulus must bring a somewhat different response. Feeding activities represent only a part of the repertoire. The same may be said for the behavior of the manubrium, and for that of the umbrella. These parts are not only bound together metabolically (*cf.* p. 39) but also are caused to cooperate in behavior according to the "strength" and the origin point of nerve-net impulses conducted to them. The coordination of parts in the coelenterate is thus dependent upon the manner in which external stimuli happen to excite the nerve net, and upon the sequence of excitation of body parts as determined by their spatial relationship. Control over behavior is centralized in no portion of the conduction system but depends upon the relative irritability of parts, upon their accessibility to stimulation,[1] and upon the play of external stimuli.

For the sea anemone there is evidence that the normal irritability of the parts may be temporarily altered through "adaptation" of receptor cells, and that during its survival the change in sensitivity may influence behavior. Further, there is evidence that a section of the neuromusculature may maintain a posture (*i.e.*, a given condition of contraction) for a time. These are primitive forms of "learning," since they are the product of the animal's experience as it temporarily alters the irritable and contractile tissues, and since the nature of behavior is altered during this interval. However, such changes lack a related continuity and one is not built upon another, because the coelenterate returns to its original pattern of behavior after each change. The evidence we have cited makes the reason for this statement quite apparent. The facts show that a nerve-net system *connects* the bodily parts but does not *control* them except in a limited and temporary sense, and that the manner in which this type of conduction system transmits energy from part to part may be altered only briefly. The behavior of an animal thus constituted is not significantly above the psychological level of the protozoan.

[1] Because of their position on the margin of the bell, the tentacles normally are stimulated first, and therefore control most of the special responses.

Suggested Readings

Gee, W. 1913. Modifiability in the behavior of the California shore-anemone *Cribrina xanthogrammica* Brandt. *Jour. Anim. Behav.*, **3**, 305–328.

Jennings, H S. 1906. *Behavior of the lower organisms.* (1923 ed.) Columbia University Press. Pp. 188–216.

Parker, G. H. 1919. *The elementary nervous system.* Philadelphia: Lippincott. Chaps. II and III (*Porifera*), Chaps. XI and XII (*Coelenterates*).

Romanes, G. J. 1885. *Jelly-fish, star-fish and sea-urchins.* New York: Appleton. Chap. V.

CHAPTER III

A TRANSITION STAGE IN SPECIALIZATION AND IN BEHAVIOR CONTROL: PHYLUM ECHINODERMATA[1]

Echinoderms are inhabitants of the sea and many of them are common along the seaside, slowly moving about at night and resting in crannies during the day. The psychological status of these animals is subject to question. They are radially symmetrical, and it is significant that radial symmetry appears to be the primitive structural pattern so far as its importance for behavior is concerned. In few animal groups other than the echinoderms are the peculiarities of bodily form and equipment as directly and strikingly effective in determining the nature of behavior. The best known representative of this phylum, the starfish, has five rays arranged symmetrically about a vertical axis and displays a corresponding radial symmetry in its sensory, conduction, and action equipment. These characteristics suggest that the animal is qualified to occupy a psychological position closer to the coelenterates than to the bilaterally symmetrical animals of higher groups. In other words, the echinoderm bodily organization of parts about a central axis (radial symmetry) may limit and determine the nature of functional organization in behavior much as it has in the invertebrate groups which we have studied. This possibility will be tested by an examination of the manner in which echinoderms move about and adjust to typical environmental difficulties.

Factors Determining the Direction of Locomotion and of Related Behavior

A Limited Permanent Difference in the Dominance of Parts.— Although the echinoderms share with the coelenterates the character-

[1] The phylum *Echinodermata* includes radially symmetrical animals with five equivalent body sections and a body cavity. Class *Asteroidea* (starfish, Fig. 19) is characterized by five rays and by tube feet as blind endings of the water-vascular system; class *Ophiuroidea* (brittle star, Fig. 15) by slender arms marked off from the disc and by the absence of tube feet; and class *Echinoidea* (sea urchin, Fig. 16) by a ball-like body bearing plates with movable spines and without arms.

TABLE 5.—ECHINODERMS

RECEPTOR EQUIPMENT	FEATURES OF SENSITIVITY
Chemical. Specialized cells in clusters ("buds") or in pits. Thickest at ends of rays, on ventral surface of disc near mouth.	Tentacles, tube feet, and mouth region especially sensitive. General body surface sensitive, but ventral surface somewhat more so than dorsal.
Mechanical. Primary receptor cells in epidermis, generally, hairs projecting to exterior. Tube feet and tentacles have concentrations of these, as have bases of spines.	Tube feet very responsive to contact, this important for locomotion. Contact is basic for spine responses of sea urchin. Local contact arouses sensitive pedicellariae.
Static. Bipolar cells in muscles of body wall, may be aroused by muscle contraction, hence proprioceptors. Starfish, brittle star, sea urchin lack statocysts proper; some echinoderms have them.	Starfish walks freely upside-down on surface film; righting response of overturned animal initiated by stimuli from tube feet. Sea urchins cluster on under side of rocks. Stalked echinoderms (crinoids) maintain upright position mainly because of fluid content of tissues in stalk, muscular tension also important.
Light. Starfish has pigment-lined cups of sensitive cells at ends of arms, each cup covered with clear lenslike structure. Others, *e.g.*, sea urchin, have pigment spots. Skin generally contains sensitive cells.	Starfish much reduced in sensitivity to light when "ocelli" removed, skin sensitivity accounts for remaining effect of light. Echinoderms in general feed in dim light, do not remain long in bright light.

CONDUCTION

Central nerve ring in disc, containing ganglion cells. Radial nerve strand extends from the nerve ring into each ray. Nerve net predominates in body wall, closely associated with tube feet, pedicellariae, and layers of large muscles.

ACTION SYSTEM

Clumps of pedicellariae, tiny nippers, on body surface of most echinoderms. Movable spines on surface of sea urchin and others. Starfish moves by tube feet, hollow bulblike extensions of the water-vascular canal, arranged in rows on ventral surface of each ray. Brittle star "walks" on rays by action of the large ray muscles; sea urchin gets about by movement of tube feet; when excited, by spine movement. In feeding, object is held by tube feet in starfish, while rays often contract under disc, and the stomach is everted around it.

istic of radial symmetry, their locomotion typically is not in line with the principal axis of symmetry (*i.e.*, with top of disc forward) as the medusa moves with the top of its bell forward. In contrast, the starfish moves at right angles to this axis, with one or two of its arms in advance. To be sure, among the coelenterates, the sea anemone appears to fulfill this condition also by crawling on its base with any part of the margin ahead. Experiments show, however, that in echinoderm locomotion one marginal zone is capable of taking the lead more frequently than do others. We must ascertain whether this permits an organization of activities which, in part at least, overcomes the handicap of radial symmetry. This is an important point for animals which have a specialized dorsal side, and a specialized ventral side which is typically in contact with the substratum, as is the case in this and in higher phyla.

Preyer (1886) reported that certain starfish advanced more frequently with given rays forward, and Jennings (1907) found a certain pair of rays, that on the side of the madreporite (the entrance to the water-vascular system), most active in the righting movements of an overturned starfish. Cole (1913) undertook a study of normal movement in order to test the matter. The starfish *Asterias* was held on the finger tips for at least 30 sec. to insure quiescence, then was carefully placed ventral-side-down in the center of a dish. On each trial a different ray was held nearest the experimenter, to equalize the effects of tactual stimulation. A record was made of the ray or pair of rays that led each time in locomotion. The rays were lettered with respect to the position of the madreporite, as Jennings had done, *a* being the ray on the right of this opening, *e* the ray on the left. Tests with ten individuals gave the following results.

Ray or rays in advance	*e*	*ea*	*a*	*ab*	*b*	*bc*	*c*	*cd*	*d*	*de*
Number of times in advance	86	61	55	35	40	40	33	43	64	43

Ray *e* led most frequently, either alone or in combination with ray *a*, and the rays *b* and *c* took the lead least frequently of all.

The results may be expressed more concisely by adding to the separate total for each ray one half the number of times that ray led in combination with adjacent rays. The frequencies then become: *e*, 137; *d*, 107; *a*, 103; *b*, 77; *c*, 75.

One might expect to find a stable anatomical and physiological difference among the rays which is responsible for the demonstrated ability of certain ones (*e*, *a*, and *d*) to lead movement most frequently. Cole found these rays on the average somewhat longer than the others, and we may believe that a certain difference in sensitivity and in conduction is also in their favor. That the madreporite side of the disc possesses a metabolic superiority which contributes to the dominance of adjacent rays is indicated by Crozier's (1920) observations on the Bermuda starfish, in which, following division (asexual reproduction), short rays next the madreporite were observed to lead more frequently in locomotion than did longer rays elsewhere.

However, Cole found a tendency for the lead to rotate among the rays on successive trials, owing, he thought, to the effects of muscular fatigue. In some individuals given rays retained their dominance for longer times than in other cases. While this stable difference among the rays is an important factor in determining which rays shall dominate movement at a given time, it cannot be of final importance in determining the pattern of the animal's behavior. So far as dominance at a given time is concerned, there are other factors to be considered as well as this one. An understanding of the extent to which these additional influences may supersede in importance the one already identified should tell us much about the psychological limitations of the echinoderm.

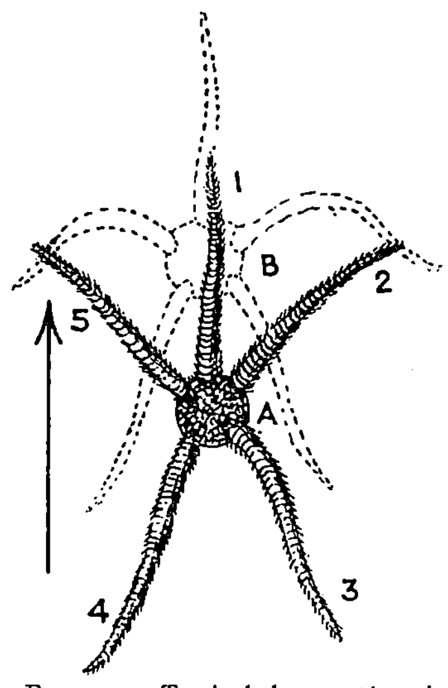

Fig. 15.—Typical locomotion in the brittle star. Ray 1 is held stiffly ahead, rays 3 and 4 are being dragged behind, while rays 2 and 5 are moving the animal. In position *A* they are extended forward, in position *B* they are stroking backward, thus pushing the animal ahead. (*Redrawn from Romanes.*)

Relative Condition of the "Contending" Rays at a Given Time.—Although, in the starfish, rays adjacent to the madreporite dominate somewhat more frequently than do others, observations over a period of time do not suggest that this difference is a marked one. In locomotion and in general activity dominance is exchanged among the rays, and under appropriate conditions any ray or pair of rays may be temporarily in the lead. It is characteristic of the starfish that it may move for a considerable time with any given ray in advance, and the same is true of the brittle star (Fig. 15).

Survival of the Effects of Recent Stimulation.—The stimulation to which the animal has been subjected in the preceding interval of time has much to do with the manner in which the various rays participate in activities, and therefore is a major determining factor in behavior. An experiment by Cowles (1910) on the brittle star shows that a difference in the action of stimuli upon the arms may greatly influence the subsequent course of behavior.

On the seashore the brittle star moves about fairly rapidly capturing food at night, but remains out of the light in the daytime. If its sheltering rock is overturned, the animal generally moves in the direction toward which the covering swings. Suspecting that the preceding contact with the rock in such cases has much to do with the movement of the exposed brittle star, Cowles duplicated the situation in the laboratory by placing the animal in a square dish with one arm in contact with a corner wall. After a time, when brought into the center of the dish and stimulated to move by the light, the animal generally moved off with the previous-contact ray in the lead. Even when it chanced into a darkened corner of the dish the subject did not remain long unless that ray was tactually stimulated. For this reason it did not stop when it entered a darkened area in the center of the dish. Any one of the rays, provided it had been the one in contact with a surface during the preceding period of rest, could lead in movement and thus determine the position in which the brittle star would come to rest. Since Cowles reported no evidence that the arms adjacent to the madreporite have a superiority in such behavior, that factor cannot be too effective in this echinoderm in comparison with the one under consideration.

Influence of Preceding Activity upon the Righting Response.—Let us take a further example of the effect that previous stimulation or the previous condition of activity may exert upon behavior. A brittle star is able to right itself when turned upon its back. Success in righting depends, as in the starfish, upon the possibility that one or a pair of rays will be more active in righting than are others. Cowles noted that a brittle star which was overturned after continuous contact with a wall was generally righted by movements in which the previous-contact arm was most active. Jennings (1907) suggested that the direction of locomotion before overturning might be an important factor in the case of the starfish. This was proved by Cole (1913). In a thorough test, Hamilton (1922) established the shift from the rays which were anterior before overturning as 0.6 interradius for 64 observations. Thus, as an approximation, a temporarily dominant ray would be displaced by a neighbor every two trials. He concluded from this and other findings that the previous pattern of locomotion or the state of activity otherwise (quiescence, general contraction) is directly responsible for the manner in which an upset starfish becomes righted (pp. 68*ff*.). The survival of a condition of activity, or of the effects of previous stimulation, is a most important factor in echinoderm behavior.

Relative Dominance of Rays May Depend upon Strength of Stimulation.—It is a basic property of behavior in these radially symmetrical animals that an intense stimulus from any direction produces movement away from the stimulated part of the body margin. Thus when the brittle star in Cowles' experiment was overturned in light of low intensity, the animal was righted through the dominant activity of its previous-contact ray, whether this swung the body toward the light or away from it; but when overturned in strong directionalized light, the animal was righted away from the illuminated side in any case. Jennings had reported a similar observation from starfish behavior, but he also observed that once the righting reaction had started, a shift in the direction of intense illumination had little or no effect upon the activity. This reminds us that in the determination of any response all three of the principal controlling factors summarized below must be involved.

We have found that the direction of starfish (brittle star, etc.) activity at a given time is determined by (1) a permanent physiological difference among the rays, (2) preceding stimulation, previous condition of activity, and (3) the strength of the present stimulus. As in the case of *Amoeba*, the third factor is most important when the stimulus is intense, but normally the interrelations of these factors make for a highly variable behavior. The constant shift in the dominance of rays indicates the importance of factors 2 and 3 and suggests the improbability that, in the ray which is most frequently in dominance, there exists a special controlling center which is absent from other rays. This presents the question of what mechanism *does* unite the parts in echinoderm behavior.

The Problem of Organization in Echinoderm Behavior

Local Activities and Their Combination.—In the first place, certain of the local activities are of no more than secondary and accessory importance for general behavior. The best examples are found in the sea urchin (Fig. 16), an echinoid which has mobile spines over the horny surface of its ball-shaped body as well as various types of pedicellariae (movable stalks bearing little pairs of jaws). The distinctive and independent manner which characterizes the responses of these structures to stimulation led von Uexküll (1890), after a detailed investigation of sea urchin activities, to call the animal a "community of reflexes." When he cut from the animal a single spine or a single pedicellaria, the structure responded to stimulation in its characteristic manner. The spines typically bend toward a spot which is mechanically stimulated, but if the stimulus is intense they bend away. In the strongly stimulated

FIG. 16.—A sea urchin in the last stages of its righting response. Note tube feet and spines. (*Modified from Romanes.*)

animal the spines consequently are involved in locomotion, stroking vigorously back and forth, although normal movement depends upon the tube feet (p. 67) alone. Impulses are transmitted from the stimulated point through a dermal nerve net to these structures; therefore, as we should expect, the stimulus acts most strongly upon the spines in the immediate locality, weakly upon distant spines. This depends upon conduction with a decrement (p. 50) in the nerve net.

Romanes (1885) cut away the central nerve ring from a sea urchin without changing the local responses of spines and pedicellariae in the least. But the spines of such operated animals were not involved in locomotor responses to strong stimuli, as they were in the intact animal. After a circular section of tissue above the shell had been isolated by cutting, stimuli within this area could cause a reaction of its spines and pedicellariae without exciting surrounding parts at all.

In the sea urchin von Uexküll (1889) found that the three types of pedicellariae are subject to excitation by different degrees of stimulation. For the starfish, Jennings (1907) studied the action of these structures in some detail. This echinoderm has large pedicellariae which are called into play by weak stimuli and small pedicellariae (grouped into "rosettes") which respond to stronger stimuli. For the latter type very weak contact usually is ineffective, but if repeated, as by the movements of a small organism along the back of a ray (summation), the stimulus causes the rosettes to rise and the jaws of the pedicellariae to open. The jaws close when further contact is presented. A very strong stimulus, however, causes the rosettes to shrink at once. Because of structural peculiarities of the pedicellariae, the struggles of an object upon which one of them has closed only increase its grip.

From an excited locality impulses spread through the nerve net to other pedicellariae, increasing their muscular tonus (readiness to contract) so that they close at once when the writhing prey contacts them. Jennings ascertained that a pedicellaria, which was partially aroused through nerve-net excitation from a distant point, would respond much more readily when touched than would structures more distant from the point first stimulated, a decrement-conduction phenomenon. After closing upon an object the pedicellariae often remain contracted for some time, and when they open through muscular fatigue, the released (by this time, paralyzed) capture is "stepped" toward the mouth by the tube feet (p. 65). The latter structures, little "suckers" on the ventral surface of each ray (Figs. 17, 18), typically are called into play by chemicals which diffuse from the crushed prey, and they pull it from the pedicellariae.

The mechanism of the tube-foot action which transports small objects to the mouth (in the center of the disc on the ventral side) is much like that of the disc tentacles in the food taking of a sea anemone. If the stimulus is strong and has extent (*e.g.*, as when a clam, or a fish, is captured) the large arm muscles are called into play and the arm bends. In certain cases observed by Jennings (1907) the pedicellariae held a captured crab until finally, evidently through summation, the ray or rays holding the object bent toward the mouth. This raised the central part of the animal, and after five minutes, in one case, the ray had brought the prey close to the mouth. The stimulation of the mouth region excited the protrusion of the soft stomach lobes, these pressed around the object, and digestion began.

The parts of the starfish, *i.e.*, pedicellariae, spines, tube feet, rays, and stomach, thus are capable of functioning together in behavior. But their cooperation is a sluggish one, and interference among them is frequent, as when pedicellariae tear out tube feet when both types of structure have taken hold of an object, or those of one ray catch and hold the tissue of another ray.

With some understanding of the local activities we are ready to study their organization in general behavior.

Starfish Locomotion: A Study of Coordination.—Dominance of one part over others, however brought about, is the manner in which organization is introduced into the action of tissues in the behaving organism. We have ascertained three factors which are instrumental in bringing a ray or a pair of rays of the starfish into dominance over other parts of the animal at a given time. To repeat, these are (1) the stable though slight superiority of one particular region, when stimulation is equal on all parts; (2) the surviving effects of previous stimulation or activity; (3) the differential effects of present stimulation. If a part gains temporary ascendancy through a given combination of these factors, the dominance may be impressed upon other parts so that a measure of organization appears in the general adjustment of the animal.

A study of the manner in which the starfish comes to move toward a weak stimulus which has acted upon one ray, with the stimulated ray in advance, should throw light upon the nature of this organization. *The animal is moved by the concerted action of tube feet on its five rays.* We approach the difficult problem of how this is brought about by first examining the response of the ray which is initially stimulated and thus leads the movement.

Action of Tube Feet on the Leading Ray.—In his detailed study of starfish locomotion, Hamilton (1921) carefully suspended a normally sensitive starfish by strings and lightly touched the tip of one ray. The tube feet[1] near the end of this ray first extended toward the ray tip, then bent toward the base of the ray, then re-extended toward the tip and remained in that position. This action of the individual tube foot Hamilton called the "step-reflex" (Fig. 17). Meanwhile the excitation passed along the arm from tip to base, carried by the radial nerve (Romanes, 1885), exciting tube feet along its course.

[1] The tube feet are elongated hollow bulbs, blind endings of the water-vascular (circulatory) system which extend ventrally through a longitudinal groove in the ray. In the individual tube foot the longitudinal muscles which shorten the member work against the circular muscles and the pressure of fluid in the water-vascular system, which extend it.

These structures successively gave their step-reflexes, so that shortly all tube feet on the ray were extended toward its tip.

This type of response depends upon the structure of the stimulated arm, its radial nerve and tube feet. If all five rays of the suspended starfish are stimulated lightly at their tips, the above response occurs independently in each ray (Hamilton, 1921). The tube feet do not move as in normal locomotion—something is lacking.

When the starfish rests on the substratum, tube feet on a stimulated ray receive tactual stimulation when they extend toward the ray tip after having executed their initial step-reflexes. They attach and step again, each is tactually stimulated when it re-extends, and step-reflex activity thus continues. The step-reflex activity of the tube feet *pushes* the ray in the direction of its tip. Hamilton (1921) demonstrated that *the continued stepping of the tube feet on the ray*

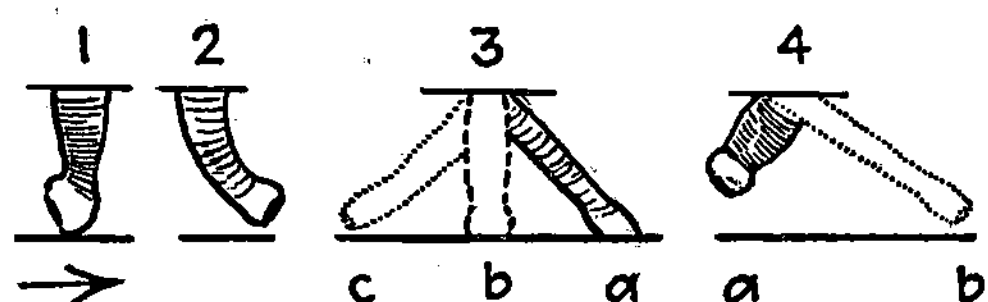

FIG. 17.—Schematic representation of a tube-foot "step-reflex." (1) A tube foot on a "following" ray of a starfish, pulled in the direction of the arrow by another ray, is stimulated by contact with the substratum. (2) This tube foot extends toward the side of its disc which was tactually stimulated. Having extended, the tube foot attaches to the substratum (3*a*), bends back and in the latter part of its arc (3*b*-*c*) pushes against the substratum, releasing with a slight kick (3*c*). Then it retracts (4*a*) and extends in the direction of locomotion (4*b*).

which is first stimulated depends upon the manner in which they contact the substratum when they extend. An excised and mounted tube foot stepped toward the side on which its terminal disc was stimulated. In another test, a starfish was suspended by threads and the end of each ray was touched with the inner surface of a hanging glass tube. The tube feet at the ends of all rays gave an initial step-reflex, then bent back toward the tips of their respective rays, but continued their step-reflexes when the re-extension brought their terminal discs into contact with the glass surface. This "walked" a tube onto each arm, and following step-reflexes brought it further toward the disc, so that tube feet successively nearer the base (which meanwhile had stepped once and re-extended) were brought into play. This process occurred on each arm, independently so that all five arms were at length provided with tubes, as sleeves.

This shows that the arms are capable of independent movement, as does Hopkins' (1927) finding that excised arms are capable of locomotion. Indeed, Hopkins was able to make the tube feet of rays on opposite sides of the disc move the rays toward their respective tips so that the specimen was "locked" for a time, and Hamilton's experiment was thus duplicated in the crawling starfish. Crozier (1920) observed that the Bermuda starfish commonly divides in this manner.

But when the weakly stimulated ray of the normal animal has begun to migrate toward its tip through tube-foot action, soon the rest of the animal is following. In this process tube feet on the "following" rays do not step toward their respective ray tips, but toward the side of the leading ray. What governs this coordination among the rays?

Cooperation of the Rays in General Locomotion.—Once forward locomotion has begun, tube feet on the leading ray (or ray pair) execute their step-reflexes in the direction of its long axis, while tube feet on other arms work in the same general direction regardless of their position on the animal. Thus, in giving the starfish additional impetus in the direction of the leading ray, tube feet on the "following" rays work at angles to the long axes of their respective rays. The tube feet continue to work in the direction of general locomotion, even though the rays to which they belong may bend variously during movement, as Mangold (1908) reported for the slender-rayed starfish *Luidia*.

Although it is generally believed that the stepping of the tube feet in a common direction is somehow controlled through the central nervous system, there is no good evidence for this, and there is no conceivable way in which such a complex nervous coordination could be effected in this animal. Following Romanes (1885) and Preyer (1886), writers on the starfish (Jennings, 1907; Mangold, 1908) nevertheless have accepted the theory of a specific nervous control of tube-foot action, on the strength of somewhat doubtful evidence. Romanes reported that the cutting of radial nerves at the bases of the rays destroyed "coordination among the rays," but he and other experimenters (*e.g.*, Hamilton, 1922) nevertheless have observed slow locomotion with coordination of tube feet following such operations. The difficulty is that the operated starfish usually does not move, because of an induced rigidity and a general contraction of tube feet. Mangold (1908) reported that after the radial nerve had been removed from a given ray, local tonus was greatly increased and the tube feet of that ray were very unresponsive.

Hamilton (1921, 1922) has presented evidence against a direct nervous control of tube-foot coordination. He concludes that *the concerted action of tube feet depends upon the manner in which tactual stimulation is externally furnished them.* As we have noted, the excised and mounted tube foot executes the step-reflex when tactual stimulation is furnished, and steps most readily toward the stimulated side. The conditions and results of Hamilton's glass-tube experiment will be recalled (p. 65). According to his results, when a weakly stimulated ray is moved tip first by its tube feet, the animal is pulled along so that tube feet on other arms are stimulated more strongly on one side of their terminal discs by contact with the substratum (Fig. 17). Hence the locomotion of the ray (or ray pair) which takes the lead (pp. 64*ff*.) determines the stepping of tube feet on the other rays. Hamilton (1921, 1922*a*) offers evidence that these tube feet receive their greatest tactual stimulation on the side of the leading rays, and hence step toward that side regardless of the angle that this direction of progress makes with the principal axes of their respective rays. It is a convincing fact that during general locomotion the tube feet do not extend, attach, and step in unison; rather, at a given moment, all possible phases of the reaction will be found among them. This militates strongly against the possibility that the common direction of their step-reflexes is nervously determined.

We may summarize as follows the mechanism of the response which is given by the normally active starfish to a weak, localized stimulus: (1) Tube feet on the ray or pair of rays first stimulated extend toward the ray tip, and contact with the substratum causes them to begin their step-reflexes. The animal is thus pulled toward the stimulated side. (2) Tube feet on other rays are probably *aroused* by nervous impulses from the stimulated ray, so that they extend randomly. (3) Tube feet on the "following" rays are forced, by the manner in which they are stimulated by their contact with the substratum, to step in the direction in which the animal is being dragged by the ray which leads at the time.

Coordination and the "Negative Response."—In the characteristic withdrawal of the starfish from the source of strong stimulation we find another opportunity to study the mechanism of movement.

When the tip of ray *a*, let us say, is strongly stimulated, a wave of nervous impulses quickly contracts the tube feet of this ray. Spreading to the other rays through nerve ring and radial nerves, the excitation also contracts their tube feet. However, by virtue of conduction with a decrement (pp. 50*f.*) the nerve impulses reach rays *c* and *d*, farthest from the point of origin, in diminished strength. Tube feet on these rays are not so strongly contracted as are those near the intensely stimulated ray or upon it, and hence these more remote tube feet are the first on the animal to re-extend after their less increased tonus has waned. These tube feet soon extend toward the tips of their rays and begin their step-reflexes upon contact with the substratum; as a result the animal is pulled in the direction of rays *c* and *d*. When the tube feet on the intensely stimulated ray *a* and the adjacent rays *b* and *e* recover from their strong contraction and extend, they are forced by substratal contact to fall into line with the movement already under way: pushing toward rays *c* and *d*. Therefore, because of the structure of the tube feet and of the individual rays, and because of the effectiveness of nervous conduction "with a decrement," the starfish may reverse its direction of locomotion when the advance point is strongly stimulated (*e.g.*, nipped by a crab). Although the persistence of the general contraction of tube feet prevents locomotion for a time and makes this reaction a sluggish one, fortunately for survival of the animal a stimulus which is sufficiently intense to induce this stage will also call the spines and pedicellariae into play.

The Righting Response and Coordination. *The Typical Pattern of Righting.*—The righting response of the starfish has been studied by many scientists, and has been the subject of much controversy. Loeb (1900) described the act as follows. When a starfish is turned onto its dorsal surface,

> . . . soon the tips of one or more arms turn over and touch the underlying surface with their ventral side. The tube feet of these arms attach themselves to this surface and the animal is then able to turn a somersault and regain its normal position. For this reason it is essential that all five arms do not attempt simultaneously to bring the animal into the ventral position. . . . In normal starfish having five arms, not more than three begin the act of turning; the other two remain quiet.

As Loeb points out, for successful righting it is essential that not all of the rays bend around and attach their tube feet. The standard case and the primary object of study is the animal which is moving with two rays definitely in the lead when it is overturned. In this case the previously leading rays, *a* and *e*, let us say, are the ones which bend dorsally (toward the substratum) and twist so that their ventral surfaces face each other, thereby permitting their tube feet to attach and right the animal (Fig. 18).

The Origin of the Stimulus which Initiates Righting.—Various explanations have been offered to account for the bending of these rays, the first phase of the righting response. Mangold (1908) suggested that the stimulus originates from the dorsal tissues, which are not ordinarily in contact with surfaces. This cannot be the case, however, since Preyer (1886) observed righting in an animal hung upside down by threads attached to its rays, and Hamilton (1922*b*) obtained normal righting when an inverted animal was balanced on the rim of a glass beaker. Irritation of the delicate gill structures has been suggested as the cause, but the gills may be removed without significantly altering the response. Fraenkel (1928) has made a thorough experimental test of the matter. His records showed that the starfish *Astropecten irregularis* normally rights within 30 to 120 sec., and that overturning is followed by a measurable latent period which intervenes before bending of the rays begins. Fraenkel became convinced that this latent period is dependent upon the condition of the sensory mechanism responsible for the arm bending, and that variations in the value of the interval under different experimental conditions should disclose the source of the initiating stimulus. Removal of dorsal tissues failed to alter the value of the latent period, as did removal of the digestive glands; hence these tissues were excluded from consideration. However, when the tube feet were removed, the latent time was greatly increased or the righting response disappeared altogether; so these structures in some manner provide the impulses which initiate righting.

Hamilton (1922*a*) had observed that when overturning interrupts oriented locomotion, the tube feet on the leading rays (and usually those on the other rays as well) extend in the direction of previous locomotion (*i.e.*, toward the tips of the dominant arms) and remain in that position (Fig. 18*A*). It is conceivable that the muscular activity involved in this bending of tube feet releases energy which reaches the large muscles of the rays through the nervous system, and thus starts the righting response.

This directionalized extension of tube feet in the overturned (previously oriented) animal is the first phase of its reaction. The tube feet remain extended until they receive tactual stimulation, whereupon they begin their step-reflexes. Thus, as Hamilton (1922*b*) maintains, righting is a special case of normal locomotion.

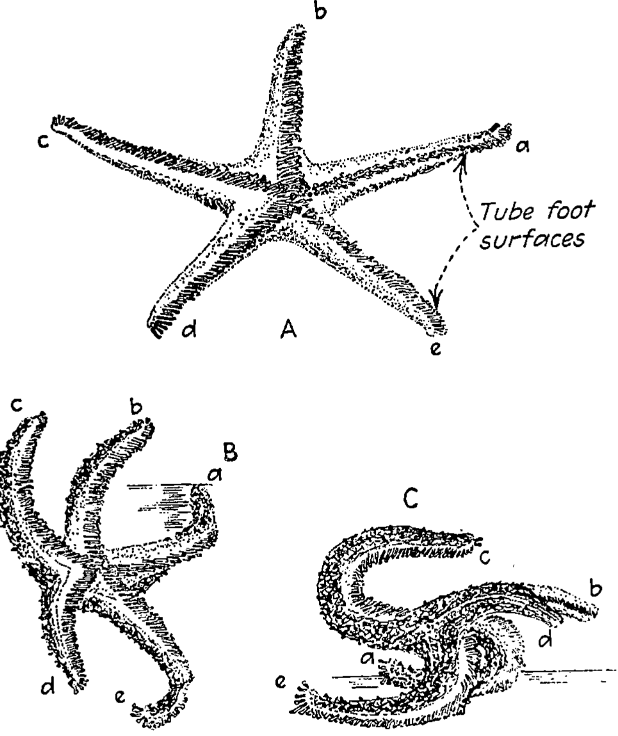

FIG. 18.—Successive stages in the righting of an overturned starfish. (*A*) Viewed from above, the animal is seen dorsal side down, with tube feet extended in the direction of previous locomotion. In *B* (from above) rays *a* and *e* have attached. In *C* (from side) the animal is almost overbalanced. (*Redrawn from Hamilton*, 1922.)

Action of the Dominant Rays in Righting.—As Hamilton describes the typical righting response, the initiating stimulus from the extension of tube feet causes the tips of the previously leading rays to bend dorsally and to twist inward so that their tube-foot surfaces face each other (Fig. 18*A*). These surfaces touch the substratum, permitting the tube feet on them to attach and begin their step-reflexes. The tube feet twist the rays still more as they step, and apply leverage to the rest of the animal so that it is raised from the substratum (Fig. 18*B*). As more of the tube feet on these two arms are attached by virtue of the twisting, and begin to step, the rays are doubled back further under the disc (Fig. 18*C*). The animal is thus finally overbalanced, falls flat upon its ventral surface, and righting is completed.

The Action of the Subordinate Rays in Righting.—Hamilton (1922*b*) maintains that it is the bending of tube feet on subordinate arms toward the bases of these arms (*i.e.*, in the direction of previous locomotion) which stimulates the arm tips to bend ventrally (*i.e.*, upward). Moore (1910) as we shall see, favors the operation of nervous impulses from the dominant arms as a cause of this response. At any rate, the upward bending of the subordinate arms continues, and tends to overbalance the body in the direction in which it is being somersaulted through the lever action of arms *a* and *e* (Fig. 18*B*, *C*), thus assisting the righting.

It is Hamilton's conclusion that starfish righting is a modified kind of locomotion which is no more dependent upon a complex nervous coordination than is normal locomotion. The many special modifications of the righting response (*e.g.*, the six varieties described by Jennings, 1907) he attributes to the action of a single previously dominant ray, or to the fact that the animal was quiescent or was not moving in a definite direction when overturned. It is perhaps unnecessary to stress the important fact, for which evidence has been cited, that the ray (or pairs of rays) which dominated previous locomotion commonly is the one which rights the starfish. Hamilton's theory of locomotion and righting has been followed in the main, because it is in keeping with the principal facts and because it is consistent with the evident limitations of this animal.

The Function of Nervous Conduction in Righting.—Nervous conduction appears to play a secondary role in the activity, a role not clearly understood at present. Moore (1910) has accepted Loeb's (1900) hypothesis that nervous impulses arising through the activity of the dominant arms are conducted with a decrement through the nervous system and inhibit the attachment of tube feet on the other arms. He relies on the results of nerve-cutting experiments (*cf.* p. 66). If arms *e* and *a* dominate righting, a cut through the nerve ring at a point between arms *a* and *b* will, he reports, prevent the inhibitory impulses from reaching arm *b* directly (from *a*). Arm *b* therefore continues in its activity, and attachment of its tube feet interferes with the righting. In this instance, Moore believes, *b* may be nervously inhibited only from the arm *e* by a long route around the nerve ring, which greatly reduces the strength of the impulses.[1]

These results do not oppose the general conclusion that the primary mechanism responsible for coordination in starfish locomotion and righting is that outlined in Hamilton's theory. Nervous conduction may well have a subordinate role in controlling activity, (1) by maintaining a condition of tonus in the tube feet (local nerve-net activity); and (2) by means of conduction of impulses "with a decrement" around the nerve ring from the arms which first attach. Such conduction may furnish impulses which facilitate the bending of the subordinate arms away from the substratum, a response which is primarily aroused by local tube-foot activity.

Nervous Conduction and Locomotion in Other Echinoderms.—In echinoderms which lack locomotor tube feet the role of nervous conduction appears somewhat more essential to locomotion than in the case of the starfish. For instance, the normal locomotion of the brittle star is effected through the stroking of four of the five rays. von Uexküll (1905) worked out the mechanism of the most frequently observed action of the rays (Fig. 15). Adjacent rays strike toward a strongly stimulated point between them, in response to impulses directly transmitted to them through the nerve ring, and in flexing they lift the body away from the stimulus. This determines the pairing of the arms in the resulting locomotion. If the *tip* of a ray is stimulated, the rays on each side bend toward it and thus are paired, so that the stimulated ray is "unpaired" and drags behind in the ensuing movement. However, if the stimulus is applied *between* two rays the unpaired ray is the one on the opposite side of the disc, and this ray is held ahead in the movement in which it is not an important participant.

Generally the brittle star is moved by the stroking of the arms adjacent to the stimulated point, but sometimes the second pair actively participates in the locomotion which follows stimulation. Let us theoretically consider such a case. Nervous excitation spreads through the nerve ring in each direction from the stimulated point, but in the transmission it loses strength (decrement conduction), and consequently it arrives at the farther rays so weakened that it contracts the extensor muscles.[2] Hence these rays first lift away from the stimulus.

[1] It is conceivable that this result was caused, not by the prevention of an inhibition otherwise exerted by arm *a*, but by *currents of injury* from the cut surface of the radial nerve which activated the large muscles of arm *b* so that this arm bent and twisted excessively. In another experiment by Moore (1919, *Jour. Gen. Physiol.*) a starfish treated with strychnine, when overturned, was not righted normally, since all arms became very active, all attached, and hence no arms maintained the necessary dominance. The action of strychnine is to increase greatly the readiness of nervous conduction.

[2] It is probable that the extensor muscles of the ray have a low threshold of excitation and are thus contracted by weak nervous impulses. The flexor mus-

In the ensuing locomotion the members of each pair work together, while the two pairs of rays work oppositely (*e.g.*, in Fig. 15, rays 2 and 3 are paired and work oppositely to rays 5 and 4), and for this result two factors appear responsible: (1) The stroking of a given ray, in which flexion and extension alternate, is attributable (von Uexküll, 1905) to the *reciprocal innervation of the antagonistic muscles* which bind together the overlapping skeletal elements of this ray. (2) The flexion of a ray furnishes a strong nervous excitation which reaches the adjacent member on that side, then finishing its extension, and *reinforces the local process of reciprocal innervation*, thus flexing this arm. We may assume that a *difference in threshold of excitation of the antagonistic muscles of a ray* is responsible for the importance of this summation effect.

In the brittle star, coordination among the rays thus appears attributable to: (1) antagonistic action of the muscle sets in the individual ray, which is primarily attributable to a difference in their excitation thresholds; and (2) conduction with a decrement through the nervous system, into contracting muscles (von Uexküll, 1905), which permits the rays to function in pairs.

Evaluation of Echinoderm Organization.—Considering the echinoderms in general, the evidence suggests that the factors which are primarily responsible for coordination of parts in locomotion and related activities operate in essentially the same manner, whichever part may dominate behavior at the time. Our survey thus has revealed a very simply constituted animal, so far as its psychological status is concerned, an animal which does not really evade the limitations which radial symmetry imposes upon the organization of its activities. But in connection with the last point a further question remains to be considered.

THE BASIS AND LIMITATIONS OF "MODIFIABILITY" IN ECHINODERMS

Experiments on "Learning" Ability.—Various experiments have been performed to test the possibility of permanent modification in echinoderm behavior. In one of Preyer's experiments (1886) a rubber collar was slipped over one of the brittle star's rays. This collar soon was removed by the adjacent rays.[1] To Preyer it appeared that when the test was repeated, both the required time and the number of variable movements gradually decreased. However, Glaser (1907) was unable to confirm this observation, and

cles have a higher threshold, and respond only to stronger impulses. In the light of this consideration, it is interesting to note the opposite effects produced initially by a strong stimulus on the nearby and on the more distant rays.

[1] The adjacent rays strike toward a ray which is strongly stimulated, and push against it; then bend back, but repeat the movement if the stimulus persists. We have already treated the mechanism of this response (p. 72).

shortly afterward Jennings (1907) had the same experience. Another of Preyer's tests required a starfish to escape from a circle of broad-headed tacks driven into the board on which the animal lay, one tack in each interradial space. The starfish always escaped after much pulling, twisting, and squeezing of its rays. In this experiment also Preyer judged that there had been a reduction of required time and of excess movements in further trials, but again a careful repetition by Jennings revealed no evidences of improvement.

The overturned brittle star is righted by the twisting and flexing of temporarily dominant rays. Glaser (1907) repeatedly forced righting in an individual, but found no reduction in time and no significant change in movements. However, he suggested that the animal might have the capacity to "learn" without displaying it in such tests. Because of the variety of ways in which it may be righted (or may remove a collar), he thought, the chances of its establishing one and eliminating others are slight. To take an analogy, if water flows from 10 identical holes in a dike, unless some are plugged, the holes will be worn equally and will continue to let through equal amounts of water.

Improvement in the Righting Response.—Jennings (1907) observed that in successive righting responses of the starfish the rays *a* and *e*, adjacent to the madreporite, initiated and dominated the response more frequently than did any other arms. In view of Glaser's suggestion, he set out to test the possibility of training the animal not to use these arms in the righting response. On each trial the animal was overturned as gently as possible in a shallow aquarium. When righting had begun, Jennings released with a glass rod the tube feet of arms *a*, *d*, and *e* (see Fig. 18*A*) whenever they attached, in an attempt to limit the lead to arms *b* and *c*. Ten trials were made each day, usually with 2-min. intervals between them.

In five tests after 18 trials with one animal, ray *b* took hold first in each case, sometimes with *c*, but rays *b* and *c* acted together "very awkwardly." It was still difficult to restrain arms *a* and *e*. Two days later, 10 test trials showed that the "effects of the training had been lost." After 180 trials, during 18 days of training, arms *c* and *d* worked together fairly well, but it was almost impossible to get *b* and *c* to twist ventral surfaces toward each other, and consequently they paired very poorly. After a 7-day rest period from the last "training" experiments the change was evident during 10 test trials. Arm *e* was used 5 times; *c*, 5 times; *b*, 4, and *a*, 2 times. The pair *b*–*c* was used once. In general these results held for other individuals, although there were certain failures, particularly with "old, stiff specimens."

In this experiment the normal dominance frequency of the rays in righting was somehow altered and the change favored rays which before the experiment had dominated less frequently. This represented a reduction in the initiating tendency of the normally superior rays. Jennings concluded that the results proved "habit formation" in the starfish, and stressed that the alteration was evident even after a 7-day rest interval. Before evaluating this phenomenon, however, we should ascertain the nature of the change upon which the behavior modification is dependent.

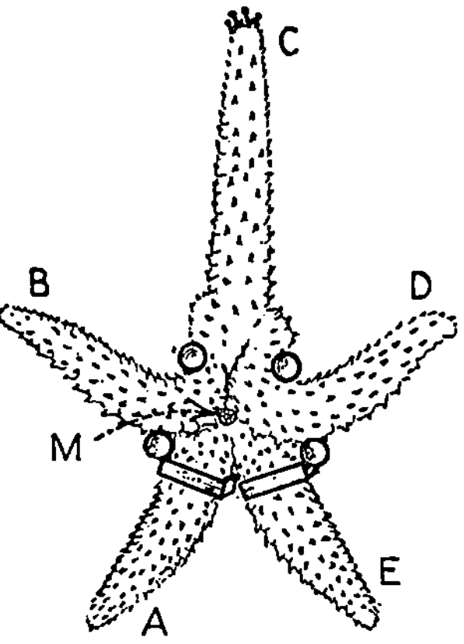

FIG. 19.—Starfish restrained with pins and rubber collars, as in Ven's experiment. *M*, madreporite. (*Modified from Preyer*.)

The Actual Basis of Jennings' Results. Judging from Jennings' procedure, local changes in the tissues, and not changes in the pattern of nervous conduction, appear to have been responsible for his findings. This interpretation is strongly supported by one of Moore's (1910) tests. When *N*/10 acid was applied to rays *a* and *e*, or when the tube-foot surfaces of these rays were rubbed with a glass rod (*cf*. Jennings' procedure), rays *a* and *e* thereafter figured less prominently in righting and the untreated rays consequently took the lead more readily than before the experiment. This change could be brought about by two or three such treatments a few minutes apart. Jennings' results thus may be attributed to local changes produced by rubbing and irritation of the tissues: reduced sensitivity of tube feet, changes in their muscular tonus, and perhaps changes in the tonus of large arm muscles as well. Similar local changes are responsible for the temporary survival of behavior modifications in the hydroid coelenterate, organic changes which are essentially peripheral in nature and hence cannot be accepted as instances of more than "primitive learning."

Ven's Test of Starfish Learning.—Ven (1921) has reopened the question, and his results confirm the above conclusion in an interesting way. The starfish was placed (ventral side down) on a board suspended in the aquarium. Pegs were driven into clay which covered the center of the board, one in the angle between each two rays (*cf.* Preyer), and loops were fastened down over arms *a* and *e* (*cf.* Jennings) so that the arms could slip out backward but could not lead the animal forward through the pegs (Fig. 19). The results were interpreted by the experimenter as proving "learning," "an experience acquired by the entire animal." A critical inspection of the proffered evidence suggests that this conclusion is not a little misleading.

To be sure, the reported results for two specimens show an abrupt initial decrease in escape time, which when plotted resembles superficially the learning curves for some higher animals. This feature was based upon a rather sudden decrease in the dominance of arms *a* and *e*. At first these arms readily took the lead, but constant irritation of surfaces from rubbing of the arms back and forth in their collars and from pulling and scraping of their tube feet (the first trial of one animal lasted 17 min., that of the other positive case lasted 78 min.!) promptly took their toll. "Trials well begun," in which rays *b*, *c*, or *d*, first took hold, apparently showed no significant change after the sharp initial decrease in time. The best specimen averaged 7, 2.5, 3.5, 5, 4, 3, and 3.5 min. for its escape times on successive series of such trials (in 10's). In contrast its average times for successive series of "trials poorly begun," in which arms *a* and *e* took the lead at first, were 12.5, 4, 6.5, 4, and 5 min. An increase in escape time during the middle of this series was attributed by the experimenter to a "bad state of health" in the starfish, evidenced by more prompt fatigue and a greater stiffness in the arms.

We recall that when Moore placed acid on one arm of a starfish, the other arms took the lead more promptly than before. Closely paralleling this, arms *a* and *e* in Ven's subjects were rubbed by collars and the tube feet were subjected to much pulling about and rubbing; in addition to which, when the specimen lay sluggishly in the apparatus without moving, a pointed stick was applied to these arms, often repeatedly on a given trial. It is virtually certain that whatever changes appeared in the behavior of the starfish during this experiment were based upon local changes in the rays *a* and *e*, causing them to lose dominance more readily than did the other arms which apparently were not altered in any important way.

Recapitulation

The echinoderm is bound by a radially symmetrical body pattern to a psychological level which is not significantly above that of the coelenterate. None of the parts is constantly dominant, although a certain metabolic difference exists among the five otherwise equivalent body regions. This difference is not crucially significant for the overcoming of structural limitations on behavior, as we have shown. The starfish and the brittle star may move with any given ray in advance, according to the manner in which conditions determine a shift in dominance.

The possession of a fairly well developed nervous system with a circumoral ring which contains ganglion cells, is an improvement over the coelenterate system, and some corresponding advances in behavior are detectable. However, the nerve ring is not a *controlling center:* it makes for more efficient transmission of impulses among the rays, much as the radial nerve of a given ray facilitates direct transmission of excitation in that part.

The form of movement and of related activities is locally determined in the parts. In the starfish it depends upon the tube feet in particular, with nervous conduction secondary in importance. In the brittle star and some other echinoderms nervous conduction is more primary in controlling locomotion, the pattern of movement depending upon conduction "with a decrement" through available paths and reciprocal innervation in the antagonistic local muscle systems. There is no evidence for the complicated coordination in nervous control which frequently has been attributed to the echinoderm.

Although the component bodily parts are well equipped for independent action, with nerve-net transmission its foundation, they are welded somewhat more efficiently into a behaving whole than is typical in the coelenterates. Generally, however, unified response to stimulation is not very prompt, and frequently it is not very complete.

There is evidence for the temporary survival of a general condition of excitability (*e.g.*, an active brittle star as compared with a sluggish one), a fact which is important for adaptation. There arise local changes in sensitivity and in muscle tonus which may alter behavior for a time by reducing or by increasing the dominance of the local rays over others. But after a time, psychologically speaking, the animal returns to the starting point. The results of Preyer, Jennings, Glaser, and Ven show these changes to be local in character, and rather limited in the scope of their effect upon behavior. The dominance of one ray or a pair of rays may be altered for some time, but apparently it is not possible to reorganize the behavior of the entire animal permanently about the newly increased dominance of a given part.

Therefore it must be said that in the echinoderm the action of each component part is dependent upon the constitution of that part, and that only temporarily and secondarily do the parts cooperate with one another in behavior. This appears to be an inevitable consequence of radial symmetry in bodily structure. In the bilaterally symmetrical animals, which we are next to consider, the permanent organization of functions about the dominance of one strategically located portion tends to overcome these limitations and develops new possibilities in adaptive behavior.

Suggested Readings

Hamilton, W. F. 1922. Coordination in the starfish. III. The righting reaction as a phase of locomotion (Righting and locomotion). *Jour. Comp. Psychol.*, **2**, 81–94.

Jennings, H. S. 1907. Behavior of the starfish, *Asterias forreri* de Loriol. *Univ. Calif. Publ. Zool.*, **4**, 53–185.

Parker, G. H. 1927. Locomotion and righting movements in echinoderms, especially in Echinarachnius. *Amer. Jour. Psychol.*, **39**, 167–180.

Romanes, G. J. 1885. *Jelly-fish, star-fish and sea-urchins.* New York: Appleton. Pp. 267–319 (Chap. X).

CHAPTER IV

NEW AND SIGNIFICANT ADVANCES IN ADAPTATION: THE WORMS

The Simplest Bilaterally Symmetrical Animals: Phylum Platyhelminthes[1]

Advances Important for Behavior.—In the flatworms there first appear certain characteristics which are of the greatest importance for the organization of behavior. These advances are evident in *Planaria* (Fig. 20), an elongated ribbonlike animal which moves with its specialized head end in advance. From anterior to posterior there is a decrease in the rate of metabolism as measured along the central axis, a *physiological gradient* (Hyman, 1919). The anterior pole of the body is most richly supplied with sense organs, and sensitivity decreases toward the posterior. In close relationship with this, the *sensitivity gradient*, there is a centralization of ganglion cells at the anterior end in the cephalic ganglion of the nervous system, which receives impulses from the sensitive elements of snout and head. The head end leads in behavior, and through its nerve-ganglion impulses originating in head receptors are conducted in two principal nerve strands and in their branches to other parts of the body. The upper (dorsal) and lower (ventral) aspects of the body differ in sensitivity and in other characteristics, a fact of great stabilizing influence in behavior.

The right and left halves of *Planaria's* body are mirrored images of each other. This condition, *bilateral* symmetry, is the typical body pattern in the flatworms and in higher phyla. Bilateral symmetry is a development which brings with it important changes in the pattern of behavior, outstanding among which are improvements in orientation. A survey of flatworm locomotion will suggest the significance of these developments.

[1] Phylum *Platyhelminthes* includes bilaterally symmetrical animals with a flattened body, three layers of cells in the body wall, and a blind gastrovascular cavity (Planaria, Fig. 20).

Locomotion and Flatworm Organization. *The Basis of the Simplest Planarian Locomotion.*—In fresh-water Planaria (Fig. 20) the simplest form of movement is *gliding*. The flatworm moves forward in a track of mucus secreted by its ventral tissue. A local portion of the ventral surface is raised (through contraction of its longitudinal muscles), pulled forward and reattached, and this activity passes along the body in the form of successive waves. This form of movement depends upon conduction through peripheral nerve-net tissue, since it appears in a piece cut from any part of a planarian or in a worm deprived of its cephalic ganglion. Strong stimulation on the posterior end generally causes the planarian to move forward more rapidly in a "humping" gait, in which more extensive contraction of longitudinal muscles causes the body to loop upward in the center, with subsequent advance and attachment of the anterior end. This form of locomotion depends upon more rapid conduction of excitation through longitudinal nerve strands and nerve net, but does not require the cephalic ganglion.

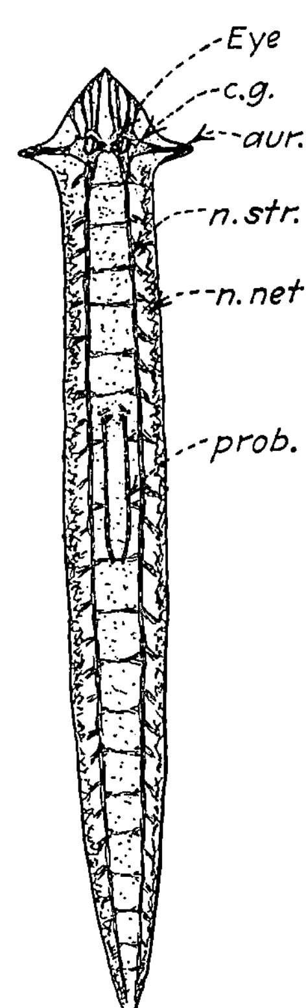

FIG. 20.—*Planaria dorotocephalia. aur.*, auricular appendage; *c.g.*, cephalic ganglion; *n. net*, representing peripheral nerve net; *n. str.*, nerve strand; *prob.*, proboscis, on ventral side. (*Redrawn from Child.*)

More Complex Movement, Dependent upon Nervous Centralization. Olmsted (1922) has described more complicated forms of movement in certain marine flatworms. In *ditaxic locomotion*, a common form, waves of muscular contraction pass alternately along the two lateral margins of the body from front to rear. When Olmsted split the head of *Leptoplana* and destroyed the right cephalic ganglion, the animal circled toward the right side, since only the left side of the body moved ditaxically. So each one of the two cephalic ganglia, conducting impulses from its side of the sensitive anterior end, controls movement on its side of the body. Transmission of nervous impulses from anterior to posterior through the body depends upon the longitudinal nerve strands, since when Olmsted

TABLE 6.—FLATWORMS AND ANNELIDS

RECEPTOR EQUIPMENT	FEATURES OF SENSITIVITY
Chemical. *Planaria:* Cells on paired ciliated areas, sometimes pits, especially concentrated on anterior end, also in proboscis walls.	Anterior end of planarian most sensitive, severed proboscis responds to chemicals as when attached.
Annelids: In earthworm, cells scattered in epithelium; marine worms have aggregations in palps, tentacles, other anterior parts.	Concentration of sensitivity, body wall also sensitive throughout. Head end, tail end, middle of body, sensitive in that order.
Mechanical. *Flatworms:* Fine anterior bristles in many; body surface sensitive in general, especially snout. Tentacles, other parts of annelid anterior contain clusters of receptor cells. Parapodia, in particular, have cells at base of bristles. Some flatworms, many annelids, have anterior statocysts.	Most annelids very sensitive to contact, substratal vibration. Order of body parts in sensitivity same as for contact. Righting response attributable to contact sensitivity. Not certain that contact, chemical, light-receptors distinct in earthworm. Contact of anterior segments important for burrowing, avoiding of objects in crawling.
Light. Paired anterior ocelli in some marine flatworms. Earthworm has only receptor cells scattered in body wall. Marine worms have one or more pairs of ocelli.	Breeding behavior elicited by dim light in some annelids. Light keeps most worms in burrows or in hiding during daytime, greatly influences mode of life.

CONDUCTION

Planaria: cephalic ganglion near anterior end, contains ganglion cells; two nerve strands pass posteriorly from it, discharge into nerve net near body wall.

Annelids: better developed bilobed cephalic ganglion, well-developed nerves from eyes, tentacles, when present. Ganglionic mass in each segment, these longitudinally connected.

ACTION SYSTEM

Flatworms: stimulated proboscis extends from near middle of planarian, ingests food. Gliding: controlled by nerve net; ditaxic locomotion in marine forms controlled by brain, nerve cords.

Annelids: Extensible pharynx in many marine forms, used with strong jaws. Vermicular movement described (pp. 93 *f.*): alternate waves of segment-thickening and of segment-thinning, setae of parapodia preventing backward-slipping. In some marine annelids paired parapodia (specialized setae) serve as swimming organs. Setae in earthworm: paired spines on ventral side of each segment.

cut these through at a point anywhere behind the cephalic ganglion, only the anterior part of the animal was capable of ditaxic movement. *Swimming*, a faster ditaxic locomotion which occurs only when the planarian is excited during ditaxic movement, similarly depends upon the cephalic ganglia and the longitudinal nerve strands.

The brain is necessary for ditaxic locomotion and for swimming because it transmits impulses from the sensitive anterior end which set the muscles of locomotion into action. Moore (1923) deprived the flatworm *Yungia* of its head (and cephalic ganglion), and was then able to elicit swimming movements by stimulating the forward ends of the longitudinal nerve strands either mechanically or chemically. In a medium containing a chemical which increased nervous irritability (*e.g.*, sodium chloride), swimming movements were readily obtained from the headless animal; but magnesium chloride, a nerve depressant, prevented swimming.

So while gliding is controlled in a primitive manner, the specialized forms of movement show the importance of cephalization (head dominance), of centralization in the nervous system, and of conduction from the specialized nerve center through nerve strands, as related to bilateral symmetry.

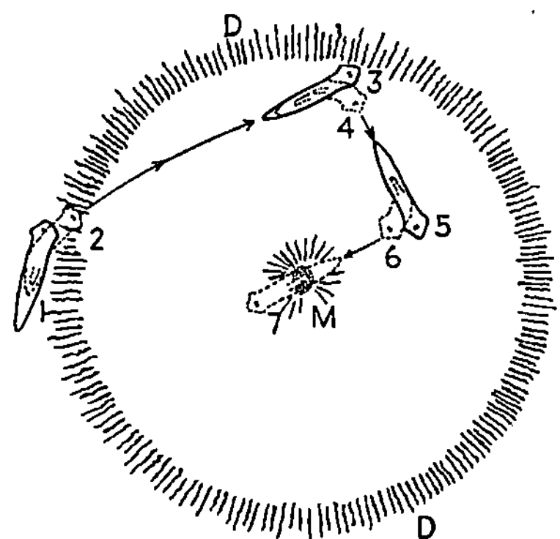

FIG. 21.—A planarian at 1 is stimulated by juices which have diffused from a crushed piece of meat at *M* (zone of diffusion marked with the circle *D*). In orienting with respect to the source of diffusion, *M*, the planarian successively takes the positions 1, 2, 3, etc. (*Modified from Pearl*, 1901.)

Cephalization and Orientation. *Orientation as Controlled by Chemical Stimuli.*—When a planarian is excited by diffusing meat juices, the head is lifted and is moved from side to side. Forward movement at first is somewhat variable, but if it brings the animal close to the source of stimulation the course becomes a fairly direct one. Variable movements ("trial reactions," Jennings, 1905) appear when stimulation of the anterior end is approximately equal on the two sides. The direction of ultimate movement depends upon which side of the sensitive anterior is stimulated more strongly by the chemical, as is illustrated by the case in Fig. 21.[1] Koehler (1932) found that *Planaria* which were deprived of the skin on the right surface of the head turned to the left when placed in a diffuse solution of meat juice. However, if the chemical (acid, sodium chloride, etc.) was strong, such operated animals contracted sharply, then turned toward the right side.

[1] A planarian which reaches a chemically stimulative object usually crawls upon it (Fig. 21). Meanwhile the long tubular proboscis (the opening of the digestive tract) elongates, extending from its attachment near the center of the ventral surface of the body, touches the object and slowly engulfs it (Kepner and Rich, 1918).

Orientation as Controlled by Light.—Planaria typically moves away from light of any intensity. If the light is diffuse, the animal is excited and continues to move until it is in darkness. When light comes from above the animal wavers much more in its course than when it is headed away from horizontally directed light (Parker and Burnett, 1901). In an experiment by Taliaferro (1920), *P. maculata* turned abruptly and moved directly away from the source of strong light, and repeated this behavior each time the direction of the light was changed. In contrast, the animal turned from weak light in a gradual curving course.

Thus the planarian is capable of rather specific adjustments to diffusing chemicals and to directed light. It is unnecessary to dwell upon the fact that it is cephalic dominance and bilateral symmetry in sensitive and locomotor equipment which give these orienting adjustments their character. Lest the efficiency of the flatworm's behavior be overestimated and the nature of the causal factors consequently misunderstood, we should be reminded that the activities of this animal display a considerable amount of variability.

Conditions Which Make for Variability in Behavior.—Unpredictable fluctuations in external stimulation and in internal condition account for a marked changeability in the flatworm's behavior. Bohn and Drzewina (1928) report that the marine flatworm *Convoluta,* in its general behavior, varies in delicate relation to the chemical conditions of the water in which it is placed, the moisture content of the air, the stage of the tide, and other factors. Certain typical influences may be described for the planarian. This flatworm is much more responsive to light after a period in the dark, because of certain chemical changes in its photoreceptors (dark adaptation). Again, well-fed animals are sluggish, remain under objects in the dark, and are not very responsive to stimulation; while the starved animal is very active and responds readily to tactual stimulation (Pearl, 1903). In Hyman's tests (1919), starved *Planaria* and young *Planaria* consistently showed a higher metabolic rate[1] than did well-fed or older animals. Thus the animal's reactivity is a function of its sensitivity at the time, and its sensitivity is a function of its general metabolic condition.

[1] As indicated by higher rate of oxygen consumption, the greater effect of killing agents, and by other signs.

The most responsive condition may be induced also by repeatedly stimulating the animal. Thus excited, the planarian moves energetically ("humping" instead of gliding), and generally does not react specifically to tactual stimuli which cause the unexcited worm to withdraw (Pearl, 1903). The manner in which the condition of the animal influences nervous conduction and thus affects general behavior is shown in an experiment by Moore (1919). After the marine flatworm *Bdelloura* had been placed for a time in sea water containing strychnine sulphate, a tactual stimulus applied to the anterior end brought "extension and activity" instead of the normal "stopping and shortening" response. In this experiment, the drug produced changes similar to those we have described for the starved animal or for the repeatedly stimulated animal.

An Advance in the Capacity for Modification of Behavior. *Learning a Primitive Capacity of Organisms.*—There is no point in the animal series at which the capacity to learn is first recognized, for learning, in a more or less primitive form, is a capacity of all living organisms (Hering, 1896). In the *Protozoa* certain protoplasmic changes are accompanied by altered behavior (pp. 33*ff.*), but such changes are very generalized by virtue of the nature of the substance in which they are brought about. In the multicellular animals thus far studied, experience may temporarily alter the form of behavior by inducing local tissue changes (in specialized receptor tissues or in specialized contractile tissues); but such changes are wiped out by subsequent events, and have no permanent altering effect on the nature of stimuli to which the animal may respond or on its repertoire of activities.

Hovey's Demonstration of Improved Modifiability in the Flatworm.—With a better developed and differently organized conduction system, the flatworms may improve upon the condition described for lower animals. An experiment by Hovey (1929) tests the possibility. Light excites the marine flatworm *Leptoplana* to movement (photokinesis); but contact with an object causes the anterior end to contract for a time, after which forward movement is resumed. After 12 hr. in the dark, a worm brought into the light would begin to creep, and Hovey endeavored to prevent movement by touching the animal on the snout as often as was required. Each 5-min. experimental period in the light was followed by a 30-min. period in darkness, until 25 test periods had been obtained for each of 17 subjects.

During the first period of 5 min. the animals required an average of 100 contacts. This number fell off abruptly in the second and third periods, then increased somewhat; but after the seventh 5-min. period there was a steady decrease in the number of required contacts. Toward the end of the series it was seldom that an animal would require the tactual stimulus, and successions of two or three consecutive motionless periods were the rule. A 10-hr. rest in darkness was then given, after which an average of only 20 contacts was required to keep the worms motionless during a 5-min test. In the next 5-min period one or two contacts sufficed, and the former low point was maintained in the following test periods.

That this ability to inhibit movement in the light did not depend upon sensory adaptation to light was shown by a control experiment in which 12 animals were exposed to light during the test periods, but were not tactually stimulated. After the 21st period the contact routine was begun. The subsequent record of these animals was much the same as the initial part of the record for experimentals. Apparently the behavior modification depended upon the conduction system; and if so, this represents a decided advance over the local type of change with which we have dealt in lower phyla.

PHYLUM ANNELIDA[1]

A Definite Advance beyond Flatworms.—The advantages of bilateral symmetry as we have sketched them for the flatworm reach a higher development in the true worms. The anterior end is more developed in sensitivity, with concentrations of specialized receptor cells (Table 6) from which excitation passes through the cephalic ganglion of the nervous system and is discharged to bodily parts. In this phylum there are animals of many habitats. The terrestrial worms and many marine worms typically bore into the ground, and leave their burrows only when certain stimulating conditions prevail. It is possible that the effect of these conditions may be modified through learning, as we shall find.

[1] Phylum *Annelida* includes animals with a three-layered, bilaterally symmetrical, and elongated body composed of longitudinally arranged segments. Our attention is confined to the class *Chaetopoda*, which includes annelids with largely similar segments, each segment with setae.

The basic annelid behavior pattern is related to the segmentation of the body and to its elongated form. We may demonstrate this by surveying the function of some of the principal controlling factors.

Sensitivity as a Determinant of Annelid Behavior. *Moisture a Basic Controller of Earthworm Activity.*—Moisture is an important factor in the behavior of terrestrial worms. At night, or after a rain, the earthworm emerges from its burrow and crawls about, mainly under the influence of moisture. When a dry spot is entered movement is checked by contraction of the anterior part of the body. This swings from side to side until dampness is encountered, and then the earthworm crawls off in a new direction.

Parker and Parshley (1911) demonstrated that the essential receptors are located on the first few anterior segments.[1] A worm with anesthetized anterior end would creep readily over dry surfaces, but in one or two days after the effect had worn off, the normal behavior reappeared. Conversely, moisture has a facilitative effect on movement, bringing the worm from its burrow after a rain and influencing its behavior in other ways. Smith (1902) observed that an earthworm crawling across a dry surface would accelerate somewhat when its anterior end contacted a moist spot. Since its anterior end recoiled each time it probed into the dry border zone, the worm would enter the moist area and remain there.

The Role of Chemical Sensitivity.—Chemical sensitivity is of basic importance in annelid activities. The anterior end of the annelid is usually most sensitive, and the middle segments of the body are less sensitive than are those at the posterior end. In most marine worms the chemoreceptors are concentrated not only in the body wall of the anterior segments but on special appendages as well. Gross (1921) showed that the reaction time of *Nereis* to acids and other chemicals was markedly increased when tentacles and other anterior appendages were removed.

Copeland and Wieman (1924) investigated the manner in which the marine worm *Nereis* (Fig. 22) locates its food. Placed in a glass tube simulating its mucus-lined burrow, the worm moved its body in a rhythmic undulatory manner, thereby circulating water through

[1] The experimenters likened the probable stimulating effect to that produced by drying of the mucous membrane of the human mouth. Dry air extracts moisture from the tissues, which coagulates (or concentrates) the local protoplasm, thereby releasing nervous impulses.

the tube.[1] This suggests the manner in which meat juices are carried into the tide-pool burrow of *Nereis* with the stream of water, exciting the worm and causing it to appear on the surface. Upon its emergence the worm holds to the burrow with its posterior segments and stretches its anterior end out toward the source of the chemical stimulus. It appears able to locate small organisms and to catch them in its formidable jaws. When the experimenters placed identical cheesecloth packets 5 cm. from the burrow opening of *Nereis* and 1 cm. apart, one packet containing snail meat and the other a pebble, the worm extended accurately toward the meat. Evidently it turned toward the side of greater stimulation, and

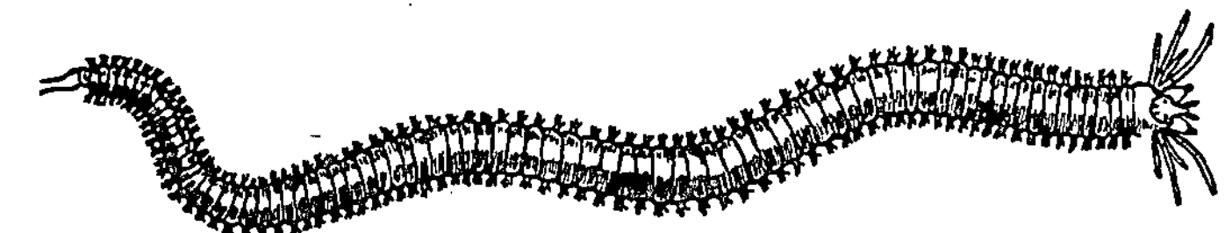

FIG. 22.—The marine worm, *Nereis*. (*Modified from Shipley and MacBride.*)

extended forward when the stimulus acted equally upon the two anterior sides. In a flat circular dish of water-covered sand *Nereis* located food less precisely than it did in open sea water, probably because of irregular diffusion of chemicals through the water in the dish (*i.e.*, as deflected by the walls). When chemicals from nearby food reached it virtually in equal amounts from all sides under these conditions, the worm could reach the source only by chance in its excited bending about.

Light as a Controller of General Activity.—Light is an important stimulative factor in the control of most annelid activities. This is particularly true in the case of the earthworm, which seldom appears from its burrow during the day. This annelid is commonly cited as an example of a *photonegative* organism, since it moves away from strong light, but the nature of its response actually depends upon the stimulative intensity of the light, as was shown by Hess (1924). After a period in the dark, an earthworm was placed on moistened blotting paper and stimulated from one side with light of a given intensity. As Table 7 shows, the subjects practically always moved away from the source when stimulated by light above 0.313 meter candle in strength. When intensities successively lower than 0.313

[1] In all probability this behavior is controlled by oxygen content of the water.

meter candle were used, locomotion was more variable and movements toward the source occurred more frequently. This helps us to understand why the earthworm ordinarily does not appear from its burrow in the daytime unless the light is dim and increased surface humidity has exerted a stimulating effect.

Organic Condition and the Determination of Behavior.—We have repeatedly stressed the necessity of knowing the protoplasmic condition of an organism at a given time if its behavior is to be understood.

TABLE 7.—VARIATIONS IN THE DIRECTION OF LOCOMOTION OF EARTHWORMS STIMULATED BY LATERALLY DIRECTED LIGHT OF DIFFERENT INTENSITIES (HESS, 1924)

Condition of the worm	Intensity of light in meter candles	Frequency of movement away from the source	Frequency of movement toward the source
Dark-adapted	785.00	139	5
	0.313	136	8
	0.0618	99	45
	0.00118	63	81
Light-adapted	0.313	65	79

For a number of invertebrate animals it has been shown that the action of a stimulus depends upon the condition of the organism at the time. This fact accounts for much of the variability in annelid behavior.

Behavior Varies According to the Annelid's Sensitivity.—Since variations in the condition of receptors are important for the nature of response, for an interpretation of behavior the stimuli effective during the preceding period should be known, as Hess (1924) has demonstrated. He subjected some earthworms to the action of bright light for one hour, then tested the response of each to a laterally directed light of 0.313 meter candle. These worms moved toward the light source much more readily than did dark-adapted worms in laterally directed light of the same intensity (Table 7). In fact, the light-adapted worms responded to 0.313 meter candle much as dark-adapted worms responded to light of 0.00118 meter candle. It cannot be concluded that the change was entirely restricted to the photoreceptor cells, since the proportionate number of turns toward light also increased in earthworms that had been subjected to strong thermal or tactual stimulation.

Condition of Muscles and Variation in Behavior.—Sometimes, because of surviving muscular tonus effects, a response is influenced

TABLE 8.—TYPICAL CONDITIONS OF THE EARTHWORM (JENNINGS, 1906) AND THEIR INFLUENCE UPON RESPONSE

Condition of earthworm	Manner or locality of stimulation	Typical responses of earthworm
State of rest (*e.g.*, after feeding)		Does not react readily; stimuli of low intensities not very effective
Moderately active	Anterior end:	Moves backward
	Posterior end:	Moves forward
	At side, near anterior end:	Turns away from stimulated side, as a rule, unless stimulus weak
State of excitement	Repeated stimulation	Persists in movement previously begun, with greater vigor as stimulus is reapplied
Greater excitement		Hastening of movement, regardless of locality of stimulation
Very great excitement (*e.g.*, long-continued stimulation)	Anterior end	Right-about-face reaction: Sudden contraction of one side, jerking anterior end around; contraction of opposite side then jerks posterior end around as well. If worm greatly excited, repetition of stimulus causes raising and waving of anterior end, often alternating with sudden turns
	Posterior end	Vigorous movement forward

more by the preceding stimulus than by the stimulus which is next applied. If the anterior end of a worm has swung to the left in response to a moderately strong stimulus presented to the right side, the next response may be a swing toward the right, even if a strong stimulus is presented to that side. The involvement of *muscular tonus* (readiness to contract) and *muscular fatigue* (chemically conditioned inability to contract) must be reckoned with in most of the activities which these highly muscular animals perform.

Previous Stimulation and Variability in Behavior.—Other temporary variations in behavior are attributable to what we must call the *general state of the animal.* From the observations of Jennings (1906*b*) we may select certain typical conditions under which the behavior of the earthworm may change. These results are presented in Table 8. In the various "conditions" described, the application of the same stimulus elicits quite different reactions. Summative effects are of evident importance; in fact, it is possible to take an earthworm which is in the *resting state*, and by stimulating repeatedly with light contact to place it successively in the different "conditions" represented in the table, so that it gives a succession of different responses to the same stimulus. Without doubt one important factor in the variability of behavior is the readiness with which the nervous tissue transmits impulses.

Other Conditions Responsible for Variability in the "Level of Activity." A worm may vary in its general condition not only according to external stimulation or previous activity but also because of alterations of its internal chemistry in dependence upon digestive condition and the like. This has already been shown for Planaria (pp. 83*f.*). Internal chemistry depends to a considerable extent upon external temperature, a fact which Crozier (1926) and his collaborators have shown to be of no little importance for the behavior of many animals. Kribs (1910) ascertained the mean concentrations of certain chemicals which would just bring a response from the worm *Aelosoma.* At a temperature of 20°C. responses to these threshold contractions greatly resembled those elicited by moderate concentrations of the same chemicals at 15°C. That is, at higher temperatures these chemicals were much more effective as stimuli because of a difference in the animal's condition.

These are conditions which contribute to variability in the activities of annelids. For an understanding of their relation to the standard pattern of annelid behavior we must consider some basic properties of the animal's inherited equipment.

Characteristic Activity as Dependent upon Neuromuscular Equipment. *The Annelid Nervous Pattern in Its Importance for*

Behavior.—The relation of the nervous system to behavior is probably better known for the earthworm than for any other invertebrate animal.

Nervous Pattern.—The most anterior part of the nervous system is a mass of ganglion cells, the *cephalic ganglion* to which well-developed nerves carry afferent impulses from the highly sensitive peristome and anterior part of the body (Fig. 23). The infra-esophageal ganglion is the important motor center of the anterior end. In each of the more than one hundred remaining segments of the body there is a local ganglion from which three pairs of nerves reach the periphery of that segment. The segmental ganglia are joined into a linear chain by longitudinal connectives. All parts of the ganglionic chain, except the cephalic ganglion, lie near the ventral wall of the body. This is the typical position of the nervous system in the invertebrate animal.

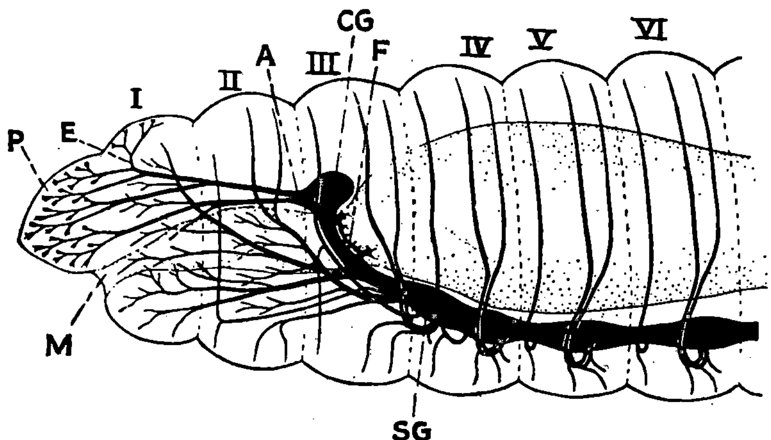

FIG. 23.—Brain and anterior portion of the earthworm's nervous system, seen from the left. *A*, *E*, nerves conducting motor impulses from the circum-esophageal connectives to muscles of anterior segments. *CG*, cerebral ganglion; *M*, mouth; *P*, prostomium; *SG*, subesophageal ganglion. (*From Rogers, after Hess.*)

The cephalic ganglion (Fig. 23) is the afferent center of the foremost segments, a function of great importance for the assertion of anterior dominance in normal locomotion. The forward part of a sectioned worm crawls with accelerated pace (just as though the intact worm had been touched strongly on the posterior end) while the posterior part remains on the spot, writhing about in an uncoordinated manner. When the cephalic ganglion is removed, no important normal motor activity disappears, but there are behavior changes which are attributable to the failure of sensory impulses from the sensitive anterior end to reach posterior segments. In response to external stimuli, the direction of locomotion in the operated worm is changed in a much more sluggish manner than it is in the normal worm. The worm readily enters dry areas after the operation, and moves toward bright light. According to Focke (1930) the operated worm is still able to burrow into the ground, but burrows much more slowly than does the normal worm.

The cephalic ganglion of the marine annelid is better developed than is that of the earthworm and conducts impulses from more sensitive and specialized sensory structures (p. 81). When this ganglion is removed from *Nereis*, the animal is somewhat more handicapped than is the "decerebrated" earthworm, but in a similar way, as the following results from Maxwell's experiments (1897) show. The operated *Nereis* crawls about restlessly and does not burrow unless its anterior end is covered with sand. The presence of this special tactual stimulation, however, causes it to burrow much as usual, and the cuticular cells secrete the mucus which lines its tube. Maxwell observed that in an aquarium the worm does not turn readily when it crawls into a corner, but pushes for some time against the glass. These features of behavior in the operated worm are attributable to the fact that tactual and other stimuli are not transmitted from the anterior end, and hence anterior stimulation does not interfere with or modify forward movement as is normally the case (pp. 86*ff.*). It is evident that while the cephalic ganglion is the most important center for normal activities and for normal anterior dominance, the characteristic properties of behavior do not depend upon this or upon any other one center of the nervous system. *Primitively, the principal centralized part of the nervous system functions as a direct transmitter of impulses from the most sensitive locality of the body, the anterior, to other parts.*

The Typical Vermicular Locomotion.—In advancing toward a weak stimulus, the anterior end of the earthworm first thins and lengthens. The backward directed spines (setae) on the ventral side then catch into the ground so that as the anterior end next thickens and shortens the rest of the worm is pulled forward. In forward movement, which begins in this way, the thinning of a few adjacent segments passes in wavelike manner from anterior to posterior through the segments, and each wave of thinning is closely followed by a wave of thickening.

The thinning of segments is produced by the contraction of circular muscles which lie just beneath the skin; the thickening of a given few segments is attributable to the deeper lying longitudinal muscles (Fig. 25). In normal vermicular movement these two muscle systems function reciprocally, *i.e.*, antagonistically (Garrey and Moore, 1915).

The backward directed setae of each segment hold it in place until it is next pulled forward by the thinning of segments anterior to it. These peristaltic waves of muscular contraction originate at the sensitive anterior end, but the manner of their conduction through the worm is a difficult problem. Recently Prosser (1934) has summarized the evidence in an able manner.

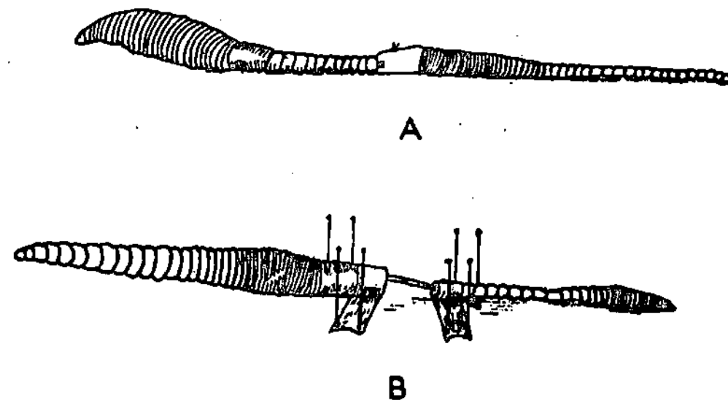

FIG. 24.—(*A*) Friedländer's experiment: coordination of posterior and anterior parts of an operated earthworm through mechanical pull upon muscles. (*B*) Biedermann's experiment: coordination of anterior and posterior parts through nervous impulses (see text). (*Modified from Jordan.*)

Factors in the Control and Maintenance of Vermicular Locomotion.—The factor of first importance is the conduction of waves of nervous impulses through the ventral nerve cord. In close connection with the nervous transmission is a wave of muscular activity which passes posteriorly at the rate of approximately 25 mm. per second during locomotion. Bovard (1918) repeated Biedermann's experiment, removing the muscles from the central segments of a worm and pinning down the skin on both sides of the open stretch of nerve cord (Fig. 24*B*), and he, too, found that after the peristaltic wave had passed through the anterior section of the worm it resumed in the posterior section. In this preparation, application of the drug stovaine to the exposed nerve cord prevented the posterior section from following the anterior section in contraction, hence the impulse must be nervous in character. But *nerve conduction*, although of major importance, is not the only factor, since the excitation does not cross a muscle-free section of worm as readily as it passes through the normal worm. The pull of one section upon the segments just posterior to it sets up *proprioceptive stimuli* which locally facilitate the passage of the neuromuscular wave (Moore, 1923). Friedländer (1894) cut a worm in two and connected anterior and posterior halves by a thread hooked into the tissues (Fig. 24*A*). When the forward movement of the anterior part caused the thread to pull upon the posterior segments, the latter also showed coordinated muscular contractions. More recent experiments show that *contact with the substratum* is a source of stimuli which facilitate passage of the peristaltic waves, a third factor in the control of movement.

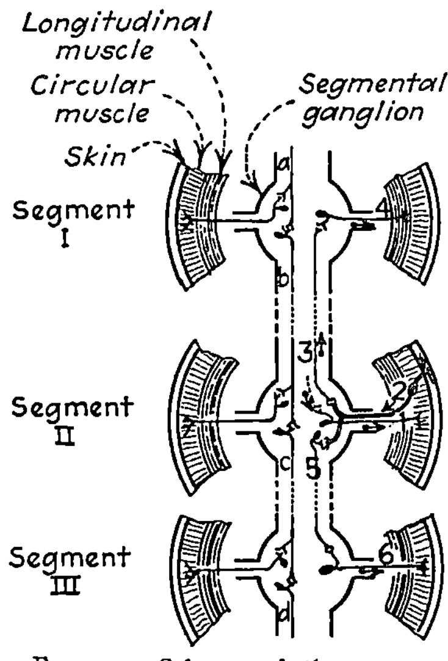

FIG. 25.—Scheme of the nervous control of earthworm locomotion. (*A*) (on left side). An impulse relayed from neurone *a* to *b*, thence to *c* and *d*, etc., is conducted to motor neurones in segments I, II, and III, and thus successively excites the circular muscles of the respective segments. (*B*) (on right side). The contraction of a given segment (*e.g.*, II) exerts traction in skin and other tissues and thus excites a sensory neurone, 2, from which impulses pass up the cord (via neurone 3) and down the cord (via neurone 5) to reinforce contraction in adjacent segments. (*Modified from Herrick*, 1924.)

The nervous excitation of normal vermicular movement is probably conducted along the cord by relatively short neurones (Fig. 25), the functioning of which is facilitated by contact and proprioceptive stimuli in each segment. The basis of movement therefore is the segmental structure of the annelid. Its elongated body form predetermines the manner in which the segments may be called into play under the influence of the major antero-posterior gradient in excitability and in conductivity. The individual segment is the unit in annelid behavior.

The Giant Fiber Tracts and Reaction.—There is another nervous mechanism of great importance for behavior. When an earthworm is extended from its burrow at night, the posterior segments remaining in the burrow and the anterior part stretching here and there, a moderately strong stimulus causes the anterior end to contract suddenly backward. Again, if the worm is crawling freely about, the anterior part may either pull strongly backward when intensely stimulated or the two ends may swing together promptly, then apart. Such sudden contractions of the entire worm depend upon the rapid conduction of excitation from one end to the other. This is made possible by the five giant fiber tracts, which run the length of the nerve cord and give off collateral fibers in each segmental ganglion (Stough, 1926). These conduction elements apparently have a much higher threshold of excitability than do the shorter neurones which take care of most normal behavior, and they conduct at a much more rapid rate (ca. 1,500 mm. per second).

The Two Principal Muscle Systems and Adjustment to Stimulation. The antagonistic action of the two major types of muscle (longitudinal and circular, Fig. 25) is fundamental in behavior. Extension of the anterior end and movement toward a weak stimulus depend upon the fact that the circular muscles (thinning of segments) are first called into play by stimuli of low energy value, and dominate the response. In contrast an intense stimulus produces a strong pull backward, or swinging about of the front end, because of the fact that strong energy first calls the longitudinal muscles vigorously into play.

We have cited experimental evidence for the dependence of a response to light or to a chemical stimulus upon the energy value of the stimulus (pp. 87*ff.*), and the above facts suggest the reason for that important relationship.

The Problem of Orientation in Annelid Behavior. *The Basic Problem of Orientation.*—A discussion of movement raises the problem of orientation, which concerns the manner in which an animal's spatial position is related to the surrounding field of energy. The basic problem of orientation is that of the "sign" of response, *i.e.*, whether the animal typically passes away from or toward the stimulus. A specific orienting response to a localized stimulus, a movement which tends to orient the animal, may be called a "tropism." If movement typically is toward a light source (or if the animal comes to rest in the light), the animal is said to be *positively phototropic.* If the animal typically moves away from the light (or comes to rest in the dark), it is said to be *negatively phototropic.* The adjectives *negative* and *positive* are applied in the same way to *chemotropism,* to *thigmotropism* (contact), to *geotropism* (*gravity*), and to *galvanotropism* (orientation in an electrical field),

Conditions Which Determine "Sign" of Orientation.—In the subject of orientation the words *negative* and *positive* are not to be used in an absolute sense. Hess (1924) has shown that the negative phototropism of the earthworm depends upon the intensity of the light, for worms move with greater frequency toward a source of light the lower its intensity (pp. 87*f.*). The same is true of responses to chemicals (Kribs, 1910): as concentration decreases, the worm is more likely to approach the source than to move from it. With stimuli of medium intensities behavior is more variable, and the result is more difficult to predict than at the extremes of intensity. These facts lead us to anticipate the solution of the problem of orientation only from a close study of the effect exerted by energy upon the animal.

The reason why the earthworm (other things equal) moves away from the source of fairly intense light is to be found in the action brought about by light in the typical physiological processes of the animal. We cannot treat that problem here[1] but may show its weight by recalling the important fact that an unusual physiological condition of the animal is likely to bring a change from its typical orienting behavior. The fact that a light-adapted earthworm turns much more readily toward light than does a dark-adapted worm (Table 7) shows the importance of knowing the effect that a given intensity of light exerts upon the photoreceptors of an animal. In this respect, the receptor may be regarded as a "filter" which lets through energy (more or less, according to conditions) to be conducted to the action system. Further, variations in the general state of excitement may cause orientation to vary from the form which is typical of the species by influencing the conductivity of nervous tissues and the readiness of muscles to contract (*i.e.*, their tonus). Jennings (1906) pointed out how spatial reactions vary with the general level of the earthworm's excitement (Table 8). A dried-out worm turns much more readily toward light than does a normal worm. In one of Loeb's experiments (1900) the normally photonegative marine worm *Polygordius* became photopositive after 1 gm. of sodium chloride had been added to the medium.

Regulation of Orienting Movements by a Localized Stimulus.—A problem closely related to that of "sign" concerns the manner in which the stimulus directs the animal's movements (*e.g.*, the process

[1] See Bayliss, Chap. XIX.

involved in the earthworm's turn from strong light) and which eventually brings the animal into position with anterior end directed away from or toward the light source. Two principal explanations have been offered.

The *trial and error theory* (Holmes, 1905; Jennings, 1906*a*) holds that random movements of the anterior portion of the worm's body occur when light reaches it from one side, and that movements which carry this end toward the light are interfered with, while movements which carry it away from the light are "followed up." The latter movements are thus "selected," and eventually orient the animal. However, it is precisely this point (*i.e.*, why certain of the random movements are "checked," while others are "followed up") that the theory should explain, and does not explain. Further, as Harper (1905) made clear, the "random movements" (the wavering of the anterior end from side to side) are seldom observed except in the orientation of a worm which is stimulated by *weak* light. For other fields of sensitivity as well, such movements may be looked for only if stimulation is weak and diffuse. Hence the trial and error theory merely describes a feature of orientation in less intense illumination, but does not point out the factors which are essential to *any* orienting adjustment.

The most adequate account of the essentials of orientation is given by the *forced-movement theory* of Loeb (1918), which is developed in the following manner: (1) In the majority of animals, the receptor, conductor, and contractile tissues are arranged in a bilaterally symmetrical manner. (2) When symmetrically arranged receptors are stimulated equally, as by a light directly behind the earthworm, equally strong nervous excitation is conducted to the muscles of the two sides, through symmetrically arranged nerve tracts. The symmetrically placed muscles of the two sides therefore contract with equal strength, and the earthworm continues to move forward directly away from the light. (3) When the receptors of one side are stimulated more strongly, nervous impulses reaching the muscles of one side (typically the opposite side, as in the earthworm) cause them to contract more vigorously. The result is a turning movement, a "forced movement," the extent of which is dependent upon the degree of inequality in the stimulation of the two sides. These forced movements reorient the animal, since they continue until the animal is again stimulated symmetrically or is moved into a zone of nonstimulation.

In weak light other stimuli (*e.g.*, proprioceptive stimuli) cause the anterior end to swing toward the light as well as away from it, but owing to the action of the light the muscles that pull the anterior end toward the source are somewhat lower in tonus than muscles on the other side of the body. The more diffuse the light (*i.e.*, the poorer its directionalization) the longer is the time required by the worm to reach its oriented position. Although the worm is excited by the light in such cases, the nearly equal action of light on the receptor cells of the two sides makes the anterior wavering a pronounced feature of locomotion. However, when laterally directed light is intense, the greater stimulation of one side maintains the muscles of the opposite side in a state of strong tonus, thus turning the animal in a continuous curve from the light. In this way, as we have shown for the planarian (pp. 82*ff.*), bilateral symmetry accounts for the appearance of a very effective mode of orientation.

Modifiability in the Annelids

Variations in behavior attributable to sensory adaptation, to a general change in level of excitement, or to a change in muscular tonus, are evident in the annelid adjustments which we have described. But there is evidence that modifications of more lasting nature than these, and somewhat more stably based, are possible in this phylum.

Establishment of a Conditioned Response in a Marine Annelid.—Copeland (1930) flashed on a light each time a *Nereis*, stimulated by clam juices diffusing through the water, was attracted to the end of a glass tube. Within some 60 trials the animal had been trained to appear when the light alone was presented. Moreover, as trials were continued, the required appearance time decreased from the initial value of nearly 30 sec. to less than 5 sec. Later, the same subject was similarly trained to appear when the tube was darkened. The performance of the planarian in Hovey's experiment (pp. 84*f.*) resembles this in its general conditions, but appears simpler in that it involved merely the inhibition of a general movement rather than the making of a given movement under new sensory conditions.

Learning of the "T Problem."—The type of behavior modification described above was only one feature of a habit first demonstrated for the earthworm by Yerkes (1912). The experiment has been repeated by Heck (1920) with substantially the same method, but with certain modifications in conditions and with a number of subjects. Heck used the T apparatus of Yerkes (Fig. 26) with the alleys made very narrow in order to minimize variability in movement. When the cover was removed from the starting alley, light caused the worm to move forward. At the junction, a turn to one side (*e.g.*, to the left) brought the worm into contact with electrodes, while a turn to the opposite side brought it into a dark moist chamber at the end of the other arm. A number of animals learned the problem, some within 150 trials.

The average record of eleven worms and the results for one of the better subjects are given in Table 9. Three stages may be distinguished in the learning. I. Following the early trips, the

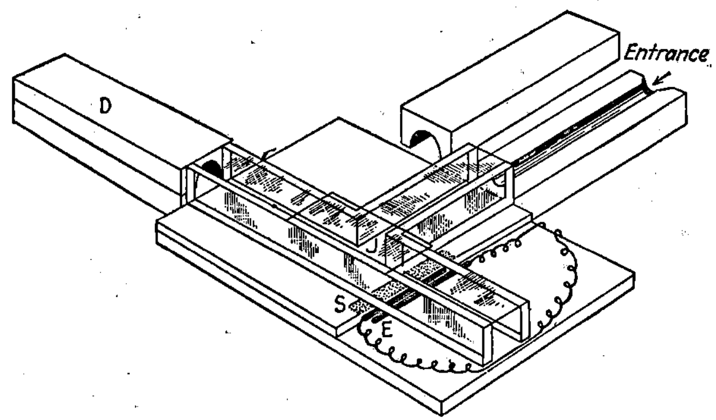

FIG. 26.—Yerkes' T apparatus for studying earthworm learning. *D*, dark chamber; *E*, electrodes; *J*, junction of alleys; *S*, sandpaper. The alley walls are of plate glass; a removable piece of blotting paper covers the floor. (*Redrawn from Yerkes*, 1912.)

animal began to turn to the open side of the apparatus directly after receiving its first shock on a given trial, without hesitantly returning to be shocked more than once, as it did at first. II. With increasing frequency the worm, when it turned "incorrectly," contacted about hesitantly in the neighborhood of the electrodes, and then turned toward the other alley without having touched the wires. III. The direct turn into the open alley became more and more frequent, and was executed with increasing precision.

The attainment of *stage I* as described above indicates that the worm has learned to respond to the shock with a specific movement (turning its anterior end to the opposite side and moving in that direction) rather than with irregular movements or with a repetition of the movement which had led to the shock. *Stage II* suggests a definite weakening of the tendency to turn toward the electrode side. Without question this change was hastened by the worm's learning to respond to certain incidental ("advance") stimuli furnished by the electrodes: possibly heat, or chemical stimuli from ionization of air by the current. To term this a "substitution of stimuli" serves to emphasize the changes that took place as the subject progressed in inhibiting the "false" turn. Some writers (*e.g.*, Fischel, 1931) insist that the entire act is nothing more than a "position habit" (Chap. XV). This, however, artificially isolates *stage III* and makes the habit solely depend upon the fact that attainment of the dark chamber after each "correct" turn somehow facilitated the making of this movement on later trials. The last fact, that the consequences of an act may influence the efficiency of its repetition later on, presents a difficult theoretical problem which will be treated in later chapters (XV to XVII).

TABLE 9.—THE LEARNING OF THE YERKES T PROBLEM BY EARTHWORMS (HECK, 1920)
(In Terms of the Number of Turns toward the Electrode Alley)

<table>
<tr><th colspan="2">A. General results</th><th colspan="4">B. Record of one subject</th></tr>
<tr><th rowspan="2">Trials</th><th rowspan="2">Average for eleven worms</th><th colspan="2">Initial record for worm I</th><th colspan="2">Worm I, after electrodes changed to "correct" side</th></tr>
<tr><th>Trials</th><th>"Errors"</th><th>Trials</th><th>Turns to side now "incorrect"</th></tr>
<tr><td>1– 40</td><td>17.4</td><td>1– 40</td><td>19</td><td>161–165</td><td>4</td></tr>
<tr><td>41– 80</td><td>14.0</td><td>41– 80</td><td>8</td><td>166–175</td><td>7</td></tr>
<tr><td>81–120</td><td>13.0</td><td>81–120</td><td>4</td><td>176–185</td><td>7</td></tr>
<tr><td>121–160</td><td>9.0</td><td>121–160</td><td>5</td><td>186–195</td><td>7</td></tr>
<tr><td>161–200</td><td>4.0</td><td></td><td></td><td>196–205</td><td>5</td></tr>
<tr><td>201–240</td><td>4.0</td><td></td><td></td><td>206–215</td><td>3</td></tr>
<tr><td></td><td></td><td></td><td></td><td>216–225</td><td>2</td></tr>
</table>

After the problem had been mastered Heck exchanged the positions of electrodes and dark chamber so that the learned turn now led into the electrode alley. The actual complexity of the learned habit was shown by the fact that the turn which had been inhibited through learning now was established in definitely fewer trials than had been required to learn the original problem. Representatively this learned reversal of the original habit was accomplished within 65 to 75 trials (Table 9, *B*). This certainly shows that the learned change was not principally dependent upon the nervous control of any particular group of muscles.

Yerkes (1912) reported that a worm which had learned the consistent turn toward one side was able to perform the habit after removal of the first five body segments, containing the cephalic ganglion. Heck confirmed this result. Twenty-four hours after the removal of the cephalic ganglion under narcosis, a worm repeated the previously learned habit with consistency. Untrained subjects were not prevented from learning the habit by the removal of the cephalic ganglion. The nervous changes brought about by the training must have involved nervous ganglia of the body segments as well as the cephalic ganglion. The worms may have been handicapped by a reduction in anterior sensitivity, owing to the loss of the cephalic transmitting center, since Heck observed a certain abnormal stiffness and hesitancy in their movements, but this did not appear to retard their performance of the learned act.

Concluding Remarks

The worms have a definite advantage over animals of lower groups, in the possession of a specialized anterior and the capacity for head-first progression. These and related improvements make for adjustment to an environment which is correspondingly more complex. Such advances are accompaniments of a permanently elaborated anteroposterior gradient in sensitivity, in conduction, and in activity. The general threshold of sensitivity is lowered, which makes it possible for the animal to locate the source of less intense stimuli. Orientation therefore increases in precision, while both the approach to weak stimuli and movement from strong stimuli are brought about more efficiently. Bilateral symmetry in structure and in function is primarily responsible for these more precise spatial adjustments.

Although nervous centralization has not progressed far in the flatworms, we find these organisms overcoming the principal limitations of radial symmetry. When local superiority in sensitivity and anterior concentration of nervous tissue make possible consistent progress with one pole of the body in advance, the learned inhibition of forward movement under given general sensory conditions is also possible. Centralization progresses in the annelid nervous system, making possible a greater complexity in normal adjustments and advancing the nature of possible modifications in behavior.

These advances correlate the parts of the body more closely in general annelid behavior, so that the independence of the peripheral mechanism is subjugated more definitely to unification of the localities than is the case for animals lower in the scale. The sensitive anterior end and its transmission centers, somewhat more dominant than those of other localities, determine what stimuli may cause the parts to become active. As Herrick (1924) has said

> . . . these worms illustrate the general plan of all segmented invertebrates reduced to the lowest terms consonant with a clear exhibition of the common pattern, with no highly specialized development of any part of it (page 141).

Suggested Readings

Herrick, C. J. 1924. *Neurological foundations of animal behavior.* New York: Holt. Chap. X.

Holmes, S. J. 1916. *Studies in animal behavior.* Boston: Badger. Chaps. III and IV.

Loeb, J. 1918. *Forced movements, tropisms, and animal conduct.* Philadelphia: Lippincott. Chaps. I and II.

Pearl, R. 1903. The movements and reactions of fresh-water planarians. *Quart. Jour. Micr. Sci.*, **46**, 509–714.

Swartz, Ruth. 1929. Modification of behavior in earthworms. *Jour. Comp. Psychol.*, **9**, 17–34.

CHAPTER V

ADVANCES IN THE COMPLEXITY AND VARIABILITY OF BEHAVIOR: PHYLUM MOLLUSCA[1]

General Features of Molluscan Behavior

A striking characteristic of behavior in *Mollusca* is its great variety among the groups, as the following brief survey will show. Most of the bivalves lead a very simple existence, creeping slowly through the mud or attaching to the substratum for long periods (even permanently, as does the oyster). However, bivalves such as the clam, when excited, may be propelled through the water by the rhythmic opening and closing of the two valves of the shell. In their simple food taking, fine material is sifted from a current of water drawn through the mantle cavity by ciliary action (Yonge, 1928).

The *gastropods* are more responsive to stimulation, and in general are more active than the bivalves. The snail glides about slowly on land by means of rhythmic muscular contractions that pass over its muscular "foot," and takes its food actively with specialized mouth parts. Some of the marine gastropods swim with agility by means of the wavelike movement of the thin specialized margins of the foot.

The *cephalopods* are the most specialized in sensitivity and the most diversified in behavior of all mollusks. The Octopus may creep about actively on its arms (the modified foot) or swim rapidly through the water by strong rhythmic muscular contractions which expel water from the mantle cavity. The rich sensitivity of the

[1] Phylum *Mollusca* includes three-layered unsegmented animals with modified bilateral symmetry, with a distinctive locomotor structure, the foot, and a mantle cavity. We treat the following classes: *Gastropoda*—mollusks with a head, bilateral symmetry modified by a single-pieced shell (snails); *Pelecypoda*—mollusks without head and with a shell of two lateral valves and a mantle of two lobes (clams, scallops); *Cephalopoda*—distinctly bilaterally symmetrical, with foot modified into arms usually with sucking discs, and usually bearing well-developed eyes (squid, Octopus).

TABLE 10.—MOLLUSKS

RECEPTOR EQUIPMENT	FEATURES OF SENSITIVITY
Chemical. Pits on body surface of some, containing chemoreceptor cells. Cephalopods have pitted papillae below eye on each side of head, with concentrations of cells. Two osphradia, pigmented cell-clusters, in many snails.	Uncovered surfaces generally chemically sensitive. Highly mobile tentacles, *e.g.*, in gastropods, very sensitive as rule. Sucking discs on arms of cephalopods very sensitive. Chemical sensitivity basic in food taking of most; controls actual grasping of food by octopus.
Mechanical. Primary sense cells in skin, also free nerve endings.	Two pairs of tentacles, foot region, very sensitive to contact in gastropods; arms, sucking discs very sensitive in cephalopods.

Static. Clams, snails, cephalopods, possess complicated statocysts, typically paired. Nerve endings in muscles may function as proprioceptors. Statocysts important for maintenance of equilibrium; muscle sense, for movement.

Light. Some gastropods have ocelli at ends of tentacles, others have compound eyes at bases of tentacles. Many cephalopods have lens eye, capable of accommodation (Fig. 28*B*). Nautilus and others have "pit eye" (Fig. 28*A*). Scallop has row of small ocelli around mantle edge. Clam has photoreceptor cells scattered over inner surface of siphon. Skin sensitivity to light is common. (See text for behavior significance.)

CONDUCTION

Ganglia fused into pairs, with longitudinal connectives (Fig. 29). Cephalic ganglia best developed, generally, with paired connectives to pedal ganglia and to visceral ganglia. Pleural and parietal ganglia often present. Ganglion cells present, central system predominantly synaptic. Nerve-net plexus in foot and other local parts.

ACTION SYSTEM

Foot, elongated muscular process, is organ of locomotion in gastropods, pelecypods. Many (*e.g.*, slugs) glide on foot, by waves of muscular contraction acting on a track of secreted mucus. Modified forms of pedal locomotion: *e.g.*, swimming in marine snails, jumping in some clams. Clams swim by rhythmic valve clapping. Cephalopods move on arms, or swim by expulsion of water from mantle cavity (Fig. 30). Skin coloration changes in excited cephalopods, due to chromatophore cells. Ink sac in squid and other cephalopods.

cephalopods and their varied activities are in decided contrast to the relatively sluggish and uniform activity of the bivalves.

The arrangement of receptors and conduction system varies greatly among the mollusks, but the fundamental behavior pattern is determined by structural properties similar to those already studied in the annelids. The complexity of molluscan behavior, as compared with that of lower animals, is increased by the somewhat greater domination of a specialized anterior over other bodily parts. Anterior sensitivity is thus more influential over general behavior.

The mollusks present greater complication in the structure of the conduction system. Greater diversity is apparent among the nervous centers in relation to the differentiation of local parts, *i.e.*, head: cephalic ganglion; foot: pedal ganglion; etc. This is responsible for a further increase in the variety of an individual's behavior since it makes possible the joining of bodily localities into total activities of very different form. The matter is complicated by structural factors which make for a considerable degree of local autonomy (independence in action). For these and related reasons we may say that the mollusks represent a transition group between the annelids and the highest invertebrates. A survey of the orientation process will demonstrate the point.

Orientation and the Control of Movement

Factors Influencing Maintenance of Direction and Change in Direction of Progress. *The Snail's Locomotion in Directed Light, as Influenced by Temperature.*—The light-sensitive ocelli at the ends of the tentacles in many gastropods are essential to the control of locomotion. In light which comes from one direction, the slug *Limax* turns toward the darker side. Correspondingly, with one eye removed, the animal circles toward the eyeless side when light is directed from above. This, Crozier (1929) concludes, is because the inequality in stimulation brings about differences in the contraction of muscles on the two sides of the body, thereby producing forced movements (pp. 97*f.*).

The slug's orientation has been analyzed experimentally. Locomotion in this gastropod involves the action of two principal systems of muscles: (1) the pedal muscles in the long fleshy foot which are responsible for straightforward progress; and (2) the parietal muscles of the lateral body wall which are responsible for the turning of the foot in locomotion. Normally these two systems function together, but their separate function has been experimentally demonstrated.[1]

Crozier and Federighi (1925) recorded the circling movement of the one-eyed slug under diffuse illumination, in terms of *degrees turning* in each centimeter of the path. As the temperature was lowered from 15°C., they found that the turning tendency became less evident, whereas between 15 and 30°C. the amount of turning in the path increased with the temperature. It was concluded that below 15°C. the pedal muscles are in the ascendancy, while above this temperature the parietal muscles dominate. Near 15°C. the pedal and parietal muscles are approximately balanced in controlling

[1] When strychnine is injected at 29°C. the parietal muscles are eliminated from function, and under diffuse illumination the one-eyed slug forges straight ahead (Crozier and Cole, 1923, *Anat. Rec.*).

movement, which accounts for a greater variability in the direction of locomotion under this condition. The experimenters thus demonstrated that at temperatures above 15°C. the light excites a different chemical process in the animal than is effective below that temperature, a fact which is of great importance for general behavior.

The results were fitted to the Arrhenius equation for the velocity of a chemical reaction:

$$\frac{(\text{Amplitude of turning at})\ T_2^\circ}{(\text{Amplitude of turning at})\ T_1^\circ} = \frac{\mu}{e^2}\left(\frac{1}{T_1} - \frac{1}{T_2}\right)$$

in which T_1 and T_2 represent different temperature values, e the base of natural logarithms, and μ the critical increment which expresses the velocity of a chemical reaction involving a catalyzed oxidation. In the slug, the value of the critical increment μ proved to be 16,820 for locomotion above 15°C., while its value for the reciprocal of creeping rate above 15°C. was 10,900.

Cases in Which Locomotion Is Controlled by More than One Type of Stimulus.—Such processes as those discussed above basically determine the manner in which these animals move about under the altering influence of external stimulation. An experiment by Crozier and Cole (1929) shows the manner in which two important sensory factors may be involved in controlling locomotion. The slug typically moves upward on an inclined surface, *i.e.*, it is *negatively geotropic.* But weak light coming from one side causes the animal to diverge somewhat from that side in its upward path. Thus the light opposes geotropism in its effect. For the slug crawling upward on an incline, the stronger the laterally directed light the greater is the angle of divergence from the direct upward path (Fig. 27), in logarithmic relation to the intensity of illumination. Thus normally in crawling up the side of a stone the slug's course is determined both by the position of its body and by the direction of illumination, in addition to incidental stimulus factors (*e.g.*, roughnesses in the surface).

The Manner in Which a Geotropism Is Effected.—The involvement of geotropism in gastropod orientation is easily demonstrated; but what is its sensory basis? Hoagland and Crozier (1931) found that when the inclined plane on which a snail is creeping upward is rotated through 90 deg. (so that the animal's head is now directed toward one side), upward orientation is resumed more promptly the greater the slope of the incline.[1] The experimenters concluded that it is

[1] For inclined planes up to 55 deg. in slope the time required for the snail's readjustment was found to decrease hyperbolically with increase in the sine of the angle of inclination.

the pull of the body weight upon the muscles of the two sides, a source of proprioceptive (muscle-sense) stimuli, which enables the gastropod to resume its direct upward progress after disturbance of position on a slope. Forced movements are elicited by the asymmetrical pull, until the proprioceptive stimuli again affect locomotor muscles equally on the two sides of the body. On steep slopes the "gravitational pull" is very different on the two sides of the displaced animal, hence readjustment is brought about through more prompt and vigorous forced movements than are effective on a more gradual slope.

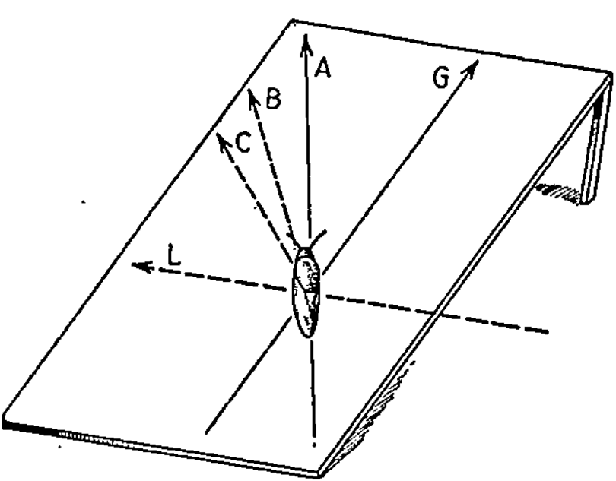

FIG. 27.—Orientation of the slug under the combined influences of "gravity" and directed light. *G*, the direct upward course when illumination is not a factor. *L*, light directed from the right side. *A*, *B*, *C*, courses taken with successive increases in the intensity of illumination. (*Modified from Crozier and Cole*, 1929.)

The paired statocysts possessed by many mollusks serve as a second important source of stimulation in the control of bodily position (Piéron, 1927, 1928). This is indicated by the asymmetrical movement of the swimming gastropod after the statocyst (Fig. 29*B*) of one side has been destroyed.

THE PROPERTIES AND LIMITATIONS OF SOME RECEPTOR MECHANISMS WHICH INFLUENCE THE PATTERN OF BEHAVIOR

One of the chief reasons for the great variety of molluscan behavior is the nature of the receptor equipment. Some cases may be cited to show how sensory apparatus is a major factor in the occurrence of important adjustments in the different classes.

The Nature of the Basic Photochemical Effect.—In dim light the clam protrudes its double siphon through which a current of water is drawn into the mantle cavity by ciliary action. Hecht (1919*a*) found that when the animal was again exposed to light, siphon withdrawal began within a definite reaction time. This interval he measured by means of a photographic record of the movements of a small mirror attached to the siphon tip. Determinations showed that the reaction time is mainly taken up, first, by a very short sensitization period during which the light acts directly upon the receptors, and second, by a latent period during which a chemical reaction occurs within the photosensitive pigment. The nature of this chemical reaction was ascertained through a study of variations in the latent period under different conditions.

When the time of exposure to light T was short and constant, the length of the latent period L varied inversely with the logarithm of the intensity of illumination I. When intensity of illumination was constant, the latent period varied inversely with the exposure time (1919*b*). The results are expressed by the equation $1/L = T \cdot \log I$. Since the length of the latent period was also found to vary inversely in geometrical ratio with the temperature, Hecht (1920) concluded that the light sets up *chemical* changes in the photosensitive substance. Since the speed of adaptation to light in *Mya* was found to depend upon the temperature, he concluded that the chemical reaction must be reversible in nature. Darkening causes the formation of a chemical substance in the receptor cells, but increase in the intensity of illumination causes the dissipation of this substance (Hecht, 1929).

Because siphon protrusion occurs in connection with one phase of the photochemical reaction, and because siphon withdrawal occurs in connection with the opposite phase, it is clear that the two phases of the receptor process set off nerve impulses which are capable of acting upon different muscle systems (see Chap. XIV). The nature of the highly adaptive siphon response thus is primarily attributable to the chemistry of the receptor cells themselves.

A Response Which Depends upon the Darkening Phase of the Photochemical Process.—The *skioptic reaction*, a more or less inclusive contraction when the body surface is shaded, was reported by Nagel (1894) as characteristic of animals that inhabit a covering of some sort or a tube from which they emerge at intervals. The snail and tube-dwelling worms are examples. For the snail, Föh (1932) obtained proof that this response depends upon light-sensitive cells distributed in the skin. Removal of the eye-bearing tentacles did not change the nature of the *shadow response*, but this reaction did vary in nature and extent according to which portion of the body was shaded. The reaction also varied according to the intensity of light to which the snail had become adapted.

It is significant that this peculiar response is characteristic of a great many animals that are subject to re-adaptation of their skin receptors to light at frequent intervals. These receptor cells are so constituted that a decrease in illumination throws them readily into the opposite phase of the photochemical reaction from that occurring in light, thereby setting up nervous impulses which arouse a specific system of muscles. In this respect the skioptic reaction is on much the same basis as siphon protrusion in the clam.

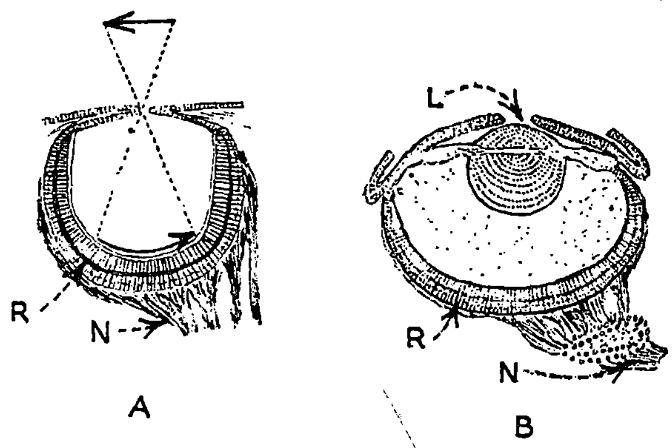

FIG. 28.—Cross sections of (*A*) the "camera lucida" eye of Nautilus; and (*B*) a typical cephalopod eye. *L*, lens; *N*, nerve; *R*, retina. (*Redrawn from Herter.*)

The Significance of Special Eye Structures for Behavior. *The "Camera Lucida" Eye of Nautilus.*—Certain principal features of specialized photoreceptors, *i.e.*, the aggregation of light-sensitive cells on a surface (retina), the development of accessory structures (iris, lens) which improve the reception of light, and the arrangement of these "eyes" on the body, are responsible for some marked differences in behavior among mollusks. For instance, *Nautilus*, a cephalopod, has a retina which lines the bottom of a spherical cavity (Fig. 28*A*). Light reaches the sensitive cells through an opening without being altered in direction, since no lens is present. It has been inferred that this eye functions on the pinhole camera (camera lucida) principle, that since the external opening is very small only a fine beam of light may reach the retina from each portion of the object. Any slight movement of the object would thus effectively stimulate the eye, since it would remove from each portion of the retina its adapted stimulus and present a different one. This may account for the notable ability of *Nautilus* to respond to the slightest movement of an object which is fairly close to its eye, a fact of consequence for the food taking of this mollusk.

The Efficient Lens Eyes of the Cephalopod.—Vision is the dominant type of sensitivity in the cephalopods, and the structure of the well-developed lens eye (Fig. 28*B*) appears mainly responsible for the great activity of the animal and the range of its behavior. The Octopus' eye is very large in relation to body size, and its retina is richly supplied with sensitive cells. The spherical hardened lens is moved backward or forward through the action of different muscles, thereby actively accommodating the eye to the movement of an object within a certain range (focusing on the same principle as a camera).

Experiments on the importance of this receptor for behavior have been preliminary in nature. Typically the Octopus springs only at moving prey. It is able to leap upon its usual prey, the crab, from a distance of 2 to 3 meters, and the release of this response may depend upon the limits of effective accommodation. Just how vision functions in the actual grasping of prey is not certain (p. 113). Giersberg (1926) reports that a crab which is hanging in the water at a distance of 25 cm. excites the Octopus so that its arms wave about and changes appear in skin coloration, but the securing of food appears to depend upon the chance that one of the arm tips with its chemically sensitive sucking discs may touch the prey.

Some Functions of Chemical Sensitivity. *Typical Distribution of Sensitivity to Chemicals.*—We have seen that in some molluscs light-sensitive organs permit the animal to respond to an object which is at a distance from its body surface. There is evidence that chemoreceptive organs may function in a similar though more limited manner. In an experiment by Jordan (1917) the marine snail *Aplysia* bent its head and grasped algae presented within 2 cm. of the anterior tentacles, but failed to respond when posterior tentacles or the parapodial (foot) margin were tested unless the food actually touched the body. Following a common practice, we may call the lappets of the anterior tentacles in this animal *distance chemoreceptors*, since they contain cells which are sufficiently sensitive to be acted upon by juices diffusing relatively weakly from a somewhat distant object. Regions like the parapodial margins, which are not effectively stimulated unless the object practically touches the body, may be called *proximate chemoreceptors*. In a given mollusk the body regions are differently sensitive to chemicals, but no sharp distinction can be made between areas which are of one or the other class. It is not always easy to distinguish between a response which is mainly attributable to the chemical effect of an object touching the body, and its contact effect as such (see note, p. 41).

Chemical Sensitivity and Food Taking.—Let us consider a specific instance of the function of chemoreceptors in the food taking of a mollusk. In some experiments performed by Copeland (1918) a marine snail was observed to extend its siphon and move forward when fish juice was released into the water from a distance, but this response was not apparent after the osphradia had been scraped away. Hence these specialized regions are the distance chemoreceptors, which are stimulated by chemicals drawn in with sea water through the siphon.

Copeland found certain marine snails able to locate objects from which juices diffused to them. In the snail *Busycon* the siphon swings from side to side as the animal glides along on its foot. If a chemical stimulus was presented as the siphon reached the leftward end of its excursion, the foot next turned in that direction; whereas if the stimulus was presented at the termination of a siphon swing to the right side, the animal turned to the right. When Copeland attached two white cloth packets to a snail, one on each side of the siphon, the animal turned in circles toward the side of the packet which contained meat. Similarly, in normal behavior the more intense stimulation of the osphradium on the side nearer the food must be responsible for forced movements which eventually result in location of the food object.

Sensory Integration in the Control of Behavior

A Basis for Evaluating the Animal's Psychological Standing.—We have been considering factors which are important in determining whether an animal may respond to given features of its environment, but which do not themselves determine the complexity of this response. Hempelmann (1926) grants the cephalopods a psychological position close to that of the higher insects, on the strength of their visual receptor which is as well developed as that of some vertebrates (Fig. 28*B*). However, it should be remembered that the use which is made of stimulative effects depends upon the organizing capacity of the animal, and this is particularly a function of its nervous system. The present question should, therefore, be approached cautiously.

The extent of sensory integration (*i.e.*, the manner in which the fields of sensitivity cooperate or check upon one another in controlling the nature of behavior) is a most useful criterion in evaluating the psychological status of an animal. In different instances of molluscan behavior it is not difficult to recognize the combined function of receptors, but it is the *manner* in which the sense departments cooperate which is crucial for the present discussion. Accordingly we may pass in review some typical interrelationships of the departments of sensitivity in mollusks.

The Readiness with Which One Type of Sensitivity Displaces Another in Controlling a Response.—Frequently in molluscan behavior the first phase of an act may be attributed to one type of sensitivity and the following phase to a different type. The first response of the scallop *Pecten* to an approaching starfish is a weak response, extension of the tentacles, which is attributable to stimulation of the numerous lens eyes arranged like decorative beads around the mantle edge (von Uexküll, 1912). If the starfish comes close, a strong response displaces the first: the scallop promptly closes its valves, or claps the valves rhythmically and swims off. The second response cannot be attributed to vision, since it does not occur if a glass separates the approaching starfish from the scallop. Dakin (1910) caused the scallop to close its valves or to swim off by releasing starfish broth near it. We may therefore conclude that the visual stimulus induces a weak local response, that of tentacular extension; then, as the starfish comes close, chemical stimulation intervenes and brings into play a stronger response. In displacing the first response, the chemically determined activity thus protects *Pecten* from its common enemy.

In many cases two stimuli may be presented together, but the mollusk may not respond to one until its response to the other has ceased. When the snail's mouth parts are touched with a lettuce leaf, characteristic "chewing" movements appear, but these are not observed when the foot is touched at the same time. The foot contraction, elicited by local contact, inhibits the mouth response while it is in force so that for a time the chemical stimulus is ineffective. In this way the foot contraction of a "rapidly" advancing snail that touches a leaf may prevent the snail from eating the food until the foot response has waned. Similarly, a snail which is gliding toward the surface under the influence of stimuli originating through the depletion of air in its lung contracts into its shell when touched, and does not resume its movement until the blocking effect of the general contraction has waned sufficiently for the viscerally derived stimuli again to become effective.

The Limitations of Sensory Integration in the Highest Mollusks.— Is it possible to find a true instance of integrated receptor control over the same response in molluscan behavior? The Octopus is visually excited by a moving crab, and is stimulated to approach and spring upon the prey. But grasping of the prey with the arms (where it is held by sucking discs) and its transport to the mouth occurs whether or not the eyes are present (Giersberg, 1926). Apparently the grasping of prey depends upon a summation of local tactual and chemical stimuli. Tactual stimuli alone are not sufficient, as is indicated by the fact that a glass tube is grasped but is promptly released by the sucking discs. Vision thus controls the general response (approaching and springing), while the contact-chemical summation controls the local response of arms in contact with the crab. In food capture, vision and the local arm sensitivity do not appear to be integrated to any extent, since neither has a modifying effect upon a response produced by the other.

This point may be illustrated by the result of an experiment which Bierens de Haan (1926) performed. An Octopus rushed directly toward a glass jar in which a crab was moving about, vivid changes in skin coloration indicating the height of its visually aroused excitement. Upon reaching the jar the Octopus continued to scramble toward the crab, persistently pushing against the glass. One of the animal's writhing arms finally *chanced* over the edge of the vessel and into contact with the crab; but the Octopus continued to push against the side of the jar. The general response to visual stimulation was not modified in any recognizable way by the local stimulation of the arm, nor was the arm detectably influenced by the fact that the Octopus could see the prey. Rather, the response of the arm holding the crab appeared to be controlled by local stimulation and to occur despite the visual stimulation.

When a crab passes over a rock out of sight of a pursuing Octopus, and the cephalopod makes its next general movement around that side of the rock on which an arm happens to touch the crab, the temporary absence of visual control apparently is responsible for the influence of local stimulation on general activity. In the glass-jar test, however, the local contact-chemical stimulation neither modified nor displaced the visually controlled response. The stimulus effects did not cooperate in the sense that they checked upon each other. (The demands of the glass-jar problem appear to lie beyond the natural neurophysiological capacity for integrated sensory control of behavior. Very probably they are also beyond the limits of cephalopod learning, although this matter remains to be tested.)

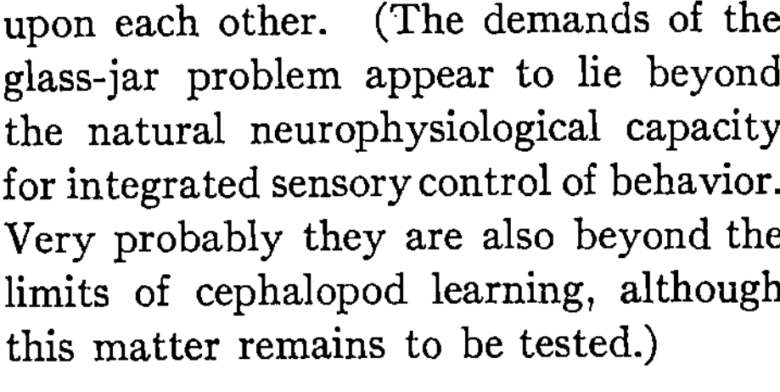

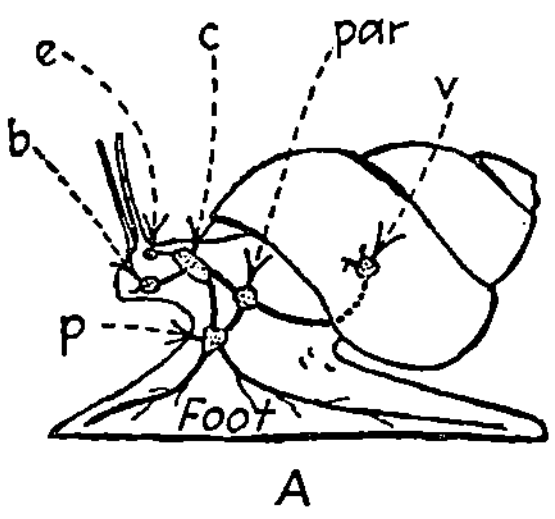

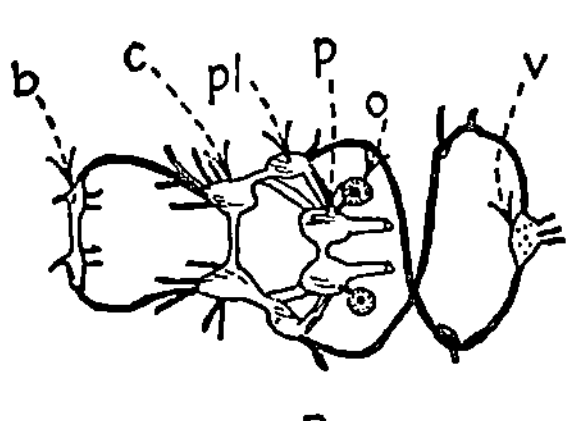

FIG. 29.—Nervous system of the gastropod *Paludina.* (*A*) schematic view from the left. (*B*) dorsal view. *b*, buccal ganglion; *c*, cerebral ganglion; *e*, eye; *o*, otocyst; *pl*, pleural ganglion; *par*, parietal ganglion; *p*, pedal ganglion; *v*, visceral ganglion. (*A*, *modified from Loeb; B, redrawn from Hertwig's Manual.*)

CHARACTERISTICS OF NERVOUS PATTERN AND FUNCTION IN THEIR IMPORTANCE FOR BEHAVIOR

Typical Pattern of the Molluscan Nervous System.—In *Mollusca* the nervous system is essentially a series of paired, connected ganglia, but those paired ganglia which are present in all groups (*viz.*, the cephalic, pedal, and visceral ganglia) vary in their positions, and to some extent in their interconnections. It is a matter of some importance that the arrangement of the ganglia and of their connectives corresponds to general bodily form, particularly to certain characteristics of the action, digestive, and respiratory systems. The cephalic ganglion is generally much larger than the others. In all mollusks it is connected directly with the pedal ganglia of the foot, and in most of them it is connected with the pleural ganglia, but only indirectly with the remaining ganglia.

General Versus Local Nervous Control in Molluscan Behavior. *Function of a Local Center in the Gastropod.*—In gastropods, the pedal ganglia locally control the rhythmic waves of slight muscular contractions which sweep over the foot (usually from anterior to posterior) in the creeping or gliding of the animal (Fig. 29*A*).

Removal of the cephalic ganglion does not prevent locomotion, but the sensory conditions under which it may be altered are reduced by the operation to local stimulation of the foot itself. The relative self-sufficiency of the pedal ganglion in controlling local activities, such as forward locomotion, is mainly attributable to the existence of a nerve net in the foot (Jordan, 1929), a plexus which is connected with the local pair of ganglia. For this reason removal of the cephalic ganglia causes the gastropod *Aplysia* to swim in lively manner through the water by the locally controlled action of its parapodium, but to stop only with difficulty.

The Unifying of Local Ganglia and Other Functions of the Cephalic Ganglion.—Fundamentally the cephalic ganglion functions as a connector of the other ganglia. For instance, when the mantle margin of the mussel is touched, the foot contracts promptly and the valves are closed together. Here the excitation (through the cephalic ganglia) has reached both the pedal ganglia and the visceral ganglia. These when excited effect their locally controlled activities, foot retraction and shell closure, respectively. Further, it is generally concluded from operative studies that the cephalic ganglia exert an inhibitory influence on the other ganglia. For example, von Uexküll (1891) found that destruction of the cephalic centers in cephalopods results in a strengthening of locally controlled activities and an increased general excitability.

General Valuation of Nervous Control in the Mollusk.—The existence of a well-developed cephalic ganglion in the higher mollusks (gastropods and cephalopods) is responsible for bringing general behavior under the control of the principal receptors. Energy reaching the cephalic ganglia from receptors may be summated there in such manner as to reinforce and steady a local activity, as Friedrich (1932) has shown for swimming in the marine snail *Pterotrachea.* The cephalic ganglia also permit the localities of the body, each of which strongly tends to act autonomously, to function together in a somewhat organized manner. Through central conduction one locally determined activity may displace another, as we noted in the preceding section (pp. 112*f.*), or one locality may dominate general behavior at the time another locality is inhibited. In the normal swimming of Octopus, energy from the cephalic ganglia paces the local control of rhythmic mantle contractions through the visceral ganglia, while impulses which reach the brachial ganglia may inhibit local activity so that the arms are held in a weakly toned or relaxed position (Fig. 30*B*).

The pattern of fiber connections in the molluscan nervous system must account for the limitations which appear in the sensory integration of these animals. Nevertheless, the arrangement of the local ganglia and the possession of the cephalic ganglia make for a greater variability and complexity in the interaction of bodily parts than we have seen in lower invertebrates.

Modifiability in the Control of Behavior

Difficulties in Ascertaining the Mollusk's Psychological Status.—Some writers judge the psychological position of the cephalopod

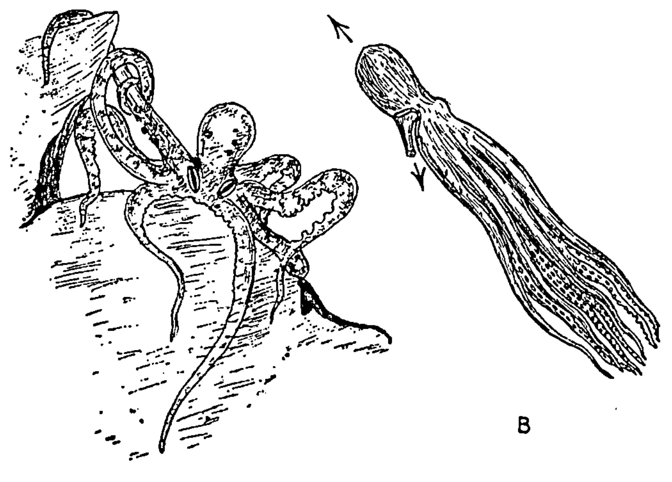

Fig. 30.—Octopus. (*A*) Locomotion on substratum. (*Modified from Beebe*). (*B*) Swimming, by means of water forced from mantle cavity through siphon (as shown by lower arrow). (*Modified from Cooke.*)

mainly on the basis of highly developed receptors, the eyes in particular, and the activity and predaciousness of the animal. For example, it has been inferred that since the Octopus frequently sits in the center of a crater of stones, it must possess an ability to "build" and have a rather complicated "sense of form." However, Bierens de Haan (1926) has shown that the structure is formed quite accidentally. The Octopus tends to pick up stones or other objects and hold them against its body at certain times, particularly after feeding. Since the stones slip to the ground when the sucking discs fatigue and lose their hold, eventually a circular heap is formed around the Octopus where it sits. Bierens de Haan concludes that the responsible factor is a temporary change in the skin which forces restless behavior unless tactual stimulation of the body surface is maintained. It is this factor which is responsible for the tendency of the Octopus to crawl into crevices in rocks, and not a desire to hide, for the animal will crawl between glass plates as readily as between pieces of slate.

Habituation ("Negative Adaptation") to Stimuli.—Evidence for habituation to stimuli is not lacking. Piéron (1909) subjected a snail to the repeated shading of its body, and after a number of trials the animal no longer gave its skioptic reaction (pp. 108*f.*) as at first. Dawson (1911) observed that a snail which had been collected from a pond or other quiet environment was easily disturbed by incidental contact stimulation in the laboratory aquarium. Slightly jostled by a swimming tadpole, such a snail would contract violently, expel the air from its lung, and drop from the surface film on which it had been crawling. Later on, the snail contracted only slightly in response to such stimuli, and then continued its feeding or forward movement. As Dawson pointed out, snails taken from running-stream habitats are already negatively adapted to incidental contact stimulation when brought into the laboratory.

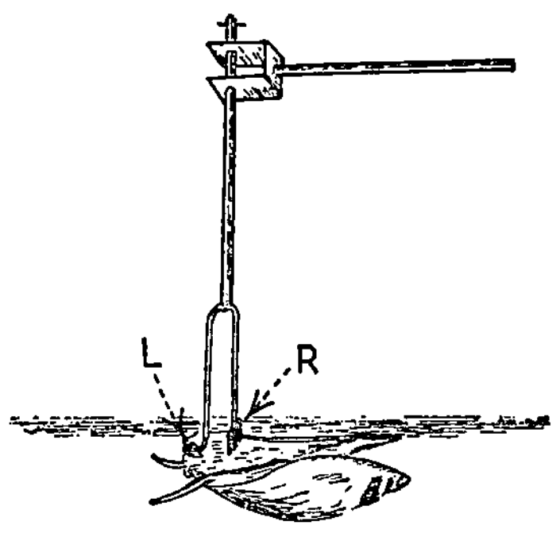

FIG. 31.—Method of applying stimuli in Thompson's experiment. The instrument is presented to a snail which is gliding, shell downward, suspended from the surface film. *R*, rake touching foot; *L*, lettuce touched to mouth parts. (*Redrawn from Thompson*, 1917.)

A Conditioned Response in the Snail.—Thompson (1917) studied a learned act of which the above type of modification formed a part. Normally, when the mouth parts of a snail are touched with lettuce, repeated chewing movements are observed. If the snail is touched on its muscular foot, the foot contracts, but the mouth parts are not visibly affected (certainly no chewing movements occur). Thompson's procedure was simultaneously to present lettuce to mouth parts and to touch the foot with a metal rake as the snail was crawling along the surface film of a dish of water (Fig. 31). On alternate days each subject was given 10 trials, with 24 hr. of starvation preceding each series. Six animals averaged the following totals of mouth movements in successive series of 50 trials: 0, 12, 82, 44, 71.

It was evident that at first the foot contact effectively inhibited mouth response to the lettuce. For instance, during the first 100 trials with a representative snail, mouth movements were observed on but one test. The buccal ganglia (controlling mouth movement) slowly recovered from the evident domination of the pedal ganglion, but even after 250 trials the average number of mouth responses to foot contact was slightly less than 1.5 per trial. In contrast, the normal number of mouth movements for stimulation of mouth with lettuce alone was more than 5.

The training thus not only overcame to a considerable extent the inhibition initially exerted by the foot response over the mouth response, but it also brought a modification in the sensory conditions under which mouth movements might be elicited. That is, after the training foot contact also would bring chewing movements. The apparent basis of the habit was a change in the manner in which the pedal and the cephalic ganglia functioned together when the foot contact was presented (Fig. 31). After 250 trials with the combined stimuli, one of the snails was given a test series in which the foot contact alone was presented. During the first eight of these test trials the contact produced mouth responses, showing that it had taken over control of the mouth response, but after 12 trials it failed to do so. The training effects had not completely disappeared, however, since the snail could be retrained in fewer trials than were needed initially. Hence the experiment also shows that training may partially overcome a natural limitation on the nature of sensory integration in the snail (pp. 112*f.*).

Learning of the T Problem by the Snail.—Thompson (1917) found little evidence that the fresh-water snail *Physa* was able to learn a T problem. In her experiment the left turn at the junction led to a vertical alley that reached the surface of the water (where the empty lung of the snail could be filled with air), the right turn was a blind alley.

Garth and Mitchell (1925) performed a similar experiment on the land snail *Ruminia*, with positive results. Excited to move by heat, the snail crawled along the approach arm of a T apparatus similar to that used by Yerkes for the earthworm (Fig. 26), and at the junction encountered two alternative alleys, each 7.5 cm. in length. In the left arm, near the junction, electrodes were in position; the right alley led to a dark box. Behavior during the first trials was marked by variability, highly irregular routes, and failures to perform. Ten snails showed a straightening of their course after some trials, but only one snail survived until a consistently direct course into the right alley had been learned. With two trials per day this point was reached in 70 trials, after which this individual made the direct turn to the right on 32 successive trials. This evidence, although preliminary in nature, shows that the snail is capable of learning the inhibition of one movement on the basis of its drastic consequences, and of establishing an alternative movement on the basis of its facilitating consequences.

Concluding Statement

We have examined the factors which account for a more complex behavior in representative mollusks than exists in lower invertebrates. This complexity is due not only to a more diversified sensory and action equipment, but in particular to further advances in the pattern of the conduction system. The local action systems, despite the elaborations which make for a certain autonomy in their functions, are brought together in behavior by the central ganglion. This permits a certain variability in the sequence of local activities in total behavior. Although sensory integration is subject to certain limitations, the ability to learn may permit certain of these to be overcome to some extent.

Having surveyed the characteristics which give the mollusks a more complex environment than that of the annelids, we are in a position to attack problems of behavior in the highest invertebrate phylum.

Suggested Readings

Copeland, M. 1918. The olfactory reactions of the marine snails *Alectrion obsoleta* (Say) and *Busycon canaciculatum* (Linn.) *Jour. Exper. Zool.*, **25**, 177–227.

Dawson, Jean. 1911. The biology of *Physa*. *Behav. Mono.*, **1**, 68–88, 107–117.

Garth, T. and Mitchell, Mary. 1925. The learning curve of a land snail. *Jour. Comp. Psychol.*, **6**, 103–113.

Thompson, Elizabeth. 1917. An analysis of the learning process in the snail, *Physa gyrina* Say. *Behav. Mono.*, **3**, 5–39.

Wheeler, G. 1921. The phototropism of land snails. *Jour. Comp. Psychol.*, **1**, 149–154.

CHAPTER VI

BASIC ACTIVITIES IN THE PHYLUM ARTHROPODA[1]

INTRODUCTION

Each of the more than 380,000 known arthropod species is characterized by certain activities which appear almost inevitably in its usual environment. These modes of behavior have been called *instincts*, because they are typically more or less stereotyped, because they appear to be influenced primarily by heredity, and because they are supposedly unlearned. But as recent work with infant vertebrates has shown, this concept has been a retarding influence in psychology, since to call a phenomenon an instinct has been accepted as an explanation and so has not encouraged research into the problem. Vague volumes have been written on the subject without advancing understanding much beyond the speculative stage. This is particularly true for the study of arthropod behavior, which has been a favorite source of incidental cases for speculation on instinct. It is our present task to see where the problems lie and, as far as possible, to treat the typical arthropod activities in terms of the factors which are most responsible for their appearance and for their form.

The Two Principal Problems of Instinct.—The form of arthropod activities in the main is characteristically dependent upon heritable equipment. The *phylogenetic problem* concerns the manner in which such equipment first appeared in the race, and how it was propagated from generation to generation, *i.e.*, its evolution or development through many successive generations. The *ontogenetic problem* concerns the manner in which the heritable equipment is related to

[1] Phylum *Arthropoda:* three-layered animals usually with more or less dissimilar segments, paired jointed appendages, chitinous exoskeleton. Includes class *Myriapoda* (tracheae, many similar legs—centipedes, millipedes); class *Crustacea* (breathe by gills, two pairs antennae—crabs, shrimps); class *Arachnida* (lack antennae, have tracheae, book lungs or book gills—spiders); and class *Insecta* (one pair of antennae, three pairs of legs, tracheae—wasps, grasshoppers, ants, bees, etc.).

behavior on its first appearance and on later appearances in the given individual.

TABLE 11.—ARTHROPODS

RECEPTOR EQUIPMENT, SENSITIVITY

Chemical. *Crustacea*, insects, have pits and sensillae containing specialized cells, on mouth parts and ends of legs (proximate chemoreceptors). Similar receptors over body surface in spiders and insects. Distance chemoreceptors present in antennules of crustacea, club and flask sensillae, and other types of cell in antennae of insects.

Mechanical. Antennae of insects (antennules of *Crustacea*) contain receptor cells. Insects have hairs, scales, etc., associated with sensitive cells on antennae and body surface. Spiders have hairs on legs and over body surface, receptor cells at bases of hairs. Special organ (Johnston's) in antennae of some insects may be important as contact and proprioceptive organ.

Static. Statocysts in *Crustacea*, usually at antennal bases, sometimes open, filled with sand grains at molting time. Statocysts not known in other arthropods, which are well equipped with proprioceptors (movement sense).

"Auditory." Tactual effects through substratal vibrations in most. Chordotonal organs specifically sensitive to air vibrations, present in many insects (*e.g.*, crickets) at antennal bases, on forelegs, or on abdomen.

Light. Spiders often have four pairs ocelli (Fig. 32*A*) on cephalothorax. Insects have 2 to 3 ocelli on top of head, as rule. Paired compound eyes typical of insects, with 1–30,000 units (*Ommatidia*). Many *Crustacea* have compound eyes on stalks.

CONDUCTION

"Ladder type" nervous system typical (Fig. 36). Large supra-esophageal ganglion in head, receives impulses from head receptors. Connectives around esophagus join brain with infra-esophageal ganglion, from which motor impulses reach anterior parts. Paired or fused ganglion in each remaining segment in simple arthropods, joined by longitudinal connectives into chain. These ganglia variously fused together in spiders, insects, crustacea.

ACTION SYSTEM

Crustacea have large pincers on forelegs; walking, swimming legs. Mobile antennae typical of insects, also antennules in *Crustacea*. Running legs in insects; wings typical in *Hymenoptera*. Many have stings, poison glands; and special glands which release gases, fluids, when insect excited. Crushing, biting, other forms of mandibles; sucking and other special types of mouth parts. Glands with air-hardening secretion, released through special tubes, typical of arachnids and many others.

General Consideration of the Phylogenetic Problem.—Although the phylogenetic problem cannot be treated within our present space, certain points should be understood. First of all, we cannot assume that the conditions which were responsible for the evolution of a given structure in the species have anything in common with the significance for behavior of the fully developed structure. For instance, in certain species of ants (*Colobopsis*) the worker individuals have peculiarly flattened heads with which they plug the small entrance into the hollow stalk inhabited by the colony. To say that such a structure is "developed for nest protection" *describes* its present function, but neither *explains* it nor tells anything about its history in the race. No student of the problem means to suggest by such an expression that some beneficent deity recognized the need of the function and supplied the appropriate structure, or that the animal "saw" the need and "grew" the structure to fulfill it. No matter how the structure first appeared in the race, there is no justification for saying that utility was a *primary* factor in the matter.[1]

The evolved structural changes may or may not alter behavior in adaptive ways, *i.e.*, in ways which adjust the species more adequately to its environment. In any case, the relative utility of an evolved structural change which may be responsible for significant alterations in behavior cannot be assumed to have any causal relationship with the factors responsible for its evolution. The matter of utility, or adaptiveness, has *survival* value but has no demonstrated value for the *origin* of a structural change.

It is generally held that any somatic change, in order to be inherited by subsequent generations and thus become the property of the race, must be correlated with changes in the germ plasm. Many attempts to bring about new and heritable structures (mutations) experimentally in multicellular animals have given ambiguous results, or have failed. However, positive results have been obtained for the fruit fly in that the action of *X*-rays has produced new features (*e.g.*, in body coloration, in bristles) which were retained in subsequent generations. Jennings (1930) has summarized the general evidence on this point.

Features Essential to a Study of the Ontogenetic Problem.—We are primarily concerned with the *ontogenetic problem*, the nature and causes of behavior in the individual organism. A scientific study of activities such as those which characterize the arthropods should take particularly into account the following points: (1) the condition of the response on its first appearance in the individual, and on later appearances; (2) the organic condition of the individual when it displays the act (*e.g.*, its life stage, its general metabolic condition); (3) the environmental conditions under which the act is typically observed, and any changes which may occur in the form of the act

[1] Utility is, of course, a *secondary* factor of some importance. A species which comes into the possession of a given evolved character which seriously handicaps its adjustment to an environment (*e.g.*, the development of "conspicuous" coloration in the wings of a butterfly edible by birds) may be forced to migrate and continue its movement until it advances into a different environment in which the maladaptive character is not a barrier to adjustment; otherwise it may become extinct.

under different conditions of appearance; (4) and possible differences in the act as it appears in different individuals of the species.

Among the lower invertebrates the pattern of behavior which characterizes a given group mainly depends upon the inherited equipment of the animal. Variations in the mode of response are attributable to the play of external stimuli and to the survival of internal changes set up by previous stimulation and activities. Our examination of arthropod activities will continue the survey undertaken for the earlier groups. The subject is more complex here, since there exist many influential factors which are interlocked in a more involved fashion than is the case in lower animals. Hence it is desirable to treat each mode of behavior (simple or complex, stereotyped or somewhat variable) according to the nature of the factor which is of central importance for its typical pattern.

Receptor Equipment and the Ontogenetic Problem

Sensitivity and the Determination of Characteristic Forms of Activity.—Of particular significance is the great variety of sensitive tissues and accessory structures possessed by the numerous groups of this phylum (Table 11). The very fact that receptors of given type are possessed by an animal accounts to a considerable extent for the almost inevitable occurrence of certain modes of activity in that animal. That is to say, an arthropod possessing a given type of receptor is forced to give a specific response under certain conditions. For this reason it will be profitable to study arthropod behavior, first, from the standpoint of the essential receptor and the importance of accompanying conditions.

Sensitivity to Temperature and Atmospheric Conditions.—Between wide limits the temperature of the arthropod's body varies with outer temperature, *i.e.*, the animal is *poikilothermous.* General activity depends particularly upon the control exerted by temperature over internal metabolism. This has been experimentally demonstrated for many arthropod functions. Given arthropod forms are to be found living in environments of the predictable mean temperature at which their physiological processes occur most smoothly (the optimum); in part because a variation much above or below this value produces restless movements which may eventuate in a change of habitat. The value of the temperature optimum for the water mite, as determined by the temperature in which the animal swims from end to end of a tube with the smallest number of irregular turnings en route, lies in the vicinity of 17°C. (Agar, 1927). The optimum temperature for many ants lies near 21.5°C.

Environmental temperature determines the form of many activities by controlling the arthropod's readiness to respond to certain types of stimuli. By influencing relative sensitivity to light, temperature figures as an important factor in orientation. It is a principal item in the control of mating activities. The winged males and females of many social insects (*e.g.*, ants) take to the air in the "marriage flight" at times which differ according to the temperature susceptibilities of the respective species. Furthermore, temperature is an important conditioner of general activity in the social insects. Members of a bee colony cluster more closely as temperature falls below 18°C. (Wilson and Milum, 1927). As the temperature drops, bees near the outside move toward the greater warmth in the center of the mass. As temperature rises above 18°C., on the other hand, the wing vibrating known as *fanning* is more and more in evidence within the hive. This response to supra-optimal values ventilates the hive and reduces its temperature, and is thus an important factor in the survival of the bee colony.

At low temperatures many arthropods bury themselves, become inactive, and their metabolism sinks to a low level at which certain physiological (enzyme) changes make possible the consumption of a limited amount of energy. This general response, *hibernation*, is essential for the fact that the female may remain with her brood, an ability which is fundamentally necessary for the development of social behavior in the higher insects (Chap. VII).

Sensitivity to humidity, to temperature, and to other conditions are major determinants of the environmental situation in which an arthropod lives (Kennedy, 1927; Talbot, 1934). These types of sensitivity indirectly influence the form of behavior, in that a different environment might well elicit different behavior from a given arthropod. Certain characteristic activities often fail to appear in a supra-optimal environment, or unusual activities may appear under such conditions.

Chemical Sensitivity and the Activity Pattern.—The diversified chemoreceptive organs of arthropods are basic for many types of activity, and particularly for food taking. In aquatic *Crustacea*, sensitive cells on the anterior moving parts of the body, especially on the antennules, are stimulated by juices diffusing from meat at a distance. The animal is therefore able to locate a source of food by turning consistently toward the more strongly stimulated side (Bateson, 1887). The antennae of insects contain numerous types of specialized nerve cells which are *distance chemoreceptors*. The less sensitive chemoreceptors of arthropods, generally located on mouth parts (although also variously present over the body surface), are brought into play only by chemical stimuli furnished by objects which are very close to or in contact with the body. Forel (1908) observed that ants could withdraw from strychnine-treated honey only after having touched their mouth parts to its surface. This suggests the importance of the role played by the *proximate chemoreceptors* in normal feeding.

Minnich (1926, 1931) has demonstrated the presence of such receptors on the tarsal segments (distal end of leg) in many insects. The red admiral butterfly extends its proboscis when a needle with a drop of sugar water is brought close to the tarsae, and removal of the antennae does not eliminate the response. Such receptors, in addition to the distance chemoreceptors of the antennae, appear of outstanding importance in the finding and taking of food by nocturnal insects (*e.g.*, moths).

Chemical sensitivity governs the manner in which mating is elicited in many arthropods, and especially in insects. A number of experimenters have shown that male moths find the sexually mature female of their species by responding to a chemical which emanates from her body. Riley (1895) reported a case in which a male ailanthus silkworm moth (marked with silk thread) reached a female over a distance of 1½ miles. Kellogg (1906) proved that the male responds to the scent-gland secretion of the female. The isolated scent gland attracts the male as does the intact female, but a female lacking the scent gland does not attract the male in any way.

The foraging and "communication" of honey bees comprise one complex activity, the form of which is greatly influenced by chemical sensitivity. Von Frisch (1923) has described the typical behavior pattern. A nectar gatherer that enters an observation hive is seen to disgorge part of her load to other bees, after which she executes a peculiar "round dance." She circles repeatedly, first to one side and then, with a sudden sweep, toward the other side, before she moves off and repeats the dance on another part of the comb. Other bees become excited and follow closely after her. Soon many of them leave the hive. If the dancer has found her food in one of several saucers of sugar water equidistant from the hive, the excited bees may visit this particular saucer in greatest numbers. For this to happen, however, it is necessary that in feeding the *finder* shall have been sufficiently excited to swell out her *scent gland*, a highly glandular pocket near the end of the abdomen from which a chemical substance diffuses onto the food place. von Frisch found that when the scent gland of the "finder" bee had been shellacked over before the experiment, the secondary bees arrived at her feeding saucer only by chance.

Chemical sensitivity may influence flower visiting in still other ways. In his early work, von Frisch (1919) demonstrated that bees may learn to respond to flowers on the basis of their different "odors." In later experiments (1923) he trained one group of bees to visit a nectar place saturated with peppermint oil, and a second group to visit a different place which was saturated with polyanthus oil. Both food places lay 9 meters from the hive, but in different directions. After a nonfeeding period of 1 hr., one of the "polyanthus bees" was permitted to feed in its accustomed place and to return to the hive. During the following hour, of 13 bees in the polyanthus group 11 visited their habituated food place, while only 4 bees in the peppermint group came to this place. Such results show that the round dance of the finder bee excites other individuals, but not in a specific way, unless the chemical (*e.g.*, polyanthus) which diffuses from her body as she dances elicits the learned response of flying out to the food place connected with it. Thus, one of the most complex and representative activities of bee life centers around the characteristics of chemical sensitivity.

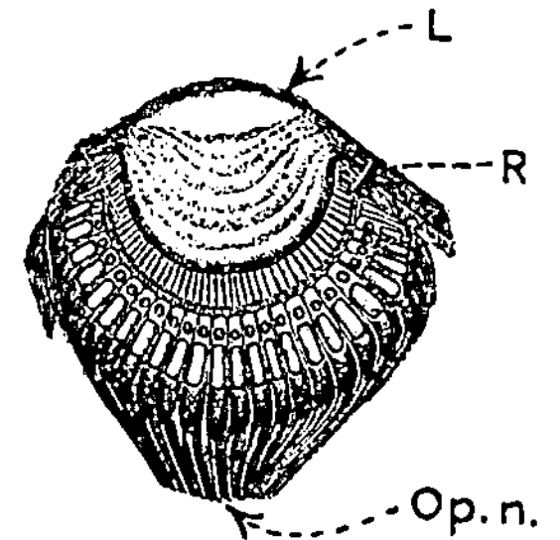

FIG. 32*A*.—Ocellus (posterior) of the spider *Epeira* (from Lubbock). *L*, lens; *R*, retina; *Op. n.*, optic nerve.

Sensitivity to Light and the Determination of Behavior.—Arthropods possess two principal types of eye, the *ocellus* and the *compound eye* (Fig. 32). The ocellus is believed to be especially sensitive to low intensities of light. The fact that most spiders have four pairs of well-pigmented ocelli on the anterior part of the body may account for the frequency of nocturnal activity among these animals. Many ants have three ocelli on the top of the head, and evidence has been presented that these organs permit sensitivity to the positions of large objects in the vicinity.

Most crustaceans and insects possess compound eyes (Figs 32*B*, 35). These receptors are generally very well developed in flying insects, and in some flies they take over almost the entire upper surface of the head. Each of the cylindrical units (ommatidia), of which there may be a few or several thousands in the compound eye, has its individual lens on the outer end and its nerve ending at

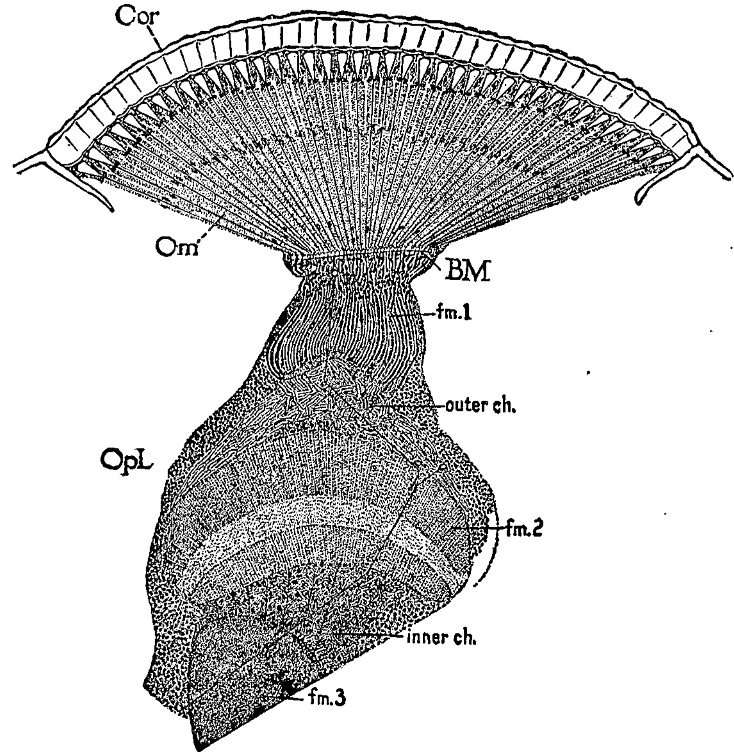

FIG. 32*B*.—Horizontal section of compound eye and optic lobe of worker honey bee. (*From Snodgrass.*) *Cor*, cornea (facets of ommatidia); outer *ch.*, inner *ch.*, crossings of optic fibers; *Om*, ommatidium; *OpL*, optic lobe of brain.

the bottom. Pigment on the walls of the ommatidium absorbs light reflected from all parts of an object, save that directly ahead of the facet. It appears that in consequence only certain well-separated parts of a motionless object could stimulate this eye. The insect is acutely sensitive to movement of an object, since even a very slight movement changes the ommatidia which are acted upon by light. The result must be a summated nervous excitation much greater in effect than the slight movement of the object would appear to warrant. The *praying mantid* does not respond to a motionless cricket, but the least motion of the prey causes a prompt head turn on the part of the mantid, and a further movement releases the pouncing response. Forel (1908) suggested that a flying insect can have more acute vision when on the wing than when at rest, since its own motion must bring a constant shift in the stimuli furnished to the compound eyes.

The nature and relative development of the visual receptor are evidently responsible for the habitat and mode of life of many arthropods. Crawford (1934) concludes that features of eye-structure are frequently of great importance in making an insect nocturnal. A "night eye" is one which is most sensitive in dim light, and which readily adapts to dim light. Welsh (1932) reports that in certain nocturnal *Crustacea* a layer of reflecting pigment readily forms behind the retinal cells when the animal is in darkness. This greatly increases the arthropod's sensitivity to low intensities of light.

The Significance of Tactual Sensitivity and of Sensitivity to Mechanical Vibrations.—Tactual sensitivity is fundamental to many arthropod activities. It permits walking arthropods to follow a surface and to avoid objects with great efficiency. The highly mobile antennae, often covered with fine hairs which amplify their sensitivity, figure prominently in the intimate control exerted by contact over movement. The side on which one antenna has been in contact with a surface beside which the animal (*e.g.*, the cockroach) is running is the side toward which a turn is made when a clear space is reached (pp. 160*f*). This characteristic behavior depends upon the general fact that arthropods tend to turn in the direction of a weak tactual stimulus (*positive thigmotaxis*). In the movement of nocturnal arthropods, or of those which inhabit quarters not reached by light (*e.g.*, subterranean ants, beetles that live under bark), contact figures more prominently as a sensory control than it does in the activities of surface forms (Banta, 1910).

Although the bodies of arthropods have a hard exoskeletal covering, the presence of specialized hairs or scales which have sensitive cells at their bases makes possible a delicate tactual sensitivity over the body surface. Many of the orb-weaving spiders are able to charge in the correct direction and seize flies which have become caught in the web. Barrows (1915) demonstrated the ability of one species to locate promptly in the dark the tip of a rapidly vibrating straw (mechanically actuated) which was touched against the web. His results indicated that the spider was responding to vibrations of the web strands which acted upon the fine hairs of the legs, by turning in the direction of the most intense web vibration.

It is apparently the sensitive condition of the soft body covering which is responsible for the hermit crab's interesting practice of living within a snail shell or a similar hollow object (Fig. 33). A crab which lacks a shell approaches and picks up small objects. The object is discarded if the fumbling walking legs do not find a cavity. Even when the object contains a hollow, the animal soon drops the object and moves away if the body is not fairly well covered (Hertz, 1933). Such behavior apparently arises from the internal disturbance which is set up when the soft and sensitive skin is not adequately in contact with a surface.

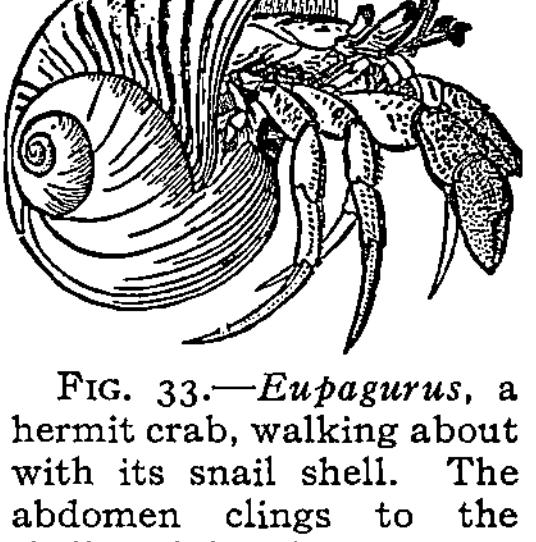

FIG. 33.—*Eupagurus*, a hermit crab, walking about with its snail shell. The abdomen clings to the shell, while the anterior part of the body is held free. When disturbed the crab quickly withdraws into the shell. (*Redrawn from Hertwig's Manual.*)

The nature of inter-individual relationships in most insect societies and major behavior modes such as communication and exchange of food depends ultimately upon tactual sensitivity (pp. 167*ff.*). Social insects arouse one another when the colony is disturbed, usually by antennal contact or bodily contact, and in many cases by vibratory disturbance. Emerson (1928) has described the manner in which the heads of excited soldier termites knock against the wooden gallery walls of the nest. This sets up a vibration which is conducted by the wood and soon arouses the entire colony, since other individuals are delicately sensitive to vibration of the surface on which they are walking. In similar manner the large-headed soldiers of some carpenter ants repeatedly knock their abdomens against the floor when disturbed, thereby exciting other ants of the colony. These acts are incidental parts of the excitement of the insects which are first disturbed, and their effect upon other individuals is also an accidental one, at least insofar as the excited individual is concerned. Nevertheless, such tactual stimuli represent one important means whereby the colony is unified.

Thus the excitement of one insect may be transmitted to others of its kind in the vicinity, through direct contact of antennae or body surfaces, through substratal vibrations, or in scattered instances (*e.g.*, in crickets) as periodic air vibrations (Regen, 1911) which stimulate specialized chordotonal organs. Frequently, vibratory effects serve to unite an insect social group, in other cases they account for the aggregation of the sexes during the mating period, and they may be of importance in other ways.

We have passed in review a number of arthropod activities which are particularly influenced in their form by the nature of receptor equipment. No other animal is so well supplied with sensory structures in relation to its size. The nature of these structures in a given arthropod makes certain activities virtually inevitable, or influences the conditions under which activities otherwise determined may appear. A consideration of arthropod orientation will afford a somewhat different approach to the general subject of the chapter.

Sensitivity and the Basis of Arthropod Orientation.—Orientation is the process whereby the movements of an organism are influenced by stimulation from a localized source, or by diffuse energy, so that a bodily position is assumed which bears some relation to the manner of stimulation. Among the lower arthropod groups (*e.g.*, crustacea) in particular the modes of orientation are fairly stereotyped. We shall study such cases first of all, in order to gain an understanding of the fundamental properties of orientation in the phylum, and in order not to be confused later by the additional factors which complicate the phenomenon in higher arthropods. For the sake of clearness, reference will be confined to orientation under the influence of light.[1]

Photokinesis.—A photonegative arthropod such as the sowbug (a crustacean), when stimulated by diffuse light or by light from directly overhead, will be found after a time in a dimly lighted zone. This phenomenon, *photokinesis*, depends upon the fact that light excites the animal so that it continues to move until a situation (darkness, if photonegative) is reached in which the disturbance is removed. Hence, although diffuse light does not specifically determine the direction of locomotion, it nevertheless may eventually orient the animal (Fig. 34*A*). However, when there exists a slight difference in the intensity of light from different quarters, the photo-

[1] It is suggested that the student review the essentials of the "tropism" question as considered in Chap. IV. Since 1935, the term "taxis" has replaced "tropism" with respect to animal orientation. For a review and discussion of the evidence see Fraenkel and Gunn (Suppl. ref., 1940, 1961).

negative sowbug reaches the darker side somewhat more promptly, since movements toward the light have slightly less vigor than movements from the light. Thus even a feeble directionalization of light may exert some direct control over orientation.

In more strongly directionalized light, the animal's movements tend to show a consistent relation to the manner in which energy reaches its receptors. Consequently, the animal reaches its region of optimal illumination more quickly than it does in diffuse light (Fig. 34*B*).

The problem of "sign" (pp. 95*f.*) in orientation is sometimes cited as unapproachable. However, the experimental evidence shows that

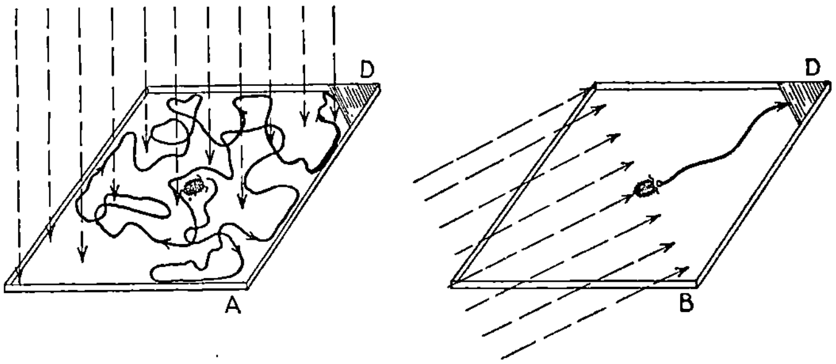

FIG. 34.—Contrast between photokinesis and directed orientation. (*A*) Sketch of path taken by a photonegative arthropod (sowbug) under diffuse illumination. (*B*) Path of photonegative arthropod in light directed from one side. Broken lines show direction of light; solid lines show path of animal. (*D*) dark corner.

this problem may be studied with profit in terms of the conditions which cause the typical sign of orientation in a given species to vary.

Conditions Which Make for Variation in Response to Light.—The most important factor in the determination of a characteristic response to light, photonegative or photopositive, is the *normal metabolic condition of the group*. Phipps (1915) discovered that amphipods from running streams moved into the brighter regions of a box which was illuminated with light which increased in intensity from one end to the other (a "light grader"). The rate of metabolism in this species was higher than that in species from rivers and ponds which moved into the darker regions when tested in the light grader. The difference in the characteristic metabolism of these two species apparently accounts for the significant difference in their typical response to light. This conclusion is supported by the fact that any effect which changes an animal's metabolic condition may also change the sign of its response to light. Phipps found that when starved the running-stream amphipods became photonegative, whereas pond amphipods when starved became photopositive. Starvation reversed phototropism in both cases.

Temperature, through its control over metabolism, is an important influence in bringing about such changes. Certain copepods studied by Loeb (1893) were positive to light in cool water, but became photonegative in warmer water. Similarly, the chemical constitution of the medium is important for the sign of phototropism in aquatic animals. By adding sodium chloride to the medium Loeb (1893) made copepods photopositive, but the addition of a quantity of distilled water made them photonegative. Chloretone, or potassium cyanide (both of which depress metabolism) were found by Phipps (1915) to cause ordinarily photonegative amphipods to become positive to light. Holmes (1905) reversed the photopositive response of the water scorpion *Ranatra* by handling the insect. Mast (1911) has pointed out that the more disturbed a moth becomes the more persistently it flies toward light, although when quiet it crawls more readily into the dark.

It may be concluded that the typical "sign" of an animal's phototropism depends upon the characteristic metabolic condition of its species, but that many conditions which tend to alter the usual metabolic state may bring about variations in orientation or may even reverse the normal response. It is unnecessary to dwell upon the fact that the sign of phototropism also depends upon the *intensity of the light*, a point which has been treated for other animals. We have also stressed the dependence of phototropic response upon the condition of the animal's receptor (*e.g.*, its adaptation to light).

These, and other factors which might be mentioned, produce variations in the phototropic response by influencing the effect which a given stimulus actually exerts upon the muscular tonus of the animal. The representative orientation phenomenon may therefore be studied from this approach.

The Basic Mechanism of Arthropod Orientation.—The lower arthropods orient fairly directly, as a rule, when strong light reaches them from a localized source, and the position which is assumed typically is related to the manner in which the stimulus acts upon the receptors. The "forced-movement" theory (Loeb, 1918) offers the most adequate account of this phenomenon. It also appears to be the best point of departure for study of more complex forms of orientation.

To review the theory (pp. 97*f.*), let us consider a hypothetical aquatic arthropod which possesses symmetrically placed light-sensitive areas, from which bilaterally symmetrical nerve tracts carry energy to the muscles of symmetrically placed legs. Let us say that this animal is metabolically so constituted that it is normally photopositive. When the two eyes are stimulated equally, the legs will beat equivalently and synchronously, sending the animal directly toward the light. If an interference causes the animal to veer toward the left, the right eye receives more light than the left. The stronger innervation which consequently acts upon the legs of the left side increases the tonus of their muscles and causes them to beat more strongly. Thus the animal is turned back toward the right. Inequality of stimulation has forced a movement which restores the conditions of equivalent stimulation and the path directly toward the light.

The manner in which forced movements are normally produced when a stimulus source is changed in position is demonstrated in an experiment by Holmes and McGraw (1913). A light was placed at the top of a paper cone so that all portions of the floor were illuminated with equal intensity. A photonegative cricket, which had its left eye covered with a light-proof mixture of lamp black and shellac, when placed on the floor of the cone turned persistently toward the left (away from the strongly stimulated eye). Photopositive robber flies circled toward the right when the left eye was coated.

Factors Which Complicate the Mechanism of Orientation.—Many experimenters are impressed by apparent exceptions or by variabilities in arthropod orientation. They maintain that such cases violate the premises of the forced-movement theory. To take one case, Rádl (1903) observed that a blowfly larva with one eye covered was able to move toward light in a path almost as straight as that taken by a normal animal. When Holmes (1905) blackened one eye of *Ranatra,* the insect at first circled toward the normal eye in progressing toward a light source, but after repeated trials the path led straight toward the light. Holmes explained this as the gradual suppression, through "learning," of movements which tended to carry the animal from the light.

Clark (1928) repeated this experiment and reached a different conclusion. One of the two lateral eyes of the water bug *Notonecta* was covered, and the animal was kept in the dark for 6 hr. When tested, the animal moved toward the source of a strong horizontally directed beam of light. At first it took an irregular course which was characterized by circus movements toward the uncovered eye. In further trials, however, the circles became larger and decreased in number, and after 43 trials the animal moved directly toward the source. After a rest period of 15 min. in darkness, the circling reappeared when a test was made, but subjects which were given a rest period of 1 hr. *in light* moved directly toward the source.

In another test, after a 50-min. exposure of the eyes to light, the direct path appeared in an average time of less than 1 min.; after a 30-min. exposure an average of 18 min. was required; whereas animals taken directly from darkness required on the average 48 min. of repeated trials in the light to attain the direct path.

The tendency to circle in an animal with one newly covered eye is attributable to a difference in stimulation of the two eyes. The above results show that circling largely disappears when the sensitivity of the uncovered eye is reduced through adaptation to light. Hence this apparently exceptional type of orientation is explained by the operation of a special factor, light adaptation, and actually serves as evidence for the forced-movement theory. We shall soon find that in arthropod orientation which is modified through learning this theory retains a fundamental explanatory value.

Another important complicating factor in the orientation of many arthropods is the fact that stimulation of different parts of the compound eye produces different effects upon movement. Mast (1923) established the fact that a one-eyed drone fly moves differently toward light according to which portion of the unblacked eye is stimulated by light. In his tests, stimulation of the anterior inner ommatidia brought forced movements toward the side of the covered eye, but stimulation of the lateral ommatidia brought turning toward the functional eye. Clark (1928) reported similar findings for *Notonecta* (Fig. 35).

These facts support other evidence that different portions of the higher arthropod's compound eye are differently connected, by crossed and uncrossed fiber tracts in the nervous system, with the locomotor apparatus of the same and the opposite sides of the body. In such animals the two eyes are much more complexly balanced in controlling movements of the two sides of the body than is the case for simpler forms. In a given position, for instance, the stimulation of a portion of one compound eye may dominate general movement for a time. Thus an animal may move directly toward one of two lights if its bodily position prevents the other light from effectively stimulating the eyes. Consequently there appears to be no necessary difference between this last case of orientation, which Kühn (1919) calls *phototelotaxis*, and the simpler case in which the animal moves toward a point between the two lights, called by Kühn *tropotaxis*.

A complex case of orientation, not uncommon among higher arthropods, involves the maintenance of a given visual field as a

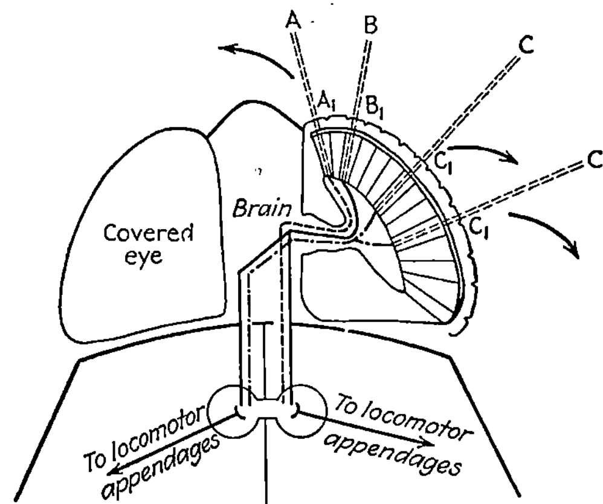

FIG. 35.—Relation between compound eye stimulation and forced movement in a one-eyed (photopositive) Notonecta. Light at *A* stimulates area A_1 of the eye and the animal turns toward the light; light at *B* stimulates area B_1 but the animal does not turn; while light at *C* stimulates area C_1 and causes a turn toward the right side. Schematic nervous paths represent the theory of H. Müller as to how this control is effected. (*Modified from Clark*, 1928.)

learned reaction. For instance, whirligig beetles collect in a given locality of their stream environment, gyrating and milling about on the surface of the water, but each animal maintains its position within a zone by swimming against the current. The movements of the beetles are visually controlled, and are not direct responses to the current, since Brown and Hatch (1929) found that in a darkened laboratory tank the insects swam about erratically and did not move against a current. The experimenters considered it probable that in the normal habitat the beetles become gradually habituated to the visual field of a given optimal (*i.e.*, as to temperature or food) locality. When an animal is carried off by the current, the visual change excites swimming movements until the former condition of stimulation again prevails. This conclusion was encouraged by the finding that beetles in a stream swarm were excited to swim (*fright reaction*) by the sudden appearance of a black spot in a field which had been plain white during the preceding few minutes (habituation time). Tested in a large laboratory tank the beetles were found capable of simpler forms of orientation: movement toward a luminous part of the visual field, or remaining within a diffusely illuminated area by swimming back whenever the shaded border zone was entered. Circumstances therefore determine whether the orientation process in such insects is to be simple or complex in nature.

The factors that may complicate the mechanism of orientation are numerous and often are so complexly interrelated as to produce variations in movement which are not easily understood. A beetle released at night flies directly toward a nearby light and may strike it, but then circles about the light, first in one direction and then in the other. It may fly past the light into the dark border zone, then turn about, become reoriented, and fly back toward the source. Each time it flies very close to the light the excessive intensity interferes with forward flight, and variable circling movements result. The observer sees the results of successive and variable stimulations of the compound eyes as the animal's position changes. Successive changes in the condition of the two eyes and in the tonus of flight muscles on the two sides are also apparently involved.

Degrees of Complexity in Arthropod Orientation.—In this phylum orientation is basically a matter of direct control of movement by environmental stimuli which owe their coercing effect to the animal's structure. In the simplest cases, feeding and other activities are incidental to the overpowering effect of orienting stimuli. A plankton crustacean feeds near the surface of its lake at night not because food is more plentiful, but because light and related stimuli have oriented it there (Ewald, 1910; Kikuchi, 1930), and migrates to the depths in the morning because stimuli have forced it to move there.

The forced-movement theory designates factors which function in all cases of orientation. In the higher arthropods, however, these factors are not so much the direct tool of external stimulation. The whirligig beetle illustrates a case in which the attainment of an optimal zone may determine the nature of the visual stimuli which control the animal's movements over a given period. In the social insects the habituation factor (learning) makes possible a much more complete dependence upon circumstances of feeding and other life activities, as improvements in receptor and in nervous equipment increase the complexity and variety of stimuli which may be involved in the control of orientation. Later (Chap. VII) we shall return to this question.

Behavior Characteristics Mainly Attributable to the Nature of Nervous and Effector Equipment

Nervous Structure and Behavior Pattern.—The activities of arthropods are basically delimited and molded by the arrangement of the conduction system. Certain important characteristics are worthy of special notice.

Head Dominance and the General Nervous Pattern.—The representative arthropod nervous system is of the "ladder type," with a pair of ganglia in each body segment and the ganglionic pairs connected into a ventral chain by longitudinal connectives and by commissures across the midline (Fig. 36). In the developed arthropod, correlated with the highly specialized receptors of the head there is an anterior ganglionic mass, the cephalic ganglion (or brain), which exerts a dominance over other nervous centers. This permits head sensitivity to play a part, usually a controlling part, in any of the numerous local activities. Consequently the form of orientation in a given arthropod is particularly dependent upon the type of head receptors possessed by that animal. To illustrate, an arthropod with poorly developed eyes and well-developed chemical sensitivity differs greatly in its orientation from one with excellent vision (p. 154).

The Insect Nervous System.—In ants, bees, and wasps, the principal social insects, the arthropod nervous system is most highly developed and specialized. The nervous system of the bee is shown in Fig. 37. The supra-esophageal ganglion (brain) in the head, which receives impulses from head sense organs (eyes, antennae, and in mouth parts), is joined to the infra-esophageal ganglion by a thick connective on each side of the esophagus. From the infra-esophageal ganglion motor nerves innervate the action parts of the head: antennae, mandibles, and mouth parts. The head ganglia are joined by connectives to the ganglia of the thorax. The local thoracic ganglia are joined by longitudinal connectives. Each of these centers receives sensory impulses from its level of the body, and innervates one pair of legs (and wings when present). The continuation of the chain joins the thoracic ganglia with the next posterior abdominal ganglia. These receive local sensory impulses, and give rise to motor nerves which innervate action parts of the abdomen (stomach, sting, etc.). There is also an important auxiliary nervous system, which mediates between central nervous system and viscera.

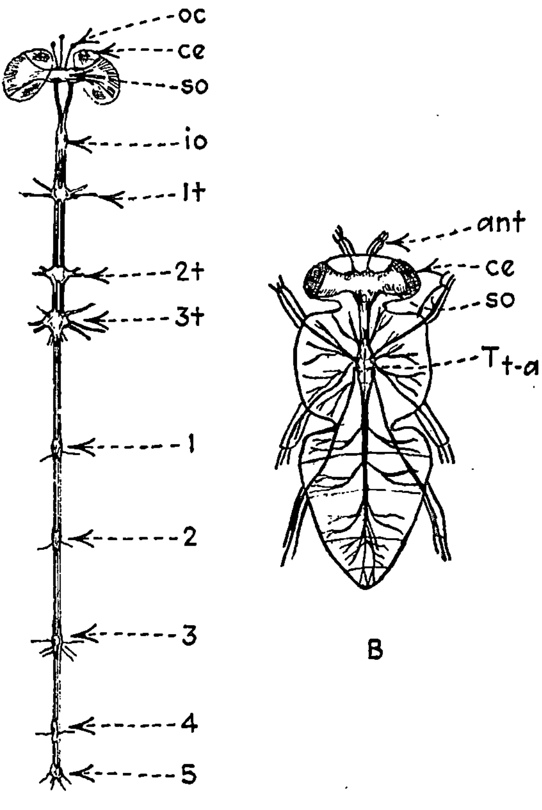

FIG. 36.—(*A*) Ladder-type nervous system of *Caloptenus* (a locust). *ce*, compound eye; *io*, infra-esophageal ganglion; *oc*, ocellus; *so*, supra-esophageal ganglion; 1_t, 2_t, 3_t, thoracic ganglia; 1, 2, 3, 4, 5, abdominal ganglia. (*Redrawn from Packard.*) (*B*) "Fused ganglion" nervous system of *Sarcophaga* (flesh fly). *ant*, antenna and antennal nerve; T_{t-a}, single large ganglion into which all thoracic and abdominal ganglia have fused during development. (*Redrawn from Herrick.*)

To a considerable extent, each part of the segmented nervous system dominates the action parts of its locality. Although the arthropods have this characteristic in common with the annelids, they have a distinct advantage in the larger measures of dominance

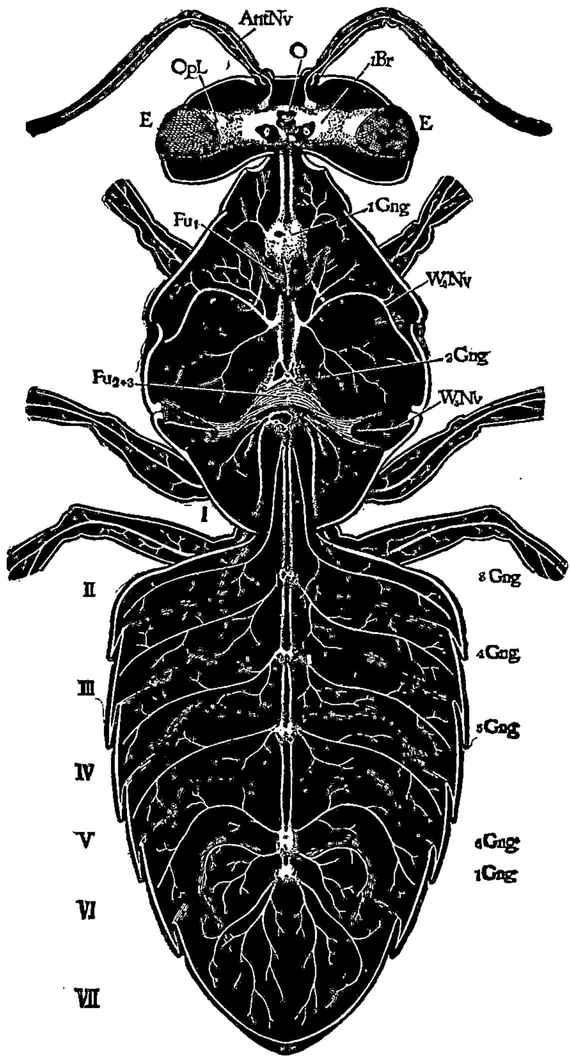

FIG. 37.—Nervous system of the worker bee. *Ant Nv*, antennal nerve; *O*, ocellus; *E*, compound eye; 1*Br*, supra-esophageal ganglion; *Op L*, optic lobe of brain; 1 *Gng*, 2 *Gng*, thoracic ganglia; 3 *Gng*–7 *Gng*, abdominal ganglia. (*From Snodgrass.*)

which is exerted by the brain over lower centers. When Bethe (1897) cut the circumesophageal connectives between the brain and the remainder of the crab's nervous system, the animal did not continue a movement unless further tactual stimuli were applied locally. The legs of the operated animal tended to swing in pendular manner, and were constantly rubbing against one another. Such results are taken to indicate the absence of an inhibition which is normally exerted over the local ganglia by the brain. In the operated animal the absence of the brain therefore permits the segments to respond independently to their respective local stimuli.

The Nervous Basis of an "Instinct."—The complicated structure of the insect nervous system makes possible many serial activities, with a locally controlled activity as nucleus and the functions of other body parts superadded. Mating and egg laying are good examples. These are virtually the only activities of the silkworm moth during its 15 days of adult life, since the animal is incapable of eating and does not fly. After mating (p. 125), the female lays her eggs, 200 to 350 in number, in rows arranged in a single layer.

McCracken (1907) performed various operations upon females, and studied the effect upon the egg-laying performance. Removal of the head (containing the brain) prevented the female from mating; or if she had mated already, made it impossible for her to lay her eggs upon the mulberry leaf, as the silkworm usually does. These shortcomings are accounted for by the fact that receptors (particularly the olfactory organs) removed with the head are indispensable for the reception of stimuli which act upon the mature female with particular potency and thus determine where the eggs are laid.

When the thorax was removed, depriving the animal of its legs, eggs were still laid when the abdominal walls were rubbed by the experimenter, but these eggs were not deposited in even rows. The abdomen, however, twisted and bent from side to side more than does that of the normal female in egg laying. This behavior caused the eggs to be laid in an irregular heap.

The abdomen of the moth, with its four pairs of fused ganglia, was the next subject of investigation. While the abdominal ganglia remained interconnected, the stimulated abdomen moved from side to side as in normal laying (although with more twisting), and the ovipositor deposited eggs. When the first, second, and third pairs of ganglia were successively cut free from the last pair, the coordination of segments necessary for the abdominal twisting was correspondingly reduced each time. When the last pair of ganglia (the pair from which nerves reach the actual egg-laying apparatus, the ovipositor) alone remained, stimulation of walls of this remnant of the abdomen caused the eggs to be laid in a heap.

Thus the actual depositing of eggs is locally controlled in the last abdominal segments. The complete act in the normal silkworm may be theoretically described as follows. The distension of the ovaries with eggs, very probably accompanied by exciting glandular effects, puts the animal into a condition which makes it respond readily to the special olfactory stimulus (*cf.* p. 146). Transmitted posteriorly through brain and nerve trunk, this excitation acts upon the susceptible abdominal mechanism, ovipositor and related structures. The action upon thoracic centers of *secondary* nervous excitation emanating from the activated abdomen causes the legs to fall into step with the abdomen in the body twisting which accompanies action of the ovipositor. Thus as Herrick (1924) points out, the response consists not of numerous discrete "reflexes" but of a complicated pattern of activity which is built about the abdominal function as its core.

Local Control and Its Limitations.—An example of local control is the movement of mouth parts in the insect larva when the mouth region is chemically or tactually stimulated. This response is a "line of least resistance" phenomenon in which the energy set up by the stimulus is conducted to the nearest action parts, which respond first and most strongly.[1] This is the basis of the feeding response in many insects, and appears subject to later modification through learning. In its elaborated state it assumes significance for the form of social behavior (pp. 171*ff.*).

A common illustration of autonomy (relative independence) in the action of local segments is the ability of many lower arthropods to walk after the head has been removed. Following such an operation, however, contact stimulation of the thorax is necessary to arouse continued movement (Rogers, 1929). This activity of the legs in walking is locally controlled by the interaction of the thoracic

[1] The relatively small size of the arthropod, in relation to the rapidity of its nervous conduction (6 to 12 meters per second in the lobster, much faster in insects), means that over very short nerve arcs an impulse is transmitted in a fraction of the time required in other animals. This suggests an intimate relationship between sensitivity and action, a major reason for the characteristic stereotypy in arthropod behavior and for the prominence of local activities.

ganglia, but normally variations in speed and in direction of progress are attributable to the influence of head receptors.

The manner in which nervous impulses from other centers influence locally controlled activities is suggested by some findings on leg activities. In the crab *Portunus*, Herter (1932) found that weak stimulation of a leg caused extension, whereas strong local stimulation caused flexion. The opposite nature of leg responses to local stimulation of high and low intensity rests upon the functioning of local ganglia and muscle systems. To study the effect of impulses normally coming from the brain, Herter cut the esophageal connectives and applied electrical stimulation to the stumps. Weak stimulation of the right connective (*i.e.*, excitation from the right side of the brain) caused flexion of walking legs on the right side and extension of those on the left side; whereas strong stimulation of the right connective brought extension of right walking legs and flexion of left walking legs. The diametrically opposite effect of locally applied stimuli and of impulses from higher centers speaks for the manner in which the dominant anterior normally alters the action of lower centers in controlling segmental activities.

A difficult problem in local control is presented by the opening and closing of the crab's cheliped or pincher, an act of great importance in the capture of prey. Tonner (1933) has demonstrated that in the crab there is no direct central nervous inhibition over the muscles that close the pincher, but that this action may be inhibited only through a local nerve plexus in the pincher itself. This represents the fact that in arthropods, as in other invertebrates, local nervous control assumes many functions which devolve on central nervous control in the vertebrates.

The segmental arrangement of ganglia in the arthropod nervous system draws inevitably together local activities which are peculiar to a given species. The fusion of two or more local ganglia (Fig. 36*B*) unites much more closely the activities which were primarily controlled by the corresponding separate ganglia. United action of the legs in initiating flight is undoubtedly more efficient in flies which have but a single large thoracic ganglion than in flies which have two or three pairs of thoracic ganglia. However, in all insects the activities controlled by the thoracic ganglia are very well coordinated. Fraenkel (1932) discovered that in flying insects generally contact of the leg tarsae with the ground somehow inhibits flight by releasing sensory impulses which reach the wing muscles through the local thoracic ganglia. Artificial removal of the tarsal stimulation (*e.g.*, by lifting of legs) led directly to flight. An exteroceptive stimulus (*e.g.*, a light) causes the tarsae to release with a spring, and this action appears to summate with the initiating exteroceptive stimulus in eliciting wing action. However, in good flyers such as bees or wasps exteroceptive stimulation is usually sufficient to produce both flight and tarsal release as simultaneous responses, the latter reinforcing the former.

Action Equipment and the Inevitable Appearance of Certain Adjustments. *Structures Related to Special Modes of Behavior.*—The

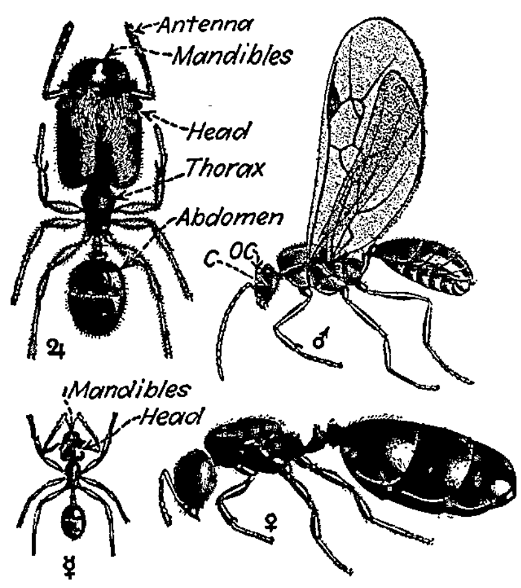

FIG. 38.—Dealated queen (♀), male (♂), major worker (soldier, ♃), and minor worker (☿), of the ant *Pheidole instabilis*. *c*, compound eye; *oc*, ocellus. (*Reproduced from Wheeler's Ants by permission of the Columbia University Press.*)

possession of given action equipment is generally a basic factor in the appearance of certain modes of behavior. Just as the nature of photoreceptors may assist in forcing an arthropod to shun the surface and lead a subterranean life, so the possession of strong mandibles or of sucking mouth parts circumscribes its food-taking behavior. In no other phylum are the action structures and the correspondingly diversified types of behavior so numerous. A simple example of the responsibility of a structural character for a mode of activity is the humping ("measuring worm") gait of the small green larva of the geometrid moth. This trait is attributable to the absence of segmental appendages in the middle of the body. When the muscles of the body wall contact, the body therefore loops upward and the posterior end is carried forward close to the front end before it reattaches. The locomotion of other caterpillars which have a full complement of legs takes the typical vermicular form.

In many species of ants there are workers which differ in size and in structure. In *Pheidole* species, for instance, certain workers have huge heads and mandibles, and other workers have smaller heads and mandibles (Fig. 38). These ants are seed eaters. The small workers forage outside the nest and carry in seeds, while the large-headed workers become able to crack the husks from the seeds with their powerful mandibles. Such *polymorphic* differences among the sexually impotent workers, with corresponding behavior differences induced by the structural characters, are of the greatest importance for social life in insects (pp. 170*ff*.).

Structural Changes in Development, and Corresponding Behavior Changes.—Some arthropods, and in particular some insects, come into the possession of new and distinctive activities when growth has made a body part functional. The early activities of the worker bee correspond in a general manner to certain changes which appear successively in its glandular structure. Later we shall find this fact of great importance for social behavior in bees (pp. 173*f*.). Emerson (1926) has described the manner in which individuals of certain species of tropical termites at first eat and digest wood and show other typical worker activities. While they have the worker structure, they behave entirely like workers. Then two molts occur in which the former body covering is lost. As a consequence these individuals look like "nasute soldiers," and they behave like soldiers. Instead of running off when the nest is disturbed, as workers do, they rush forth excitedly and squirt the sticky product of the now functional frontal gland through a specialized tube on the front of the head.

Petrunkevitch (1911) reported that the immature male "tarantula" behaves as does the female. In the daytime he remains in the burrow which he has dug (the entrance closed with silk spun by him), and at night he lingers outside the entrance and pounces upon prey which chances along. Upon reaching sexual maturity, however, his behavior undergoes a radical change. He leaves his burrow for a roving life, and at length fills the enlarged bulbs of his palpi by a special process peculiar to male spiders, involving the spinning of a *sperm web*. Then, after a day of rest he again wanders, and mates with any responsive female with which he chances to come into contact. Among the arthropods such abrupt changes in behavior, correlated with the onset of maturity (probably glandularly based) are very common.

"Instinctive" Activities Dependent upon a Complex of Factors.—We have seen that many arthropod activities owe their existence to the possession of given structures. Such activities, however, often are so complex and so intimately related to other features of behavior that the conditions of their appearance may resist analysis. For example, among *Eciton* species ("army ants") of the American tropics, raiding is one prominent activity. The ants pass out along collective chemical trails (pp. 151 *f.*), capture the brood and adults of arthropods and other small forest life, and carry the booty back to their temporary nest. Schneirla (see *Jour. Comp. Psychol.*, 1933, 1938, 1940), investigating this phenomenon, finds it dependent upon a complex of factors, such as: the conditions which exclude a stable nest site for these ants; the ability of excited foragers to lay down a trail of chemical secretions; acute chemical sensitivity enabling the ants to follow this trail; the excitability of the workers and the potency of their attacking equipment. When a great brood of active, voracious larvae is present, the daily raids are large and an emigration occurs each night. This interval, the *nomadic phase*, ends when the larvae mature. As this brood then pupates, a statary phase ensues in which daily raids are small or absent and the colony does not emigrate. These phases of the activity cycle, which are species-typical in their duration, arise through high-level stimulation (tactual, chemical) from the brood when it is larval; low-level stimulation when it is pupal. To explain this complex pattern would require "explaining" the ants themselves.

A further example may be mentioned. All spiders possess glands the secretion of which hardens into thread form in air. Having such glands, spiders spin, as do the larvae of insects, and other arthropods which have similar equipment. But the conditions under which spiders spin and the manner in which they spin depend upon many factors, and consequently there are many species differences in this activity. Certain factors are of evident importance: (1) the general mode of life of the spider; (2) the sensitive equipment and general structure of the spider; and (3) the nature of the spinning equipment.

The problem of web spinning in spiders is largely unworked, but it is necessary to recognize that for an understanding of the behavioral complex as a whole we must gain insight into the several contributory factors.

Conclusion: General Consideration of Arthropod "Instinct"

Reasons for the Appearance of Characteristic Activities.— We have treated a variety of activities, in terms of the factors which lead to their occurrence in the individual. No general law can be laid down, since each act must be studied in terms of the setting in which it may be produced, and in terms of the inherited mechanisms which make its appearance certain in the typical species environment. Although much of arthropod behavior is predictably stereotyped and consists of modes of activity which are highly characteristic of given species, the outer and inner contributing factors may be quite different in nature from case to case. The susceptibility of the highly sensitive arthropod to coercive stimulating conditions, the pattern of the nervous system, and the relative autonomy of segmental activities in connection with the possession of certain action structures, all make for the appearance of behavior which is very typical of the given species.

Many complex acts are formed by the cooperation of sensory, nervous, and motor equipment. The matter is further complicated by the possibility that continuity of the individual from one life history stage to the next (*e.g.*, from larva to adult in the insect) may be important. For instance, the fertile female of one wasp species stings a caterpillar and oviposits upon it in her prepared burrow, the female of another species stings a cricket, while another generally stings a spider. The adult wasp herself feeds upon nectar, and need not previously have performed the act of stinging a particular prey. Her response to the specific prey may depend upon the fact that the olfactory stimulus, which was constantly with her while she fed as a larva, is a particulary potent stimulus in arousing the stinging response when she is in a susceptible condition as an adult (Wheeler, 1928). Her glandular condition at the time would appear basically responsible for this *susceptibility*. This interpretation suggests a feasible experimental investigation. To say that the behavior depends upon an instinctive urge to provide for the larva is not only misleading, but shrouds the problem in mystery and does not encourage research.

Traditional Instinct Theory a Handicap.—A certain sphecid wasp captures and stings spiders which she places in her clay nest. Fabre (1918 tr.), who made himself the champion of an "inflexible instinct, unerringly serving specific ends," reported that the wasp always so directs its sting as to hit the ganglionic mass which controls leg action. According to him, the spider is stung accurately and a sufficient number of times to induce paralysis, but not to kill, "so that living food may be available for the larva." This seems to encourage the conclusion that somehow the wasp possesses an instinctive knowledge of spider anatomy. However, the Peckhams (1905) discovered that

> . . . out of forty-five species of solitary wasps only about one third kill their prey outright. Of those that remain there is not a single species in which the sting is given with invariable accuracy. To judge from results, they scarcely sting twice alike, since the victims of the same wasp may be killed at once or may live from one to six days, or perhaps ultimately recover.

It is probable that the number of stings given to the prey, and hence the chances that it will be killed, depend upon the characteristic level of excitability of the wasp species and upon the extent to which the victim happens to excite the captor by struggling. It is evident that if the wasp generally pounces onto the moving spider from behind, in curling forward beneath the body of the prey her elongated abdomen must bring the sting to the forward end of its cephalothorax. This affords excellent chances that the sting will strike close to the ganglionic mass there.

Ziegler (1910) has pointed out the futility of interpreting acts such as these in terms of their utility or purpose. The following case is typical: The female dragonfly normally deposits her eggs as she skims the surface of a stream, but she may also lay them upon surfaces such as a freshly tarred roof. In the former case the eggs sink to the bottom and develop, in the latter case they perish. As Ziegler shows, there is no justification for the statement that the dragonfly lays her eggs on the water "so that they may develop." It is much closer to the facts, and to dragonfly psychology, to say that she lays her eggs when she is susceptible and when a shining surface attracts her. If a human being were governing the act, the response to the tarred roof would be a "mistake"; but it is a dragonfly that lays the eggs and there is no mistake, only an unfortunate outcome which has no more to do with performance of the act than has the normal outcome.

We know all too little about the astoundingly diversified activities of arthropods, of "tailor ants," "diving beetles," "mason wasps," and the like. But the way in which to increase our knowledge is to investigate, and not be content with imposing labels which actually have little meaning.

Suggested Readings

Forel, A. 1908. *The senses of insects.* London: Methuen. (Tr.) Chap. V.

Hecht, S. and Wolf, E. 1929. The visual acuity of the honeybee. *Jour. Gen. Physiol.*, **12**, 727–760.

Mast, S. O. 1911. *Light and the behavior of organisms.* New York: Wiley. Chap. X.

Pearse, A. 1911. The influence of different color environments on the behavior of certain arthropods. *Jour. Anim. Behav.*, **1**, 79–110.

Schneirla, T. C. 1933. Studies on army ants in Panama. *Jour. Comp. Psychol.*, **15**, 267–299.

CHAPTER VII

SOME CHARACTERISTIC ACTIVITIES OF HIGHER ARTHROPODS, AND THE ROLE OF MODIFIABILITY

Although the behavior of all insects is markedly stereotyped, there are outstanding cases in which it is not difficult to recognize learning as an important contributant to the nature of an activity. This is particularly true of the higher insects. We shall find that learning plays a major part, although not a finally controlling part, in some of their most characteristic activities. The study of orientation in the social insects is very profitable in connection with this problem.

THE PROVISIONALLY CONTROLLED ORIENTATION OF INSECTS

In the orientation of lower arthropods there exists, as we know, a gradation from the simple, specifically determined type to more complicated types of orientation in which the animal is able to maintain a given field of visual stimulation (pp. 132*ff.*). Among the arthropods this last type of orientation reaches its most complex development in the social insects. This form we may call *provisional orientation*, since it depends upon the nature of the circumstances which prevail at a given time, and therefore results from the insect's habituation to external situations.

The Orientation of the Flying Insects. *Examples of Provisional Orientation.*—On their foraging (food-getting) trips, the flying insects (by virtue of compound eyes which are much better developed than are those of most other insects) mainly depend upon visual stimuli which are available along the route. After a number of trips between her hive and a flower patch, the worker bee takes a direct route, a "bee line." That this depends upon a visual control which is established through learning is strongly indicated by the manner in which young bees (or older bees early in the spring) first circle about the hive and fly in the nearby vicinity on short expeditions before beginning longer journeys.

The predatory wasp learns to locate her very small nest entrance quite accurately on the basis of vision. Bouvier (1922) took a flat stone from a position it had occupied for two days close to the nest opening of a digger wasp, and placed it in a new position 20 cm. away. Upon her laden return, the wasp alighted at the border of the stone in its new position and began to dig there. She returned to this place twice when driven away, but when the stone was replaced in its original position she flew to it and found the burrow without difficulty.

The nest entrance of a wasp usually is surrounded by entrances to other nests, yet she is able to find her own promptly each time she returns. Is there some specific visible feature on which she depends? In Tinbergen's (1932) experiments the bee wolf *Philanthus* was habituated to a ring of pegs which had encircled her burrow entrance during numerous departures and arrivals. Upon her return flight, the wasp would fly to the circle of pegs even if it had been moved 30 cm. from the actual position of the nest entrance during her absence. The wasp did not appear disturbed when the number of pegs was changed; neither did the grouping of the pegs into an oval instead of a circle prevent her finding the nest. It was only necessary that the pegs surround the position of the opening.

Contrast of Provisional and Directly Determined Orientation.—Such behavior as the above involves movement with reference to a habituated visual pattern, as does the swimming of the whirligig beetle (pp. 135*f*.), but there are differences which mark the wasp's performance as a distinctly higher type. The habituated visual pattern does not continuously control the wasp's movements but functions only under given conditions, *i.e.*, when she returns to the nest. Further, the wasp responds not merely in a generalized manner to the pattern, as does the beetle to the light-and-shadow field, but responds to it in a specific manner. The visual pattern is the dominant factor in the beetle's behavior which keeps it in a given locality. For the wasp, however, it is an incidental stimulus to an act which is based fundamentally upon the factors which have caused the insect to return with booty to her burrow.

Beyond this consideration, the characteristics of provisional orientation are best understood in a study of the manner in which stimuli control the movements of the terrestrial insect along an established route.

Orientation in Terrestrial Insects.—Way finding is most complex in the case of social insects that must run about on the surface and in underground tunnels. The bee, in successive journeys, readily establishes a straight course; but the ant must learn her path over terrain which presents a multiplicity of confusing stimuli and more possibilities for error. We are concerned here with the manner in which the ant's movements are controlled once she has established her route, in order to see the importance of modifiability in the act.

The first controlled experiments on animal behavior were performed to solve the problem of way finding in ants (Bonnet, 1779–1783). A number of subsequent investigators have dealt with the question of sensory guidance of the ant's movements on her journeys. The evidence warrants the recognition of two general orientation types among species of ants: (1) collective foragers, and (2) individual foragers.

The Basic Role of Chemical Sensitivity in the Orientation of Collective Foragers.—In the *collective foragers*, species that typically pass along a narrow trail which they have established between their nest and a food place, the ants are able to follow the route by virtue of a chemical which saturates it. Bonnet demonstrated this by rubbing one finger across the line of progress: the ants collected on both sides of the affected area and showed evident disturbance. Further, Bethe (1898) found that under certain conditions the trail is chemically different from end to end. Brun (1914) demonstrated this for conditions similar to those which prevail on the natural aphid-visiting trails of some ants. Using a method devised by Lubbock (1881), he permitted a *Lasius* colony to establish its trail between nest and a supply of honey, across a narrow bridge of paper strips laid end to end. A section of bridge was then turned through 180 deg., a change which caused a definite stoppage of ants on both ends of the reversed section. To produce this result, the trail on the two ends of the strip must have been chemically different. The ants apparently had learned to follow their path in terms of such chemical differences.

Under other conditions, for example, when large pieces of food are being carried along the trail toward the nest, the ants are not significantly disturbed by reversal or interchange of sections of their pathway. This Lubbock (1881) found to be true for a paper pathway over which *Lasius* ants had been carrying larvae (placed at the end of the bridge by the experimenter) back to the nest. When, however, a light which had been in place on the right side of the path was moved to the left side, the ants were disturbed, turned around, and reversed their direction of progress. Response to chemical stimuli enabled the ants to remain on the trail, but the experiment shows that the direction of their progress in this case had been chiefly controlled by the light.

Visual Sensitivity Plays the Major Role in the Orientation of the Individual Forager.—The ants of many species (*e.g.*, *Formica* species) may be called *individual foragers*, since they appear to be practically independent of one another in their orientation on foraging trips. This was demonstrated by Shepard (1911). A maze, with its long central alley complicated by various detours and blind alleys, was placed between the nest and a food place. After one ant had learned to pass through the maze without entering blind alleys, newcomers were not assisted in any way by her rapid and direct progress but had to learn for themselves.

Shepard found that the individual *Formica* maze runner was influenced by the chemical saturation of her pathway, since she was much disturbed by the reversal or interchange of cardboard alley linings which had been in place during the learning of the maze. Under natural conditions, however, as a number of experimenters (Huber, 1810; and others) have shown, the chemical factor is of negligible importance in the orientation of the individual forager. It is seldom that her path on successive trips in the open is superimposed as it is in maze alleys and produces a saturated chemical trail on which she may learn to depend. Even if successive trips are made to the same place, the ant is very unlikely to cover exactly the same ground each time, except in places where her route is canalized (*e.g.*, where she follows the top of a log or the side of a large stone).

The orientation of the individual forager is controlled predominantly by vision. In the daytime the direction of the sun's ray is an important sensory cue, although it is often subordinate to others. Shielding part of the route from direct sunlight, and reflecting light from the other side with a large mirror, as Santschi (1911) did, often confuses the individual forager and causes her to proceed in the opposite direction to that taken before the experiment. Although trips are usually made in a given general direction, booty is seldom found in the same place twice in succession. As a consequence, the ant usually must change her principal direction of progress more than once on a given trip. Generally, if food is not found in foraging about on the first stop, the ant sets out in a different direction; while if successful, she turns about and starts promptly for the nest. Laboratory experiments show that the *Formica* forager may learn to change her direction successively, with reference to the principal source of illumination, a number of times on each trip (Schneirla, 1929). She may pass through the first alley of a maze with the light

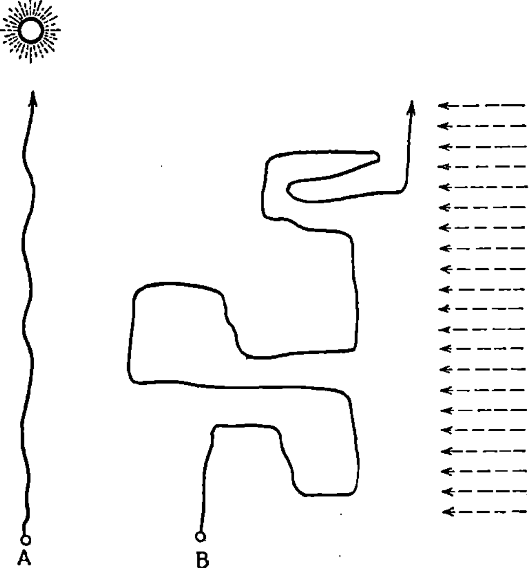

Fig. 39.—(*A*) A phototropic response to light. (*B*) A case of provisional orientation in the higher insect, in which many successive positions are assumed with respect to the light in passing through a maze (see Fig. 42).

on her right, through the second with the light behind her, through the third facing the light, and so on (Fig. 39*B*). This is provisional orientation at its best.

Brun (1914) reported evidence that frequently the orientation of individual foragers may even depend upon the position of large objects, *e.g.*, a tree behind the nest. With his finger Brun guided a booty-laden *F. sanguinea* individual through a series of changes in direction, and upon being released at any point the ant would set out directly toward the nest. Figure 40 shows a course of this kind as it occurred under natural conditions (Brun, 1916). Cornetz (1910) made many careful records of foraging routes under natural conditions, and among them such polygonal courses as the above are numerous. Brun discovered that blinded individuals, or members of a poorly visioned species, could not succeed in such a test but wandered about upon being released. It is also significant that the experiment failed with species (*e.g.*, *Lasius*) that lack ocelli and have only compound eyes as visual receptors (p. 139).

Comparison of Different Types of Orientation.—It is apparent that the relative importance of the different types of stimuli in the orientation of various ant species depends upon differences in sensory equipment. A poorly visioned species is forced to depend more

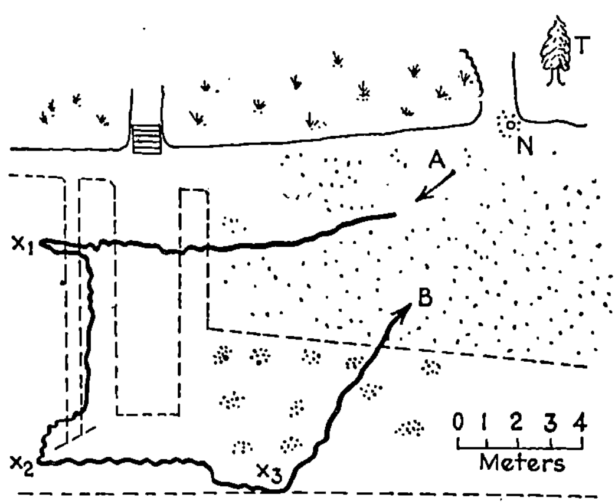

FIG. 40.—Foraging trip of a *Formica rufa* individual under natural conditions. *A–B*, path of the ant during one hour: across a sand place, through a garden, and back toward the nest. x_1, x_2, x_3, successive stopping places. *N*, nest opening; *T*, large tree behind nest. (*Redrawn from Brun*, 1916.)

extensively upon the other receptors, particularly contact and chemical sensitivity, than is a species with well-developed eyes. It is also significant that the relative importance of the fields of sensitivity in controlling the orientation of a given individual may vary according to external circumstances. The individual forager normally may learn to depend mainly upon the position of large objects in given territory, but in the event that no large objects are visible, she is forced to depend upon other sensory factors.

In the stereotyped, nonprovisional form of orientation to which most lower arthropods appear to be limited, there exists typically a specific relationship between external stimulation and the animal's movements. The manner in which external energy reaches the animal *primarily* determines its orientation, and does so in a consistent manner. The higher arthropod, however, may change its position according to a learned dependence upon the energy sources (Fig. 39). In provisional orientation, the insect maintains a specific position with reference to the stimulus source only until circumstances next in the learned order cause the assumption of a different bodily position. The movements which will be learned in a given field of stimulation depend upon special conditions, such as where the insect finds food and in what places obstacles interfere with her progress.

Experimental Studies on Arthropod Learning

Thus far we have found many evidences of the function of modifiability in arthropod life, but we have been mainly concerned with the modified activities themselves, after they have become established. By passing in review certain representative experiments, the nature and the limitations of the learning capacity in this phylum may be evaluated more satisfactorily.

The Establishment of a Provisional Orientation Habit in a Crustacean.—Many experimenters have found the crab able to learn simple modifications in its movements. Schwartz and Safir (1915) set up an experiment in which the fiddler crab was made to find its way from one end of a box to an opposite corner, from which an opening led to moist sand and the animal's burrow. With respect to the starting point this opening was located so that in order to reach it directly the animal must inhibit the normally dominant tendency to turn toward the side of its larger pincher. First the crab learned not to climb up in the corners of the box. Then it slowly learned to leave the closed corner promptly whenever it turned that way, and direct turns from the starting point toward the open corner increased in frequency. Seven crabs averaged 15.4 turns toward the closed side in the first series of 20 trials, but on the tenth day (after 180 trials) they averaged only 2.6 such turns in a series of 20 trials. The best subject made the following record in successive series of 20 trials: 14, 5, 4, 3, 4, 1, 1, 0, 1, 0. This crab perfected the habit in 180 trials. The opening was then changed to the opposite corner, and in successive series of 20 trials the animal very slowly decreased its turns toward the previously open side as follows: 19, 13, 15, 9, 5.

A check experiment showed that vision afforded the principal sensory control of the habit. Yerkes and Huggins (1903) had reported from a similar experiment that the crawfish learns under the control of vision, contact, olfaction, and muscle sensitivity. Such experiments demonstrate that a crab may learn to return to its burrow by turning with reference to its surroundings, but do not suggest an ability to master very involved situations.

The Learned Inhibition of a Shock Response in the Spider.—In testing the sensitivity of spiders to tuning fork vibrations, the Peckhams (1887) found that a strongly vibrating fork held close to the animal would cause it to fall from its web and remain for some time about 18 in. below, dangling from the end of a thread spun during the fall. A vibrating fork was presented 22 times in succession to an excitable *Epeira* female. After the seventh trial, the spider fell a shorter distance and returned more promptly to the web each time. After the twenty-second trial the animal responded by merely holding up her front legs as the fork approached. A female *Cyclosa* was subjected to daily tests throughout a month, and at the end of this period was able to remain on her web from the first trials of each new day. It may be concluded that the change in behavior was not due to sensory fatigue (*i.e.*, to an actual reduction in sensitivity), but to a learned inhibition of the "dropping" response when a specific stimulus was presented.

The Learned Inhibition of a Phototropic Tendency.—Normally the cockroach avoids the light, and unless very hungry emerges from hiding only in darkness. Szymanski (1912) sought to change this strong avoidance of light. He placed the animal in a long box, one half of which was glass and open to the light, the other half darkened. Each time the cockroach ran into the darkened end of the box it received an electric shock through plates in the floor. Violently excited, and unable to climb the walls, the roach finally would run into the lighted part of the box, where the shock ceased.

After a number of trials the animal would stop at the edge of the darkened zone, energetically clean its antennae (as insects do when startled), and then turn back into the light. Shortly afterward it might pass into the darkened end of the box after hesitating at the border, receive a shock, and hurriedly retreat into the light. Ten cockroaches finally reached a stage at which each of them was able to turn back into the light ten times in succession without entering the dark. One animal was able to remain in the light for almost an hour, after long training. Thus, even the strong factors which are responsible for the photonegativity of this insect may be inhibited through learning, at least for a time. However, the strength of the influences which must be controlled makes the habit a very difficult one to establish, and a difficult one to maintain.

Visual Discrimination and the Flower Visiting of Honeybees. *Discrimination Habits Based upon Color Vision.*—In testing the responses of bees to flowers, von Frisch (1915) demonstrated the ability of this insect to learn a visual discrimination. Two dozen marked bees were trained to visit a table on which a series of 16 paper squares (each 15 by 15 cm. square) were arranged as squares on a chessboard. The bees found sugar water in the watch glass that covered a square of "blue" paper, but the watch glasses on the other 15 squares, "grays" of different brightness values, were empty.[1] During the training period the papers were constantly changed in position, to prevent the bees from depending upon a given position of the "food color." Then a test experiment was conducted in which new papers were supplied, again one "blue" and 15 "grays," but with all of the watch glasses clean and empty. However, the bees gathered over the "blue" paper and crawled about on it, largely neglecting the other patches. Von Frisch concluded that the discrimination habit depended upon the characteristic wave length of the "blue," and not upon the intensity (brightness) of this stimulus, since otherwise the bees would have been disturbed by the presence of "grays" close to or identical with the "blue" in intensity value.

In other experiments, bees were unable to discriminate "blue" from "violet" or from "purplish-red" patches. They could, however, discriminate a "yellow" patch from the "blue" and from "grays." This was taken as proof that the "yellow" and the "blue" regions of the spectrum could be differentiated by the bees on the basis of wave-length differences. As is typical of insects in general, the bees were unable to discriminate "red" from "dark gray."

This question is not yet closed. Bees and other insects are very responsive to the ultraviolet rays which are reflected from the petals of a great many flowers, as Lutz (1924) has shown. Bertholf (1931, *Jour. Agr. Res. Wash.*) found that spectral light near 350 mμ in wave length was greatest in stimulating efficiency for the bee, and that the bee responded to rays as low as 250 mμ, far below the lower limits of the human visible spectrum. Further, although the upper limit

[1] It is preferable to name a visual stimulus in terms of its physical characteristics, *intensity* (amplitude), *wave length*, and *complexity*, rather than in terms of its effect upon the human visual system. The quotation marks will serve as reminders of this fact.

of spectral sensitivity has been found to lie near 550 mμ for the bee and other insects, which show themselves practically blind in red light, Lotmar (1933) found that some flowers (*e.g.*, the poppy) which reflect light of the long-wave values are nevertheless readily visited by bees on the basis of the ultraviolet which they also reflect.

Visual Discrimination Habits and Orientation.—The work of Opfinger (1931) disclosed a very interesting characteristic of bee learning. During the arrival flight of the bee she presented a "yellow" square beneath the glass plate of the feeding place; a "white" square was quickly inserted as the bee landed and began to feed; and a "blue" square was inserted at the beginning of the circling and departure flight. Having presented the stimuli separately in this manner during several visits, the experimenter then placed the squares together on the feeding table. Practically all of the subsequent visits of the bees were paid to the "yellow," the "arrival color." In one case, the landings were as follows: upon "yellow" 58 times; upon "white" ("feeding color") 1 time; upon "blue" ("departure color") 0 times. If the "arrival color" was not presented, the arriving bees would respond to the other patches as to new stimuli.

This shows that the visual appearance of the feeding place during arrivals in the course of training comes to control subsequent arrivals in a highly specific manner. On the other hand, the visual appearance of the place during other phases of the trip (*i.e.*, during feeding, during the departure flight) apparently does not influence discrimination on arrival flights during or after training. This militates against the common belief that the circling flight of a bee, wasp, or other flying insect on its departure from the nest is an important factor in permitting the insect to find the place again. It is much more probable that in the "orientation flight" the insect receives sensory cues which are necessary for the start on her flight *from* the nest, but function only in connection with that act.

The Learning of Terrestrial Insects.—In the case of the bee, the wasp, and other flying insect foragers, a swooping approach to the food place from the air makes possible the learning of visual discriminations which govern later arrivals. Similar visually controlled habits are learned in connection with the arrival at the nest. Although Rau (1929, 1931) has reported convincing evidence that the bee line is established through learning, the difficulties of studying orientation in flying insects have withheld knowledge as to the manner in which the direct flight is learned.[1] In any case, the learning of a route cannot be so difficult a problem for the flyer (because of the extensive scope of visual control) as is orientation for the terrestrial insect.

The Maze Method Used to Study Terrestrial Insect Orientation.—Our previous treatment of orientation in the ant has opened the problem of how the foraging route is established. That this involves learning was first demonstrated by Lubbock (1879); and Fielde (1901) reported that ants were able to straighten their path in carrying larvae to the nest through a simple maze. Shepard (1911) showed the usefulness of the maze as a means of essentially duplicating the natural foraging situation so that learning of the route might be studied with the sensory factors under experimental control. The maze was placed so that the ant must pass through it in order to reach the nest with food carried from a food place. A record was taken of the subject's errors: her returns toward the starting point, or her detours from the true (shortest) path into blind alleys. Both types of error are observable in foraging under natural conditions; but the maze method permits a detailed study of the manner in which the subject eliminates them from her maze run.[2]

The General Course of the Ant's Maze Learning.—During her first trips the ant improves greatly, particularly reducing the frequency of her returns to the starting point, back tracking less in maze alleys, and running less on walls and ceilings of alleys. The rapid decrease in the frequency of such "general errors" during the early runs accounts mainly for the abrupt descent of the "error curve" (Fig. 41*B*). In learning the maze pattern shown in Fig. 41*A*, the *Formica* individual's record after her first 8 to 10 trips is largely one of progress in avoiding the blind alleys. The relatively crude habit that was easily and rapidly learned during the early runs now

[1] Young carpenter and mining bees were seldom successful in returning to the nest when liberated at a distance of one half mile from it, whereas middle-aged bees were more successful in returning, and older bees seldom failed to return when liberated at a distance of nearly three miles from the nest.

[2] Illumination may be directionalized (from one side of the maze) or may be diffuse (equal in all parts of the maze), according to whether vision is to be an important sensory factor in the learned orientation or is to be reduced to minimal importance. The involvement of chemical stimuli may be controlled by means of bristol board linings for all alleys of the maze.

Ants other than the experimental subject are excluded from the maze. The subject is permitted as many consecutive maze trips as she will make.

becomes slowly modified, since learning to avoid a given blind alley requires a much more precise running of the adjacent true pathway than was learned initially, as Schneirla (1929, 1933*a*) has demonstrated.

The Course of Blind-alley Elimination, and Factors Which Influence It.—In studying maze learning, it is important to recognize that original behavior in the problem is not "chance" in the sense that errors are made unpredictably. Rather, the subject enters some blind alleys much more readily than others, and this difference in

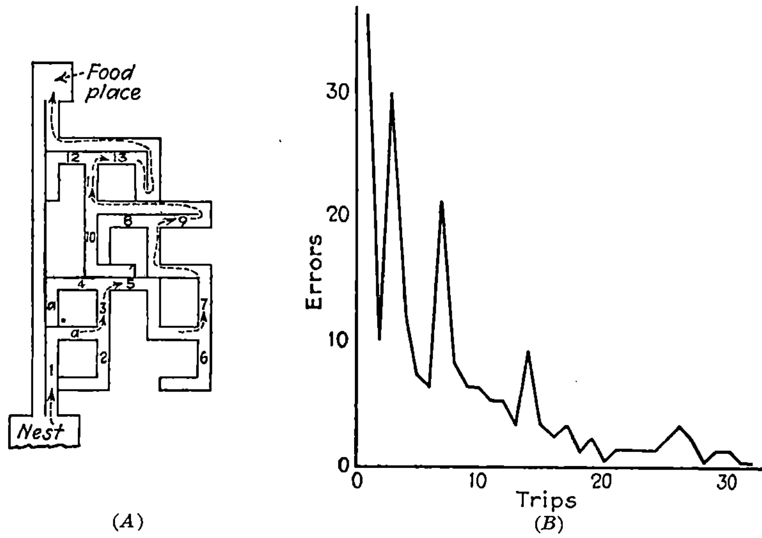

FIG. 41.—(*A*) Maze pattern learned by *Formica incerta* subjects usually within 35 trips. (*B*) Sample "error curve" representing the learning of this problem. (*Redrawn from Schneirla,* 1933.)

their "difficulty" depends upon the arrangement of the just preceding alleys in the maze. For instance, the *Formica* ant generally requires fewer than 40 trips to learn the maze pattern shown in Fig. 41*A*, which has *six* blind alleys, but in learning the pattern shown in Fig. 42, which has *four* blind alleys, 150 trips are the minimum (Schneirla, 1929). In the latter pattern, each of the blind alleys 4, 6, and 8 is preceded by a U arrangement of alleys, requiring two successive turns toward the same side (*e.g.*, from 3*a* into 3*b*, from 3*b* into 3*c*). This serves to "throw" the ant strongly toward the outside wall of the last alley (*e.g.*, toward the "4" side of alley 3*c*) so that she is forced to turn to that side at the next junction (see p. 128). This phenomenon depends upon the combined influence of momentum and of centrifugal force in running, as determined by (1) the running speed of the ant and (2) the length and arrangement of the maze alleys. Consequently it has been termed *centrifugal swing*. This is the most important factor in determining the initial difficulty of blind alleys in the maze learning of other animals as well as in that of ants (Schneirla, 1933*b*). We cannot hope to understand the process of learning unless we know the nature

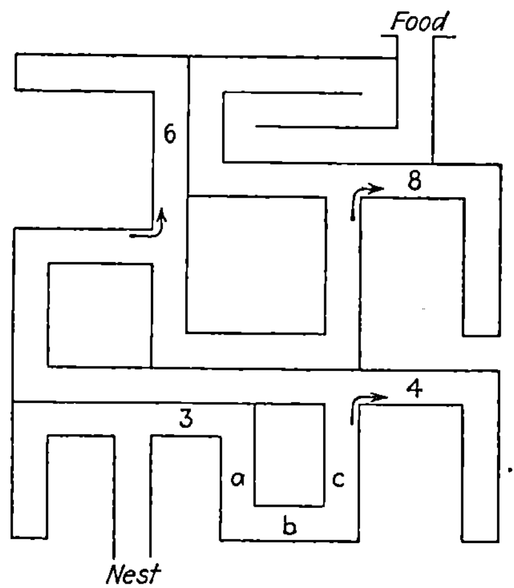

FIG. 42.—A maze problem made very difficult by the influence of "centrifugal swing." (See text.)

of the factors that oppose modification of movement in the problem (Chaps. XVI and XVII).

In learning the maze pattern of Fig. 41*A*, *Formica* ants have their greatest difficulty in connection with the elimination of blind alleys 4, 6, 8, and 12 from their route. Starting at about the tenth trip (in this case), blind-alley elimination typically follows this course. (*a*) After 15 trips the second arm of the blind alley (*e.g.*, 4*a*) is no longer entered, but the ant turns around each time upon touching the end of the first arm. (*b*) Then, on occasional trips, the ant turns around after having entered the first arm a very short distance. This behavior becomes more frequent, while activity *a* decreases. (*c*) At first occasionally, then more frequently, the ant turns directly into the true pathway alternative, without entering the blind alley at all. (*d*) Finally, the ant turns consistently into the true alley at each of the junctions.

The Unity of the Maze Habit.—As progress through steps *a*, *b*, and *c* leads to the avoidance of the blind-alley turn at a given true path-blind-alley junction, the movements of the ant in passing through just preceding true path alleys are gradually transformed in connection with the appearance of improvement *d*, the consistent turn into true pathway. This change in true pathway behavior (*e.g.*, in alleys 1*a* and 3, preceding junction 4-5) under the influence of learned avoidance of the blind alley which comes next in the maze (*e.g.*, alley 4) speaks for a close relationship between the successive parts of the maze habit. Learning of more efficient passage through true path alleys, and progress in avoiding blind alleys, shorten the ant's path toward the end point of the maze. That the carrying of food into the nest at the end of the path is responsible for the directionalization and unification of the learning in the maze alleys themselves is suggested by the manner in which the pace of the somewhat advanced learner quickens on each trip once the first half of the maze has been passed. The manner in which the conditions affecting a maze run influence behavior on subsequent runs will be treated as a special problem in Chap. XVII.

Motivation and the Demonstration of Its Function.—If one group of ants, *A*, is made to run the maze in passing from the nest to a food place, and a second group, *B*, runs a maze of the same pattern in carrying food from the food place to their nest, the *B* ants behave differently in the problem and learn it more efficiently than do ants of the *A* group (Schneirla, 1933*a*). After the eighth or tenth trip, in what we may call the second phase of learning, ants of the *B* group show consistent and efficient progress in eliminating crucial blind-alley errors, while the progress of *A* ants is less marked. The retarded error elimination of the *A* ants is much less like the standard progress described above (p. 161) than is that of subjects in the *B* group. It may be concluded that the organic condition of the ant that runs the maze (*i.e.*, her "motivation") is important for the efficiency and the manner in which she learns the problem, and particularly for the elimination of difficult errors.[1]

The Complexity of the Ant's Maze Habit.—The receptors (visual, chemical, tactual, and muscle sensitivity) function in close cooperation in directing and controlling the orienting movements of higher insects, in definite contrast with the state of affairs in mollusks

[1] "Motivation" may be defined tentatively as the totality of factors, directly or indirectly dependent upon organic condition, which impels a given activity.

(pp. 111*ff.*). The complexity of this sensory integration is shown by the ability of the ant to master a change in the sensory setting of a problem which she has learned. A *Formica* individual that has learned her way through a maze which is illuminated from one side, (*e.g.*, the east) is disturbed if the light is moved to the opposite (the west) side of the maze. However, she adapts to such a change within relatively few trips, in the course of which the unchanged sensory factors appear to play a greater part. Contact is one of the temporary substitutes for unreliable vision; the increased importance of its role being evidenced by significant features of behavior such as the more frequent touching of walls, especially at difficult turns. Once this visual change has been mastered, the source of illumination may be shifted to successive new positions (*e.g.*, to the north side, next to the south side), and each time the subject will be able to adjust to the changed sensory circumstances (Schneirla, 1929).

When a change in a learned maze pattern makes a former blind alley a true pathway turn, the ant learns to turn consistently into the former blind alley within fewer trips than were required for original learning. There is also a change in the running of the ant through the preceding sections of true pathway, in conformity with the new state of affairs at the junction of true path and blind alley. The ability to change certain parts of a previously successful act when an experimental change makes the movements "inadequate," and to learn other movements in their places, represents motor adaptability of a high order. This efficient movement control, together with a facile sensory control of the habit, indicates the possession of a complex nervous system which is susceptible to extensive changes in its function of conducting sensory impulses to the action system.

The Specificity of Sensory Control in Insect Learning.—In the food foraging and "homegoing" of the insect we find modifiability in its major role. However, there is evidence for the ant, as Opfinger (1931) has found for the bee (p. 158), that what is learned may be performed again only in the presence of the specific sensory and organic conditions of original training. Schneirla (1934) has reported that after an ant has learned to run a maze pattern in successive trips from nest to food place (returning to the nest each time through a straight pathway), the maze pattern must be learned virtually as a new problem when the ant is then forced to run through it on her trips from food place to the nest. Apparently, what is learned when the ant is in a given organic condition (that in effect during the trip from nest to food place) cannot be performed except under conditions very similar to those of original learning. These results suggest why, despite the ability of social insects to learn difficult discrimination problems and complicated orientation problems, their general behavior under natural conditions shows the typical arthropod stereotypy. As we shall see (Chaps. XVI, XX), for rather fundamental reasons this contrasts definitely with the manner in which the mammal learns.

We are now in a position to consider the most complex phenomenon in insect behavior.

Social Organization among the Insects

General Definition of the Insect Society. *Social or Nonsocial; "Society" or "Association."*—The gathering of individuals is a fairly common phenomenon among the arthropods, and indeed among all animals. Temporary groups are commonly formed by the accidental assembling of members of the species within a restricted area in which optimal conditions (illumination, temperature, humidity) prevail. Allee (1931) has demonstrated that the principles which underlie the formation of this type of animal group are of importance for the appearance of true social organization in many animals (*e.g.*, in birds). Alverdes (1931) terms this incidental type of aggregation an *association*, in order to stress the fact that the group owes its existence to extraneous factors, which also keep its members together. The point is that in such a gathering each individual is oriented primarily toward its external environment and not toward other members of the group *as such*. In contrast, a *society* shows a closer and more adequate unity in its membership.

The true social organization may therefore be defined as *an aggregation of individuals into a fairly well integrated and self-consistent group in which the unity is based upon the interdependence of the separate organisms and upon their responses one to another.* The insect social organization (the colony) typically is based upon the grouping of parents and offspring, and displays a complexity and stability of organization which are not found in any arthropod association.

However, no sharp distinction can be made between the association and the type of aggregation which we have called a society. As Espinas (1877) has pointed out, associative life does not arise fortuitously in certain groups only, but all animals form simple or complex groups at some period in their life history. Those who restrict the term *society* as we have done, stress the complexity of the individual relationships in the (insect) social group, and the important difference in its origin as compared with that of the arthropod *association*.

The Probable Phylogenetic History of Ant Societies.—The insect social organization reaches its highest condition of development and its greatest complexity in the ants, therefore our treatment will center about conditions in that group. A close study of existing insects has led Wheeler (1928) to the view that the complex social pattern of ants had its origin among a group of wasplike ancestors. The state of affairs among existing wasps presents the following gradation toward the true social organization, which is found only in the highest groups: (1) A fertile female finds her prey, stings it, is stimulated to lay an egg upon it, and leaves it on the spot (*Mutillidae*). (2) The female stings her prey, then carries it off and stores it in a natural cavity or in a cell which she digs or constructs. Then she lays an egg upon the prey, closes the burrow or cell, and does not return. Roubaud (1910) called this *mass provisioning* (*Sphecoidea*). (3) The female remains in the cell with the egg which she has laid upon the prey, and as the egg develops into a larva and the larva grows, she leaves the cell from time to time and brings in further prey. The captures are fed to the growing larva from day to day, and consequently the phenomenon has been termed *progressive provisioning* (*Bembecidae*).[1] (4) In the true social wasps (*Polistes*, *Vespa*) the female not only stays with her young during their development, but also remains with them in a constructed nest after they hatch into adults, and a rudimentary family thus comes into existence.

Two factors appear of importance in the evolution of a true social behavior from conditions in the "solitary wasps" (stages 1, 2, and 3): (1) the ability of the female to oviposit upon prey (Nielsen, 1932); and (2) the appearance of important metabolic changes which prolong the life of the female and make it possible for her to remain with her young (p. 124). Now we may study the result of this evolution.

[1] An excellent description of behavior in *Sphex* and in *Bembix*, "mass provisioner" and "progressive provisioner," respectively, will be found in a book by the Raus (1918).

Colony Founding in Ants.—First, a word about the personnel of an established ant colony. The appearance of *castes*, or different types of individual, is a feature of fundamental importance in the insect social group. Males and females exist as separate types (sexual dimorphism), but generally the males (Fig. 38) are found only during the mating season. There are two principal types of female ant: the fertile female, or *queen*, who loses her wings and becomes an egg-laying machine after she is fecundated by the male; and the *worker* type of female (Fig. 38), incapable of being fertilized and ordinarily unable to lay eggs. The worker is the active individual of the colony; she tends brood and queen, gathers food, excavates or builds, and otherwise busies herself about the nest. The workers of many species of ants are polymorphic (present two or more structural types, as do *Pheidole* species), a fact of great importance for colony behavior, since corresponding differences appear in the behavior of these individuals (pp. 173*f*.). The organization of the colony usually is further complicated by the presence of insects other than ants, as parasites or as "guests" (p. 168).

The chief methods of colony founding in insects are: (1) swarming, in which sexual forms leave the parent colony accompanied by a number of workers (as in honeybees and a few ant species); and (2) independent colony foundation, which is prevalent among the ants. We may describe the founding of a colony in the latter way.

The male ants and the fertile females are winged and develop in numbers during the spring, or later in the summer, according to the species. They remain in the nest for a time. Then, evidently in response to meteorological conditions, the entire population of the colony becomes excited on some warm day as the winged forms come out in numbers and take to the air. The males fertilize the females in the ensuing "marriage flight." The fertilized female promptly descends to the ground, now shuns the light and hides beneath an object, or makes a small closed burrow. Her wings fall off, or she bites them off. Hibernating in her burrow, she may survive the winter.

As the first eggs are laid, the queen licks them, gathers them into a packet, and also eats some of them. When the larvae develop, she feeds them upon a salivary gland secretion which contains energy derived from her degenerated wing musculature and abdominal fat-body, and from eggs eaten.

Wheeler (1933) finds a significant similarity between the behavior of the female in certain species of ponerine ants (a primitive group) and that in progressive-provisioning wasps. The female *Myrmecia* (Australian bulldog ant) remains in the cell with the eggs she has laid, but unlike other colony-founding female ants she leaves this burrow from time to time and brings back pieces of insect, although she herself feeds upon nectar. The significant point is that these females are small and lack the reserve tissues of the portly females in other species. This lack *forces* them to obtain food by leaving the burrow, as do wasps.

In attending to her first brood the queen behaves much as would a worker in the normal colony, but she ceases this for the life of a timid egg-laying machine as soon as the undersized workers of the first brood become active. These new individuals soon busy themselves with caring for further eggs, enlarging the burrow, and shortly they are bringing in food.

Particular attention may be called to the marked change in the queen's behavior after fertilization. She flees from the light, to which she is generally positive before fertilization, hides, and removes her wings. Such changes, all closely associated, appear to depend upon the alteration of her physiological processes by the chemical changes involved in fertilization. Secondly, there is the fact that during the development of the first clutch of eggs she fondles them in a packet, and later feeds the larvae much as would a worker. It is recalled here that the queen is a female, as are the workers, but that she differs particularly in having enlarged, functional ovaries. The appearance of worker behavior in the queen, if for a short time only, must depend upon structural and physiological characters which the worker and the queen possess in common. Thirdly, there is the important change in the queen's behavior as soon as the first workers begin to give *her* food.

The Factors Responsible for Colony Organization. *Trophallaxis.* Now we may summarize in chronological sequence some of the inter-individual relationships which are important in the life of the normal ant colony: (1) licking, eating, handling of eggs by the queen; (2) licking and feeding of larvae by the queen; (3) newly hatched workers feed the queen, as do other workers thereafter; (4) the workers, in turn, lick a fatty secretion from the queen's body; (5) the workers pick up eggs as they are laid, lick them, gather them into a packet, and bundle them off from excessive moisture or warmth, or from any condition which disturbs the workers themselves; (6) the growing larvae are fed by the workers with regurgitated food, and from the body surfaces of the larvae the workers lick fatty exudates; (7) adult workers are constantly feeding one another with regurgitated food (Fig. 43), licking one another's bodies, and tapping and stroking one another with antennae.

It is noted that all of these activities involve the exchange of substances between participants. This fact, as Wheeler (1923) shows, accounts for (1) the queen's care for eggs and larvae; (2) the attentions of workers to the queen, the eggs, and developing individuals; and (3) the unity of the colony in general. The process which is responsible for these reciprocal activities he designates as *trophallaxis*, a concept signifying *the exchange of food, or of stimuli related to food in their effect, among adults, between adults and developing young, or between adults and other insects present in the colony.* Concerning trophallaxis, the following points are of importance: (*a*) The exchange need not be in kind. The adult worker feeds the larva, but licks fatty exudates from the larva's body. (*b*) Trophallaxis is not a process which is merely superadded to the care of young, but is itself responsible for such activities. (*c*) Olfactory, tactual, and other stimuli which have accompanied feeding or the receiving of exudates in the past, may acquire through learning the ability to elicit the effect which is at first brought only by the substance itself. (*d*) The contribution of a participant need not be food, but may be the stimulation of sensitive zones, a stimulus which has an evident soothing (or erogenous) effect. When another colony member is "cleaning" her, the ant stands immobile, with slowly oscillating antennae as when feeding.

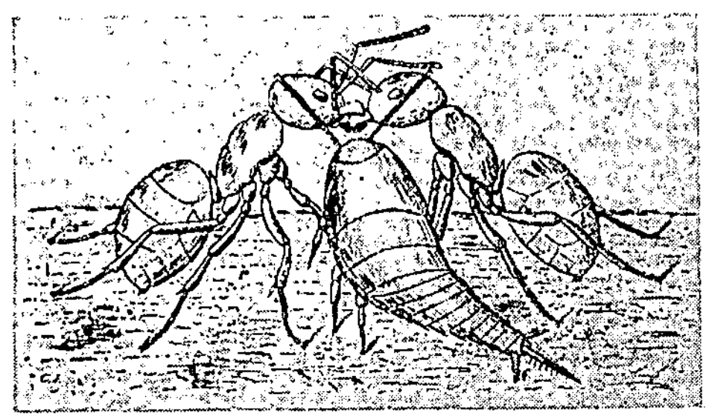

FIG. 43.—A *Lasius worker* is regurgitating a droplet of fluid to a nest mate (on the right), but a predatory beetle is about to snatch the food. (*Redrawn from Janet*, 1897, *Note* 13.)

Interindividual Stimulation and Colony Unity.—An ant touches antennae with a member of her own colony that enters the nest, but attacks an ant of another colony, even though she may belong to the same species. That this ant has learned to respond positively to the specific chemical of her own colony, a chemical which surrounds her constantly in feeding and in other activities, is suggested by the following experiments. Fielde (1904) took cocoons from the nests of different species and hatched the pupae artificially. Raised in a common nest, this heterogeneous collection of ants lived together peaceably. On the other hand, adult ants taken from the original nests and put together attacked one another at once.

An experiment first performed by Bethe (1898) demonstrates clearly enough that the differential responses to nest mates and to "enemies" depend upon the existence of a chemical which is specific to the given colony. An ant of species *A*, normally attacked by species *B* ants, was bathed in alcohol, in water, and then in the body juices of species *B*. The treated *A* ant was not attacked when introduced into a group of species *B* workers, although the latter were restless when near her. Introduced into her own nest, she was attacked and killed.

The facts suggest that the newly hatched ant, entering a situation in which a specifically stimulating chemical predominates, becomes habituated to it as a constantly encountered feature of feeding and of other colony adjustments. Since exudates and food are continually being exchanged among members of the colony, it is not difficult to see how the specific metabolism of workers, of queen, and of larvae could produce a chemical which would cling to the bodies of all. Olfaction is therefore a basic factor in the development of colony unity.

"*Communication.*"—Studies by von Frisch (1923, 1926) on honey-bees indicated specific excitatory effects by a returning food finder on other bees. Further results (Suppl. ref.: von Frisch, 1953) show that the finder bee, through the direction and rate of her waggle-dancing on the comb, somehow influences the direction and the distance of flights by aroused bees so that they are aided in finding the new food source.

The phenomenon is somewhat less complicated in ants. When a food-finding ant meets another individual upon entering the nest, there is usually an excited and rapid exchange of antennal taps on antennae, head, and thorax. If the "finder" ant is very excited, the exchange is very energetic and each ant may bring her front pair of legs into play as well. Differences in these taps have been described by some imaginative writers as conveying "information" concerning the nature, the amount, and the whereabouts of discovered food. Although popular writers take for granted the existence of an "antennal code," the practice is not based upon evidence, but upon a desire to make the ant "interesting."

In principle, communication occurs in ants much as in bees. Santschi (1930) permitted a *Tapinoma* ant (a collective forager) to find food, and then to cross an area to the compartment in which the colony was located. Other ants soon appeared and were able to cross the area in the very path followed by the first ant, so they promptly reached the food. Santschi pointed out that when excited, members of this and some other species release from the anal glands a substance which forms a trail, and that other ants in the colony are able to follow this trail. To account adequately for the results, we need only assume that the finder aroused ants in the colony by excited antennal tappings and by rushing about, and that upon emerging from the compartment these individuals were able to follow the chemical trail. Eidmann's results agree with these (1925, 1927). He found that a *Myrmica* worker that had found a large piece of food would return to the nest and excite other individuals. Exchange of the paper floor of the area for a fresh one just after the finder ant had entered the nest and before the secondary ants emerged, prevented the latter from finding the food except by chance, although otherwise they found it readily.

Schneirla has found that if the ants of a *Formica* colony have learned to get food in one certain corner of a large table, they will visit this corner in greatest numbers when a finder ant returns to the nest and excites them after having discovered food *anywhere* on the table. Thus, for ants in general, the facts show that the returning ant arouses the others, but the behavior of secondarily aroused ants depends (in trail formers) upon the existence of a chemical trail laid down *incidentally* by the excited finder, or depends (in individual foragers) upon the manner in which members of the colony have learned to forage in the vicinity.

Development of the Worker as a Unit in the Social Group. Investigations upon newly hatched wasps (Steiner, 1932), bees (Rösch, 1925, etc.), and ants (Heyde, 1924) show that the worker assumes her colony duties gradually, rather than by displaying the abilities in fully developed form at the start or by having the "teaching" of adult workers, as Büchner (1880) suggested. Heyde (1924) artificially removed a number of worker ants from their cocoons and kept them in a separate group, under observation, as they were reared free from adult influence. She found that typical activities such as fighting, caring for the brood, nest excavating, and the like, appear in these segregated ants. Any one type of behavior first appears only in very simple and limited ways, the initial activities being forced by certain local structures (see pp. 143*ff.*). The development of such activities into the adult types requires more time in the segregated callows than when behavior develops in the midst of an adult colony. In either case, incidental stimulation furnished by other individuals accounts for a sequence of changes in the responses of the new individual (the callow).

The Basic Process: Elaboration of the Feeding Response through Learning.—To understand this succession of behavior changes, let us give our own interpretation of Heyde's observations on feeding; selecting this activity because of its fundamental importance for the unity of the social group. Events are pictured as they occur within a normal colony.

1. When food is presented to the newly hatched worker, she opens wide her mandibles, and by action of the mouth parts the liquid is ingested. (It is recalled that as a larva the individual was fed by adults.) As the callow feeds, her antennae quiver and usually the front legs vibrate rapidly. This suggests that nervous excitation spreads directly through the brain to the nearest motor centers (in the infra-esophageal and first thoracic ganglia), directly arousing action parts supplied by these centers.

2. A few hours later, if she has not received food for a time, the callow moves about restlessly.[1] She soon encounters an older ant, and is fed by regurgitation. After a few such feedings, the original quivering movements of the antennae and the rapid front leg movements are to be observed *only at the outset.* Then, during the actual feeding, the antennal vibration is replaced by a slow oscillatory motion of these appendages, and the front legs remain on the ground.

3. Somewhat later, the young ant responds with quick movements of antennae and front legs *when she encounters another individual.* The same response is given by the "hungry" adult ant that

[1] It is probable that this excitement is attributable to visceral tensions set up through depletion of food in the digestive tract. By means of the pressure-balloon technique, Patterson (1933) has demonstrated the existence of rhythmic gastric contractions in the fasting crab, *Cancer.* He seems justified in his conclusion that the significance of this activity for general behavior in arthropods is much the same as that proved for higher animals.

meets a returning food carrier (Fig. 43). Without question this response, the basis of "communication" (pp. 169*f*.), is the survival form of the general "head-end movements" of the callow's feeding response.

4. Heyde found that when raised in the colony with adults, the young worker first feeds other individuals on the second day after hatching, but in the segregated group of callows feeding of others does not appear until the fourth day. Another worker persistently stimulates the young one with quivering antennae and jerking front legs, and this finally forces regurgitation from the young individual.[1] Later the regurgitative response is elicited more promptly; it has been trained. Then, too, on first occasions the callow must continue to disgorge liquid for some time once the act is begun, and thus gives up a considerable amount of food to the same ant; but in time she is able to discontinue the response shortly after it has begun. This is another mark of a learned control.

5. The young ant next begins to find food herself, and soon she is able to forage at a distance from the colony site.

We may briefly review the facts in the light of their significance for the socialization of the individual. At first the feeding behavior of the newly hatched worker is the larval response plus incidental excitation of new anterior action parts (antennae and front legs), but this reaction is changed through learning so that its later form depends upon the circumstances of stimulation. The susceptibility of the regurgitative response to stimulation from other individuals makes for a fairly even distribution of food throughout the colony, and is indirectly responsible for the fact that a foraging worker returns to the nest and gives up any food which she may have found. Thus, not only foraging, but also "homegoing" (without which, as Descy (1925) says, the insect colony could not survive), depend upon the elaboration of the original feeding response. The social training of the newly hatched worker is a complex process in which direct determination and learning closely cooperate.

The Motivation of the Socialized Individual.—The motivation of the individual worker presents an interesting problem (p. 162). It is the

[1] Regurgitation is the return of food to the mouth from the highly muscular crop when a pumping action is reflexly set into play by stimulation of sensitive contact receptors. Forel (1928) calls the crop the "social stomach," since its contents are truly *the property of the colony*. Once food has seeped through a strong valve into the "individual stomach," from which regurgitation is not possible, it becomes *the property of the individual*.

chronic hunger of the colony, as Wheeler (1923) says, which causes some of the workers to run forth and forage.[1] The readiness with which workers forage from laboratory nests is almost directly dependent upon the amount of food which has been given the colony during the preceding period. As for the acquisition of food, the act must be assisted by the immediate olfactory and gustatory stimuli involved. Such stimuli appear important for the return trip to the nest as well (Schneirla, 1934), since if the worker is carrying solid food and drops it she circles about in the vicinity, and if it is not found she makes a wandering return to the nest. In species that acquire liquid food outside the nest (*e.g.*, aphid visitors), the return to the nest must be greatly assisted by visceral tensions arising from the stretching of abdominal walls by the well-filled crop. Once in the nest, this replete worker rushes about, regurgitating food to workers she meets, actively hunts other individuals, and may even force her food upon them. In contrast, a less-filled worker gives up food only when stimulated energetically.

These considerations show the fundamental importance of Wheeler's *trophallaxis* (p. 168) for the socialization of the individual worker. The function of learning is evident at every stage of the process, but what the worker may learn to do in the social group is determined and restricted by her equipment.[1] Learning may expand her activities but cannot deprive them of their arthropod stereotypy. This accounts for the fact that the general pattern of social behavior in a given species is highly constant from colony to colony.

Division of Labor in the Insect Colony.—Structure delimits activities in the insect colony, and thus accounts for "division of labor" among the workers. This is clearly to be seen in species which have polymorphic workers (p. 143*f*.), in which a close correspondence is found between function and structural peculiarities (Goetsch, 1930).

Division of labor also may depend upon structural changes that come with growth, as Steiner (1932) has shown for wasps, and Rösch (1925) for bees. Between the third and fifteenth days after hatching, when the salivary glands are functional, the young worker bee is

[1] In an experimental *Formica* colony some of the workers appear much less frequently from the nest, and when they do, carry back little food and make poorer records in learning problems than do the workers that are regularly active in foraging (Schneirla, 1933*c*). It is possible that the poorer learners in a natural colony, less successful in learning their way about in the open, automatically settle to larva-feeding and other activities within the nest.

most active in feeding larvae. Then these glands atrophy and brood feeding ceases. From the fifteenth to the twenty-fifth day the wax glands are in their most active secretory phase and the young bee busies herself within the hive in comb building and also in receiving materials (nectar, pollen) brought in by foragers. Generally before the end of this period orientation flights outside the hive have begun. After the twentieth day and until death (at about fifty-five days) the bee is most active as a forager outside the hive. The time at which these changes in function appear differs greatly among individuals, and may vary according to hive conditions.

Concluding Remarks

Although the capacity to learn is important for the development of activities in higher arthropods, the direction in which a given activity expands is fairly rigidly determined by the equipment factors. The fact that all arthropods in a given group will have virtually the same environment contributes to much the same behavior in all cases, even though certain modes of this behavior are subject to a measure of change through learning.

To illustrate, let us consider the factors which contribute to social behavior. The evolution of certain changes in equipment permits a female to remain with her offspring. It is virtually certain that she will care for the young, because of the inevitable trophallactic relationship. All fertile females of a given species have the same type of equipment which contributes to this result; are caused by their physiological makeup to take the same type of environment after mating; and have essentially the same capacity for learning. Consequently the pattern of colony founding is highly constant for the group.

In the same sense, although learning appears to be involved in every phase of the worker's growth into the social organization, highly stable factors (stable environment; sensory, conduction, motor equipment) determine *what* may be learned and the *extent* to which modification is possible. Foraging outside the nest, which is essential to the survival of the insect colony, is made possible by the ability of the worker to learn a route. The more successful orientation of one worker, permitted by the learning of a path which represents an impossible problem for others, benefits the rest of the colony only through adding to its food supply. The manner in which such problems are solved is highly constant for a given species, since the pattern of provisional orientation depends particularly upon the worker's sensory and neural equipment.

In the arthropods, "learning" is therefore the tool of stable factors which themselves determine the nature and extent of typical activities in the given group. While, in the social insects, the well-developed capacity for learning may extend given modes of behavior to degrees that are scarcely possible in lower arthropods (not to speak of lower invertebrates), it does not change the "directly determined" pattern of behavior.

Suggested Readings

Imms, A. 1931. *Social behavior in insects.* New York: Dial. Pp. 117. Chap. II.

Rabaud, É. 1928. *How animals find their way about.* New York: Harcourt, Brace. Pp. 142. Chap. III.

Rau, Phil and Rau, Nellie. 1918. *Wasp studies afield.* Princeton: Princeton Univ. Press. Pp. 372.

Wheeler, W. M. 1923. *Social life among the insects.* New York: Harcourt, Brace. Lecture IV.

Yerkes, R. M., and Huggins, G. 1903. Habit formation in the crawfish, *Cambarus affinis. Psychol. Rev. Mono. Suppl.*, 4, 565–577.

CHAPTER VIII

BASIC FEATURES OF VERTEBRATE BEHAVIOR: CLASS PISCES[1]

INTRODUCTION

Distinctive Vertebrate Characteristics.—In the lowest vertebrate classes, as in the invertebrates, behavior is characterized by activities which are directly enforced by inherited mechanisms. In the fishes and Amphibia, in particular, the possession of a repertoire of dominantly stereotyped activities may be attributed to factors which are on the level of those outlined in Chap. VI as determining the arthropod's status. In fact, the social insects are superior to fishes and in many respects to Amphibia in the extent to which the natively determined behavior patterns (Chap. XII) may be modified through experience. The mere fact that these vertebrate classes stand "higher" in the taxonomic scale does not warrant placing them above the highest invertebrates in psychological standing.

The properties of behavior in the lower vertebrate classes bear a significant relation to the structure of the nervous system. Primarily, the nervous system of the vertebrate may be regarded as a bridge between receptors and the action system. However, the vertebrate conduction system is not dominated in its function by the segmentation of the body, as is the "ladder system" of the higher invertebrate animals (Fig. 36). On the contrary, the nervous tissue of the vertebrate forms a continuous tube which is distinguished from the arrangement of the invertebrate system by its position along the dorsum or back of the body. Such a conduction system is capable of unifying the action of parts of a larger body in a way that is scarcely possible for the ladder type of system. As the conduction

[1] The phylum *Chordata* includes animals with a dorsal nerve cord and a supporting skeletal axis or a vertebral column. The subphylum *Cephalochorda* contains Amphioxus, the lancelet. The subphylum *Vertebrata* includes the following classes: *Elasmobranchii* (fishlike vertebrates with cartilagenous skeleton,—sharks, rays); *Pisces* (true fishes); *Amphibia* (frogs, toads, salamanders); *Reptilia* (snakes, lizards, turtles, crocodiles); *Aves* (birds); and *Mammalia* (rodents, carnivores, primates).

TABLE 12.—FISHES

RECEPTOR EQUIPMENT	FEATURES OF SENSITIVITY
Chemical. Free nerve-endings in skin, generally over entire body surface.	"Common chemical sense," permits responses to fairly concentrated chemicals. Distinct nerve supply.
Grouped cells with hairlike projections, in and around mouth, on barbels, often over body surface.	Gustation (proximate chemoreception), permits responses to chemicals (acids, alkalies, salts) from objects close to body surface.
Ciliated cells in sacs opening to exterior. Connect with olfactory bulb cells, 1st cranial nerve.	Olfaction (distance chemoreception); response to weak chemicals from distant objects. Food location, response to chemical gradient in water, etc.
Contact. Receptor cells in skin, thickest at fin bases, on barbels when present, gill slits, borders of mouth.	Most fishes "positively thigmotactic" (swim against pressure), orient to tide, bottom feeders orient with reference to substratum, etc.
Lateral line. Ciliated cells clumped at intervals in mucus-filled canal along body sides, often branching on head. Swim-bladder permits responses to pressure differences.	Probably stimulated by streaming movements of water, possibly slow oscillations of water. May permit orientation with reference to objects, *e.g.*, through deflection of water from objects. Sensitivity to movement of objects.
Static. In *pars superioris* of vestibular apparatus (three canals and sacculus, latter with statoliths) lined with hair cells, contain endolymph fluid.	Direct control over position, equilibrium. Important in controlling bodily orientation. Cooperates with visual, tactual sensitivity, in controlling movement.
Auditory. *Pars inferioris* of inner ear contains poorly developed "primitive cochlea," the lagena, small pocket at base.	Some evidence that rapid water vibrations stimulate. May permit orientation to vibrations reflected from surface, from the shore.
Light. Flattened cornea, lens of fixed shape, accommodated by muscle pulling it toward retina. Large optic nerves, total crossing of fibers in chiasma. Elongated cells, "rods," characteristic in retina.	Vision predominantly monocular, owing to position of eyes, nervous limitations. Distance accommodation limited. Evidence for color vision in some (*e.g.*, dace).

CONDUCTION

Brain much like remainder of cord in *Amphioxus*, only two pairs cranial nerves. In fishes olfactory system dominates basal forebrain. Dorsal forebrain non-nervous. Ten pairs cranial nerves: afferent, efferent functions of anterior body, viscera. Well-developed cerebellum, lateral-line lobes.

ACTION EQUIPMENT

Body form, type of tail and fins, air bladder, determine swimming. Four longitudinal muscle bands in body wall. Respiration (aeration of blood) by gills. Circulation more sluggish than in higher vertebrates. Chromatophore cells in skin, coloration changes; phosphorescent organs (secretory cells) in many deep sea fishes. Thyroid, adrenal, thymus, other endocrine glands.

system increases in complexity in higher vertebrate classes, the tubular structure readily permits new combinations and types of behavior which represent extensive modification of the basic patterns.

This occurs particularly in close connection with the elaboration of one part of the central nerve tube under the influence of the sensory systems of the dominant head region. Correspondingly, significant changes appear in the external aspect of the anterior part of the

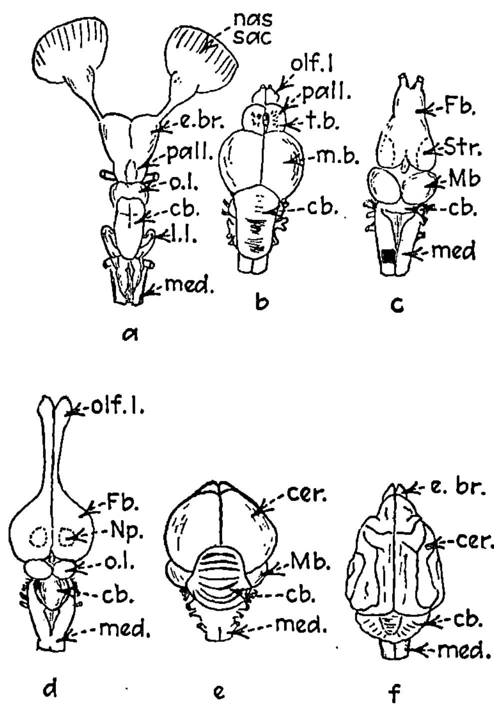

FIG. 44.—Representative vertebrate brains, sketched from above. (*a*) Dog fish, (*b*) trout, (*c*) frog, (*d*) alligator, (*e*) pigeon, (*f*) dog. *cb.*, cerebellum: *e.br.* end brain; *Fb.*, forebrain; *l.l.*, lateral-line lobe; *Mb.*, midbrain; *med.*, medulla; *Np.*, neopallium (approximate location); *olf.l.*, olfactory lobe; *o.l.*, optic lobe; *pall.*, pallium; *Str.*, striatum (approximate location). (*Modified from Herrick, Ziegler.*)

nervous system, as will be noted in an inspection of the brains of different vertebrates sketched in Fig. 44.

It is true that in the nerves which connect the central system with the sense organs and action system, a segmental arrangement is primarily responsible for some of the stereotyped behavior patterns which are fundamental in the lower classes. However, this is not a basic feature of nervous function, so far as its general contribution to psychological organization is concerned. In a general sense, the vertebrate is an animal which, by virtue of a new departure in the form of its tubular conduction system, finally is able to overcome the specific determining influence of its inherited sensory and action equipment.

The Fundamentally Stereotyped Vertebrate Behavior Pattern Represented in the Fishes. *Sensory Systems Dominate Behavior.*—In primitive vertebrates such as *Amphioxus* the lancelet, anterior dominance is relatively feeble, and in this sense behavior is on the annelid level. In the fishes, however, behavior is dominated by the exteroceptor systems, particularly by the olfactory and visual systems, and hence is basically dependent upon anterior supremacy. As Herrick (1925) has shown, the brain of the fish is monopolized by sensory systems (by different conduction systems which belong to one or another of the principal receptors). The olfactory system dominates the forebrain (Fig. 44*a*, *b*), and the visual system dominates the optic lobes in the midbrain; while the remaining exteroceptors have their particular conduction centers in other parts of the anterior nervous system. The structure of the spinal cord also makes for a direct determination of behavior, but on the primitive basis of bodily segmentation rather than on the characteristic vertebrate basis of sensory systems.

Homogeneous Environments Contribute to Simplified Behavior.—The fish has a highly homogeneous environment as compared with terrestrial vertebrates, a difference which is an important clue to some of its characteristic psychological limitations. Various fishes live at different levels of the sea which correspond to differences in their physiological makeup, and at a given depth the water medium represents an environment of minimal variability. In a somewhat more variable environment, a given species of fresh-water fishes presents a larger repertoire of activities. Among fishes in general, however, the relative uniformity of the water environment corresponds to a behavior which is simpler than is that of terrestrial animals in the higher vertebrate classes.

Species Habitats Differ Greatly among the Fishes.—Notwithstanding the above fact, we find a great variety of habitats among the more than 20,000 species of fishes. Some species live a roaming existence near the surface of the sea, others remain close to shore, others live at intermediate depths of the sea, and still others live at great depths. Furthermore, the great variety of fresh-water habitats also must be taken into account.

To illustrate one factor that may figure in the determination of habitat, species in which the air bladder generates an enormous internal pressure are forced to remain in the depths of the sea throughout life. In contrast, the herring is able to vary its environment to some extent. Owing to a peculiarity in its air bladder (an opening to the exterior) which permits the intake or release of water and hence makes possible an adjustment to different external pressures, this fish may live at the surface of the sea or at depths as great as 100 fathoms (Kyle, 1926). Although we cannot enter into a consideration of the wide variety of structural and physiological characteristics which determine differences in habitat, it must be remembered that (as pointed out for the arthropods) the inhabited environment stamps its character upon the behavior of the given species.

In the constant environment of a given species, the pattern of behavior is to a considerable extent a direct expression of the receptor equipment possessed by the fish. Hence it will be profitable to treat the animal first in terms of the importance of sensitivity for its characteristic behavior.

Sensitivity as a Primary Determiner of Behavior

The Coercive Effect of Stimuli on Action.—The close relationship between the receptor equipment and the characteristic behavior pattern of the fishes is viewed to advantage in a study of their feeding activities. Jarmer (1928) emphasizes the fundamental role of the exteroceptors in pointing out the general fact that under given commanding visual or olfactory conditions "the fish must bite." The power of a given stimulus over behavior is closely related to its intensity. Either the visual system or the olfactory system dominates the others, depending upon the species. While the commanding type of sensitivity may be supplemented at a given time by other modes, there is good evidence that in the fish the secondary systems (*e.g.*, the lateral-line system, pp. 185*f.*) do not remold the activity in cooperation with the primary system, but either support the latter or displace it according to conditions. Sensory integration in the fishes is therefore on a primitive level (*cf.* Chap. V).

The "Sensory Systems" and Species Differences in Food Taking. Fishes differ widely as to the type of food taken or captured by them. These differences correspond to the relative importance of various anatomical and physiological characters, among which are gross size, sensory acuity, swimming speed, and structure of mouth and digestive canal. An outstanding difference is that many fishes take their food more or less passively, as do the species of *bottom feeders*, while others actively hunt it down.

Species Classified According to Characteristic Food Taking.—It is apparent that species differences in the receptors and in related equipment have much to do with the determination of characteristic modes of food taking. Evans (1931) has reported differences in the structure of higher nervous centers (the medulla, in particular) which appear to be important in accounting for the occurrence of corresponding species differences in feeding activity. Therefore it is well not to limit our consideration to differences in end-organ equipment alone.

Fishes have been grouped by some writers according to their characteristic ways of taking food. In one group are placed quick and lively species such as codfish, dace, and trout, which pursue rapidly moving prey in daylight. Among other prey, these fishes feed upon flies which they snap up at the surface of the water, and upon agile minnows. Their capacity for rapid movement is important in permitting the capture of such prey, but a well-developed vision is even more essential, as Scheuring (1921) has shown.

The bottom feeders comprise a second group to which species such as the carp, skates and rays belong. These fishes generally have poor vision (*e.g.*, many of them have small eyes, placed low on the head) but have well-developed chemical and contact sensitivity. The group therefore contains many carrion feeders, fishes which can extract organic matter from mud, and many species which feed predominantly at night.

Chemical Sensitivity and Food Taking. *The Role of Olfaction in the Activities of Bottom Feeders.*—In the bottom feeders, chemical sensitivity is clearly the dominant sensory modality in fixing the nature of behavior. When the fish nears a food object, water which passes through the nasal cavity in respiration carries chemicals which stimulate sensitive cells in the epithelium lining this cavity. From these cells nervous impulses pass to the cells of the large olfactory bulbs and thence reach the forebrain, which, as we already know, is dominated by excitation from olfactory receptors. The work of Parker and Sheldon (1913) and of others shows that the olfactory bulb apparatus is a true distance chemoreceptor in certain fishes, since it mediates sensitivity to weak chemicals diffusing from objects which may lie at a considerable distance from the animal.

Parker and Sheldon kept specimens of *Ameiurus*, the catfish, without food for a few days. Starved individuals became markedly excited when pieces of earthworm were dropped into the water at a distance from them. They would seldom fail to turn toward a cheesecloth packet containing earthworms when they passed near it, whereas a packet containing waste was not molested. In contrast, catfish in which the olfactory nerves had been cut (thus preventing impulses from the nasal epithelium from reaching the forebrain) would pass by earthworm packets without any change in behavior.

In normal feeding, olfactory stimuli may not specifically orient the fish, but they may cause it to move about excitedly in the vicinity of the stimulus source. This activity eventually carries the fish close to the food, where a more specific difference in the intensity of chemical stimulation on the two sides of the body may direct its movements more effectively. That olfaction may serve in this way is shown by the manner in which a fish with one olfactory nerve severed turns toward the normal side when meat juice comes to it from a point directly ahead. We may conclude that in addition to its principal role of "general exciter," olfaction serves as an orienting sense, so that it is of essential importance in the adjustments of the fish to its environment.

Gustation and the Control of the Snapping Response.—The final stage of the feeding act, the snapping response, is touched off by the action of more intense chemical stimulation upon the *proximate chemoreceptors*, the gustatory endings or *taste buds*. When a piece of meat is brought close to or in contact with one side of the catfish's body, or in contact with one of the barblets, the animal turns promptly and snaps up the object. The response may be elicited even after contact sensitivity has been eliminated, which shows its essential dependence upon gustatory sensitivity. Herrick (1903) repeatedly stimulated the catfish with cotton wool, to which the fish ceased to respond by snapping, after a few trials. Fish that had been habituated to contact in this way would turn and snap at meat which touched the body. Strong localized chemical stimuli then presented almost anywhere on the body surface produced the snapping response independently of contact. Such stimuli were particularly effective when applied to the barblets. Sectioning of the olfactory nerves did not impair this response in the tomcod. These results show that when close to the fish, the prey stimulates taste buds which are distributed over the body surface, on barblets and on mouth parts in these fishes, thereby releasing the snap.

Vision Dominates Feeding in Many Species.—The catfish, a bottom feeder with a rich supply of taste buds in the skin of the body, snaps much more readily when these receptors are adequately stimulated than when visual stimuli are presented. Fishes of many other species, however, snap with far greater readiness at the sight of a moving object, and these are the ones that typically hunt living prey. For instance, in the food taking of the killifish, vision is definitely the dominant modality. Parker (1911) observed that a killifish would dart toward a small moving piece of meat much too quickly for chemical stimulation to have played any part. So effective is the stimulus, and so promptly does such a fish move when stimulated in this manner, that the darting approach and the snap may represent one response if the object appears close by.

The powerful and directly coercive rôle of vision in the behavior of many species is well shown by Scheuring's (1921) description of food taking in a predatory fish such as the cod. As the fish is swimming slowly about, a moving piece of food or a prey animal which appears elicits a prompt response. Simultaneously with a movement of the eyes toward the side on which the prey appears there is a quick twist of the body which turns the long axis in the direction of the object. Then a play of fins swiftly brings the fish closer to its prey, and at a certain distance the snapping response occurs, usually within a second or two after the first appearance of the object. The trout is capable of responding in this manner to small objects which move within a panoramic horizontal field of 160 deg.; whereas in the flatfish, whose two eyes lie on the same side of the head, the snapping response is elicited by stimuli from an object moving within a 50-deg. field.[1] In a given species, the snapping response may appear only when the fish is at a certain distance from the prey, this distance varying somewhat according to the size of the object. This last fact suggests that the response comes about within the limited distance of effective visual accommodation and that the booty must stimulate a certain extent of the retina before the snapping response may be elicited.

In certain species the visual control of snapping is subject to at least a small measure of learned modification. In the spring the

[1] In panoramic vision, which is characteristic of lower vertebrates, the visual fields of the two eyes are continuous and essentially different.

very hungry trout snaps at almost any moving small object; but later in the year, when food is plentiful, its feeding snap is given most readily to the movement of the particular insect which is plentiful at the time and is otherwise somewhat inhibited. That learning may reduce the readiness with which the response appears under certain visual conditions is also indicated by the finding of Buytendijk and Remmers (1923). Early in the course of their experiment fresh-water fishes would snap with equal readiness at bread and at chalk falling through the water, but after more than 170 trials the subjects became able to inhibit the response to chalk, apparently on the basis of a difference in its motion.

The Sensory Systems Compared in Their Influence upon Behavior. The group of fishes in which vision is dominant may be contrasted with the group in which chemical and contact sensitivity are dominant. In general, the visual determinant appears more rigid and specific in its control than does the olfactory. The role of the eyes, with their impulses conducted through specialized centers in the nervous system (the optic lobes and thalamic centers), is that of direct enforcer and coercer of behavior. Olfaction, which dominates the forebrain, is a general exciter which accounts for less specificity or stereotypy, although it is a modality which has a wide scope of influence in behavior. In this respect, the less specific behavior controlled by olfaction may be contrasted with the role of vision as described above, and with the effect of localized gustatory and contact stimulation. As Herrick (1924) points out, it is through the evolution of the forebrain, and hence through mechanisms primitively under the dominance of olfaction, that higher vertebrates become less subject to the specific enforcement of stereotyped behavior.

Some Important Factors in the Control of Movement

Maintenance of Position through Response to Change in a Visual Field.—Fishes that live in running streams are able to maintain a given position by actively swimming in the upstream direction. Although tactual stimuli from the current are of importance for this response in many species, this is not the principal factor in most cases. Lyon (1904) placed killifish in a long bottle filled with water and corked to eliminate the current effect, and when this bottle was moved in one direction beside the algae-covered wall of a large tank, the fish promptly swam to the opposite end. They were not oriented by a current but by "apparent motion" of the wall. Similarly, when the experimenter permitted this bottle to float in the current of a small tide stream, the fish shut within it quickly swam to the upstream end. In another test, minnows swam in the direction in which a surface of transverse black and white stripes was moved beneath the glass bottom of their aquarium, and reversed their direction when the movement of the artificial substratum was reversed in direction. Most of the species tested by Lyon showed this ability to move in such a manner as to retain a given field of visual stimulation.

The Control of Posture and Equilibrium.—Since fishes move in a tridimensional environment, the maintenance of position is a complicated matter. In experiments on the dogfish by Maxwell (1923) and others three sense modalities have been shown to cooperate in this function. Destruction of the entire labyrinth of the right ear in a dogfish resulted in depression of the right eye, elevation of the left eye, and an inclination of the body toward the right by virtue of an opposite bending of fins on the two sides of the body. Consequently, in locomotion the body rolled toward the operated side. Dogfish deprived of the semicircular canals and utriculi in the ears of both sides were able to swim toward objects, and to turn toward a moving object, although near the surface they swam without good control of bodily position. This shows that visual control is of some importance in the normal maintenance of posture. The role of contact was shown by the way in which the operated animals righted themselves promptly when they happened to touch the bottom of the tank in which they were swimming. However, the primary function of the inner ear apparatus in the normal maintenance of an upright position and in movement control was shown by the frequent failure of operated animals to compensate for disturbances of position, and by the frequency of swimming back-downward or on one side. Hence the semicircular apparatus, vision, and contact may be considered supplementary senses in the control of equilibrium and posture.

The Lateral-line Organs and Movement.—Contact sensitivity is of evident importance in the responses of fish to currents and in their reactions to other movements of the water or to a surface near which they are swimming. Movements of water also stimulate the lateral-line organs, which are cells grouped in pits at intervals in subdermal canals along the sides of the body. Hofer (*cf.* Watson,

1914) presented water currents differing in direction and force to fish deprived of lateral-line sensitivity by operation and found these animals unable to respond as did normal fish. By means of the extirpation method, and by testing trained fish, Dykgraaf (1933) has recently shown that the lateral-line organs are stimulated by weak currents and by changes in water pressure. He finds that this sensitivity not only permits the fish to localize motionless objects (from which water vibrations are reflected), so that it may avoid obstacles in swimming, but also to respond to moving objects, as in food capture or avoidance of predatory animals.

The Cerebellum, and Integration of Receptors in the Control of Equilibrium and Posture.—The preceding treatment has disclosed the fact that in the control of equilibrium and posture a number of receptors commonly function. About the semicircular apparatus as chief controller, vision, the lateral-line reception, and contact are combined. We note that while the brain (forebrain and midbrain) of the fish permits no great amount of interaction among the important sensory systems (p. 180), these systems do cooperate in a general function such as the control of posture and equilibrium. For this cooperative control the cerebellum is responsible. Herrick (1924) has compared the role of the cerebellum in general behavior with that of a gyroscope on a moving steamship. The cerebellum does not determine the *nature* of movements made nor are its connections responsible for the initiation of new movements, but rather this nervous center is essential for the regulation of the strength and timing of movements which are aroused and directed by other nervous means. Following the analogy, the quartermaster steers the ship by controlling the rudder, while the function of the gyroscope is to hold the vessel steady and on an even keel.

In the movements of a vertebrate animal the integration of sensory impulses (p. 209) by means of the cerebellum maintains the animal in a steady posture and insures a condition of activity which is gained by other means. Movements which are specifically initiated and controlled by one sense department (*e.g.*, by olfaction through forebrain centers) may be reinforced and supported by other sense departments through cerebellar connections, without thereby having their essential nature altered. The cerebellum does not combine the sense modalities in a manner that permits them to check or to counterbalance one another. It is restricted to combining or integrating them in making for postural adequacy, a fact the importance of which is shown by the comparatively large size of the cerebellum in vertebrates that adjust to a tridimensional environment (*e.g.*, fishes, birds). The reasons for this decided limitation in the psychological importance of the cerebellum may be found in its tract-and-nucleus structure. A cooperative and counterbalancing integration of the fields of sensitivity in controlling the appropriateness of environmental adjustments necessitates the functioning of conduction mechanisms which are but feebly forecast in the nervous system of the fish.

Important Modes of Behavior in Fishes

Among the fishes and other lower vertebrate classes we find modes of behavior which, like those of some of the higher invertebrates, may not express the influence of any one mechanism alone, but rather show the result which is inevitable when a constant environment plays upon an animal of complex structure. The principle, and the way in which it makes for adaptive behavior, will be illustrated in terms of two behavior modes which characterize the fishes as a group.

Migration.—Some fishes are mainly sedentary and the movements of others are very irregular, or if regular are dependent upon periodic changes in available food; but the migratory movements which are most common in the class bear a significant relation to the breeding cycle. The case of the salmon is best known.

The Typical Migratory Pattern in the Salmon.—The young Pacific salmon develop from eggs which are laid in some fresh-water tributary stream. During their second year the young salmon begin to move downstream by easy stages, and finally they pass from the main stream into the sea. As adults they live in salt water during their next two years. Commonly during their fifth year the adults become sexually mature, and now begin to move into fresh water and to migrate up the river to its tributaries. They move rather slowly at first, tarrying at the mouth of the river and dropping back for a distance at each ebb tide, but once started against the current, they move steadily toward the river's source. Many of them fail to pass obstacles such as falls and rapids which are encountered on the way, and die on the rocks. The mortality is decreased when artificial "ladders" are provided, up the stages of which the salmon may flip themselves against the current. Finally the mature adults reach some tributary of the river and spawn there, after which they float tail-first downstream and die.

Factors Involved in the Control of Salmon Migration.—Let us attempt an understanding of this behavior. Roule (1928) attributes the downstream passage of the young salmon to the fact that near the end of their second year they begin to lose pigment from their skin. As more pigment is lost there is an increase in the irritating effect of the strong sunlight which floods shallow water, and this sets up an organic disturbance which starts the fish on its way along the line of least resistance, *i.e.*, downstream. It continues to move until the

FIG. 45.—A Labrador salmon surmounting an 18-ft. waterfall in the course of its upstream migration. (*Photograph by Dr. R. T. Morris. From Jordan and Kellogg.*)

new optimum, deep water, is reached. Very probably this is not the only factor in bringing about the movement. However, it appears significant that the trout, a close relative which loses much less pigment than does the salmon, remains in fresh water.

After its life of two or three years in the sea, the adult salmon migrates into fresh water, and hence upstream, under the influence of physiological changes which come with sexual maturity. According to Roule (1928) one factor in forcing the move is that when swollen with developing eggs (female) or with sperm (male) the individual's respiratory rate is greatly increased, and the oxygen content of sea water thus falls considerably below the new optimum. The fish then responds more readily to water which has a greater quantity of oxygen mixed with it, *i.e.*, to water from a river system. Unfortunately, at present there is no direct evidence for this factor.

Other factors may be suggested. Greene (1926) has described rather extensive changes in the chemical composition of the king salmon's tissues just before and during its upstream migration. It is important that a great increase in the size of the female's ovaries comes about during the months that precede spawning. Then too, at migration time there are changes in muscle tissue and in the salt content of the blood, as well as a destruction of stomach tissues which soon prevents feeding. There are many indications that the sexually mature salmon is more highly sensitive to slight differences in salinity and temperature, a fact which may be attributed to internal changes at the time. One study suggested to the experimenters (Shelford and Powers, 1915) that the fish is oriented into the river mouth by virtue of its ability to respond to a gradient of decreasing salinity. Responses to temperature differences may be influential in controlling behavior at the river mouth, and current pressure undoubtedly is important.

It is now known that migrating adult salmon frequently enter the river from which they came as young individuals. Jordan (1907) held the opinion that, while in most cases salmon return to the same river system, the reason is that few of them ever go far from the river mouth. Consequently, in view of the known fact that a river such as the Columbia affects temperature, current, and chemical conditions of the ocean at considerable distances from its mouth, most of the salmon would be within the scope of its influence at the time when they were becoming susceptible to fresh-water conditions. Rich and Holmes (1928) reported extensive tagging experiments in which none of the salmon marked on the Columbia river were recovered as adults in any other river system, although some of them had moved far up the coast to a point near Alaska during their life in the sea.

It is possible that the factors which control entrance into the river mouth have much in common with those which operate during the upstream movement. In any case, the persistent movement of the adults upstream is an interesting and perplexing problem. Once on their way the fish swim doggedly against the current, ascending rapids and surmounting other obstacles until the headwaters are reached (Fig. 45). The upstream progress appears to be based mainly upon a response to the pressure of the water, but how this factor operates is not clear. It is not improbable that the strong tendency to move against current pressure represents a phenomenon of counterirritation; that the nervous effect of internal distension and the actual degeneration of tissues accompanying sexual maturity is counterbalanced only by strong external stimulation and hence forces the maintenance of such stimulation.

The particular route taken by the salmon in their upstream passage remains to be explained. Although there is a lack of evidence that pressure differences are important for the movement into one tributary rather than into another at places where the river branches, Ward (1921) has pointed out the great importance of temperature sensitivity in determining the upstream route. His records show that when migrating adult sockeye salmon encounter a fork in the stream up which they are passing, they turn into the tributary from which flows water of lower temperature. Similarly, upon reaching the headwaters, they are most likely to build nests and to spawn in places where the water is below 4°C. in temperature. This suggests one reason why salmon which migrate at different times of the year may differ in the routes they follow.

Although temperature is important, other factors may be involved in accounting for the fact that salmon of different species, under appropriate conditions, are able to return to the identical tributary stream which they left as young individuals. Rich and Holmes (1928) found that when the young of Chinook salmon were transplanted as eggs to a different Columbia river tributary from that of their hatching, and permitted to develop in the "foster stream," the majority of them succeeded in returning as adults from the sea to the tributary in which development had occurred. Surprisingly enough, while this held for the fall hatch of salmon, it did not hold to any extent for spring hatches. Somehow the fish may become acclimated to the specific nature of the waters in which it develops, so that as an adult it is fitted to respond more readily to the chemical effect of these waters than to that of other waters. Although definite proof that chemical discrimination is involved in this particular phase of migration is still lacking, we must not forget the ingrained importance of chemical sensitivity in the life of the fish.

It may be concluded that the upstream migration of the adult salmon depends upon a change in responses to outer stimuli when progressing sexual maturity has brought certain important internal changes. These internal changes represent the core of the phenomenon. However, it may be said that the factors which cause the young individuals to leave fresh water are responsible for the occurrence of migration in the life cycle of salmon.

Behavior Associated with Reproduction.—The breeding behavior of most fishes is so rapidly accomplished and so complicated that only by observing it many times with special attention to important items, as was done in Reighard's (1920) study, is it possible to obtain a clear picture of events. This observer carefully pieced together the following account of breeding activities in a common fresh-water fish.

Breeding in the Sucker.—The male suckers congregate on the rapids during breeding season, when they are sperm-laden. Occasionally a female appears from deeper waters and remains quietly in place until males approach her. As they come near she darts forward, then pauses with belly touching the stream bed, and swims quickly away as they reapproach. This flight is repeated a number of times, but finally the female does not swim off. Now attention shifts to the approaching male. When close to the female he stops, with spreading pectoral fins, dorsal fin coming erect, and with protruding jaws. Changes in the body coloration of the male now appear, and his eyes redden. For a second his head vibrates from side to side, disclosing the tension of his entire body.

The males now press close against the female, one on each side of her (as a rule), and all three individuals head upstream. The pectoral fins of the males are spread beneath the female, and their downward spreading anal fins press against her tail, so that she is held in a tight grip between their bodies. There now occurs a shuddering vibration of the three bodies, and although the heads move in little more than a tremor, the wide and vigorous excursions of the tails are sufficient to stir up gravel from the stream bed. This, together with the sperm-containing milt which spurts from the males at this point, clouds the water so that the eggs cannot be seen as they are shed from the female. This is the climax of the mating act, and requires only about two seconds of time. After it the female moves upstream, subsequently mating with other males.

The distinctive features of this behavior depend upon the general excitability, for which the glandular condition of the fish is responsible when sexual maturity is reached, and the increased excitement attributable to bodily contact at the time of actual mating. This pattern is typical, and although the mating of other fishes differs somewhat from it (particularly in the behavior of the male before and after the spawning act), the elemental features are much the same in many species.

The Place of Learning in Fish Behavior

The Intelligence of the Fish.—Judgments of the comparative intelligence of different species of fishes, or even of the general psychological position of the class, frequently are not based upon dependable criteria. For instance, although there would be general agreement that the trout stands higher in this respect than does the herring, the reasons for this judgment have not been clearly stated. The ability of a species to avoid traps may depend upon general agility or upon acute sensitivity, upon boldness or timidity, or upon capacity for learning. A given species of fish may be more able to learn than larger species but may become excessively timid because its small size automatically creates for it a host of enemies. Then too, special advantages oftentimes permit behavior that is misjudged easily. One writer has called the eel clever because it is able to escape from tanks, to climb slippery posts, and to wander over fields. Again, adaptation of the skin to surroundings depends upon factors which have nothing directly to do with ability to learn, but these and similarly based occurrences have been cited as proof of the intelligence of certain fishes.

There is, also the popular desire to tell a good fish story: the *anecdotal tendency*. We cull an example from the correspondence columns of the London Evening Standard:

In answer to the question, "Do Fish Think?" an experience of ours some years ago is interesting. We kept several large gold and silver fish in a glass tank, which became frozen over. One day the frost formed delicate fern-like leaves of ice, which shot down to the bottom of the tank, and enclosed the fish in frozen cells. They became so still and lifeless that we were alarmed. My father broke the ice, and, lifting the fish out, poured a few drops of brandy down their throats, replacing them in the tank in a more lively condition. Next morning, which was milder, we were astonished to find them all in a row, motionless, pretending to be frozen in.

It is evident that our question can be answered only by the use of criteria which have experimental dependability.

Learning a Simple Series of Orienting Movements.—Many fishes (*e.g.*, the trout) are able to orient successfully within a given locality of considerable area, and to return to the same places in their daily feeding. Most aquarium fishes readily learn their way about the container.

A simple test of this ability was performed by Churchill (1916). Once goldfish had learned to swim through an open tank and be fed on the farther end, two mesh-wire partitions were inserted which required the fish to swim to the first opening *A* near one side wall (Fig. 46), then downward toward the second opening *B* near the opposite side, and thence upward to reach the feeding place *C*. The openings were each 2.5 cm. square. Visual distinction of the openings from the squares in the wire mesh apparently was not involved in controlling the movement, since a fish would repeatedly swim past the door and get through only "by playing about the netting and thrusting the head through the interstices until the opening was hit upon . . . " On further trials the exploration was confined more

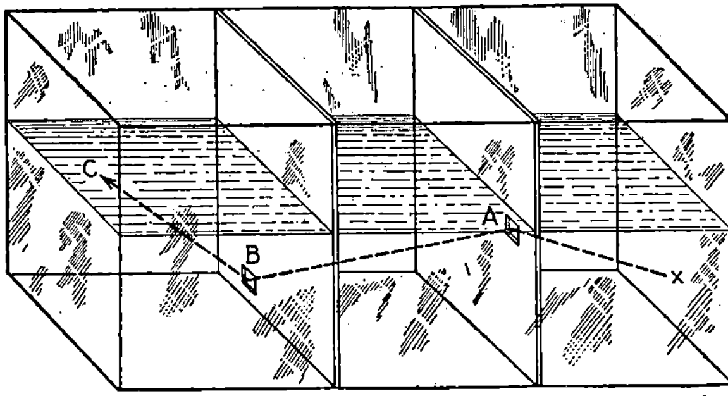

FIG. 46.—Churchill's apparatus for testing learning in the goldfish (see text). *X*, starting point; *C*, feeding place; *A*, *B*, apertures in the partitions. (*Modified from Churchill*, 1916.)

closely to the vicinity of the opening. Apparently this change depended upon an increase in the accuracy with which the fish moved toward the first partition, since at length there was much less nosing about near *A* and the fish swam directly toward the opening. The fish improved more rapidly in getting through the opening *B*, apparently on the basis of a specific orientation obtained in passing through *A*.

A second group of subjects was presented with wooden partitions in the tank, with openings cut in the corresponding places. The goldfish did not contact about the surface of these partitions, as with the mesh-wire partitions, but when a subject happened to come directly in front of the opening and an inch or two away it usually would pass directly through. This response, as well as subsequent improvement in finding the opening, apparently depended upon visual control alone.

In this study, one test shows no evidence for the modification of a visually controlled movement by tactual stimulation which the situation also affords. The other test, discussed first above, shows that tactual (probably with kinesthetic) sensitivity may dominate independently without the essential involvement of vision. The results contribute to an understanding of the manner in which the sensory control of a simple movement series may differ according to circumstances, and yet may be fairly specific for any given type of situation.

Learning Which Is Distinguished by Sensory Discrimination.—We are familiar with the results of Herrick (pp. 182*f.*) and of Buytendijk and Remmers (p. 184), which demonstrate the ability of certain fishes to inhibit snapping in sensory circumstances which at first elicit the response. A number of experiments have shown that for each of the major sense departments it is possible to train an individual to respond specifically to a stimulus which at first did not produce such response. This has been demonstrated by coupling the (at first) ineffective stimulus with one which forces the response in question from the outset.

Establishment of Conditioned Responses.—Froloff (1925) employed the strong "starting response" to electric shock. Five seconds before the shock was given, a light would flash in front of the fish. Within 5 to 30 trials a subject would give the starting response when the light was presented alone. Training thus attached a response to light which had previously been limited to other stimuli. Fewer trials were required to bring the response when the shock was preceded by the actuation of a telephone receiver under the water than when the new stimulus was the ringing of an electric bell above the surface. Bull (1928) used a somewhat different method, and successfully trained the blenny to rise to a feeding place in response to a momentary increase of 4°C. in the temperature of the water, or to an increase of 3/1,000 in the salt content of the water. When these changes were introduced after training, the fishes swam up to the feeding place even though no food was presented.

Sears (1934) used a current of water as "unconditioned" stimulus, the typical response to this being an easily identified movement of fins and tail. When this stimulus was presented together with light as the "conditioned" stimulus, the response was conditioned to light in nine subjects within an average of 50 trials (see Chap. XV).

In Sears' tests electric shock proved very undependable as an "unconditioned" stimulus. After 45 shocks had been applied in series to one subject the electrical shock response could be readily elicited by presenting either a "white" light or a vibratory stimulus alone, although neither of the latter stimuli had been applied before that time. This phenomenon, not a "conditioned" response, was named by Ufland (1929) "a dominant." The apparent explanation is that repeated response to the relatively strong electric stimulus so lowers the animal's threshold of reaction that stimuli of other types, although presented then for the first time, will bring that response rather than their own usual reactions.

A Discrimination Habit Based upon Visual Differences.—Somewhat more complicated is the change produced when the fish is trained to react to one sensory feature but at the same time is trained not to react to a similar feature, *i.e.*, the "discrimination habit" (Chap. XIII). Reeves (1919) obtained evidence for color vision in the fish by means of this method.

Two colored filters were placed in lamp boxes so that light passing through the filters would fall into spaces separated by a partition in the aquarium below (Fig. 47). The light from each of the filters passed through a slit which could be varied in width in order to control the intensity. A white surface placed at an angle in each compartment reflected the light from the filters so that a fish swimming toward the compartments would be exposed to two lights of different wave lengths, one on each side: a "blue" light (wave lengths between 460 and 524 mμ) as against a "red-orange" light (wave lengths above 598 mμ).

Food fastened to a rod was lowered into the water when the fish swam into the compartment illuminated by "blue" light, but no food could be obtained in the "red-orange" compartment. Since the "correct" light was sometimes to the right and sometimes to the left, it became necessary for the fish to learn a discrimination between the lights if food was to be obtained consistently.

Horned dace required about 400 trials and sunfish about 900 trials to learn to turn toward the "blue" stimulus and to avoid the "red-orange." To test whether the discrimination was based upon a brightness difference, Reeves began to change the intensity of the "red-orange" light. After this light had been reduced to a certain intensity value the discrimination broke down and the subject made a chance score. However, training was continued and after a time the ability to discriminate reappeared. Apparently the fish had first learned to discriminate on the basis of intensity, but when changes were introduced so that the "red-orange" was sometimes more and sometimes less intense than the "blue," the fish learned to react on the basis of wave-length difference alone. Other tests in which "white" versus "blue" and "white" versus "red-orange"

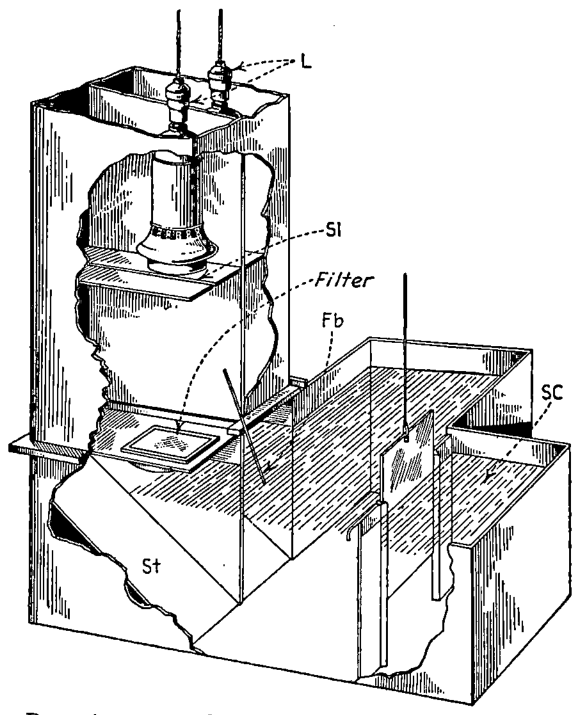

FIG. 47.—Reeves' apparatus for testing color vision in the fish. *Fb*, food bar, now down in right compartment; *L*, light sources; *Sl*, slit for controlling intensity of light; *St*, stimulus plates, from which light is reflected into starting compartment *SC*. (*Modified from Reeves*, 1919.)

were used as stimulus pairs substantiated the conclusion that the animals were capable of learning to discriminate on the basis of differences in the wave length of lights.

Modifiability and Conduction Equipment.—Although fishes are capable of simple modifications in their responses to stimuli of a given sense department, and show by an acquired distinctiveness in response that certain energy effects act upon the receptor in different ways, we find no indication of important learned integrations among the sensory systems. This is a fact of some importance, in view of the apparent fact that the basic structure of the conduction system contributes to a virtual independence of the major exteroceptive departments in controlling the nature of behavior. Some experiments involving destruction of nervous tissue are of interest in this connection.

Nolte (1932) trained sticklebacks to discriminate a given patch of "colored" paper from other patches, after which he bilaterally destroyed part or all of the *forebrain* in given subjects. He reported that after the operation these fish reacted just as efficiently to the training stimuli as before, avoiding the nontraining patches and snapping only at the training patch or at those close to it in wavelength value.

However, in the nervous system of the fish the forebrain is preempted by olfaction, which was not involved in the control of this learned response. The visual system dominates the optic lobes; and it is there that we should test for traces of whatever physiological changes were produced by this training. Destruction of optic lobe or thalamic tissue, but not of forebrain tissue, might be expected to impair or remove the habit. Unfortunately Nolte did not operate upon the visual complex.

We have mentioned the conditioned response to light which Sears (1934) established in the goldfish (pp. 194*f.*). He destroyed a large part of the optic-lobe tissue (in the outer cell layers) without impairing ability to acquire this conditioned response. Animals subjected to the operation after the conditioning had been established were able three days afterward to equal the performance of normals. There was some evidence that the habit was less stable in operated cases than in normals, but in general it was concluded that the extensive (although not complete) destruction of optic-lobe tissue interfered neither with the establishment nor with the performance of this conditioned response. However, it is by no means certain that these operations would not have impaired other types of visuomotor habit which make greater demands upon the animal (Chap. XIII).

Conclusion

The fishes are subject to limitations prescribed mainly by a virtual independence in the control exerted by the major exteroceptive systems over behavior, since there appears to be a lack of any important intersensory correlation in the qualitative control of adjustments to the environment. For this reason in particular, the behavior of the class is characterized by directly determined activities and by a lack of plasticity. The discovered capacity for modification of behavior apparently contributes only to alterations in given specific activities (*e.g.*, response to a light of given wave length or to an opening in a certain position) rather than to the broad recombination and selection of activities which appear in the highest vertebrate groups.

Segregation of the major sensory systems in the qualitative control of behavior thus is a primitive vertebrate character, which is somewhat improved upon in the Amphibia.

Suggested Readings

HERRICK, C. J. 1924. *Neurological foundations of animal behavior.* New York: Holt. Chap. XII.

KYLE, H. M. 1925. *The biology of fishes.* New York: Macmillan. Pp. 396.

REEVES, CORA. 1919. Discrimination of light of different wave-lengths by fish. *Comp. Psychol. Mono.*, 4, 57–83.

ROULE, L. 1928. *Fishes, their journeys and migrations.* New York: Norton. Chaps. III and V.

WATSON, J. B. 1914. *Behavior, an introduction to comparative psychology.* New York: Holt. Chap. XIII.

CHAPTER IX

IMPROVED INTEGRATION IN THE CONTROL OF STEREOTYPED ACTIVITIES: CLASS AMPHIBIA[1]

INTRODUCTION

The Importance of a Shift in Typical Environment.—The amphibians mark a crucial transition in vertebrate life, the change from an aquatic to a terrestrial environment. Although most of them develop in the water, the larger part of the adult's life typically is spent on land. The frogs and toads are significantly different in their sensory and other equipment, as compared with an aquatic amphibian such as the salamander. The terrestrial amphibian possesses the vomeronasal organ as an auxiliary to the olfactory apparatus. This development evidently increases sensitivity to chemical stimuli received from food in the mouth and may make possible a control of the snapping response through learning. Certain significant changes appear in the conduction system in correlation with this receptor.

Another difference is notable: terrestrial amphibians lack the lateral-line organs which they possessed in the aquatic (tadpole) stage of their life history. This makes for an important difference in the feeding of adult anurans and urodeles. Scharrer (1932) showed the outstanding role of the lateral-line receptors in the food finding of the larval salamander. *Amblystoma* larvae were deprived of both vision and olfaction by operation, but could still readily locate small moving prey by responding to water currents from the objects, which were then snapped up.

The possession of lung respiration in the terrestrial stage, instead of the gill type of breathing which is characteristic of aquatic animals, is significant for the rate of metabolism and therefore makes for a more active life. An important related possession (which, however,

[1] Amphibia are vertebrates distinguished by their smooth, slimy skin which typically lacks scales, and by gill breathing in the larval form. The two amphibian orders with which we deal are the *Urodela*, with long cylindrical bodies and flattened tails (*e.g.*, salamanders, Fig. 49); and the *Anura*, which in the adult stage have broad flattened bodies without tails (*e.g.*, frogs, toads).

represents a similarity to aquatic animals) is the moist skin of the terrestrial amphibians, which makes them delicately sensitive to

TABLE 13.—AMPHIBIA

RECEPTOR EQUIPMENT	FEATURES OF SENSITIVITY
Chemical. Free nerve endings over body surface.	Moist skin sensitive to concentrated chemicals. Reaction time differs for acids, salts, alkalies.
Taste buds in mouth, pharynx; none on skin.	Gustation of some importance for feeding of terrestrial forms.
Ciliated cells in epithelium of nasal cavities, which open by nostrils near snout end, connect with mouth at back.	Olfaction functions in aquatic forms much as in fishes. Tracking of chemicals demonstrated for newts. Anura much less sensitive.
Contact. Receptor cells over body surface, especially head. Moist skin sensitive to humidity changes. Separate temperature endings probable.	Important reflex responses controlled tactually, *e.g.*, respiration, clasp response. Determines male's carrying eggs. Humidity conditions determine habitat; place where eggs are laid.
Lateral line. Organs present only in aquatic Amphibia.	Sensitivity to movements of prey animals. Similar to fishes.
Static. Somewhat better developed semicircular apparatus than in fishes.	Aquatic amphibians similar to fishes. Head orientation in terrestrial forms attributable to utriculi.
Auditory. Tympanum embedded in skin of head, middle ear present (Anura). Lagena improved.	No evidence for auditory sensitivity in aquatic forms. Effects general bodily changes in terrestrial forms. Important in mating.
Light. Lens fixed in shape, eye focused by pulling of lens forward from retina. Fovea, central area, absent. Skin receptor cells in some.	Sensitivity to movement important for feeding. Typically orient toward light, but gather in dark places. Skin coloration controlled by brightness of background, temperature, humidity. Scattered evidence for color vision.

CONDUCTION

Olfactory lobes smaller than in fishes. Thick-walled forebrain, dorsal part nervously developed (archipallium). Well-developed optic lobes, thalamus. Small cerebellum. Ten pairs cranial nerves. In terrestrial Amphibia spinal cord most extensively developed on level of fore and of hind limbs.

ACTION SYSTEM

Extensive development of reciprocally innervated muscles (flexors and extensors) attached to parts of bony skeleton which are moved as levers. Hind-legs furnish motive power, front legs mainly steer in locomotion. Webbing of rear legs in some important for swimming. Front of tongue attached in *Anura*, tongue quickly extended in feeding. Sound production: two vocal cords across larynx, inflated throat as resonator. Chromatophores in skin of some. Secretions of mucous glands keep skin moist. Skin respiration important; adults breathe mainly by lungs, larvae by gills. Oviparous: Eggs fertilized after extrusion from female.

humidity changes in air. To instance one importance of this characteristic, upon their appearance from hibernation in the spring, frogs, toads, and salamanders move toward ponds and streams where they breed. These migrations are particularly frequent after heavy spring showers, owing to the fact that enlargement of the sex glands (gonads) has made the animals very responsive to any sudden increase in humidity (Cummins, 1920; Blanchard, 1930). Amphibians also possess, as do fishes, the common chemical sense, a general skin sensitivity to acids and other chemicals in fairly strong concentrations. In higher vertebrates this type of receptor generally is restricted to interior mucous-membrane surfaces.

In the action equipment of terrestrial amphibians, the possession of legs is responsible for certain important developments in the organization of local activities in behavior and also accounts for an increase in the number of specific modes of action (burrowing, swimming, leaping, hopping). To cite another difference in action equipment which is responsible for departures from the piscian behavior pattern, the amphibian tongue (absent in aquatic forms) functions in connection with the snapping action of the jaws and thus provides the basis for a modification in the feeding response.

The differences we have noted are accompanied by some important changes in conduction equipment and forecast an advance from the dominance of specifically enforced and stereotyped activities. This suggests an inquiry into the adequacy of sensory control in amphibian behavior.

Advances in Sensory Control.—The external environment of the amphibian largely commands its behavior, and hence the animal displays a fund of rather specific and distinctive activities. Although the number of items in the amphibian's behavior repertoire appears greater than in the fish, the separate activities of the amphibian more readily lend themselves to organization, and the manner in which a given response appears is somewhat more influenced by other responses. Hence the amphibians appear superior to the fishes in the unity of their behavior.

Although the control of the sensory systems over behavior is similar to that in fishes, it is difficult in any amphibian to find one system that dominates consistently, in the manner that olfaction controls the behavior of some fishes and vision the behavior of others. Then, too, a given sensory system more readily brings another system into play. Riley (1913) observed cases in which a toad, previously quiescent in continuous weak light, was aroused by sudden contact with another individual to respond definitely to the light (usually hopping toward the source).

However, at any given time the amphibian's behavior is directly dominated by one of the exteroceptive fields; most frequently by vision (except in cave-dwelling forms). The frog is so strongly compelled to snap at any small moving object that in an experiment by Abbott (1862) an individual severely injured its jaws by repeatedly lunging and snapping at a fly pinned in the center of a ring of needles. In such cases the fly does not stimulate as "food," but as a small moving object, a visual stimulus of powerful coerciveness. There are, however, certain features in the feeding of other amphibians, and even of the frog, which suggest some improvement over this stereotypy.

We have pointed out that the primitive vertebrate characteristic, specific sensory control over behavior, is prominent in the feeding of these animals. As Schaeffer (1912) put it, a frog might starve to death through its failure to snap at motionless flies strewn around it. For most amphibians, as for fishes, the object must be small and must move in order to elicit the snapping response. However, in certain members of the class sensory dominance is somewhat reduced in rigidity and specificity. For one thing, the control of the feeding response may shift with some readiness from one sensory system to another when conditions force the change. Ordinarily, *Amblystoma* larvae depend predominantly upon vision in locating food, but blinded animals in Nicholas' (1922) experiment located moving objects readily on the basis of olfactory stimulation and through water vibrations (lateral-line sensitivity). Through olfaction they were able to locate motionless food objects, as were normal animals in darkness. Other examples might be cited.

The amphibian snapping response is more complex than is its homologue in the fish; and this greater complexity is paralleled in its sensory control. Hempelmann (1926) points out that the summation of stimuli is an almost indispensable feature of the frog's snapping response. The first slight movement of a fly causes the frog to move its head; if further stimulation comes there is a turning of the trunk; another visible motion brings a lowering or lifting of the head according to the position of the booty object, and a subsequent movement of the fly brings snapping. This response consists of a simultaneous opening of the jaws and throwing out of the sticky tongue, followed by tongue withdrawal and closing of the jaws.

If the frog gets nothing from the first snap, the stimulus must be repeated much as before to elicit another snap (Edinger, 1908). While the feeding response of the frog is somewhat more stereotyped than that of certain other amphibians (*e.g.*, the toad), the functioning of nervous summation (by delaying the final act) permits a certain variability in it that is not seen in the fish.

The specificity of the stimulus conditions, and the stereotypy of this response, should not obscure the fact that the animal is oriented toward the stimulus source over a longer time than is possible in the feeding of a fish. This not only makes the feeding response more complex, but also permits a somewhat greater measure of flexibility in its sensory control. Similar and even more significant features appear in the food taking of other amphibians. Before proceeding with the treatment of this problem, it is desirable to survey the amphibian's repertoire of movements.

The Nature of Directly Determined Activities in Amphibia

The Control of Local Activities. *A Local Activity Essentially Similar in Various General Responses.*—Animals of this class are said to have a "large fund of reflexes," but this statement is somewhat misleading. Noble (1931) has pointed out that the hind legs act very much alike in total activities which appear to be quite different, as in leaping and swimming. The difference is an incidental one attributable to the different resistance of the medium in the two cases. In diving, hopping, righting, and swimming movements, the rear legs function as *pushers*, an action which is primarily dependent upon their specific muscular makeup. The action of the front legs varies somewhat more in these responses, but even in their case the variation is not marked from activity to activity, as Holmes (1906) noted.

Reasons for Variation in Rear-leg Action.—Whatever variation there may be in the activities of front or of rear legs depends largely upon how other parts of the body act in the different circumstances, or upon the strength of local stimulation. In the "spinal frog," for which a cut through the lower medulla frees the spinal cord from the influence of higher centers in its control over trunk and legs, different local stimuli produce the following responses: weak contact applied to a rear leg causes it to extend; pinching typically causes the leg to flex (bend toward body); a stronger pinch causes the leg to "kick" (to flex and extend repeatedly); and a still more intense pinch brings "jumping" (very energetic flexion and extension of the leg). These differences in response fundamentally depend upon the manner in which stimuli of different strengths furnish different impulses to flexor and extensor muscles of the leg. In the intact animal, such differences would be highly crucial for the nature of total behavior. In principle, it is not important whether the impulses of varying strength (Chap. XIV) come from higher centers through the spinal cord tracts, or originate through local stimulation as in the experiment.

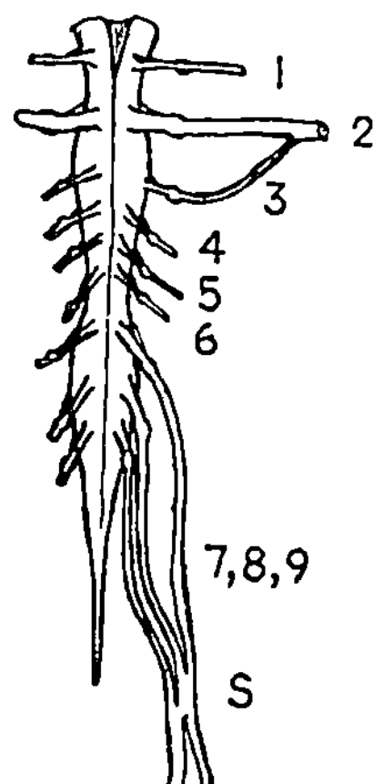

FIG. 48.—Spinal cord of the toad. Numbers on figure refer to spinal nerves. 2 and 3 unite to innervate the pectoral level of the body; 7, 8, and 9, form the sciatic plexus (*S*) from which the pelvic level of the body is innervated. (*Redrawn from Smallwood.*)

The different action of the rear legs in various general activities, for instance their action in leaping as compared with reaching an acid-stimulated spot on the skin, may be understood in terms of the strength of stimulation (Foley, 1923) and the locality of stimulation. In the classic spinal-frog experiment, when a piece of paper moistened with weak acid is laid on one haunch of the animal, the rear leg of that side comes forward in a thrashing motion which may wipe away the paper. If the chemical is more concentrated, both hind legs become active, and if sufficiently concentrated, the chemical causes movement of all four legs and twisting of the trunk.

Control of Appendages in Behavior Viewed as a "Segmental" Process. *Spinal-cord Organization Based on Limb Activities.*—The vertebrate conduction system fundamentally has a segmental pattern, and spinal and cranial nerves are arranged in pairs along the tube. However, as we have already pointed out, in fishes and higher vertebrates the upper centers of the nervous system are "segmented" only on the basis of the principal sensory systems. With the acquisition of legs, the terrestrial vertebrate also develops a modified segmentation of the lower part of the nervous system. In the anuran, which may be thought of as the "adult" amphibian, it is as though the spinal-cord portion of the nervous system had two important segments, one controlling the front legs (the pectoral region) and the other controlling the rear legs (the pelvic region). So far as general behavior is concerned, the activity in this spinal cord is mainly organized about the levels of the two pairs of legs as discharge centers (Fig. 48). This view will assist in understanding the function of the front and rear appendages in different types of behavior.

A Specific Front-leg Response.—That the front legs are capable of highly specific activities is shown by the appearance of the "clasp reflex" at mating time, shortly after the male anuran awakens from his winter hibernation. Gonadal secretions (*cf.* p. 250) so increase the irritability of the pectoral level of the cord that when the skin of the thorax or the inner leg surfaces is rubbed, the front legs clasp strongly together and hold tightly to the object.[1] Goltz (1868) cut away the upper and lower body, leaving the pectoral level with arms and a short section of spinal cord, and found that the clasp reflex still could be elicited by stimulating the skin of the body section remaining. This shows the dependence of the response upon the local stimulation and special excitability of the pectoral level itself.

Elicitation of the Clasp Reflex and Its Role in Mating.—An exteroceptive control over the clasp response might be suggested by the fact that male frogs and toads seldom clasp other males. However, vision and olfaction are not influential, since Goltz (1868) eliminated these exteroceptors operatively without impairing the ability of a male to avoid other males in mating. First of all, certain incidental selective factors appear important for this "discrimination" in most frogs and toads. One such factor is the ability of the male's croaking to excite the female so that she hops toward the source of stimulation. Then, too, the excited hopping of a male during the breeding season makes him much more difficult to hold than a female. Further, in certain American frogs and toads the males avoid other males, or release them promptly, because of their croaking or chirping, but retain the silent female once she is clasped (Noble, 1931). The "discrimination" of females by males therefore may be termed an indirect process.

It appears that an important reason for the usual failure of males to retain their hold upon other males accidentally clasped also depends upon certain differences in local (tactual) stimulation. Noble and Farris (1929) found that wood-frog males, injected with water until

[1] Steinach (1910, *Zentrlbl. Physiol.*) abolished the clasp reflex by removing the testicles of the male. After injection of testicular extract into the dorsal lymph sac, the reflex reappeared in the operated animals within a few days.

they were the size and firmness of egg-swollen females, were grasped and held just as were females in the mating act. Hinsche (1926) found that male toads would grasp and retain for some time pieces of rubber sponge and other objects that gathered fairly well into the arms. He noted also that the jolting of the female in carrying the male, and her breathing movements, provided additional stimulation which served to maintain the male's grasp. Altogether, there is no evidence that the clasp response and the dependent mating act are called out directly through exteroceptive control. Rather, the act of mating seems to depend upon the manner in which circumstances, more or less accidentally, furnish the adequate and necessary local stimulation of the male's pectoral region.

Exteroceptive Control of Spinal Segments in Behavior.—It is important that in general behavior, the pectoral level of the cord (controlling front-leg action) is much more widely influenced by higher nervous centers than is the pelvic level (controlling rear-leg action). To illustrate, in the toad's locomotion rear-leg action varies little from stage to stage except in the strength of kick, but between hops the position of the front legs is shifted by means of little movements which turn the front end in response to exteroceptive stimulation. In swimming, the synchronous stroking of the front legs assists in propelling the body directly forward, but asymmetrical exteroceptive stimulation causes the front legs to differ in their beat or to stroke alternately, thereby changing the direction of progress. The exteroceptive control which produces these changes, if olfactory, acts upon the pectoral level through the forebrain, or, if visual, it acts through midbrain and thalamus.

Stimuli produce variations in locomotion by acting upon head receptors, but the actual maintenance of locomotion is basically a function of the two "leg levels" themselves. Langendorff (1877) removed the optic lobes of the frog, and after shock effects had passed, he found that jumping and swimming activities occurred much as in the normal animal. However, in other operative cases stimulation of the exposed optic lobes brought a reduction in locomotor activities of the limbs. This suggests the controlling effect normally exerted upon these levels by impulses coming to the spinal regions from higher nervous centers. When Langendorff applied electrical stimulation to the exposed left optic lobe, the reduction in leg action was seen predominantly on the right side, and vice versa.

Development of the Individual and Control of Behavior

The Appearance of Locomotion and Related Activities in Amphibia.—In Chap. XII we are to undertake a special study of the appearance of basic activities in the vertebrate individual. At present we are interested in the early background of certain adult amphibian adjustments.

Terrestrial Locomotion Develops from Swimming.—We have already treated locomotion in the adult anuran. In his studies on a urodele,

Fig. 49.—Amblystoma tigrinum, a typical urodele amphibian. (*From Noble.*)

the salamander *Amblystoma* (Fig. 49), Coghill has shown that terrestrial locomotion develops from the swimming of the aquatic life stage. A report on this work will be offered in Chap. XII.

Our present treatment begins as the salamander takes to land at the end of its initial period of aquatic life. At this time its front legs are capable of supporting the anterior part of the body. Each leg is at first called into action only in company with the trunk muscles of its side when the body bends, since at this stage neurones that innervate the body wall also innervate the adjacent leg. When the trunk bends from side to side, the legs therefore stroke alternately.

Consequently, as Coghill puts it, walking is "swimming activity at reduced speed" and develops from swimming. Somewhat later, the hind legs go through a similar process, lagging behind the front legs in development. As walking improves, once the salamander takes to land, the S-movement (swimming) of the trunk becomes less evident until finally it becomes a minor part of terrestrial locomotion.

Although Carmichael's (1926) experiments show that the development of aquatic locomotion is not dependent upon stimuli from the external environment, it is not certain that external stimulation may be dispensed with in the improvement of *walking*. Growth changes in nerve tracts in receptors and in muscles make walking possible, but improvements in leg action in passing over irregularities and around obstacles suggest that in the case of terrestrial locomotion the changes forced by environmental stimulation are of importance. The rise of complex forms of movement such as this may forecast that type of close cooperation between "maturation" through neuromuscular growth and the molding effect of the external environment which gives rise to "basic activities" in birds and mammals (Chaps. XI and XII).

The Appearance of the Feeding Response.—In *Amblystoma* the feeding response develops from a movement which is closely related to terrestrial locomotion. At an early stage the animal lunges quickly forward when the snout or head is touched. At this time, Coghill finds that a tract of nerve fibers (the median longitudinal fasciculus) has pushed down from the midbrain region of the front-leg level of the spinal cord. Hence stimulation of the snout directly discharges impulses to the front legs, and these members respond just as though the impulses had arisen through local stimulation.

Shortly after its appearance on land the young salamander is able to lunge and snap at objects which move a few millimeters in front of its eyes. This ability depends upon the fact that axones from retinal cells (in the optic tracts) have meanwhile grown into the midbrain region. Upon reaching the midbrain, optic-nerve impulses therefore discharge directly into the median longitudinal fasciculus which carries the energy to the shoulder region and thereby touches off the developed lunging and snapping response as though the snout had been touched. We should not leave out of account the possibility that with repetition the snapping response (a complex combination of trunk, neck, jaw and tongue activities) is improved through learning.

An inspection of this act in the adult suggests that modification through learning is important. The newt (a urodele) slowly approaches an object which excites it visually and then may lunge and snap in response to further movements of the object or to olfactory stimulation (Copeland, 1913). Matthes (1924) reported that this animal may be aroused to respond to movement of prey through visual, olfactory, or lateral-line (water-vibration) sensitivity. A tree frog, excited by a movement, hesitates for a time and, as the prey approaches, finally leaps, evidently when the object has moved within the range of effective visual accommodation. These and similar features of the act are susceptible to learned modification, as we shall find. With a general understanding of the early development of locomotion and related activities in their importance for feeding, let us resume our study from a somewhat different angle.

Factors Making for Plasticity in Stereotyped Behavior

The Problem of Sensory Integration.—We have given reasons for the statement that the sensory systems are more adequately integrated in the control of amphibian activities than is the case for fishes. A judgment on this point should not be based upon sensory acuity as such, since acuity depends upon the receptor and is a factor of secondary importance in this connection. It is not merely the fact that a particular receptor is differently aroused by more or less slight differences in energy (*i.e.*, its acuity), but rather it is the closeness with which the related sensory system cooperates with others in behavior control that makes for adequate sensory integration. Let us consider some results which on their face demonstrate acuity alone, but which actually lead into a more important problem than that.

Stimuli May Have a General Arousing or "Tensing" Effect.—Yerkes (1905) presented different auditory stimuli to a frog harnessed to a saddle in an aquarium. Some stimuli (*e.g.*, shrill whistles, the ringing of a bell) brought an evident decrease in the rate of respiration, other stimuli (*e.g.*, splashing water) brought a slight decrease, and still others brought no detectable change in the animal. The results showed that the stimuli had different effects upon the receptor, but that none of them elicited more than a respiratory change. It is significant, as the experimenter pointed out, that overt responses were not called out by the stimuli used.

Nevertheless, as further results showed, such stimuli may indirectly influence overt behavior. If a "sound" such as that of splashing water was shortly followed by a visual stimulus, the animals jumped sooner in response to the visual effect than would have been the case without the auditory stimulus. In another test, one member of a group of silent male frogs was stroked and responded by croaking. This caused the others to straighten up from their squatting positions, and when the first frog croaked again the others presently joined him in the noise making. Other results show that a frog may be "tensed" by an auditory stimulus so that its response to a following stimulus of different type may be influenced. A frog, which Yerkes stimulated with a tuning fork "tone" and then with the movement of a small "red" card, jumped very promptly at the card, but frogs which were not first given the auditory stimulus required a few seconds to respond to the movement of the card. This evidence agrees with other findings in showing that the animal may be so acted upon by one stimulus that it responds more readily to one that follows. We have mentioned Riley's (1913) observation on the manner in which a quiescent toad was aroused to respond to light by another individual colliding with him. In this case, although visual stimulation readily displaced contact in controlling the animal's behavior and brought a specific response, the contact stimulus had prepared the way by bestirring the animal. However, there is no evidence that in the frog the "tension" produced by the first stimulus specifically *alters the nature* of the response to the second stimulus. Certain other amphibians are somewhat superior to the frog in this respect.

Time Relations and the Integration of Stimuli.—The time interval between the stimuli may be of significance for the effect of the first stimulus upon the response to the second. Yerkes (1904*b*) found that for the harnessed frog, a bell sounded more than 1 sec. before a tactual stimulus (hammer stroke on head) had no effect on the leg kick elicited by the latter stimulus. If the bell preceded the tactual stimulus by an interval between 0.35 and 0.9 sec., the leg kick was reduced below its normal amplitude, but if the interval was less than 0.35 sec., the bell stimulus increased the amplitude of the leg kick. Bruyn and Van Nifterik (1920) found in a similar experiment with the frog that the maximum time interval between stimuli, for augmentation of response, was 0.2 sec. Using the *toad*, however, they found that the leg kick was strongly reinforced, even when the sound (a telephone-receiver actuation of short duration) preceded the tactual stimulus by 0.38 sec. The trend of the results suggested that the interval between the stimuli might have been even greater than 0.4 sec. in the case of the toad, without totally removing the augmenting effect of the auditory stimulus upon the reaction to the tactual stimulus.

Significant results were obtained when the auditory stimulus was continuous, *i.e.*, remained in force until the tactual stimulus was presented. Under these conditions Yerkes found that sounds lasting for *more* than 2.0 sec. before a tactual stimulus was applied had the same effect upon the *frog's* tactual response as did a sound of 2.0 sec. duration. In contrast, Bruyn and Van Nifterik found for the *toad* that differences in the duration of the continuous sound within 10 sec. of the tactual stimulus were of importance for augmentation of the response to the latter stimulus.

Integration of Stimuli and "Preparation" for Response in Toad and Frog.—These differences assume greater significance when we consider the first stimulus one which "prepares" the animal for a response to the second. In this light, the results suggest that the toad may be prepared over a somewhat longer time than the frog. The toad is a more successful forager than is the frog and leads a more complicated life. One reason for its superiority as a food getter may be its ability to frequent regularly profitable feeding places, but another reason certainly is an ability to be more adequately tensed or "prepared" by stimuli which typically precede the appearance of prey. Further, the toad gives its feeding response to a wider range of stimuli. Franz (1927) reported that frequently a toad was seen to turn and snap up a crawling mealworm which had happened to touch its posterior quarters. This ability was not observed in frogs.

It is characteristic of the toad that a slight rustle of leaves may tense it so that when the insect moves within view, the capture is readily made. In the normal feeding of the toad and other amphibians, the approach represents a more adequate adjustment to the stimulus, and the snapping response is held in abeyance until a suitable sensory setting is gained. This represents a more efficient sensory control than serves for the rather inflexible response of the fish, and rests upon a definite superiority in sensory integration. The perfection of this mechanism may be the product of training, but the ability to delay a response and keep it under control (to "hesitate") depends upon the nature of the animal's conduction mechanisms.

Nervous Mechanisms Contributing to an Increased Plasticity in Behavior Control.—There are known superiorities in the amphibian nervous system, as compared with that of the fish, which make possible the more facile control of behavior. We are indebted in particular to the studies of Herrick (1924, 1933, 1934) for a large and well-organized body of facts concerning conduction patterns in these lower vertebrate classes.

This evidence may be summarized in the general statement that between the lower part of the amphibian forebrain (olfactory system)

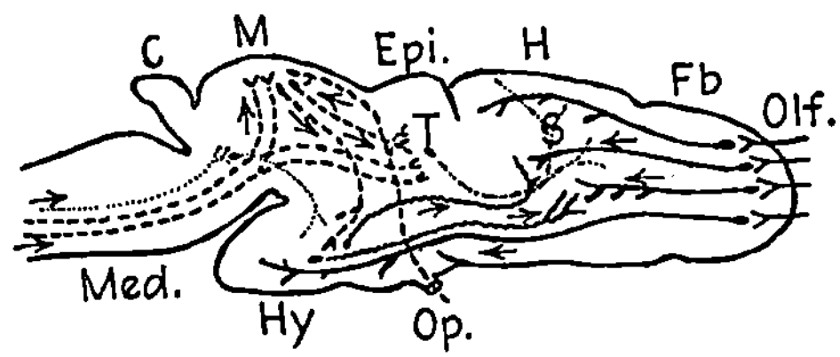

FIG. 50.—Major lines of conduction in the brain of the lower vertebrate, as based by Herrick on the salamander.
——————Olfactory impulses.
- - - - - - - - - -Impulses from optic, auditory, and somesthetic (*e.g.*, contact) receptors.
—··—··—··—··—Thalamic radiations to the cerebral hemisphere.
~~~~~~~~~Hypothalamic radiations to the cerebral hemisphere. *C*, cerebellum; *Epi.*, epithalamus; *Fb*, forebrain; *Hy*, hypothalamus; *M*, midbrain; *Med.*, medulla; *Olf.*, olfactory nerve; *S*, striatum; *T*, thalamus. (*Redrawn from Herrick*, 1926.)

and centers in the between-brain and midbrain (dominated by the visual system, connected with gustatory and tactual systems) there has developed a closer relationship than exists in the fish. (See Fig. 50.) Because of its connections with the thalamus, the lower forebrain of the amphibian may be said to have recovered somewhat from olfactory domination, but this relationship is held to the level of stereotyped control by the fact that from the thalamus the combined control works directly on the action system and is thus restricted in its scope. Despite this fact, the described advance in nervous structure makes for an improved cooperation of the exteroceptive systems in the control of behavior.

Although the dorsal forebrain of Amphibia has a complex structure, it is dominated by the olfactory system, and has limited connections with nonolfactory centers in the between-brain and midbrain. Hence this tissue is called *archipallium* (primitive olfac-
~~~~~~~~~

tory cortex). Neopallium (true cerebral cortex) is not found in the amphibian brain.

The Limitations and Advantages of Learning in Amphibia.—We have considered the status of sensory integration in the control of amphibian activities. The evidence in general supports the findings of the comparative neurologist, cited above, to the effect that the improved sensory integration functions only in the control of stereotyped (directly determined) activities. Let us see whether learning changes this condition in any important respect.

Learned Improvement in the Sensory Control of the Feeding Response. There is evidence that through learning the frog may somewhat improve the sensory control of its feeding response under given conditions. Schaeffer (1911) observed a frog which was presented with a hairy caterpillar, after having snapped up such prey on a few previous trials, on each of which it had promptly gotten rid of the mouthful. The frog now took two short hops toward the source of visual stimulation, then lowered its head and brought it close to the moving caterpillar. After a period of not more than 2 sec. the head jutted forward a little farther and the tongue was very slowly extended until it barely touched the hairy surface. Then the tongue was quickly withdrawn and there was no further response. This animal had learned to delay the snapping response in circumstances which at first were adequate for its arousal.

In this phenomenon, apparently a visual stimulus became more closely integrated through training with a tactual or gustatory stimulus. Thus on later trials the visual stimulus brought the response (ejection of object from mouth) which was at first produced only by the gustatory and tactual stimulation, instead of the response it first enforced (snapping).[1] The nature of the effective visual stimulus furnished by the caterpillar in this experiment was not ascertained, but very probably it was some peculiarity in movement. Other experimenters have found that such training is very uncertain, and also short-lived. In general, it may be said that the frog learns

[1] In the mouth of the amphibian there is a well-developed organ, the vomeronasal, or Jacobson's, organ, which is a chemoreceptor (p. 231). On the basis of its well-developed structure and the extensive midbrain connections correlated with it, we may postulate that the organ permits a gustatory effect so powerful (*e.g.*, when an acid-secreting ant is taken) as to force immediate expulsion of the object from the mouth. Apparently it is the response controlled by this system which comes partially under the control of visual stimulation in the described modification of feeding.

with great difficulty to inhibit its snapping response, as Abbott's (1882) experiment suggests. For instance, it appears impossible to train the animal not to snap at prey which moves behind glass.

Haecker (1912) succeeded in training the larval Amblystoma to react differently to a (moving) piece of meat and a (moving) piece of wood of the same size and dimensions. In the absence of an experimental control, the probability is that the animals learned to snap only when adequate olfactory stimuli accompanied the moving visual stimulus; a good illustration of a learned sensory integration. It is important for experimentation on this problem that Amphibia may differentiate rates of movement and sizes of objects in a general way, without training. Toads snap at small moving objects, but generally hop away from large ones; apparently on the primitive basis of intensity of stimulation. However, there is no evidence for the discrimination of form (outlines) of moving objects in normal feeding. Franz (1927) found that the toad and the frog did not snap at a motionless mealworm even after months of training with moving mealworms. The nature of the amphibian's responses to visual stimuli, and of modification in their sensory control, deserves a more thorough experimental investigation than it has received.

The Learning of a Series of Orienting Movements.—The toad is able to hide in a given place during the day, and to return to this place after nocturnal feeding expeditions which take him to habitually frequented places (*e.g.*, under a nearby street light). In a series of observations, many labeled toads and pond frogs were found able to return over a wide stretch of ground to their own pool from one to which they had been transferred, although many obstacles and other pools lay between (Breder, Breder and Redmond, 1927). Distances as great as one quarter mile were traversed in some cases.

Yerkes (1903) tested the green frog in a simple maze problem. Placed at *A* in the box (Fig. 51), the animal must avoid the successive blind alleys *B* and *C* (the latter closed by a glass plate) in order to reach a tank of water *T* at the opposite end. The animal received an electric shock when it entered a blind alley. At first the frog entered the blinds frequently on each trip, but the number of entrances decreased noticeably within a few trials. After 100 trials one subject passed directly through ten successive times without error.

During the training, "red" cards were at *R* and "white" cards at *W*. That the frog had learned to respond to these in making the first turn was shown by the persistence with which it entered *R* when the experimenter exchanged the cards. Apparently the first turn was learned mainly in dependence upon visual control. Also, the movement from the first section may have guided the animal into the second section of the apparatus, much as the fish in Church-

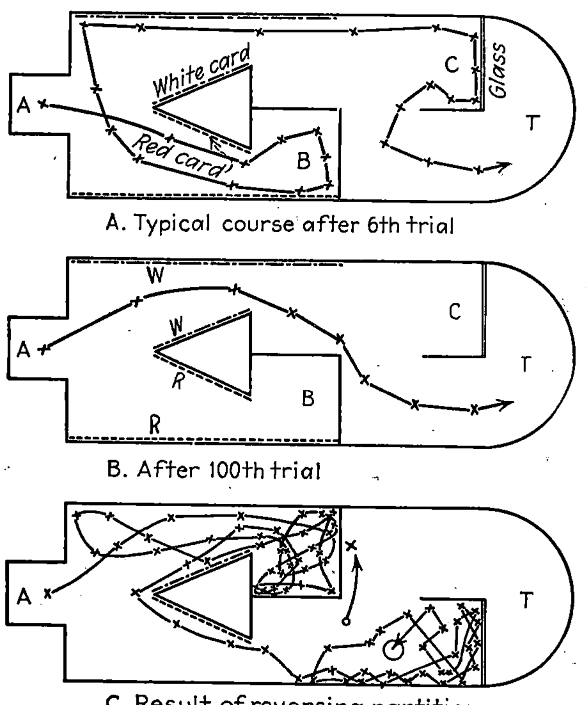

FIG. 51.—Typical records from Yerkes' test of learning in the frog. The solid line marks the general course taken by the subject, and the separate hops are marked by *x*'s. (*A*, starting point; *T*, tank.) (*Redrawn from Yerkes*, 1903.)

hill's experiment (pp. 193*f.*) was better able to learn its way through the second partition.

The stereotypy of the frog's habit was shown by the great difficulty one animal experienced in getting through the problem after Yerkes exchanged the partitions at the first "choice point" so that the learned turn became incorrect. Apparently the fixity of the habit contributed to good retention, since after 30 days a subject made few errors in its first series of ten trials, and no errors in its second series.

Comparisons Wait upon Further Analytical Studies of Learning.—The evidence on learning in this class is too fragmentary and incomplete to permit a comparative study. Analytical studies of learning are needed before the meaning of results such as the following will become clear. Burnett (1912) presented normal frogs with a simplified form of the Yerkes maze problem which was largely mastered after 20 trials. In contrast, decerebrated frogs showed no improvement after 100 trials in the same problem. The results prove that the frogs which lacked forebrain had great difficulty with the problem, but they do not show us why this was the case, since little is reported on the sensory control which such habits demand of the frog and the manner in which learned changes may come about.[1]

In addition to knowing that an animal is able to learn a problem, we must know what the problem demands of the subject and the process whereby the situation is learned, before the results become of much value. As will be shown in Chap. XVI, it is not always possible to compare learning ability on the basis of results from different problems. For instance, the ant has been shown capable of learning a maze with many more choice points (blind-alley junctions) than in any problem thus far mastered by a fish, amphibian, or reptile, but this does not justify the conclusion that all members of these vertebrate classes stand below the social insect in learning ability. The importance of sensory acuity, and the ease and quickness of movement, may not favor the lower vertebrates as compared with the insect. Hence it is not possible precisely to evaluate the amphibian's learning ability in comparison with that of other animals.

Recapitulation

In this class of vertebrates a shift is made from an aquatic to a terrestrial mode of life. As between aquatic and terrestrial *Amphibia*, or as between aquatic and terrestrial stages in the life

[1] The contradictory nature of reports from operative studies on lower vertebrates is partially accounted for by the lack of thorough behavior study both before and after the nervous destruction is made. The meaning which various experimenters attach to the expressions "spontaneity" and "initiative," which are freely used in comparing behavior, is far from being standardized. Further, Herrick (1934) has pointed out that post-mortem histological studies of the specimens are not adequate, since it is seldom certain whether any of the important nervous tissues adjacent to the forebrain (thalamic nuclei in particular) are injured along with the forebrain tissue.

history of the same individual, the latter presents significant advances in sensory, in motor, and particularly in nervous equipment. In the present chapter we have shown the importance of some features in equipment which contribute to the evident psychological superiority of the terrestrial amphibian over the fish. Certain general indications of a superior sensory integration are evident in the amphibian's adjustments to its environment. This improved sensory integration controls in a more plastic manner the stereotyped activities (attributable to the nature of bodily equipment) which constitute the repertoire of activities in this vertebrate class.

The improved cooperation among the sensory systems does not noticeably change the form of the directly determined activities but *does* make possible such advances as the following: (1) a more ready shift from one sensory field to another in controlling a given activity, which permits greater plasticity in the conditions under which the stereotyped responses may appear; (2) more than one exteroceptive field may be involved in controlling the *nature* of a given activity; (3) a closer relationship between two stereotyped activities, one of which follows the other by an interval of time; (4) the ability, under certain forceful sensory conditions, to inhibit or to delay the completion of a stereotyped act. This last advantage makes for greater complexity in the activity itself, and for certain important changes in the conditions under which it may be aroused. In consideration of these improvements, it may be said that while learning does not remove the amphibian from the level of stereotyped and specifically determined behavior, it makes possible an environment which is more complex in its psychological demands than is that of the fish.

Suggested Readings

Hargitt, C. W. 1912. Behavior and color changes of tree frogs. *Jour. Anim. Behav.*, **2**, 51–78.

Nicholas, J. S. 1922. The reactions of *Amblystoma tigrinum* to olfactory stimuli. *Jour. Exper. Zool.*, 35, 257–281.

Noble, G. K. 1931. *The biology of the Amphibia.* New York: McGraw-Hill. Chap. XVII.

Yerkes, R. M. 1903. The instincts, habits and reactions of the frog. *Psychol. Rev. Mono.*, 4, 579–638.

CHAPTER X

MODIFIABILITY AND THE PARTIAL ALTERATION OF DIRECTLY DETERMINED BEHAVIOR: CLASS REPTILIA[1]

INTRODUCTION

A general consideration of the reptile as an adjusting organism discloses variability and complexity in both the sensory and motor aspects of behavior. These features make the capacity for modification of behavior much more prominent here than it is in lower vertebrates. In this class, behavior attains a quality which is largely attributable to the fact that the sensory systems may either cooperate or may exchange dominance with facility according to conditions. In the case of a response which is controlled by vision, for instance, a change in circumstances may bring olfaction into dominance, quickly forcing vision into a subordinate role. Although in this case the change makes vision a secondary factor for the time, this sense may nevertheless influence the nature of the response in cooperation with the dominant sensory system. (It will be recalled that these characteristics were virtually absent from the sensory control of fish behavior.) In sensory control such advantages account for efficient integration, and for the variability in integration which we may call *plasticity;* on the motor side of behavior they make possible the animal's adaptation to an environment which is correspondingly more complex than that of lower vertebrate animals.

[1] The reptiles are characterized by a body covered with scales, frequently with bony plates, but lacking dermal glands; and by a larva with an amnion membrane. We shall deal with the following orders: *Testudinata,* which have a short stout body encased in a bony capsule, and jaws without teeth (*e.g.,* turtles); *Crocodilini,* which are lizardlike in form, but with jaws extended into a long snout ending in a single nostril, and with a thick leathery skin (*e.g.,* alligators); and *Squamata,* reptiles usually with horny epidermal scales (this layer of the skin being cast off periodically), and with movable quadrate bones in the skull (*e.g.,* snakes, lizards).

Directly Determined Activities and Reptilian Adaptation

In reptilian behavior there is a plenitude of activities the nature of which is more or less directly determined by the properties of

Table 14.—Reptiles

Receptor Equipment	Features of Sensitivity
Chemical. "Common chemical" sense endings absent from external surfaces.	Have sensitivity of internal head membranes to certain strong gases.
Taste buds not well developed in mouth cavity.	Gustation unimportant in snakes and alligators, more important in tortoise.
External nares lead to olfactory epithelium, generally well-developed.	Olfaction important in feeding of all. "Sniffing" of food is common.
Jacobson's organ in blind sac opening into mouth on each side.	Sensitivity apparently intermediate between gustation and olfaction.
Contact. Fair supply receptor cells in skin (except turtles). Tactile hairs often present near scales.	Contact basically important for but few (*e.g.*, snakes). Limited by presence of scales. Substratal vibrations figure.
Static. Well-developed semicircular apparatus.	Snakes hold head straightforward while crawling; turtle, alligator, keep head horizontal while swimming.
Auditory. Tympanum at end of external auditory canal. True cochlea with short basilar membrane.	Some (lizards) apparently have auditory sensitivity; others (snakes) apparently do not.
Light. "Camera eye" typical, lens accommodated by changes in thickness. Central area in retina, often a fovea. Excess of cones in most retinae, rods lacking in some. Eyeball rotates slightly.	Generally the dominant type of sensitivity, most important in food hunting, orientation. Mobile head on muscular neck compensates for lack of eye movement. Panoramic vision. Some evidence for color vision.

Conduction System

True cortex present in small amount in dorsal forebrain (Fig. 44). Corpora striata (basal forebrain) and thalamus become dominant parts of nervous system. Cerebellum well developed. 12 pairs cranial nerves.

Action Equipment

Limbs when present have five digits each, usually clawed. Snakes have unusually muscular body wall, special forms of movement. Fang-and-venom apparatus in snakes. Snapping response well developed in general. Blood temperature somewhat variable according to surroundings (poikilothermic). Breathe by lungs. Inflatable anterior parts in many. Many are oviparous (*e.g.*, turtles), laying eggs on ground, usually in simple excavation. Some viviparous: young born alive. Many have melanophores in skin: skin coloration controlled by brightness of background, temperature, humidity, through glandular secretions and sympathetic nervous system.

inherited equipment. Certain examples of this type of behavior may be considered in terms of their importance for the animal's adjustment to its environment.

The Movement of Newly Hatched Loggerhead Turtles to the Sea. Young loggerhead turtles, when they attain a vigor of activity which forces an increase in the scope of movement, dig from the nest in which they have developed and promptly pass toward the sea. Seldom do they fail in this act, even when the sea lies beyond the immediate range of vision. Hooker (1911) found that it is not the water content of the air which determines this movement, nor is it the "smell" of the sea. Of some importance is the fact that the young turtles move downward when placed on an incline, since the nest is usually located near the top of a rise which slopes toward the water. However, this cannot be the most important factor, since the young animal sets off toward the sea, even though the movement takes it up a slope. Hooker concluded that the animals tend to move toward any broad expanse of blue, and thus toward the sea.

Parker (1922) verified the suggestion of Hooker that the essential factor must be visual in nature, since turtles liberated on a wharf at night did not move toward the sea but did so in the daytime. However, the direction of the sun's rays does not control the movement, since the animals moved away from the sun's position when the sea lay in the opposite direction.

When Parker placed young turtles in a large flat-bottomed pan near a hedge beyond which lay the sea, they moved away from the hedge (thus away from the water) and toward an open field on the opposite side. They moved away from a clump of bushes although the water lay beyond this growth, but moved toward two well-separated bushes between which a stretch of clear sky showed. In all cases, when set down the young turtle soon raised its head, circled about, and then set off toward the open side. So Parker concluded that retinal sensitivity somehow determines the normal migration of the turtles by forcing a movement away from the direction in which objects interrupt the visual field and toward an open expanse of horizon (whether of water or of land) which affords a uniform retinal stimulation. In the normal environment the seaward side is most likely to afford uninterrupted retinal stimulation and a downward slope. Therefore it is the nature of the animal's image-forming eye, and of its retinal sensitivity, which mainly is responsible for the movement.

"Defensive Activities": Their Basis and Control in Behavior.—In vertebrates below reptiles, "defensive activities" are direct consequences of the animal's excitement. They appear in much the same form regardless of the fact that upon different exciting agents they have quite different effects. There is no reason for attributing them to "attempts to defend from" or "attempts to frighten away" the source of excitement, although not a few writers have implied that such is the case. Reptiles present a variety of activities of this nature.

Special Protective Devices in Snakes and Lizards.—The specific nature of many defensive responses in the reptile is attributable to a particular feature in the action system. For instance, in some snakes (*e.g.*, the hooded cobra) the anterior region of the body dilates as part of the animal's response to a disturbing stimulus. This device depends upon the presence of an elastic tracheal lining which is blown out by a sudden and considerable increase in pressure in the air tube of the excited animal, and which is supported by the spreading of the long anterior ribs. This type of response (see Fig. 52) is often called a "warning" if the snake is venomous as is the cobra, but "mimicry" or "bluff" if the same response occurs in a nonvenomous animal such as the American hog-nosed snake. Some readers at least would take this to imply that the harmless snake "knows how to defend himself by imitating the response of the venomous snake." However, as concerns an understanding of the act in the individual snake, such descriptions are very misleading and fallacious. The matter should be put somewhat as follows: "Snakes which possess *this* distinctive tracheal and anterior body structure behave in *this* manner when strongly excited, whether they happen to be venomous or not."[1]

FIG. 52.—The frilled lizard, *Chlamydosaurus kingi*. The more excited the animal becomes, the higher its frill is raised and the wider its mouth is opened. (*From Boulenger's Reptiles and Batrachians by permission of E. P. Dutton & Co.*)

[1] As we have pointed out (Chap. VI), a clear distinction should be made between the ontogenetic (or individual) behavior problem and the phylogenetic problem (the evolution of characteristic species structures).

We may consider further cases. In many lizards the tail is weakened by the presence of a septum in the vertebrae, and separates readily from the body when it is grasped by a would-be captor. However, the evidence does not suggest that the lizard learns either to depend upon or to control this very fortunate part of its equipment. Likewise, the buzz of the excited rattlesnake is not an "attempt to warn trespassers," as popular writers are fond of stating, but is one direct accompaniment of his excitement. Possession of the "rattle" is an incidental consequence of the periodic skin-shedding process. The vibration of the tail, incidentally, also accompanies excitement in numerous species of snake that lack the rattle.

"Tonic Immobility" versus "Death Feigning."—Lizards, other reptiles, and many animals in other vertebrate and invertebrate groups typically become immobilized for a time when sufficiently excited by a suddenly presented visual stimulus or a repeated tactual stimulus. This response is frequently called *death feigning*, or by the fancier name *letisimulation* (which means the same thing); often with the implication that the animal "plays dead in order to escape its enemy." Some believe that certain reptiles are able to abandon the immobile condition as soon as the exciting object moves away.

Hoagland (1928) found that the chameleon *Anolis*, which one may immobilize by turning on its back and pressing with the hands against both sides of the thorax, may be somewhat influenced by external stimuli when emerging from the state. However, the duration of the immobility is definitely dependent upon the nature of the physiological change induced by the stimulus. In *Anolis* the period of immobility was found to vary in length according to the temperature, in the sense of the Arrhenius equation (p. 106). This indicates that the immobile state is based upon a chemical reaction of fixed duration. The facts show that the animal may emerge from the state only when the physiological processes controlled by this chemical reaction have run their course.

Hoagland fitted to the facts the assumption that stimuli such as maintained pressure (summation effect) or intense shocks cause the release into the blood stream of a chemical substance (possibly adrenin) which prevents the action of higher nervous centers but which permits lower centers to pass impulses to muscles which maintain a condition of contraction (tonicity) as long as the chemical acts. If the inducing stimulus is repeated when the animal is coming out of this condition (as when the molesting animal is seen once more), re-immobilization occurs; but in the absence of further stimulation of this kind the animal is free to get up and run off. There is no evidence that in the condition of "tonic immobility" (*i.e.*, while it is drugged) the reptile watches its aggressor and that it shakes off its stupor and scurries away as soon as the enemy withdraws. It is closer to the facts to say that the withdrawal of

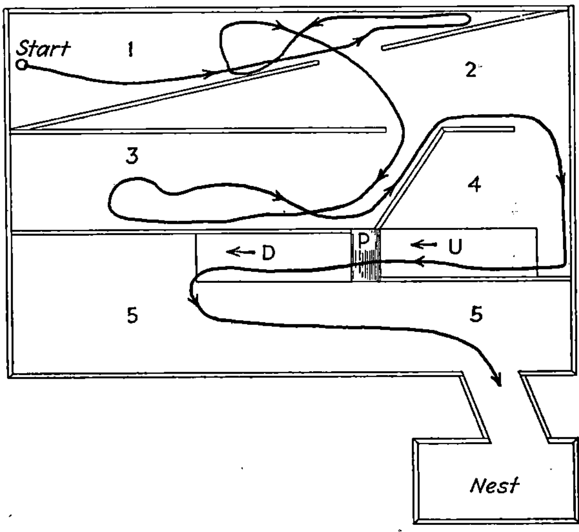

FIG. 53.—Yerkes' turtle maze, showing the course of the animal on its fifth trip. Slopes of the incline are indicated. (*Redrawn from Yerkes*, 1901.)

the source of disturbance prevents the induction of further sieges of immobilization when the first one has run its course.

SIGNIFICANT CHARACTERISTICS OF REPTILIAN ADJUSTMENT

A Suggestive Feature of Maze Learning.—Some reptiles cover a considerable amount of ground in their food-hunting and other activities, which suggests that learning must be of importance in their orientation. Yerkes (1901) tested the ability of the speckled turtle to learn maze problems. The subject was introduced into corner 1 of the apparatus (Fig. 53) and could escape only by reaching the diagonally opposite corner from which an exit led to a darkened nest containing damp grass. The box was divided into compartments by partitions 10 in. in height.

On its first trip through this situation the turtle wandered extensively, taking those turns which were the easiest. After the second trial there was little of the crawling against walls and the stopping in corners which was observed at the outset. Improvement on further trials may be represented by the subject's time record (Table 15). The major errors (returning repeatedly into certain

TABLE 15.—REPRESENTATIVE TIME RECORD OF A TURTLE IN LEARNING YERKES' PROBLEM

Trial	Time
1	1 hr. 35 min.
5	16 min.
10	4 min.
20	4 min. 5 sec.
30	3 min. 20 sec.
40	4 min. 20 sec.
50	4 min. 10 sec.

blind alleys) were largely eliminated during the first ten trials. The turtle's pace quickened and excess wandering soon disappeared. After the twentieth trial the course was mainly a direct one, and errors were infrequently made. Yerkes reported that the learning performance of the turtle was very superior to that of the frog in a simpler problem.

An interesting change in behavior became apparent after the turtle had practically mastered the runways. Incline *U* in the apparatus led from compartment 4 to a raised platform *P*, from which incline *D* led down into compartment 5. The animal first learned to run up incline *U*, then slide all of the way to the bottom of incline *D*, and turn toward the nest in compartment 5. Soon it began to turn toward the nest before reaching the bottom of *D*, toppling or jumping from the edge of the ramp to the floor below. Finally the turn was made even before the turtle had started down incline *D*; since on the platform it would turn toward the nest, then would fall or jump over the side, right itself, and run into the nest. Passing down the incline had brought the turtle to the nest and had not obstructed its passage as had blind alleys in the maze itself, and yet the animal learned an improvement upon this behavior also.

An Important Advance in Nervous Equipment.—In lower vertebrates the pattern of the conduction system makes for the dominance of stereotypy in behavior. In the reptile there are indications of a partial escape from this condition, as we have suggested. The improvement in nervous structure upon which this is apparently based is most evident in the forebrain and thalamus (*cf.* p. 212), the parts more directly related to the sensory control of adjustment.

In Amphibia, areas in dorsal forebrain include archipallium (primitive olfactory cortex), nervous tissue which represents a forecast of true cortex and a partial removal of the forebrain from olfactory domination. In one region of the reptilian forebrain there is a thin layer of neurones, arranged near the outer surface, the existence of which makes possible more advanced and extensive integrations among the sensory fields as the animal's experience grows. Into this center all of the important sensory systems discharge energy (Crosby, 1917), and through it they are able to control behavior in a new and more plastic manner. No one sensory system dominates this correlation center, the *neopallium* (true cortex, Fig. 44*d*), and hence plastic patterns of sensory control represent its proper function. In other words, its use is not specifically determined by its spatial relationship with other nervous centers (the primitive determinant of function in a nervous center) but depends upon dynamic patterns which are set up within it according to the experience of the animal. It is very suggestive that *this improvement has evolved on the receptive side of the nerve arc (in connection with the closer relationship of sensory systems) rather than on the motor side.*

Sensory Discrimination and the Appearance of New Behavior. *Learned Differentiation of Relatively Slight Spatial Variations.*—The reptile shows an ability to develop new movements in connection with crucial sensory differences in its environment. This rests first of all upon improvements in the receptor which increase visual acuity: a retina which has a central area or *fovea,* and a lens which may be changed in shape by muscles, thereby accommodating the eye to the distance of an object. The following experiment shows how the turtle may learn to respond distinctively in dependence upon a relatively slight difference in the spatial characteristics of two visual stimuli.

Casteel (1911) studied visual discrimination in the painted turtle. From an entrance compartment the subject passed up a sand-covered incline and through one of the two doors at the top (Fig. 54). Each of these doors was cut in the front of a tall box, a cardboard with alternate black and white stripes. In the first experiment the stripes differed in width and in direction on the two sides.

Upon passing through one door (*e.g.*, that in the box with vertical stripes) the turtle found meat, but behind the other door it received an electric shock. (The two boxes—with the cardboards—were exchanged in position from trial to trial.)

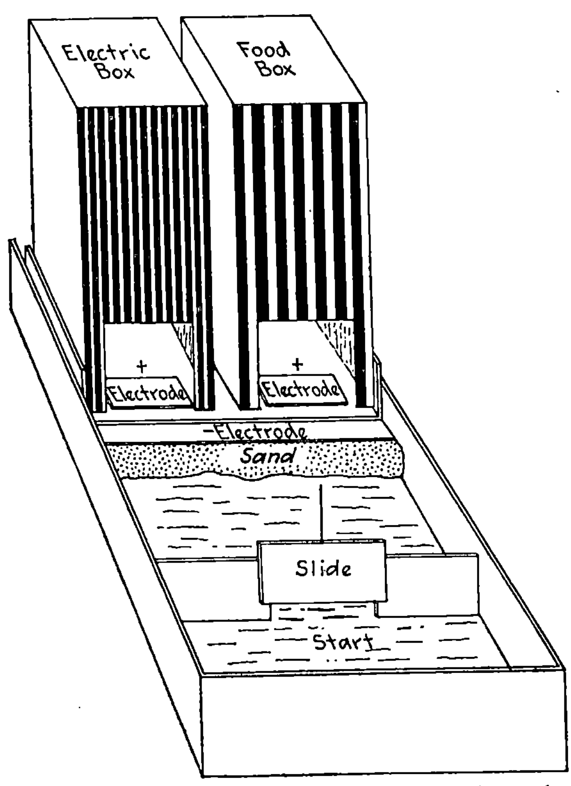

FIG. 54.—Casteel's apparatus for testing visual acuity in the turtle (see text). (*Modified from Casteel*, 1911.)

One turtle learned within 150 trials always to turn toward the card with vertical lines, as against horizontal lines. This animal succeeded in learning the vertical-horizontal discrimination when the lines were only 2 mm. in width. Then four turtles were presented with cards on which the lines ran in the same direction but differed in width. The above mentioned subject mastered 8 mm. versus 1 mm. in 170 trials, and finally reached 90 per cent success in discriminating 3-mm. from 2-mm. lines. This last combination was more difficult than the preceding ones (*e.g.*, 4 mm. versus 2 mm.), and appears to have been close to the turtle's "difference threshold."

One purpose of this experiment was to test the visual acuity of the turtle by discovering the least difference in the width of lines (the "difference threshold") which could be responded to overtly by the animal. If x (Fig. 55) is the least difference in strip width

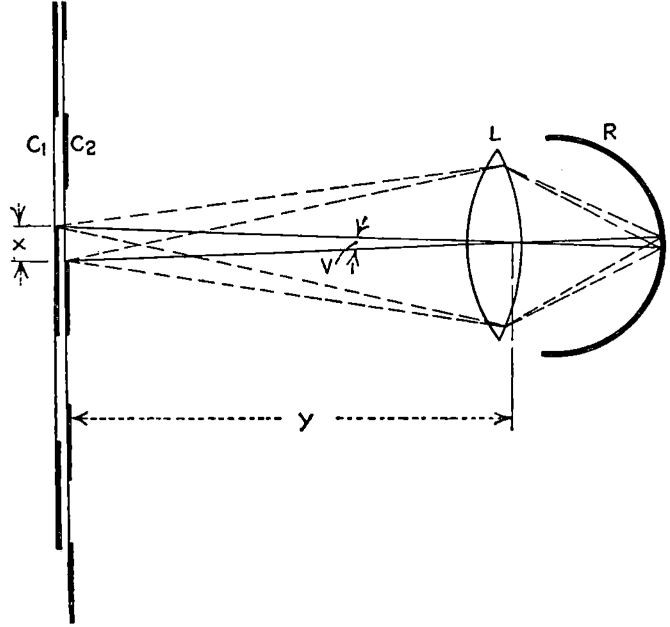

FIG. 55.—Schematic illustration of the visual angle, a representation of visual acuity. V, visual angle; R, retina; L, lens; y, distance from eye to discriminated cards; x, least detectable difference between strip widths; C_1, C_2, the discriminated cards.

which can be responded to (*i.e.*, below which the strips are reacted to as though they were the same in width), and if y is the distance of the animal's eye from the cards when the discrimination is made, the *visual angle V* may be calculated. The visual angle represents the acuity of the animal's receptor in comparison with that of other animals as determined on the same type of problem.[1] Casteel's results show that a visual difference may be slight, and still provide the basis for a differential response by the turtle.

[1] However, this value cannot be determined from Casteel's results, since the distance y varied from trial to trial in the experiment.

Such experiments tell us how readily the animal may learn a differential response to stimuli which act differently upon its receptor. Acuity depends upon the efficiency of the receptor, and sensory discrimination is limited by the animal's acuity as well as by its ability to learn. The results suggest that visual acuity may be of considerable significance for a reptile's adjustment to its natural environment.

As Herrick (1924) has said, the psychological standing of an animal "is measured not by what the organism can receive from the environment through its sense organs, but by the capacity for diversity of adaptive responses and adjustments." This capacity advances in step with increasingly efficient nervous organization of sensory impulses and with the dependent ability to respond with movement which is fitted to circumstances. In this respect, as well as in the range of its responses and in its ability to eliminate excessive movements in improving a learned act, the reptile is definitely superior to the fish and the amphibian.

Feeding Behavior and Reptilian Psychology.—Although there is available no experimental analysis of the manner in which the superior qualifications suggested above operate to lead reptilian behavior farther from the stereotyped pattern of the lower vertebrates, some results concerning food taking suggest the need for investigation of this problem.

As some writers have pointed out, reptiles are capable of "actively hunting" their prey, but the meaning of this expression has not been made clear. The greater "alertness" which is granted most reptiles over lower vertebrates is not merely a matter of agility, or of quickness in the sense that the fish is quick. The psychological superiority implied by such descriptions is suggested in the mode of life of a reptile such as the alligator (Kellogg, 1929), marked by overland trips on the trails between its haunts and by "lying in wait for prey" by the river bank. The lizard takes its prey with considerable facility, adjusts readily to differences in the position of prey, and may overcome obstacles in getting to it. It is not uncommon in the tropics to find a swarm of marauding army ants accompanied by lizards, which remain away from the bites and stings of the ants until an object of prey (*e.g.*, a grasshopper) is driven from cover by the swarm; then dart among the ants, quickly seize the booty, and move out with lightning rapidity. A given lizard may remain with the ants until he is satiated, moving along as the swarm advances. This type of phenomenon suggests the assistance of learning, and involves a more ready adjustment to varied sensory circumstances than is to be observed in amphibian feeding. The toad's approach to the light under which he feeds is a response which is forced from the beginning and inevitably produced under given sensory conditions, but the food response of the reptile appears susceptible to control by sensory conditions which are not at first effective, and thus lends itself to more extensive modification.

Some observations by Honigmann (1921) are suggestive. He presented turtles of various species with two equidistant white bags, one of which held meat and the other sand. The bag containing sand was not reacted to in any case, but the turtles would all bite at the bag which contained meat when they first came close to it. This, of course, was a direct response to chemical stimulation. However, after a few snaps and in some cases only two or three, all but a few of the animals failed to react further to the bag.

In another experiment, Honigmann brought out two sealed glass beakers, one of them empty, the other containing fish meat, the regular laboratory food of the turtles. Practically all of the animals struck repeatedly at the glass containing meat, and struggled against its side, but there were no such responses to the empty glass. One turtle bit seven times at the meat beaker, then stopped; but when the beaker was removed to another place further responses were given. Other turtles struck only a few times at the beaker, and a few "noticed the food behind the glass, but did not bite."

These results suggest the prompt inhibition of a response which was at first readily elicited by one type of exteroceptive stimulus (chemical, in the case of the bags) or by another type (visual, in the case of the beakers). This represents a significant modification of the feeding response. A repetition of the experiments, with better control of stimulus presentation and of hunger, would be valuable.

The Alteration of Directly Determined Activities in the Food-getting Pattern.—As we have suggested, in the capture of food by reptiles there are definite indications of a modification according to circumstances. Although the typical form of reptilian food capture is fundamentally dictated by features of inherited equipment, the role of learning must be taken into account. We may consider this matter in connection with evidence on the food getting of certain snakes.

Snakes feed irregularly, some of them at long intervals, and they actively forage only when the last meal has been digested. Apparently it is stomach contractions (Patterson, 1933) set up at this time which arouse the snake to begin hunting. The nature of the prey differs widely among the groups, mainly in dependence upon such factors as size and aggressiveness of the snake concerned, and the snake's typical environment. This foraging activity may be properly called "hunting" largely because it is so frequently characterized by activity in those places where the prey is likely to be found. For instance, tree snakes and others which are able to climb trees display a noteworthy ability to find birds' nests from which they take eggs. The "cat snake" (*Tarbophis phallax*) of southeastern Europe owes its common name to the manner in which it "stalks" the lizards upon which it feeds (Boulenger, 1914). Baumann (1929) reports that in nature a viper may hunt for the burrows of mice, or after tracking a mouse to its burrow may remain in the vicinity (a response to chemical stimuli) until the prey appears. It is certain that accident is not a major factor in the finding of prey by reptiles.

The Sensory Control of the Strike.—Most snakes approach and "investigate" things which stimulate them visually. They respond in this manner to moving as well as to motionless objects. Sudden movement of an object is most effective in causing the snake to approach and take the erect striking position before it. Wiedemann (1932) reported a field observation on an adder that approached within 10 cm. of a sitting frog and then suddenly struck at it, suggesting that the breathing movements of the frog may have held the snake's attention during the approach. Usually, unless the object appears suddenly and moves rapidly, the snake is likely to raise itself and to pass its darting tongue over the object before the striking response is given.

The visual circumstances play a major part in determining whether or not the snake strikes, and how promptly it strikes. This appears true even in snakes for which vision is not the outstanding type of sensitivity. The act of striking generally is not elicited if the moving object is small and can be taken without difficulty, *e.g.*, objects such as newborn mice are swallowed without being struck. This suggests the influence of learning. Baumann (1929) observed that the viper always approaches a living mouse quickly and strikes promptly, but that a dead mouse is slowly approached and leisurely eaten without the strike being elicited.

In the European viper, although vision is not the dominant sense modality in food hunting, visual stimulation may figure in the original location of the prey. However, when the prey is approached and the erect position assumed before it, the darting of the forked tongue over the object is a more evident preliminary to the strike than it is in the dominantly "visual" snakes. The tongue is not of importance for the tactual stimuli it receives (since it seldom actually touches the object) but for the fact that chemical substances volatilize to it and are then carried in solution to the sensitive Jacobson's organ, a

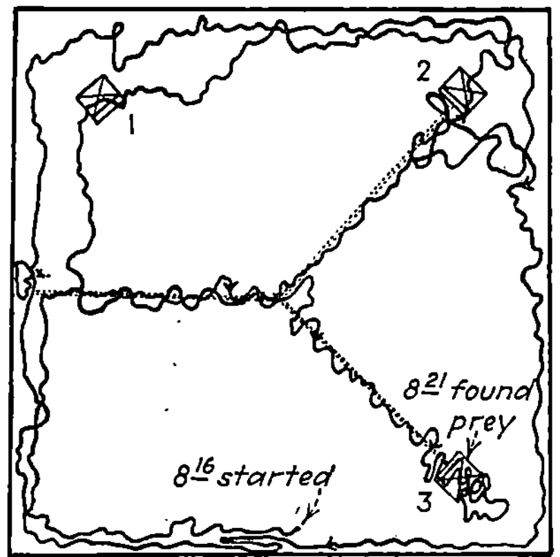

Fig. 56.—Tracking experiment with viper. A viper was permitted to bite and venomize a mouse, which was then dragged over the sand floor of the area as shown by the double broken lines. At 8:16 the snake was released, and took the course shown by the solid line. At *x* it responded to the chemical trail while following the side wall. At 8:21 it located the prey in box 3. The absence of a trail from box 1 to the center constitutes a control. (*Redrawn from Baumann*, 1929.)

chemoreceptor in the mouth (see note, p. 213). The short period of tongue play over the object that precedes the strike in most snakes permits the reception of chemical stimuli which check upon the visual, in controlling the striking response. A certain amount of tongue play before the strike is observed even in the predominantly visual back-fanged snakes (*Opisthoglypha*).

A Special Modification of Food Capture in the Viper, and Its Basis. The behavior of the snake after it has struck the prey shows features that are even more significant. The viper, after having struck, opens its jaws and draws back, and the released prey runs away. But the viper has not finished with its catch: a very hungry individual pursues it at once, but a less hungry individual may remain on the spot for a time before it starts on the pursuit.

Bauman (1929) tested the trailing behavior of the viper by dragging a freshly bitten (venomized) mouse across the sand floor of a terrarium as shown in Fig. 56, and then introducing a viper which had struck prey a few minutes before. Such an individual showed "searching movements" which were much more active than were those of a snake which had not struck prey just before the test; it moved with a characteristic to-and-fro bending of the anterior part of the body (not a feature of normal crawling) and there was a constant darting play of the tongue close to the ground. In this test, as a rule the snake soon hit upon one of the trails in the course of its movements about the experimental area, and followed this trail with energetic meandering movements and with marked tongue play. Eighty-six per cent of the snakes which had struck prey before the test finally located and ate the mouse which had been hidden at the end of one of the experimental trails (Fig. 56). Snakes which had not struck prey just before the test also followed the trail in many cases, but less than half of them located the mouse.

Kahman (1932) proved that this trail-following depends upon the sensitivity of the Jacobson's organ. Cutting away this receptor, or the forked tongue, which carries chemicals (by a pumping action) to it, removed almost completely the ability to trail prey. Bauman reported that a mouse which has been struck is more efficiently tracked than is an uninjured mouse. This indicates that the addition of venom to the chemical complex increases its stimulative value. He also found that a snake which had just struck prey would take up a trail and follow it more actively and efficiently than would a snake which had not struck prey for some time. This may mean that the act of striking has increased the snake's excitability, and that in this condition it responds most readily to the chemical which it has vividly encountered just before the test.

Conclusion

The activity pattern which we have just inspected may be taken to represent others in the reptilian behavior repertoire (mating, nest building, and the like). It is not representative because of its specific form, but rather because of the fact that its matrix (mainly furnished by the determining effect of inherited mechanisms) is overcast by the capacity for modification according to circumstances.

It cannot be said that all similarities in the food getting of different individuals in the same species may be traced directly to the influence of inherited mechanisms. When the characteristic equipment of the species (*e.g.*, those features which make for a given temperature optimum) causes the members to frequent much the same environment, it is to be expected that the learned adjustments of the different individuals will be very similar. For this reason, "universality," or presence of a behavior trait throughout the animal group, is not a feature which tells us dependably whether or not a given act is natively determined in the main.

In reptilian behavior, there are decided indications of a trend from "direct determination" toward the modification of stereotyped behavior in accordance with circumstances. Learning plays a much more recognizable part in shaping behavior in this group than is true of the Amphibia. Acts such as striking at prey may be reasonably attributed in the snake to the possession of the fang-and-venom apparatus and glandular characters which make for easy arousal of this apparatus when intense stimulation also elicits the natively determined lunge. However, the conditions under which this response is called into play appear to influence constant variations in its form and to mold it to the "demands" of the environment.

It is therefore by no means certain that the tendency of the viper and of other snakes to drop the booty after striking it, and to wait for a time before taking up the trail, can be traced directly to the influence of inherited mechanisms. The viper represents snakes that possess venom which kills prey within a few minutes. These snakes also have a receptor (Jacobson's organ) with accessory apparatus (the tongue) which permits the finding of prey by responding continuously to a stimulus (the "booty chemical") which accompanied the striking act. Such snakes might well gradually work into the habit (learned act) of tracking after a few occasional escapes of a prey object which is more easily eaten when dead than when still alive and struggling. It is quite probable that this fact causes the viper to learn to respond to the struggles of the prey by releasing it, instead of tightening its hold.

In definite contrast, snakes which are predominantly visual and not very sensitive chemically (*e.g.*, back-fanged snakes such as the coral snake) are the ones which hang to their prey and swallow it at once after having struck it, despite its struggles. In case the prey happens to escape, the hunt is generally fruitless for these snakes.

In all probability the shortcoming in chemical sensitivity is mainly responsible for the failure of such snakes to learn this modification of their feeding act.

Thus we find the reptiles capable of a type of learning which is wider in its scope and more plastic than is the learning of which lower vertebrates are capable. Modifiability therefore plays a more important part in shaping reptilian behavior than is the case in lower classes, and correspondingly this capacity reduces the dominance of "direct determination." This advancement becomes much more evident in the behavior of mammals.

Suggested Readings

Barbour, T. 1934. (rev. ed.) *Reptiles and Amphibia, their habits and adaptations.* Boston: Houghton Mifflin. Chap. II.

Boulenger, E. G. 1914. *Reptiles and batrachians.* New York: Dutton. Chap. IV.

Parker, G. H. 1922. The crawling of young loggerhead turtles toward the sea. *Jour. Exper. Zool.*, 36, 323–331.

Reese, A. M. 1915. *The alligator and its allies.* New York: Putnam. Pp. 358. Chap. I.

CHAPTER XI

MODIFIABILITY IN A COMPLEX REPERTOIRE OF DIRECTLY DETERMINED ACTIVITIES: CLASS AVES[1]

INTRODUCTION

Birds possess an extensive repertoire of highly stereotyped activities, the content of which varies greatly from species to species in correspondence with a variety of heritable structures. A very specialized action equipment inevitably limits and at the same time guides the occurrence of behavior in the particular environment of a given species, and is thus basic in the production of characteristic modes of behavior. For instance, the type of food taken and the manner of food capture are particularly dependent upon the relative power of flight, the structure of bill and foot, and upon visual acuity. The possession of very light bones and a rapid and efficient respiration (based upon a high rate of metabolism) insures the occurrence of flying in most birds which have strong wings. Further, the form of wing (and tail) in a given species directly influences the nature of flight activities. The capacity for flight greatly increases the scope of behavior; and in relation to this it is a matter of great importance that in governing bird behavior visual sensitivity is the dominant modality. This fact depends not only upon the excellence of the receptor but also upon an extensive specialization of the visual conduction system.

In the bird there appears a new and important physiological character, the maintenance of a constant temperature, or *homoio-*

[1] Birds are homoiothermic (warm-blooded) vertebrates with feathers, and usually with an adaptation of the forelimbs which permits flight. It is convenient to organize the many orders of birds in two major groups (Warden, Jenkins, and Warner, 1934): the division *Ratitae*, large, flightless birds such as the ostrich and cassowary; and the division *Carinatae*, containing passerine birds such as lark and sparrow, fowl-like birds (*e.g.*, domestic fowl), surface-swimming birds such as duck and swan, ocean flyers such as tern and gull, divers such as loon and penguin, waders such as heron and crane, and birds of prey such as the eagle and owl.

thermism. Changes in the environmental temperature act upon sensitive skin receptors, and within limits physiological adjustment may occur; but beyond these limits the bird is forced to migrate,

TABLE 16.—BIRDS

RECEPTOR EQUIPMENT

Chemical. Free nerve endings in mucous membranes of mouth, nose, throat.	Sensitivity to strong gases, similar to "common chemical" sensitivity in fishes.
Taste buds fewer in number than in lower vertebrates, as rule. Horny tongue partly responsible.	Gustation poorly developed in most. Of some importance to grain and insect feeders, in discarding mouthed food.
Scanty supply nasal epithelium, small olfactory bulbs in most.	Poorly developed as rule, important in few (*e.g.*, some marine birds, carrion feeders).
Contact. Specialized free nerve endings in skin, plentiful at bases of feathers, base of bill, on head. Specialized temperature endings.	Orientation to air currents during flight (equilibrium), etc. Bill contact important for feeding in many passerines. Delicately sensitive to external temperatures.
Static. Well-developed semicircular apparatus. Forced movements of head, body, when apparatus removed from one side.	Controls head position during walking, flying. Bird cannot fly when both organs removed. Maintains tonus during flight, cooperates with vision in postural control.
Auditory. True cochlea, with less than one turn. Short basilar membrane, organ of Corti present. No external pinna; highly mobile head.	Pitch, intensity, discrimination habits learned; sound localization accurate. Auditory control of singing; indirect means of unifying social group.
Visual. Rapid accommodation by changes in lens shape, contraction striated musculature. Fovea well-marked, sometimes two or three present. Rods and cones in retina of most; nocturnal birds only rods.	Panoramic vision (complete crossing of fibers in each optic nerve). Limited eyeball movement compensated by flexible neck. Visual acuity much greater than lower vertebrates. Evidence for color vision good.

CONDUCTION SYSTEM

Short, broad, rounded brain. Olfactory lobes small in most species. Corpora striata, optic lobes, thalamus, very well developed, account for decided dominance of visual control in stereotyped activities. Cerebellum well developed.

ACTION EQUIPMENT

Close relation between form of bill, feet, and feeding; between form of wings, tail, and relation of flight to other activities. Webbed feet—swimming; clawed feet—digging during feeding, catching prey on the wing, perching, etc. Syrinx with vocal cords at base of windpipe in most birds (absent in ostrich), characterizing male. Homoiothermic: blood temperature not truly variable with external temperature. Oviparous: eggs develop outside mother's body.

or to alter its manner of living if it remains in the locality. This makes the bird a much more active animal than is the reptile, and constitutes a second factor which greatly broadens its environment.

In addition to highly specialized sensory, nervous, and action equipment, behavior is elaborately influenced by visceral factors.

The glandular secretions are basically involved in the appearance of activities which distinguish the sexes. Periodic internal changes are very important for seasonal variations in behavior which determine the general pattern of bird life. The state of excitement of the bird and its pattern of viscerally determined activities (emotions) vary delicately in dependence upon external stimulation, as Rouse (1905) and others have shown. The patterning of this emotional activity as it may direct behavior toward certain objects at given times (*e.g.*, fixation of the young during incubation) makes motivation a much more important factor in bird life than in that of lower vertebrates. While these influences bear upon behavior in a complex manner, a sufficient amount of experimental work has been performed to make their nature and mechanism far from obscure.

Vision as a Dominating Factor in Behavior

The very extensive action repertoire of the bird is predominantly controlled by vision. The manner in which birds such as eagles and gulls swoop directly upon their prey from a height, with seldom a miss, and the manner in which flight is dexterously controlled with reference to objects, speaks for a great precision in visual control.

The Receptor and Visual Dominance.—First of all, the bird has an exceedingly well-developed eye. The retina always has at least one central area or fovea, and in many birds there are two such areas (as in ducks, one horizontal and one central fovea) or even three (as in the swallow). The great precision of this receptor is mainly attributable to accessory apparatus which makes possible a very rapid accommodation to the distance of an object. In particular it is important that the shape of the lens may be quickly changed by means of striated muscles. The ability to shift from "far" to "near" accommodation with facility and rapidity vastly increases the scope and variety of behavior in birds which are capable of flight.

Panoramic vision (see note, p. 183) is characteristic of birds. With the eyes on the sides of the head, most birds have an extensive visual field. Accuracy of response according to the relative distance of objects is made possible by the quick shifting of the head on the mobile neck (so that first one and then the other eye fixates the object in rapid succession), by sensory (kinesthetic) cues from the delicately adjustable accommodation apparatus, and by differential responses to size and to brightness.

Size discrimination was demonstrated by Breed (1911). Two chicks learned (in about 80 trials) to pass through the smaller of two openings to the living cage. The smaller opening then was made brighter than the larger without disrupting the habit, showing that the discrimination did not depend upon brightness differences. Bingham's (1913) chicks discriminated a circle of 6 cm. from one of 5 cm. diameter after extensive training, and also learned to discriminate between a circle and an upright triangle of equal area. Since the latter habit broke down when the triangle was inverted after learning, Bingham concluded that such habits in the bird depend upon a very specific retinal effect given by the "shape" of the object in a constant position.

The accessory equipment of the eyes and their position on the head account for a high degree of sensitivity to movement. The side to side turning of the head as the bird specifically responds (attends) to the source of a slight movement within its wide visual field suggests that while one eye is accommodated so that the object stimulates its central area, response to stimuli acting upon the other eye must be inhibited.

Great precision in the visual control of movement, as in the pecking of the adult bird or the landing of the skillful flyer, speaks for a high degree of visual acuity. The work of Johnson (1914) showed the chick markedly inferior to monkey and human (Chap. XIII), but for the pigeon Gundlach (1933) found evidence that the bird discriminated lines subtending an angle of 29 sec., as against a difference threshold of 50 sec. for the human subject. However, sensory acuity depends upon many factors and the meaning of such comparisons is not too certain. The evidence shows that birds have great visual acuity, but a more satisfactory technique must be devised if the problem of acuity value is to be investigated for many species, as is desirable.

Nervous Structure and Visual Dominance.—The importance of vision in bird behavior depends to a considerable extent upon a very specialized conduction apparatus. The cerebrum of the bird is much larger with respect to the size of the remainder of the brain than is that of any lower vertebrate (Fig. 44*e*), *but* this fact is attributable to a great increase in the size and fiber complexity of the basal part (striatal area) and the lateral wall of the forebrain. Neopallium is minimized in the bird brain (Herrick, 1924). Comparative neurology has shown that *the extent to which stereotyped,* *directly determined activities bulk in the repertoire of a vertebrate corresponds directly to the development of these basal areas of the cerebrum.*

In the bird, vision is the sensory field which principally supplies impulses to this dominant part of the brain and hence almost from the beginning takes the lead in behavior control. This, together with the extensive development of those parts of the thalamus through which visceral sensitivity discharges into the anterior part of the nerve tube, accounts for the two most basic properties of bird behavior: (1) the refinement of specific visual control and (2) the strong influence of motivation. In our further treatment this point will find no lack of confirmation.

Olfaction Minimized in Importance Except in a Few Bird Species. In decided contrast to vision, olfactory sensitivity appears of negligible importance for the behavior of most birds. Although the vulture has been considered an exception among land birds, in the experiments of Audubon (1835, see Watson, 1914) vultures passing overhead did not find meat and carrion covered with brush, although they flew down to a stuffed deer skin on the ground, and to the painting of a sheep. However, Chapman's (1929) tests with the turkey buzzard, a close relative of the vulture, show that under appropriate conditions olfaction may figure prominently in the activities of these birds. Buzzards repeatedly visited a shack at times when putrid meat was within, and when such booty was hidden beneath grass these birds soon alighted at a distance from it on the *leeward* side, walked about, found the spot, and tore away the covering to seize the morsels.

Homing, an Activity Made Possible by Extensive Visual Control. A great power of flight and keen vision permit most birds to roam over considerable areas. In no other animal are the natural territorial limits of the environment so wide, a fact for which the foregoing advantages are mainly responsible. For a number of reasons, homing is more prominent in the life of certain birds than of others. The greatest amount of attention has been given to the abilities of the homing pigeon.

The Training of a Homing Pigeon.—The motivation factor is basic in homing. In training the homing pigeon, the bird is first thoroughly attached to his cote through having received food and having mated there. Although the trained bird may roam from the place in which his mate and nest are located, soon he is impelled to return by the disturbance which arises through absence of the emotionally attached objects. Hunger is also important.

In connection with the second problem of homing, the nature of the stimuli which control the return, let us limit our attention first to the pigeon, since that bird is best known in this respect. Fanciers know that a pigeon must be trained to "home" and that the work should begin during youth, that is, after the bird has become well habituated to his cote. Representatively, the birds are started in the spring when about five months old and are first permitted to fly about in the vicinity of their cote. Then a given bird is liberated repeatedly at short distances from the home site, always in the same direction at first, and is permitted to fly back each time. Soon the distance is gradually increased; for example, the bird may be successively liberated at 3 miles, 8, 15, 25, 40, 60, 90, 130, 200, 300, 400 miles, and then at greater distances. Some pigeons do not make good homers, but even a good learner may be spoiled for homing if the distances are increased too rapidly at first.

How Orientation Is Controlled on the Return.—What does the bird learn during training that makes possible its later return from new territory and over great distances? Schneider (1905) reached the conclusion that the bird learns the grouping of prominent landmarks (houses, trees, hills) around the home site and its vicinity, and that training distances may be increased safely only as this visually mastered area enlarges. When the home cote of trained pigeons is transferred to a strange locality, the birds generally are unable to return successfully unless they are able to see the cote from the point of release. Visual stimuli, and possibly other exteroceptive stimuli, certainly furnish the major control in this learning (Warner, 1931). In general, the number of successful returns is fewer and the return time is longer the greater the distance of the point of release from the home site. Birds may be lost in numbers if the weather is stormy and visibility poor, and Gundlach (1932) has shown that the amount of time required for the return varies according to general visibility and atmospheric conditions that influence it. In a record flight, one pigeon returned 500 miles in 10 hr., another required three days to return 617 miles, and a third bird released at 1000 miles required nine days to return. This suggests that the disproportionately long time for return over greater distances is due to a considerable amount of excess flying and to extra stops.

How does the bird set off in the correct direction, upon his release? It is very probable that unless a familiar landmark (*e.g.*, a mountain, far distant but very prominent) or some other constant exteroceptive effect (*e.g.*, prevailing direction of the wind) directly orients the bird, it must fly about until it becomes oriented. A reasonable suggestion is that the bird flies in ever widening spirals from the release point outward, and at some point in its course comes within the range of habituated territory and becomes specifically oriented with reference to a visual landmark. The number of pigeons that fail to return from distant release points (stopping at strange cotes), and the disproportionately long time required by the others, agrees with the body of facts that supports such an assumption as the above. Homing thus appears to be one of the activities made possible by the dominance of vision in an animal which possesses great range of activity.

The experiments of Watson and Lashley (1915) are among the few tests on homing in other birds. Three of 5 terns liberated at Cape Hatteras returned to their nests on Bird Key, one of the Tortugas off Florida. The birds returned a distance of more than 1,000 miles within about 5 days. Five of 12 terns returned successfully from points near Galveston, 855 miles from Bird Key. The writers do not commit themselves to an explanation, but the facts they report are very suggestive. Gundlach (1932) has taken up one of their points, that the birds may have flown along within sight of the shore line in either direction from the release point. Since terns feed on minnows (he points out), they would not be likely to fly far inland, and since they nest on land, they would not lose sight of the shore for long. If they followed the coast to the eastward from Galveston, they would reach the tip of Florida within sight of the Keys; if along the Mexican coast, they would eventually reach Yucatan, territory which must be on their line of migration and perhaps familiar to them.

Stereotyped Activities and the Manner of Their Appearance

Among the numerous and varied stereotyped activities of birds, the pecking response of the fowl is representative. During the early part of the present century, when psychologists were beginning to feel dissatisfied with the instinct concept as an explanatory device, one of the first problems to be attacked was the origin of this response in the chick.

Experimental Studies on the Pecking Response of the Chick.— As a rule, the pecking response appears infrequently on the first day after the chick has hatched. Then the chick begins to peck clumsily and at random at bright objects, such as the eyes of other chicks, drops of water, and the like. The ability to peck at food soon develops. In studying this process, Breed (1911) graded the act as follows: (1) striking at, but missing; (2) hitting, but not seizing; (3) catching in bill, but failing to swallow; and (4) the complete act, ended by swallowing. He found that three-day old chicks (after about 24 hr. of pecking) averaged 30 four's in 50 trials, and that after 12 days of pecking the average score was 40 four's in 50 trials.

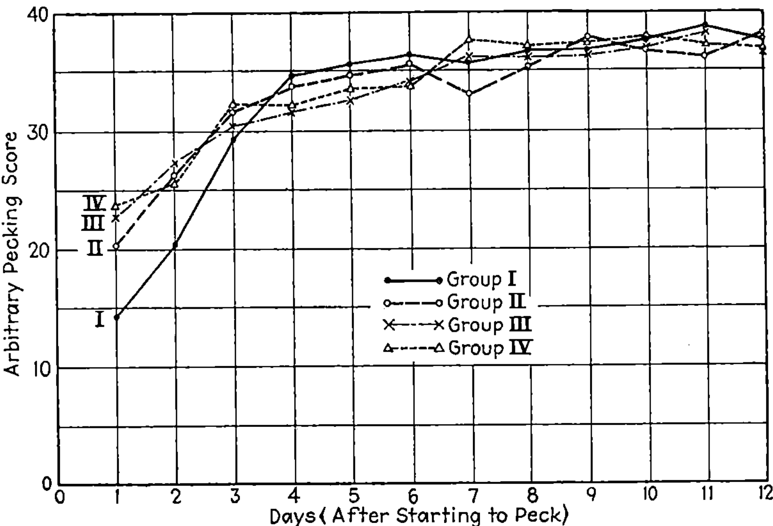

FIG. 57.—The pecking performance of chicks in four groups delayed for different periods after hatching (see text). Age at first test: I, 18 hr.; II, 30 hr.; III, 52 hr.; IV, 72 hr. Daily tests thereafter. The "arbitrary pecking score" in the figure is obtained by counting 1 point for each hit and 2 points for each "swallow" in a group of 25 sample pecks. (*Adapted from Bird*, 1933.)

His results agree with those of later investigators in showing that pecking accuracy improves at a rapid rate during the first three days, and at a slower rate on later days (see curve I, Fig. 57).

In order to discover the nature of the factors which contribute to this improvement, Shepard and Breed (1912) kept chicks in the dark after hatching, to prevent them from pecking; watering and feeding them by hand. Chicks delayed for different intervals (1, 3, 4, and 5 days, respectively) were successively brought into the light and tested for pecking. The chicks that had been delayed in the dark soon became as efficient in pecking as those of the same age which had received practice from the first day. The experimenters concluded that the rapid improvement which normally comes during the first two days of pecking may be attributed to practice. They stated the additional conclusion that the following period of slow improvement is mainly attributable to the influence of maturation (improvement through growth), and expressed the belief that maturation sets a limit on how much the bird can improve with practice at a given age.

The difficulty of interpreting results obtained from experiments involving the "delay technique" appears from the studies of Bird (1925) and Moseley (1925). These experimenters delayed chicks for periods ranging from 3 to 11 days after hatching, and made daily tests of pecking accuracy after the respective delay periods. They found that the accuracy of pecking improved more gradually the longer the delay period. The result was attributed to the setting up of "contrary habits" in the delayed chicks. These chicks had learned to swallow from a spoon, and the longer they were artificially fed, the more difficult it became for them to feed by pecking. Padilla (1930) delayed chicks for 14 days after hatching, and then found it impossible to teach them to peck grains except by an indirect and time-consuming process of training.

In a recent study, Bird (1933) has endeavored to remedy the defect in the method of previous experiments. He delayed separate groups of chicks for 18 (Group I), 31 (Group II), 52 (Group III) and 72 (Group IV) hr., respectively, and during the delay periods gave the chicks water from a medicine dropper but did not feed them. The chicks remained in the dark between the daily tests. On its first test, Group I averaged 0.65 complete pecks (four's, in Breed's terms) per group of 25 pecks, and on its second test at 31 hr. of age it averaged 2.91 complete pecks. In contrast, Group II, with no practice before its first test at 31 hr., averaged 2.62 complete pecks on this test. The experimenter attributes this improvement in Group II without the benefit of practice to the influence of "maturation" (bodily growth or development). He points to a marked improvement in postural activities disclosed by his observations on the chick during the first 30 hr. after hatching.

At first the chick mainly lies flat or sits low; at the eleventh hour after hatching "high sitting" is more noticeable, and a semi-erect position is assumed during the pursuit of grains; whereas after the seventeenth hour and before the thirtieth hour there is a greater frequency of high sitting and of standing. Only after 30 hr. of age is the chick able to stand while pecking at grains.

Bird questions the importance of practice during the first 30 hr., as noted above, in pointing to the fact that Group II in its first test at 31 hr. made virtually the same score as did Group I at that age in its second test series (Table 17). Further, he does not believe that maturation is a factor after 30 hr., since at 31, 52, and 72 hr. of age, respectively, Groups II, III, and IV scored practically the same on their first tests. On the contrary, he concludes that practice accounts for improvement *after* 30 hr. of age.

TABLE 17.—THE AVERAGE NUMBER OF "SWALLOWS" PER SERIES OF 25 TRIAL PECKS FOR DIFFERENT GROUPS OF CHICKS DELAYED IN PECKING (BIRD, 1933)

Group	Average age at testing, hr.				
	18	31	52	72	96
I	0.65	2.91			
II	....	2.62	5.6		
III	....	....	3.58	5.0	
IV	....	....	....	4.63	4.63

It is not easy to reconcile with this conclusion the fact that the "swallowing" score on the first test of Group III was slightly higher than that of Group II (granted that a difference of 0.96 swallows is significant), while the score of Group IV on its first test was higher than that of Group III. The reason why Group IV made the same score on its first and second tests also remains obscure.

The influence of the maturation factor (*i.e.*, of advantages attributable to growth and development of tissues) cannot be distinguished with any readiness from the influence of the learning factor in these results. However, the function of practice is clearly shown in the following comparison of the average "swallowing" scores made by Bird's experimental groups at the age of four days:

Group I:	three 8-hr. periods of practice	= 14.0 swallows
Group II:	two 8-hr. periods of practice	= 11.9 swallows
Group III:	one 8-hr. period of practice	= 8.5 swallows
Group IV:	50 practice pecks only	= 4.6 swallows

Bird's conclusion that maturation is effective at first, practice later, conflicts with that of Shepard and Breed (pp. 242*f.*). In any case, the evidence from these studies is not too conclusive. Experimenters may have been attempting the impossible, in their efforts to consider the roles of practice and maturation separately. That these factors are both of importance becomes evident from the results, but their relative importance at different ages must be conjectured. These experiments do not lay open to us the essential features of the mechanism by means of which the pecking response normally develops. Perhaps this is because the experimenters have neglected the fact that the chick does not begin its behavior at hatching time.

The Embryonic Activities of the Chick in Their Relation to the Pecking Response. *Significant Changes in Embryonic Behavior.*— The most extensive study of embryonic activity in the chick is that made by Kuo (1932, *a-e*). It was Kuo's purpose to find how embryonic movements first arise, to trace their form through fetal life, and to follow them into the behavior of the chick after hatching. A disc of shell was chipped from the larger end of each egg and the underlying membranes were coated with vaselin, so that the embryo and its movements could be observed. In all, 3,000 embryos were studied. It will be interesting to survey Kuo's findings in their significance for an understanding of the pecking response.

1. First (at 3 days or before) the alternate rising and falling of the thorax wall due to the heart beat cause the head to lift and drop rhythmically, since at this time it rests bent down against the thorax. This may be called *passive head bending.*

2. The head movements are at first slight and passive until the appearance of more vigorous nodding after the fourth day indicates that the embryo is being gradually trained to the movement. Now the head bends "actively" when it is *touched,* in addition to bending when it is *pushed.*

3. Head bending is induced by a general change in body position, and this modifies the nodding movement. After the sixth day the head moves in response to special tactual stimulation, such as yolk-sac pressure and incidental scratching of toes against the head during leg flexion.

4. Next, the beak alternately opens and closes when the head nods, evidently through nervous excitation furnished by the head activity. Control of the beak improves as these movements increase in frequency and extent.

5. After the ninth day the bill movement brings fluid into the mouth, thus forcing swallowing. This practice in swallowing may combine the act to some extent with bill opening and bill closing.

6. On the twelfth day, even though head movement is greatly reduced by the position of the embryo, it is observed that any head movement is followed by bill activity.

7. Meanwhile, growth has strengthened neck and bill muscles. On the seventeenth day, *when the head* (which is now mainly free and has room to move) *occasionally lifts and thrusts forward, this movement is accompanied by opening and clapping of the bill.* As Kuo points out, it is through the strong lifting and forward thrusting of the head that the stretched membranes are torn and the shell chipped by the bill so that hatching occurs.

Four principal factors cooperate in producing the foregoing changes: (1) anatomical changes in size and weight, and in sensory, conduction, and motor tissues; (2) the position of the embryo in the egg at a given time; (3) stimulation from growth changes within the embryo; and (4) the pattern of stimuli furnished by the medium, particularly by structures such as the amnion membrane and the yolk sac. The principal influences bearing upon changing behavior thus are embodied in embryonic developmental processes and in coercive or stimulative effects from the environment. As Kuo concludes, it is not possible at any stage to distinguish the functions of these factors apart from each other, since they are inseparably coalesced in influencing behavior.

This study shows that the newly hatched chick possesses the head (lunging), bill (opening and closing), and throat (swallowing) components of the pecking response. Not only that, but as Kuo points out, these activities have been combined to some extent through the influence of repeated stimulation. Two such combinations are clearly indicated: the head lunge combined with bill opening (in response to a touch on the head) and bill opening combined with swallowing. Swallowing occurs first in response to fluid in the mouth; later, very probably as a response to nervous excitation furnished through bill opening.

We are now in a position to study the posthatching act of pecking, with some understanding of its basis and an appreciation of the manner in which its components come into being in the individual.

Reconsideration of Pecking in the Newly Hatched Chick.—Shortly after hatching, as we know, the chick strikes at bright objects, giving its embryonic head lunge. However, this response is given to *visual* stimulation, whereas in embryonic life *tactual* stimulation alone had come to elicit it. We have already considered a shift of this sort, in connection with the sensory control of the amphibian's snapping response (p. 208). The two phenomena appear essentially similar: In both cases sufficiently strong visual impulses, upon arriving in midbrain, spread most readily into an adjacent nerve tract which is already under the control of head contact. Thus vision fairly promptly takes over control of the head lunge, a response previously controlled only by contact. That a certain amount of training is involved in effecting this transfer is evident from observations of the first pecks given by chicks.

During the first 30 hr. after hatching, as Bird (1933) has shown, progress in pecking is retarded by the weakness of the animal, but improvement becomes evident as first neck, then trunk, and finally leg muscles acquire strength. The chick may peck more and more readily as it becomes stronger; in that respect the function of maturation is apparent. But practice cannot be excluded at any time. Some practice in supporting and moving the body undeniably is obtained as muscles become stronger, and this practice certainly contributes to the advancing of the pecking activity of the chick, although perhaps not by increasing the number of "swallows" during the first stages. The difficulty in our understanding is due to the fact that the crudity of the usual criterion of progress, the number of "swallows" alone, gives a distorted picture of improvement in the act. This is an act of skill, and the acquisition of such an act cannot be measured accurately in terms of "bull's eyes" alone.

Normally, as improvement in postural activities (both through increasing strength and through practice) permits the chick to strike more frequently at bright objects, the only way in which its exciting hunger (empty stomach, providing "hunger contractions") may be removed is by chancing to get into the mouth, among objects struck at, certain edible morsels. Some of these are swallowed, and soon the chick has learned to discriminate them visually from other objects. In this way normal circumstances force the combination of the embryonic components, head lunging (with bill opening) and swallowing.

After this stage, the closer coordination of the components striking-at, catching, and swallowing, into a single sequence, is limited mainly by the amount of practice received by the chick (Fig. 57).

Circumstances in the normal environment thus bring about the combination of the embryonic components in the manner described above. However, if circumstances differ from the normal, the components are differently incorporated into the posthatching behavior of the chick, as the "delay" experiments abundantly prove. If the artificial training is exceptionally thorough, as it was in Padilla's experiment, it is virtually impossible to retrain the chick

FIG. 58.—Forced feeding under natural conditions. A female king bird feeding her young. *(From Francis Herrick's Home Life of Wild Birds, by permission of G. P. Putnam's Sons.)*

into a coordination of striking and swallowing.[1] In artificial feeding, swallowing becomes connected with a pattern of stimuli (sight and touch of the experimenter's hand) which is unrelated to the act of pecking, while pecking is practiced in connection with visual stimuli (bright inedible objects, etc.) which have nothing to do with feeding.

The act of pecking at food may be considered a special modification and recombination of embryonic head, throat, and body

[1] In this connection Fig. 58 reminds us of the fact that the period following hatching, during which young birds of many species are fed from the beak of the parent (forced feeding), may account for the failure of pecking to appear with any prominence in the behavior of these birds as adults. In contrast, the pecking response is typical in races of domestic fowl, in which the young are not fed by the parents during the brooding period.

activities in connection with specific external stimuli to which the chick learns to respond after hatching. The pecking response thus has a continuous history from the time the embryo gives its first movement.

Activities Associated with the Reproductive Cycle

The complex of activities related to propagation of the species actually furnishes the framework of bird behavior, since most of the characteristic activities of birds are associated with it. Among these are migration, courting and mating (with song and other motor accompaniments), nest building, brooding, incubation, and the care of young. We shall survey the successive phases of the reproductive complex in a general manner, observing the chronological order of their occurrence from the beginning to the end of a season.

The Initiation of the Reproductive Cycle: Migration.—Migration is the recurrent, usually annual movement of birds between two alternative habitats of which one is the breeding locality. Although some species of birds do not migrate regularly, and others migrate but short distances, the onset of the breeding cycle is accompanied by a movement of some kind in almost all of the birds. We are concerned here with the ontogenetic problem of migration; its origin in the race is quite another problem.

The Two Phases of Migration.—In general, birds of the northern hemisphere engage in a northward movement in spring or early summer, preceding breeding activity, and a southward movement in the autumn. Our first problem concerns the stimuli which set off these movements. The autumn movement to the southern locality has been attributed to one or a combination of the following seasonal factors (Wetmore, 1926): (1) a decrease in available food, which forces out most of the birds (*e.g.*, purely insectivorous species); (2) a decrease in temperature, furnishing a gradient to which the birds are capable of responding, more or less promptly according to their sensitivity or their powers of resistance; (3) a decrease in available light, and especially in ultraviolet.

It is the northward movement in spring (or early summer for some species) which is closely related to the present general subject, since this movement precedes the mating and breeding activities which occur after arrival. The theory that environmental changes occurring at this season of the year set up internal changes which lead to the northward movement suggested an experiment to Rowan (1931).

Changes Responsible for the Onset of Migration.—At the time when crows normally move southward from middle Alberta (August 20 to September 10) Rowan captured large numbers, which he divided into two groups. The aviary containing the control group was not artificially illuminated, so that the birds were subject to the regular autumnal decrease in the daylight period. For the experimental group, on the other hand, the aviary was illuminated electrically during an additional period after dusk, so that the "daylight period" of each successive day was 5 min. longer than that of the preceding day (as occurs in spring, in the southern habitat). The groups received equal amounts of food, and neither aviary was heated.

By the ninth of November the gonads (testes of the male; ovaries of the female) of the control birds had become very small, while those of the experimental birds were greatly enlarged, having increased regularly in size during the test period (as they normally do in the spring). On this date, 69 experimentals and 14 control birds were released. The control birds did not readily leave the vicinity. Most of them had to be driven from the aviary, and apparently none of them went northward. In contrast, the experimentals flew off with greater readiness, and many of them flew northward. The results given in Table 18 probably would have been more decisive had the test been performed in a warmer locality (farther south), in which a better reproduction of springtime migration conditions would have been achieved.

The results strongly suggest that in the southern environment the daily increase in radiation which comes in the spring sets up internal changes, particularly in the sex glands, the secretion of which becomes greater as the size of the glands increases. This augments the rate of general activity so that the bird is more and more ready to move away.[1] The departure time of a given species may be expected to occur within a given few days, a fact which may be traced to the one unfailingly regular springtime change, the gradually increasing length of the daylight period.

[1] Just as female mammals that are confined at the height of the estrous cycle are unusually active, the bird that is caged at migrating time exhibits a pronounced restlessness and consistent hyperactivity.

Factors Inflencing the Direction and the Route of Migration.—The springtime migration of temperate-zone birds of the western hemisphere is northward. What determines the direction of this movement? It is a reasonable hypothesis that the growth of the gonads

TABLE 18.—GENERAL RESULTS FROM AN EXPERIMENT ON THE CONTROL OF BIRD MIGRATION (ROWAN, 1929)

Groups	Movement of released birds				
	Remained in vicinity		Departed		
	Caught near aviary	Killed locally	Killed to north	Killed to south	Unaccounted for
Controls (14) (small gonads)	6	2	0	4	2
Experimentals (69) (enlarged gonads)	15	12	8	8	20[1]

[1] Many reports of crows flying in northern Alberta were received from residents unaware of the experiment. This was an unusual observation for that time of year in Alberta. Most of these experimental birds must have flown northward, since such reports were not received from southern parts of the province.

and constant increase in their secretions bring about changes in skin circulation which cause the birds to move in the direction in which an equilibrium with environmental temperature is better approximated. In some species young birds set off before the adults leave, a difference which may be related to the fact that the circulatory adjustments to environmental temperature are not so delicately made in the young bird as they are in the adult. The direction of migration thus would appear to be physiologically determined by the organic changes which constitute the migratory drive.

While it is probable that the restriction of migration routes to narrow well-defined paths has been exaggerated, bird-banding experiments indicate that the general paths of many species are fairly constant, year after year (Lucanus, 1921). River valleys which lie in the north-south direction are followed by some species, the coast lines are followed by many, and mountain ranges by others. Figure 59 shows some general routes which are taken by many North American birds. Differences in the characteristic routes of bird species suggest that once the migratory drive has set off the movement, and the birds are headed northward (or southward), their particular course is determined by factors such as the nature of sensory equipment (particularly visual) in the species, their manner of feeding and sleeping, and their power of flight. It is conceivable

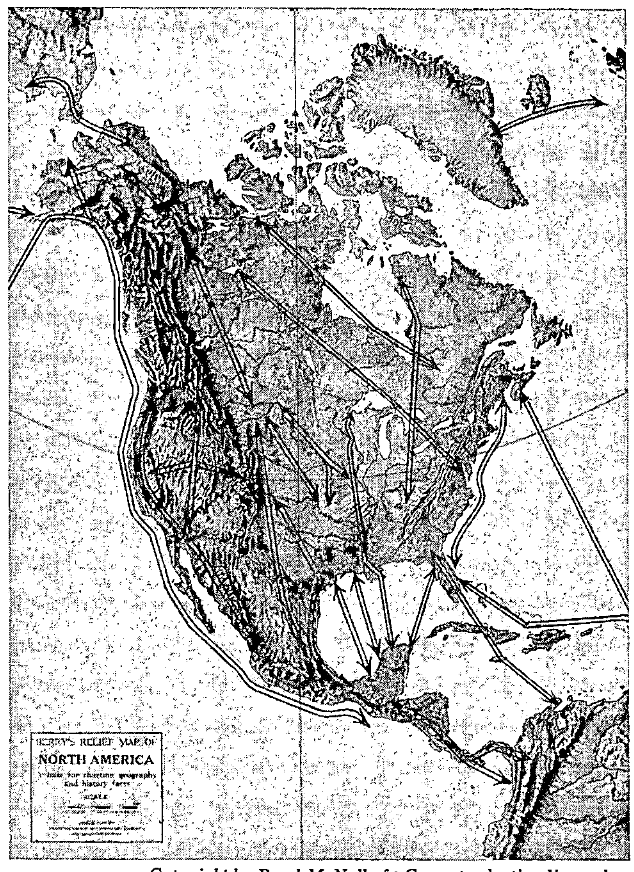

Copyright by Rand McNally & Co., reproduction licensed.

FIG. 59.—Some of the principal routes taken by various species of migrating birds in North America. (*Modified from Wetmore.*)

that migrating birds respond to prominent visual features which lie consistently in a given direction which is determined by other factors (see above), without having previously seen the territory.

Migrating birds commonly fly at heights of 1 to 2 miles, as ornithologists have ascertained by training their telescopes against the sun (or against the moon at night), and at such a height a visual field 150 to 200 miles in diameter may be commanded in clear weather. Scattered evidence suggests the importance of feeding activities in influencing the migration route. Rowan (1931), for instance, notes a correspondence between the movement of the adult golden plover (a shore bird) southward along the Atlantic coast of Canada (rather than through the central part as the young plovers move earlier in the fall) and the plentiful berry supply at that time in the eastern coastal region.

The fact that young first-year birds in many species leave before the adults do, makes learning of doubtful importance in the determination of the migration route proper. The "ancestral memory" conception contributes nothing to the answer. Sometimes birds nest in the same locality more than once, which suggests that when they chance into the region in which they have nested in previous seasons, the migrants may become oriented through learned cues much as a pigeon finds his cote. It is important that the same species may go and come by different routes. For instance, the golden plover migrates southward from northeastern Canada via the Atlantic coast, the West Indies and the Caribbean, along the South American coast and across the eastern part of the continent to the Argentine. But after six months the plovers pass northward by an entirely different route, via western South America, the Central Americas, then across the gulf to Mexico, and up the Mississippi valley toward Canada.

The Reproductive Cycle Proper. *Initial Behavior of the Male.*— Mating preliminaries begin soon after the birds have reached their breeding ground. In most species the male arrives first. He spends the greatest part of his waking time alone, lingering in prominent places, at first for short times only, but soon he takes for himself a regular "calling station." He quarrels increasingly with other males as time goes on, remaining within or near the "territory" which he has preempted (Howard, 1920) and driving off other males. The increasing tendency to fight other males is one feature of his behavior; on the other hand the male sings with greater vigor as time goes on, and sits more prominently in his station.

The active singing of the male is a major factor in attracting a female to his station. Her behavior also, not quarrelsome like that of a male, is responsible for her being received rather than being driven away. The male now sings less, and quarrels more vigorously than before with other males, with much wing activity and a variety of flying both before and after each encounter. According to Howard's (1929) analysis of his observations on buntings and other birds, the excitation of the male through glandular agencies has been increasing from the time of his arrival, and he would be capable of mating at any time after the female comes. Although she is not particularly responsive to him at first, her presence excites him further, so his surplus energy must take the form of excessive overt activity. This, in various birds, takes the form of "strutting," "display" of feathers and other parts, inflation of throat pouches, and similar activities.

Events Leading to Mating.—After a time, upon awakening in the morning, the posturings and stretchings of the male are followed by the "sexual flight." The female has become somewhat responsive to the male, but not sufficiently to mate; and so she flies from his rough advances, keeping out of reach and yet remaining near him. The increasing excitement of the female is also indicated by her occasional picking up of fibrous roots or other materials while feeding, but after some manipulation in the bill these are dropped (abortive building?). The advancing internal condition of the female is next shown by her remaining in place as the male approaches. This stimulates him to further posturing and to special activities (*e.g.*, inflation of throat pouches in the prairie chicken) which vary in nature according to the species. The male postures, is stimulated to approach closer, and mating then occurs.

Whitman (1912) describes the mating of the pigeon in the following manner. "As the period of consummation approaches, the composition of the activities changes with the addition of new elements. Along with bowing, there is billing and fondling of each other's head, hugging or necking, jumping over the female without any attempt at mounting, opening the beak by the male, inserting of the female's beak in his, and often the shaking or rattling of the crop as if the male fed the female. The female stoops with lowered head, the male mounts with a jump, the female raises her wings and lifts her tail, while the male reaches back, moving the tail from side to side until contact (*i.e.*, of cloacal surfaces) is effected."

The appearance of the final stage seems to depend upon the behavior of the female, in which the vibration or spreading of wings and the raising of the tail are major components in most species. The wing vibration occurs as well outside the mating period, as a consequence of any sufficiently intense emotional excitement. However, the tail raising, so important for the actual mating act, is specific to this period and may depend upon a local irritation of cloacal surfaces for which the condition of the female's genital tissues is responsible.

That mating depends upon the attainment of a particular stage in the reproductive changes of the female is also indicated by the fact that mating time is specific for a given species, although the courting phase varies in dependence upon the time of the female's arrival at the breeding place.

Nest Building.—Although the female may toy with materials in the preceding period, nest building proper usually does not appear until after mating. The female now picks up coarse material, and may deposit it in different places before successive trips are made to the same place. The "platform" of a nest may next be made, but then abandoned for a beginning elsewhere. After this two or three "shells" may be constructed successively, in different places, before the female finally goes on to complete a nest. The male may or may not assist, depending upon the species, but in any case his presence is an important source of stimulation for the female (Craig, 1908). However, it is the condition of the female which determines the appearance of building activity at this time.

Francis Herrick (1901, 1911) has made the most thorough study of nest building in birds, principally by means of observations from "blinds." He found that this activity varies according to the state of the weather, the nature of the site, the character of the general environment, and the condition of the bird. The height of the nest above ground depends upon the characteristic flight of the bird, and the site may also depend upon the mode of feeding and sleeping, and upon the relative timidity of the bird. The nature of the materials is determined largely by availability in many cases but is always influenced by the bird's sensitivity and motility. Usually, coarser materials are worked with first, then finer materials, and this rule is seldom violated. Herrick pointed out that it is the least adaptive birds (*e.g.*, the Baltimore oriole) which show the greatest uniformity in the nature of the materials used. The birds which build "statant" nests (supported mainly from below) go through the molding-and-turning movements in a fairly uniform and stereotyped fashion, with trips at intervals for fresh materials, and thus the growing nest is gradually and automatically adjusted to its site. For this reason (usually) the cup is finally horizontal regardless of the inclination of the supporting surface.

Brooding, Incubation, and Care of Young.—The nest is completed as the female nears egg-laying time; usually before this she has begun to "brood" (*i.e.*, to sit upon eggs, incubating them, or in the bare nest if eggs are not present). For this activity no eggs are necessary; any uneven material beneath her ventrum will do.[1] It would appear that changes which come at this time in the skin of the ventral body surface (decrease in the fat layer and increase in capillary circulation of the skin) are of basic importance for the tendency of the female to brood.

[1] Whitman (1919) observed that in the pigeon brooding begins a day or two in advance of egg laying. Females with nonfunctional ovaries never show a tendency to brood (Pearl, 1914, *Jour. Anim. Behav.*).

Once the young have hatched, the female (and often the male as well) supplies the nestlings with food at regular intervals, spreads her wings over the young, and removes excrement from the nest or occasionally swallows it. In view of the importance of these acts for survival of the young, it is surprising that for a good part of the period brooding and feeding of the young appear to be disconnected phenomena, so far as the stimuli which elicit them are concerned. Howard (1930) substituted eggs for the nestlings of a bunting, and placed the young in a different nest close by. For some time the female would sit upon the eggs in her own nest, or even upon the nest when it was emptied, but she did not brood her young in the nearby place. At intervals she returned with food, but gave it to her young only in response to their chirping. Only after a time was she able to sit upon her young in the nearby nest.

Feeding of the young and brooding of the young take turns in many birds. Each act has its normal duration before the other displaces it; and the food getting and feeding phase does not appear to be shortened in favor of brooding, no matter what the temperature may be. Hence the mortality rate of nestlings is high. It is also interesting that the female appears impelled to carry something away after having fed the young. If no excrement is visible, other waste material is picked from the nest, or failing that even part of the nest structure may be taken.

Factors Controlling the Sexual Cycle.—In many birds the female becomes subject to the beginning of a new sexual cycle when the young are not much more than 10 days old. The female bunting, according to Howard, demonstrates this by engaging in "sexual flight" with the male, by abortive building activities, and by paying less attention to the young. A few days later she ceases to respond to the young altogether. Howard (1929) cites this and other facts as evidence that *during the entire cycle the female's internal condition* (increase in gonadal secretions, maturation of eggs) *determines the nature of the activities in which she may engage.* The male is a secondary factor in determining the progress of this behavior cycle, except for the important matter of settling upon a "territory." The sequence in which mating, nest building, and brooding appear depends principally upon the progressing condition of the female.

Carpenter (1933) observed the sexual behavior of male pigeons that had been subjected to complete or to partial castration at the age of one month. The frequency of copulation was reduced almost to zero by the operation. Charging behavior was next readily eliminated, but billing and preening (in that order) could be elicited in castrated birds providing that social stimulation was sufficiently intense. The cyclic character of the male's normal reproductive behavior, Carpenter concludes, is largely determined by the changing behavior of his mate and by special stimuli such as nest, eggs, and young as they appear.

Craig (1914) studied the appearance of the separate responses bowing, cooing, preening, and mouthing in four male doves which were reared as isolated individuals. The acts appeared in response to various external stimuli: particularly in response to the hand that fed them, to the human foot, and to other objects. These responses were transferred only gradually to a female when one was presented later. This shows, as do Carpenter's findings, that normally it is previous companionship with other birds of the species that mainly accounts for the great potency of the more submissive female as a stimulating object to the sexually excited male.

A close correspondence exists between the cycle of internal changes in the mature female bird and the sequence of activities which she is able to perform. The manner in which the male and specific objects or situations in the environment function in the elicitation of these activities will be better understood when the interindividual activities of birds have been studied in more detail. We have cited results which suggest the importance of learning for directionalization of the male's viscerally supported and structurally determined activities of the sexual period. For suggestions on this problem we turn to an examination of the evidence on the social activities of birds.

The Significance of Social Organization in Bird Life

Factors Which Promote Social Organization.—We have the suggestion that the nature of the environmental stimuli to which the bird responds during the reproductive period may largely be determined by the effects of his previous experience; perhaps as a nestling with parents, fellow fledglings and the nest, and perhaps as an adult with others of his kind. Most birds have a social background of some kind, if only that of their nest life when young, or of their contacts with a mate at breeding time. But true social organizations are common among birds. By a true social organization we mean a group of two or more individuals, more or less permanent, in which there is an organization based primarily upon the interstimulation of the individuals (*cf.* pp. 164*f.*). Allee (1931) has pointed out that birds may come together on the basis of family groups, or incidentally as separate individuals which at first react to the same common external condition. This condition may be a common feeding zone; may arise by means of their sleeping activities; or through their response to available nest sites (*e.g.*, flamingoes). Since the forming of a flock depends upon such matters, some species rarely aggregate, or do so only irregularly.

Once they have flocked, the birds may remain in a group for a considerable time. This depends upon the persistence of the condition which brought them together in the first place, or upon the rise of habitual interindividual adjustments which would tend to keep them together.

Taylor (1932) maintains that "gregarious" birds are different only in degree from characteristic "solitary" species, and that any species must form aggregations under certain conditions (see above), to which some species are naturally more susceptible than others. He describes the formation of an integrated group when many pairs of pigeons are introduced into a limited space one pair at a time. The first pair nests in one place and takes over the rest of the area for strutting and other activities. A pair of newcomers are treated as enemies, and must fight for a place. In time the newcomers get as much unchallenged space, in addition to their nest, as they are able to defend. It is less if they are small, and dwindles in size if they are taken sick. Each pair now tolerates the other, and there is no fighting unless a common zone between the territories is crossed. Further pairs must struggle with those which are already established if they are to settle, and consequently the territory of each pair decreases somewhat as the flock grows. Conversely, Taylor found that if successive pairs are removed, the remaining ones "divide" the vacated space among themselves in the same indirect but effective manner. In this way, *the pigeons are forced to become gregarious, to adjust to one another in a group.* Much the same process occurs during the establishment of the "territories" of male birds (*e.g.*, robins) at the beginning of the mating season.

As the group forms, then, the birds become adjusted to one another. They learn to respond to one another's movements when in flight, each taking a position in the flock which depends upon his power of flight or his condition at the time, *e.g.*, fatigue. They learn to respond to sounds made incidentally by others when frightened or when food is discovered, and to similar stimuli.[1] In this way, some birds are forced by conditions to live in close proximity to others of their kind, and may learn interdependent adjustments as well as toleration.

Individual Behavior and the Social Organization.—The nature of organization in the bird group has been disclosed by the studies which Schjelderup-Ebbe (1922) and others have made. Schjelderup-Ebbe has described the development of a fairly stable "peck order" in a flock of domestic fowl in terms of the manner in which a given individual responds when others peck at it or "threaten" to do so. The rooster has the "peck right" over all of the hens. Some hens stand above others in this respect, threatening to peck or actually pecking them (in food competition) without being pecked in return, but those which submit have the "peck right" over still others, and so on to the "bottom" of the group. This peck order, fairly permanent once it is established, usually represents an approximately continuous decrease in ascendency from top to bottom of the group: *A* pecks *B*, *B* submits to *A* but pecks *C*, *C* pecks *D*, and so on (Fig. 60*B*). Sometimes, however, the organization is complicated by the fact that certain individuals lower in the general order may have the peck right over certain ones which otherwise occupy a superior position. Newcomers must struggle for food, roosting places, and so on, and gain a "position" on the basis of the outcome.

Murchison (1935) has reported an investigation of the establishment of a "dominance hierarchy" in a group of six young roosters, studied in terms of physical combats and other encounters arranged at various times under experimental conditions. An accurate measure of dominance was furnished by the testing of respective pairs of roosters in the "social reflex" runway. In this test, time and distance records were made of the manner in which any two individuals of the experimental group rushed toward each other and quickly met at some point in a narrow experimental corridor after having been released simultaneously from opposite ends.

[1] This should not be termed *communication* until it can be shown that the "sentinel" sounds the "alarm" because it will cause the other birds to fly away, *i.e.*, that the sound is anything more than the incidental product of his own excitement without regard to the effect it has upon the other individuals. The evidence appears to be negative so far as birds are concerned.

At intervals between the sixteenth and thirty-sixth weeks of age Murchison found that the order of dominance among the birds was subject to change, and that only one individual, I, was able to maintain his superiority throughout the period. Straight-line dominance (*i.e.*, each individual dominating all below him, while dominated by all above him) was finally achieved in the group (Fig. 60*B*). It is Murchison's conclusion that straight-line dominance "is a function of adjustment over a long period of time within an isolated social group," but that the required time would be longer in larger groups, or the condition might never appear in very large groups.

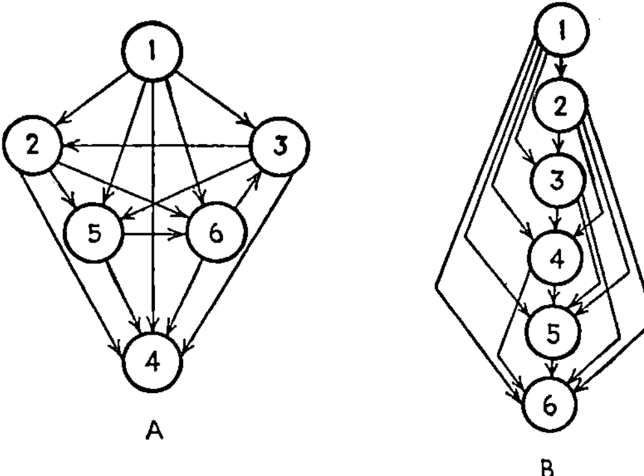

Fig. 60.—The order of dominance in six young roosters studied by Murchison (1935). (*A*) Dominance relations at 16 weeks of age, and (*B*) straight-line dominance hierarchy at 36 weeks of age. (*Modified from Murchison*, 1935.)

In the establishment of a "dominance hierarchy" in a given group, each bird gradually learns to respond to every other bird as to a distinctive visual pattern which has been accompanied by specific consequences in previous competitive situations. Where the results are drastic, the individual soon learns to withdraw from a certain other one as from an electric shock. Finally the "dominant" bird needs merely to appear or to begin its belligerent approach (to "threaten") and the others draw off without fighting. Strength is not the only factor which makes for dominance, since Katz and Toll (1923) found a correspondence between learning ability in laboratory problems and position in the pecking order of the group.

Social groups in other species of birds have been found to owe their unity to a similar basis in individual behavior. The most aggressive and active become *leaders*, since other individuals learn not to push ahead of them. The timid birds take less exposed positions in sleeping groups, and therefore the group is aroused ("warned") by the cries of the dominant birds which are first to become excited by an intrusion or disturbance. Observation of bird aggregations suggests that while a given bird is eventually forced into a certain position in the dominance-submission order of the group, its behavior throughout group life is dependent solely upon an individualized motivation.

MODIFIABILITY IN BIRD ADAPTATION

In many types of bird behavior (homing, migration, the reproductive cycle, feeding activities) we find evidence that learning is involved. Learning influences the course of development of activities which are inevitable because of inherited equipment, but this capacity appears to occupy a subordinate position in general behavior. For an understanding of its properties and its limitations, certain laboratory evidence may be surveyed.

The Visual Discrimination Habits of Birds.—Since vision is by far the dominant sense modality in practically all bird behavior (pp. 237*ff.*) one would anticipate the series of experiments which have been performed on the bird's ability to learn visual discriminations. The habits appear to be formed very slowly, as is characteristic of the bird, but their retention value is relatively great. Kroh (1927) found that a chick which required 95 trials to learn a simple visual discrimination, relearned the habit in 12 trials after a no-practice interval of 11 months. A chick that needed 72 trials to establish the same habit performed correctly after 4 months without further training. The great importance of visual discrimination in bird life has been touched upon. The general topic of visual discrimination is to be given special treatment in a later chapter (Chap. XIII).

The Learning of a Maze Problem.—Hunter (1911) tested pigeons on the maze problem. The birds were first trained to hop across the open space from the starting point to the corner in which they were fed, and then their progress was obstructed by the wooden partitions, leaving a series of six alleys (5 in. in width) to be avoided. On the average, errors did not decrease much before the seventh trial, and it was unusual for a consistently low score to be made before sixty runs had been completed. Although this maze problem (which would not be difficult for the ant or the rat) was learned very slowly, tests that followed original learning by an interval of 29 days showed only a negligible decrease in efficiency of performance.

The Learning of a Problem Box.—Porter (1904, 1906, 1910) permitted birds to become used to feeding within a small mesh-wire puzzle box which was left for a few days with open door in the center of their living cage. The door was then closed and could be opened only by the pulling of a string which was connected with the light door latch and which stretched horizontally along one side of the box near the top.

The birds busied themselves about the box, and particularly about the door, which they had been trained to enter. The number of successive periods of effort ("starts") which preceded the successful entrance of a given bird into the box decreased abruptly during the first few trials, once the bird had learned to confine its efforts to the vicinity of the string. A sparrow needed more than ten minutes and many "starts" to enter the box on its first trial, but on the twenty-fourth trial succeeded within 5 sec. and with but 3 "starts" in rapid succession. One bird learned to push with his head against the string, later with his bill. Another learned to hop out from the upper edge of the box, turn partly around, and catch the string in his claws. Pecking at the string was common with sparrows, more so with blue jays, and especially with a woodpecker.

In another experiment (1910) two horizontal trip strings were used, one slightly above the other. One subject mastered the problem by alighting on a string. Following original learning, Porter moved the strings somewhat farther apart. This caused the subject to fail, since it now alighted between the strings and did not press either of them. The bird had not learned to "open the door" nor had it learned to "press a string," but merely to make the specific response of alighting on the box in a visually distinctive place. Thus pressing the strings (often incidental to another act), then entering the door, were disconnected responses which had been learned in a well-practiced succession. It was usual for a bird to pull repeatedly at the string even after the door had opened.

The results speak for a fixity in sensory control and for a high degree of stereotypy in the responses learned by the bird. Apparently such features contribute to good retention, since after 30 days birds performed the learned act much as before, and even after 120 days one cowbird showed unmistakable signs of the original training.

Concluding Statement

We have found the behavior of the bird very complex. Many of its varied modes of activity have an important emotional facilitation, and vary in close relationship with visceral condition. Learning may be responsible for the extension of a specific type of activity and may make possible its elaboration within certain limits. However, the bounds within which learning may function appear to be established by the heritably determined factors (*e.g.*, dominance of visual control, well-ordered visceral cycles, extensive development of the striate area in the brain, etc.) which themselves are responsible for the basic stereotypy of the activity. Therefore, while the bird is enabled to live in a very complex environment by virtue of mechanisms which have evolved to a high degree of elaboration, it is these possessions which are responsible for its psychological limitations. It is otherwise with the mammals, which will be studied in Part III.

Suggested Readings

Bird, C. 1933. Maturation and practise; their effects upon the feeding reaction of chicks. *Jour. Comp. Psychol.*, **16**, 343–366.

Herrick, F. 1911. Nests and nest-building in birds. *Jour. Anim. Behav.*, **1**, 151–192.

Howard, H. E. 1929. *Bird behavior*. Camb. Univ. Press. Chap. III.

Porter, J. 1904. A preliminary study of the psychology of the English sparrow. *Amer. Jour. Psychol.*, **15**, 313–346.

Wetmore, A. 1926. *The migration of birds*. Harvard Univ. Press. Chap. I.

EQUIPMENT OF MAMMALS[1]

TABLE 19

RECEPTOR EQUIPMENT

Chemical. Taste buds generally plentiful on upper tongue, few on roof of mouth, on throat.	Variously important in feeding, for rejection of food, etc., especially if tongue habitually first applied to food.
Free nerve endings in mucous membranes of mouth, nasal passages.	Sensitivity to strong gases (*e.g.*, camphor).
Ciliated cells segregated in special areas in upper part of nasal chamber. Extensive in lower mammals, poor in primates.	Efficient tracking of prey in many, sniffing of food in ground (*e.g.*, pigs), important during mating. Degenerate in primates.
Contact. Numerous types of (corpuscles, etc.) specialized nerve endings in skin, especially at distal ends movable members. Free nerve-endings, plentiful at hair bases.	Vibrissae important in orientation of many. Sensitive snout in location of food, general skin sensitivity (or hair bending by air currents) in orientation. Related to manipulatory specialization of forelimbs in higher mammals.
Static. Well-developed semicircular apparatus in most: two vertical canals and one horizontal, sacculus and utriculus.	Functions in close cooperation with vision, muscle sensitivity, contact, in maintaining balance and postural activities.
Auditory. Well-developed ear: movable pinna in most, middle ear with drum, three bones; inner ear with cochlea (between 2½ to 5 turns).	Localization of sound generally good, intensity discriminations learned, but pitch discriminations very difficult for some lower mammals.
Visual. Lens shape changed by ciliary muscles. Poorly developed fovea in many lower mammals, well-developed fovea in primates. Rod-cone retina in most; rods only, in few. Gradual increase in "crossed" fibers at optic chiasma (virtually none in rodents, ⅓ to ½ in primates).	Vision not of major importance in many lower mammals (*e.g.*, rodents), but dominant in higher orders. True convergence of eyes present in primates only. Stereoscopic vision develops in primates. True panoramic vision in ungulates (*e.g.*, deer); lateral position of eyes responsible.

CONDUCTION EQUIPMENT

Developed cortex covers surface of cerebral hemispheres, typically convoluted. Ratio of brain weight to body weight: elephant (5000 gm.) 1/560; cat (30 gm.), 1/110; chimpanzee, (400 gm.), 1/75; man (1400 gm.), 1/35. Corpus callosum connects hemispheres across midline, less extensively in lower mammals. 12 cranial nerves.

ACTION EQUIPMENT

Variety limb types, of varied importance for locomotion, manipulation. Mouth, teeth, used in manipulatory activities of lower mammals, in conjunction with forelimbs. Varied forms digging activities in shelter obtaining, food getting. Many anatomical specializations, (*e.g.*, regurgitative digestive apparatus in ruminants). Vocal cords subject to variety of specialization in adaptation, depending particularly upon development of cerebral cortex.

[1] To be used in connection with Parts II and III.

PART II

PREREQUISITES TO BEHAVIOR DOMINATED BY MODIFIABILITY

In approaching the study of mammalian behavior one finds a great shift of interest in the problems investigated. The behavior of lower animals depends largely upon their structural characteristics and the immediate stimulating conditions, but that of mammals is largely conditioned by previous responses. The study of mammalian psychology is therefore characterized by an analysis of the process of modification in behavior through experience. This does not mean that structural characteristics and immediately present stimuli cease to be important in mammalian behavior. Nor does it mean that modification in behavior is absent in lower animal forms. Rather the interest shifts because modification in behavior reaches its height in mammals. Since modifiability plays a dominant role in the representative adjustments of this vertebrate group, it can be divorced from certain structural characteristics and studied as a general problem. From this it does not follow that all mammals have the same capacities but that generalizations regarding modification are specific to the problem and not to the animal. The following analysis will therefore be characterized by the problem investigated rather than by the animal.

Before the problem of modification by experience is attacked one must have a thorough understanding of the mechanisms underlying behavior. To begin with there are certain *behavior patterns* (fixed responses to stimulation) that are not subject to modification. Again, there are behavior patterns or at least partial behavior patterns (*i.e.*, tendencies for behavior) which serve as a basis or starting point upon which modifiable behavior is built. In such cases one must know what has been modified and to what degree the behavior has been extended. The material for such a treatment must be largely determined by the problem; and investigations which are pointed to isolate general principles in native behavior will therefore be selected.

Modification also depends upon the adequacy of the *receptive mechanisms* since these necessarily set a limit upon the degree of modifiability. Often great advances in adaptive behavior arise in connection with increases in the complexity of the sensory processes. In reptilian behavior, for example, modification is largely through complex sensory integrations. In other forms, such as in fishes, sensory development plays a minor role in modifiable behavior. Since we are concerned with the relation between sensory capacity and modification and since, as we shall see, methodology is most important, the material for this treatment will be chosen from studies which deal with highly developed sensory capacities and from studies which develop adequate and general methodology.

Modification in behavior is dependent upon changes in the *central nervous system*. Because animals which have well-developed cortical tissue are especially capable of modification in their behavior, an understanding of modifiable behavior presupposes a knowledge of cortical neural mechanisms. Material for this treatment will therefore be selected from studies which throw light upon the function of the mammalian cortex.

These three topics are therefore treated in Part II. Their importance in the behavior of lower forms has already been considered. Since modifiability dominates mammalian behavior, its prerequisites require a separate treatment. This does not mean that the subject matter of Part II is secondary to that of Part III. Rather it is fundamental to an understanding of complex behavior.

CHAPTER XII

NATIVELY DETERMINED BEHAVIOR

Introduction

In Chap. VI the "instinctive" behavior of arthropods was analyzed in detail. It was found that the nature of their responses depended upon the efficiency of their sense organs, the nature of their nervous systems, and their fund of action equipment. These in turn were dependent upon the animal's inheritance. Insect behavior contains almost unlimited examples of complex "instincts," a few of which have been carefully analyzed. Similar analysis is desirable in mammals, but complications arising because of their more elaborate nervous system (in particular, their well-developed cortex) make this difficult. Nevertheless what has been said regarding the psychology of "instinctive" behavior in arthropods holds for higher forms. In addition new problems arise. The necessity of taking modification by experience much more fully into account offers many difficulties. The complex behavior patterns of higher forms are also more variable than are those of lower forms, which makes it difficult to analyze the behavior. Although much of the work is not convincing, as a whole it is quite enlightening. The reader must, however, keep in mind that more important work is still to be done before adequate theories can be formulated.

The effects of heredity and environment on behavior are intimately connected with problems concerning the genesis of behavior, especially in higher forms. Sometimes they are regarded as being entirely different in their effects, sometimes as being closely related.

Heredity and Environment as Determiners of Behavior.—Those who uphold the *preformistic theory* maintain that the germ plasm possesses the properties which largely determine the nature of an individual in both structure and behavior. The inherited traits of an individual are thus alleged to be found in its ancestry. Evidence in support of this point of view as concerns behavior is presented in terms of correlations between traits in related individuals. Thus Stone (1932) found that wild rats are more likely to have wild offspring than tame rats.[1] In a similar manner Tolman (1924) has shown that rats which easily learn mazes are more likely to have offspring that also learn easily, than are parents which learn mazes with difficulty. Although wildness and maze-learning ability are not specific behavior patterns such as we are interested in analyzing, it is not difficult to point to cases of specific responses. Breeders of bird dogs have observed that good pointers are likely to be descendants of good rather than poor pointers.

Behavior patterns due to environment, stressed by the *epigenetic view*, show correlations between traits and environment. If, for example, different rats have different amounts of experience on mazes, there will be a correlation between the maze performance and the experience in the maze. For the same reason good tennis players are likely to be found among people who have been exposed to the game.

More careful examination, however, reveals that our problem is not so simple. Child (1921) has presented convincing proof that the pattern of an organism is dependent upon the *physiological gradient*. Thus a developing egg shows different rates of metabolism in different parts. The differences in metabolic rate form lines of diminishing and increasing activity, and these lines of activity form an environment which is fundamental to the development of the pattern of the individual. The nervous system as well as other structures are not predetermined but can be changed if the gradients are altered. Because the gradient is an environment and because the environment (temperature, chemical constitution, etc.) in which the developing organism is embedded modifies the gradients, it is not adequate to distinguish between hereditary and environment as separate influences. Rather, development must be considered as an interaction between undifferentiated germ tissue and an environment. The environment must be understood to include not only the surrounding conditions but neighboring living cells as well.

To conclude that development is a resultant of both a germ plasm (the nature of which is determined by previous ancestors) and of a most complex environment does not solve the problem. There is still the possibility that behavior patterns may differ in origin.

[1] Since Donaldson (1928) found the adrenal glands of wild rats to be roughly twice as large as those of albino rats, it is possible that the inheritance of wildness is explicable in terms of the inheritance of large adrenal glands.

There are responses which an animal obviously must acquire whereas others seem to be a natural or innate part of it. One must teach a setter to shake hands, but need not teach him to point at birds.

Responses Appearing at Birth.—It has been contended that responses which are present at birth are due to growth or maturation, whereas those which appear later are either acquired through learning or at least have been somewhat modified by learning. This distinction is, however, not clear cut. Various animals are born at different stages of development. The guinea pig is rather complete and self-sufficient at birth, whereas the rat, its close relative, is in an embryonic condition and most helpless. Chickens shortly after hatching are capable of feeding themselves, whereas robins cannot do so. It is obvious from these illustrations that the same amount of growth has not taken place before the time of birth. In no case is growth complete at birth, and as a consequence many responses dependent upon growth do not appear until later. Sexual behavior is a good example of this.

Further, one cannot be sure that modification of the sort we understand as learning, may not take place before birth. Holt (1931) describes how the "grasping reflex" may be learned. A random impulse traveling to the muscle of the hand of a fetus causes it to close. The closure in turn, necessarily sets up tactual and proprioceptive impulses which travel to the brain. These impulses set up an activity in the sensory area of the brain, shortly after activity of the motor area in the brain has caused the hand to close. Because of their contiguity these active areas become associated. Later, at birth, tactual stimulation of the palm causes activity in the sensory area which may spread and arouse the associated motor area. The aroused activity in the motor area results in hand closure, *i.e.*, the so-called grasping reflex.

A response to stimulation appearing at birth therefore need not be due to growth and maturation, but may be due, at least in part, to prenatal learning.

Responses Which Are Common to the Species.—To argue that responses which are common to the species are of a different origin from those which show species difference likewise has limitations. The fact that a response may be characteristic of a species proves only that it depends on common conditions and these may be either hereditary or environmental. Generally speaking, hereditary conditions are more alike than environmental, yet prenatal environmental conditions are very much the same.[1] If learning is possible in the fetus, then responses common to the species might be learned. Furthermore this distinction does not include responses which are not common to the species, but which might be inherited. Finally the word "common" is not sufficiently definite. How much alike must two responses be before they may be regarded as common?

Natively Determined Behavior as Unlearned Responses.—Since our knowledge of learning is more definite than our knowledge of what constitutes natively determined (or directly determined) behavior, it seems desirable, at least tentatively, to define natively determined behavior in terms of learning. In this case behavior patterns or parts of behavior patterns which are not dependent upon previous contact with the stimulus situation for their origin and development would be regarded as natively determined. An animal learns to avoid one arm of a maze and to go to another because of previous contact with that situation, but it does not have to learn to recoil when given an electric shock for the first time.

This definition serves to assist one in setting up an approach to the study of natively determined behavior. It does not assist one in recognizing such behavior when it is displayed, since learning may have taken place under unknown conditions. This means that in order to study natively determined responses one must separately analyze each behavior pattern and must exclude all possible sources of learning. In many cases one may suspect that there has been no opportunity for certain responses to be established through learning and may classify them as unlearned, but in such cases one must bear in mind the lack of conclusive evidence. The fact that chicks experience no visual stimuli before they are hatched suggests that a response to visual stimuli is natively determined. This does not mean that the complete pecking response is unlearned. The pattern as observed in the adult animal may have been modified by the first pecks. To determine what is native and what is acquired, the *nature* of the pecking response must be analyzed. Such an analysis has already been made in Chap. XI.

Unfortunately the changes in the nervous system produced through experience are unknown. The histological appearance of a brain taken from a rat which has learned many mazes cannot be distinguished from one taken from a rat that has not acquired such

[1] Child (1924) emphasized the importance of the highly constant environment of fetal mammals and its relation to the later uniformity in their behavior.

learning. If, however, observable changes in the growth of the nervous system can be found which account for the unlearned responses, a good basis for a positive characterization of natively determined responses will be had.

Since the definition of native behavior in terms of learning is useful only as it is suggestive of experimental approaches, it should be abandoned as soon as such behavior can be independently characterized. However, our first problem is to determine whether or not unlearned behavior is a scientific fact since this has actually been questioned.[1]

Evidence of Unlearned Behavior

We found the lower animals capable of rather complex behavior patterns, yet in many cases unable to profit by experience. The mass of their behavior thus constitutes evidence of the existence of native patterns. In cases where some degree of learning is present it constitutes a relatively unimportant part of their behavior. *The nature of their responses is thus primarily the product of genetic factors and certain environmental conditions (temperature, moisture, etc.) under which they develop and does not depend upon modification through previous exposure to stimulus situations.*

The existence of native behavior can also be demonstrated in higher forms. An animal such as the dog is incapable of either learning or retaining what has already been learned after destruction of its cerebral hemispheres. These structures are apparently necessary for both learning and retention. Nevertheless after such destruction the dog is capable of a variety of responses. Vital functions continue, and many responses to stimulation, though somewhat modified, remain. Irritation of the skin, for example, produces scratching. It seems reasonable, therefore, to believe that the remaining responses (reflexes) are natively determined. Whether or not such destruction also eliminates some responses which may have been natively determined, cannot be stated, but it is probable that at least those remaining are native in origin.

The most enlightening studies which are in point have been performed by Coghill (1929). He was interested in correlating the

[1] Kuo (1924) makes a strong case in which he points out that the structure of the animal facilitates certain patterns of behavior and that these are built upon through learning. He questions any neural basis for the unlearned behavior.

development of the nervous system with that of behavior. Amblystoma (see Fig. 49), because of its elementary nervous system, and because it showed striking behavior patterns at various stages in development, proved to be a most suitable animal for this purpose.

Behavior Stages in the Development of Amblystoma.—In developing the swimming response the salamander passes through the following distinct behavior stages: (1) The nonmotile stage during which it is unresponsive to sensory stimulation—yet has fully developed anterior muscles which respond when stimulated directly. (2) The early flexure stage during which a response to sensory stimulation is shown. This reaction develops gradually. At first only the muscles of the neck region opposite to the side stimulated contract and thus turn the animal's head away from the stimulus. As the animal grows in age, the muscles involved extend farther and farther down the trunk. An increase in age of about 36 hr. produces a response which involves the entire trunk. (3) The coil stage which is characterized by a complete contraction of one side of the body which is sufficient to produce a tight coil. The coil may be to the right or left and often forms first one way and then the other. In all cases the coil begins at the head end. Accompanying the development of the coil reaction there is an increase in the speed with which the coil is executed. At first the contractions are slow and sluggish, but by the time the coil is perfected considerable speed has developed. (4) The S stage in which a distinct change appears in the response. A contraction begins in the head region and spreads downward as in the coil, but before the contraction has spread all the way down, the contraction in the head and neck region reverses itself. Thus a new "coil" begins in the opposite direction before the previous one is completed, and like the first it spreads downward. This results in two flexures at different levels of the body, both spreading tailward. A succession of such contractions beginning at the head and spreading caudally characterizes the S reaction which is illustrated in Fig. 61. (5) The swimming reaction is characterized by an increase in the speed of the movements which constitute the S reaction and hence produces locomotion.

Neural Changes Which Accompany the Behavior Stages.—Having isolated these stages and the ages at which they appear, Coghill carefully examined the nervous system of animals at the critical ages. He found that in the nonmotile stage both sensory and motor mechanisms were present, but that there was no bridge between them. By the time the flexure stage was reached, however, a third type of cell appeared which bridged the gap between the sensory cells of one side and the motor cells of the other. These cells, known as *floor-plate cells*, are located in the brain stem and the

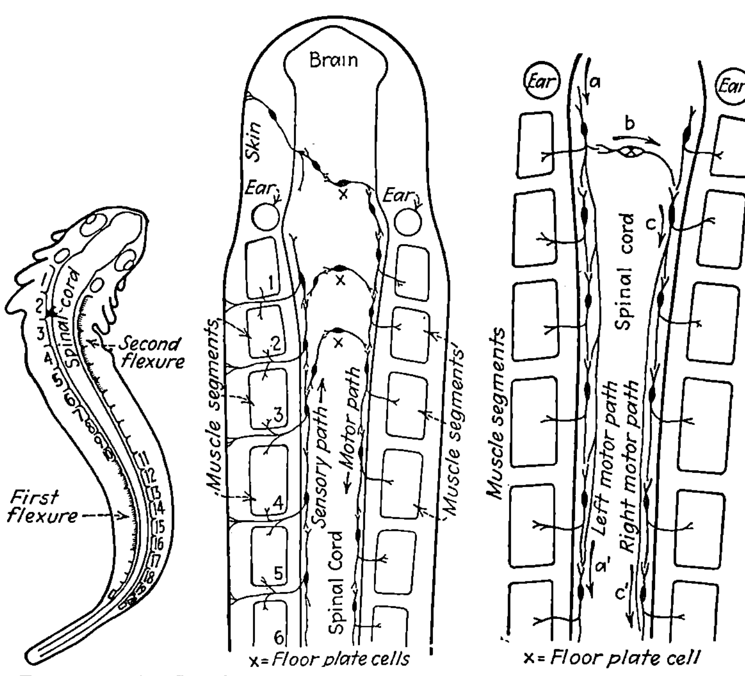

FIG. 61.—The S response in *Amblystoma*. The first flexure has passed tailward and the second is beginning in the anterior region. (*From Coghill*, 1929, *p*. 8. *By permission of The Macmillan Company*.)

FIG. 62.—The mechanism for the "coil" response. The floor-plate cells are indicated by x; the motor chain is shown on the right half of the spinal cord; and the sensory chain on the left half of the cord. The connections between sensory and motor cells with skin and muscles are also shown. (*From Coghill*, 1929, *p*. 12. *By permission of The Macmillan Company*.)

FIG. 63.—The mechanism for the swimming response. The sensory chain is omitted in this diagram. Both motor chains with the newly developed collaterals are shown. The arrows indicate the course of the nervous impulse. A disturbance at a spreads down both chains, but reaches a' before it reaches c'. (*From Coghill*, 1929, *p*. 16. *By permission of The Macmillan Company*.)

upper part of the cord. In Fig. 62 these cells are indicated by x. The growth of these new cells explains why the flexion suddenly appears at a certain age, why it is toward the side opposite that of stimulation, and why the flexion always begins at the head end. The only way stimulation of the tail can spread to the muscles is by way of these intermediate cells in the head region. Hence stimulation of either head or tail produces the same kind of flexion. The coil is executed when the chain of motor cells (shown in the right half of Fig. 62) is fully developed. Development begins at the head end and continues tailward. The degree of flexion is thus dependent upon the spreading of development toward the tail.

The reversal of the coil is produced by sensory cells (shown in left half of Fig. 62) which have connections both in the skin and in the muscles. Thus a contraction of muscles excites a second coil in the same manner as did contact with the skin, and as the contraction is on the opposite side to stimulation the succeeding contraction wave is opposite to the side previously contracted.

The S response becomes possible when the motor cells of each side make contact with the sensory cells of the same side. This is shown in Fig. 63. In this figure the sensory chain has been omitted and both motor chains are included. An impulse which now enters the brain spreads down the chain of motor cells of the same side as well as that of the opposite side. Since the impulse from the sensory neurones reaches the motor cells of the same side first, the wave of contractions at that side begins first. It is shortly followed by contractions of the opposite side which are set up as soon as the impulse has traveled the extra distance and passed over the floor-plate cell. Thus two waves of contraction are set up at this stage. The S reaction gains speed and so becomes the swimming response, through the development of collaterals in the motor cells of the neck region. These motor cells send large axones down the whole length of the animal and thus make contact with all of the motor cells in the chain. These collaterals are shown in Fig. 63. Before the appearance of the collaterals, the impulse could spread tailward only by way of a chain of neurones. Their growth, however, offers a direct route for the impulse, making the passage through separate cells unnecessary. In this manner the speed at which the contraction wave spreads tailward is greatly increased.

Neural Mechanisms Underlying Unlearned Behavior.—Coghill's study gives us a clear picture of characteristic behavior patterns which arise as soon as a certain growth stage has been completed. The accompanying changes in the nervous system are apparent and adequately account for the behavior. No degree of exposure to stimulation which is favorable for learning could have produced these changes. These experiments therefore not only give us evidence for the existence of unlearned behavior, but what is more, they give us insight into the mechanisms underlying such natively determined responses.

We may therefore regard the growth in the nervous system as a most important determinant of certain patterns of behavior. Through growth certain connections are formed and the nature of these connections determines the response. There is no evidence which shows that learning produces similar changes and it is probable that changes in the nervous system produced by learning are of a different nature.

Behavior which depends upon the formation of nervous connections of the sort described suggests a characteristic which we might expect in it. If a response depends upon the contact of neurones, it should appear rather suddenly. Thus if an animal at a definite age suddenly develops a new pattern of behavior it is quite likely that this behavior is directly determined. It is necessary, however, to observe the behavior the first time it is executed because, as we shall see later, learning may modify natively determined responses.

The Individuation of Specific Reactions

Previous to the work of Coghill it was believed that behavior involving the whole organism was made up of more primitive specific responses. Nest building, for example, was regarded as a chain of reflexes. Instincts were regarded as complexes or chains of reflexes. Coghill (1929) demonstrated that specific reactions such as reflexes are not primary, but rather the total response of the organism is the first to mature. All muscle groups are supplied by relatively few motor neurones and as a consequence the response of the whole musculature is synchronous. For example, the limb movements of the salamander first accompany the movements of the trunk; and only later, with the development of separate innervation of the limbs, are these capable of independent or individuated movement. A reflex movement of the limb is only possible where its muscles have special motor neurones. Further, in order that a particular set of muscles may act in isolation, the inhibition of the total pattern of movement is necessary. A reflex act thus involves the further maturation of motor neurones plus such nervous development as is necessary to produce inhibition of total movements. Because of these factors development follows the course of generalized to specific behavior patterns. As is the case of all development, the course of individuation is from head to tail. For this reason the fore limbs acquire individuation before the hind limbs.

The Individuation of Responses in Mammals.—In order to demonstrate the process of individuation in mammals, studies of fetal behavior become necessary. Fetal animals may be made available to stimulation by operating the mother, and with the placenta intact and the circulation maintained, placing the litter in a saline solution of the proper temperature and concentration. In this way the embryos may be kept alive long enough for experimental purposes. Knowing the exact age of gestation, embryos at various ages may be obtained.

Angulo y Gonzàlez (1932) studied the responses to stimulation of 643 rat fetuses at various ages. He found that the first appearance of motility is about 15 days and 18 hr. after insemination. At this age stimulation with a horse hair produces a slow lateral turning of the head opposite the side stimulated, the activity being confined to the muscles in the neck region. In older fetuses more and more of the trunk movements are present because of the more matured motor chain.

Strong stimulation (porcupine quill) according to Angulo y Gonzàlez (1933) produces a turn of the head toward the side stimulated. This response appears in the nonmotile stage as well as later and differs from light stimulation also in that contraction is maintained. This suggests that strong stimulation acts directly on the muscle rather than by way of the nervous system.

Limb movements though present at this stage are never made independently of the trunk until the seventeenth day. It may therefore be concluded that movements begin as mass reactions of the trunk and limbs. Only later do lesser mass movements, such as head extension, appear. Gradually the movements become more and more limited until specific reflexes appear. The appearance of specific reactions does not seem to be due to a breaking up of the total mass movements, because these total responses often reappear to dominate over the secondary movements. Rather the specific responses seem to depend upon the development of an inhibitory action by means of which the total mass movements are kept in the background. The gradual release of reflexes from the dominance of the total mass movements explains why reflexes vary from time to time. The inhibitory state existing at any time determines what the specific reaction will be.

On the nineteenth day the fetal response is specific so that the reaction is dependent upon the site of stimulation. On the twentieth day reactions of the embryo are similar to those of a newborn, except that they are less vigorous. By the twenty-second day the animal is born.

Similar results have been obtained by Coronios (1933) working with fetal cats. For cats the gestation period is about 62 days. The earliest responses to stimulation appear in 23-day-old fetuses. Early behavior is "diffuse, massive, variable and unorganized." Gradually this behavior is replaced by relatively individualized, consistent, and organized responses. Individuation, as in the rat, begins at the head end, and in the case of limb movements, the joints nearest the body are the first to move separately. Later the other joints become freed from the total arm movements. Many of the more complex patterns of behavior, such as crawling, seem to be due to the organization of individuated responses.

The Restriction of Sensory Zones.—The process of individuation is not merely the progressive reduction in the extent of muscular reaction, but involves also, as Coghill (1930*a*) has pointed out, the gradual restriction of the zone of stimulation which is adequate for producing the reaction. Coronios also observed the restriction of the sensitive zones in the cat.

Pratt, Nelson, and Sun (1930) obtained striking evidence of this in their study of human infants from birth to the eleventh day. In the case of a specific response such as sucking, the sensitive zone at birth covers the lips, the areas above and below the lips, the tongue, and the interior of the mouth. Stimuli applied to these parts in the form of touch, temperature, and taste all produce sucking. As the infant progresses in age, tactual stimulation of the lips becomes more and more effective, whereas the stimulation of other parts becomes less likely to produce sucking. Likewise temperature and tastes (except for "sweet") become less effective. "Sweet" tastes increase in effectiveness. Thus the zone becomes limited to tactual stimulation of the lips, and to gustatory stimulation in a specific form. It is not clear, however, whether this restriction is due to learning or to maturation. The infants in these studies were repeatedly tested and thus may have changed because of previous stimulation.

Raney and Carmichael (1934) found in the rat fetus an increase in the specificity and accuracy of responses to tactual stimulation on different parts of the body as the normal birth age approached. This indicates that the perception of space in the tactual sense is largely a process of maturation and is not the product of learning as is often supposed. Gestalt psychology has stressed the importance of development in perception. Traditional psychology seems to have overemphasized the importance of learning in space perception.

It appears now that much of the organization in behavior is due to genetic development and is not altogether acquired from past experience.

Concluding Remarks.—That development is a function of a maturing nervous system is well supported by experimental evidence, but that the nervous system is the sole factor must be questioned. In Chap. XI we have described the experiments of Kuo on embryonic chicks and found strong evidence supporting the belief that the development depended upon muscular maturation, relative size of various organs, the elaboration of sensory surfaces, and the animal's position in the embryonic state. These factors cause certain forms of behavior to develop. As in the case of arthropods, the behavior of an embryonic chick is the product of an interaction between its structure and the environment. Before the relative importance of the development of sensory and motor structures and nervous maturation can be known, further studies must be made, and the possible influence of each factor must be studied. Kuo was concerned with the problem of finding evidence which demonstrated the role of structural development in unlearned behavior, whereas Coghill and his students planned their experiments to study neural development. The plan of an experiment determines which facts will be brought to light; it does not determine the nature of the facts. One must recognize that no one experiment brings out all of the processes at work, and it is therefore fruitful for scientific progress that opposing points of view exist among the research workers.

The Method of Isolation in the Study of Native Behavior

Experiments on Swimming.—The role played by maturation in determining behavior patterns may be studied by controlling the conditions which allow for learning. If a particular stimulus situation calls out a behavior pattern in an adult, it is possible to isolate young animals and prevent exposure to this source of stimulation. In an isolated state maturation may progress normally and at the same time prevent learning.

In a study of the swimming response Carmichael (1926) divided frog and salamander eggs into two groups. One group developed under normal conditions, the other group was drugged with chloretone which allowed for normal growth, but caused the animals to remain inert. When the tadpoles attained the age at which the normal group had been swimming for five days, the drugged animals were placed in fresh water. After half an hour in fresh water, the swimming behavior of the drugged animals could not be distinguished from that of the normal. With a marked difference in the amount of experience in swimming no observable difference in this response was present, indicating that the response is a matter of maturation. However, it might be contended that only a small amount of experience is necessary, and that the 30-min. recovery time required by the drugged group was actually the acquisition of experience sufficient to perfect the swimming response. To answer this question Carmichael (1927) repeated his experiment and in addition reanesthetized the experimental animals after they had had experience in swimming. On again releasing the animals from the drug the same amount of recovery time was required. This shows that the gradual appearance of behavior after release from the drug was due to a gradual recovery from the anesthetic and not to an inability to perform because of a lack of learning.

These experiments show that previous experience in swimming is not essential to coordinated swimming, but that this pattern of behavior is due to normal growth, as was demonstrated by Coghill with his method.

The Pecking Response of Chicks.—The studies of the pecking response in chicks are studies of natively determined behavior by the method of isolation. Since this response has already been treated in Chap. XI, it is only necessary to remind the reader of the conclusions which are pertinent here.

1. Native mechanisms are present in the chick which cause it to peck at, to grasp with the bill, and to swallow small objects. Taste undoubtedly supplements the visual stimuli to bring about swallowing of food objects.

2. The behavior pattern is imperfectly coordinated on hatching, but maturation, either muscular or nervous, improves this coordination somewhat.

3. Practice supplements the natively determined pattern, primarily by improving the coordination.

4. Long delays (14 days) in which the chick is kept in darkness and is artificially fed, deprives the animal of its tendency to peck. This loss is perhaps due to an interference of habits. The animal has been fed artificially and has acquired tendencies to eat in response to contact, rather than in response to visual stimuli. Eating by means of pecking

must then be developed by establishing a connection between tactual and visual stimuli.

The pecking response found in the adult chicken must therefore be regarded as the product of both maturation and learning. The animal develops certain neural and muscular mechanisms during its embryonic life, and it is known that the latter continue to develop after hatching. To what extent the neural mechanisms continue to develop after hatching cannot be easily determined since muscular development is necessary to utilize what neural patterns are present. This makes it difficult to isolate the two mechanisms from a study of behavior. That development of some sort occurring after hatching improves pecking, and that practice further perfects the response is, however, demonstrated.

Space Perception.—As has already been pointed out, the experiment of Raney and Carmichael (1934) indicates that through development the skin on specific parts of the body of the fetal rat gradually acquires specific responses. The animal apparently has a native equipment for differential responses to various spacial relationships. Further evidence in support of space perception of native origin has been obtained by Lashley and Russell (1934) by the method of isolation. They compared rats reared in darkness with normal rats in the ability to regulate the force of a jump according to the distance to be jumped in reaching a food platform. Both groups used more force to jump the longer distances, and the accuracy of adjustment was nearly equal for both groups. Since the control of the jump depends upon the visual perception of distance, considerable organization between the visual and motor aspects of the behavior must be present without learning.

Sexual Behavior of the Rat.—Stone (1922) segregated male and female rats at weaning time (21- to 25-day-old rats). From time to time males were placed with females and careful observations were made of their behavior. Copulatory activity appeared at a certain age in a rather complete form. The average age for the first appearance was between 70 and 85 days and varied to some extent with nutriment. The basic behavior of the male seemed to be largely an inherited pattern, since it appeared suddenly, and since it was little modified by experience. Stone had difficulty in distinguishing the behavior of a young rat from that of an experienced adult. Similar behavior, but somewhat more stereotyped in nature, has been described in Amphibia (pp. 205*f.*) and in birds

(p. 254). In rats two types of differences were noted. These were differences dependent upon structural limitations and those dependent upon experience.

Structural difference in young and adult males resulted in differences in location of forepaws on the female and in the force used. Failure to grasp the female properly sometimes resulted in unsuccessful copulation. Lack of experience caused young males to be less aggressive and they failed to mate with resisting females. They sometime showed fear reactions in response to licking, which is the characteristic behavior of a female in heat. Lack of aggression on the part of young rats, however, might be a matter of their lesser strength as well as their inexperience. A study of males isolated until full strength was acquired would answer this question.

More characteristic deficiencies due to inexperience is the slight disorganization in the pattern. This results in failure to make definite clasp, mounting at head, failure to make pelvic movements, etc. Although in many cases, the observable behavior is rather complete in young rats, it is likely that in many cases complete copulation was not effected.

Stone also determined the external stimuli which were necessary for effecting the response. He found that young rats which had their vision, smell, and taste destroyed, behaved as normal animals in their sex activity, both as to the original appearance of the sex act and the frequency of copulation. Partial deafness as well as the removal of vibrissae also had no effect. It was found that the behavior of the female was the most essential feature of the environment in arousing the behavior. Females which responded to the sniffing and nosing of the male produced more excitement than did less responsive females. Females not in heat, but which were made sensitive by having their sex zones painted with a weak acid solution, became effective stimuli. The quick jumps made by a small guinea pig also proved adequate for arousing inexperienced males. It seems that the quick jumps made by females in heat or by other females made sensitive, are essential. This behavior of the female is communicated to the male by means of touch receptors and perhaps any others which are intact.

The physical and physiological conditions necessary for sex behavior are sufficient size, strength, maturation of nervous system, and glandular development. Stone (1927) found that 67 per cent of the male rats still copulated one month after castration, but at the end of six months only 9 per cent retained the ability. In the case of females, Stone (1926) found that sexual behavior appears about 50 days after birth. The pattern of behavior is fairly uniform and is similar in females raised in isolation and those raised with males. The necessary somatic conditions and the oestrous cycle seem the necessary internal conditions.

Rat Killing in Cats.—The most recent study of the genesis of the behavior of cats toward rats is that of Kuo (1930). During their development in isolation the behavior of kittens toward rats was observed periodically. Of 20 kittens raised in such isolation only 9 killed rats. However, 18 of 21 kittens raised in an environment in which they were exposed to rat killing by other cats, killed rats at about the age of 4 months. Only 3 out of 18 cats which were raised with rats as cage mates killed rats, and none of these killed their cage mates. By placing nonkilling cats in a rat-killing environment 9 out of 11 became rat killers.

He also found that the degree of hunger had no effect on the cat's responses, but cats raised on a vegetarian diet failed to eat rats, although their tendency to kill rats was not affected.

These results strongly indicate that the cat's behavior toward the rat is acquired. The construction of the cat is such that it readily makes certain responses, but the integrated pattern of behavior must be acquired through experience. Kuo points out that it is just as natural for the cat to love rats as to hate them. Having a particular structural equipment and being playful, it is easy to see how a tendency to attack rats may develop. That the rat can be eaten as well as played with must also be learned. As the response normally develops, it seems to depend upon a tendency to be excited by, and thus to pounce upon, a small moving object, the eating of which is due to accidentally tasting blood when the claws and teeth are sunk into some such object.

There are undoubtedly numerous other examples of behavior acquired in a similar manner which appear to be of native origin. Carpenter (1934) in his extensive study of howling monkeys describes many behavior patterns which are characterized by their structure. For example, their feet and tails make them well fitted for living in trees.

Since structural equipment plays such an important role in determining what will be acquired, Kuo (1924) believes that it is unnecessary to suppose that there are native behavior patterns. There is no doubt but that in many cases supposedly native patterns are the results of the structure of the animal, but the evidence already presented in this chapter strongly suggests the existence of some native neural patterning.

The Interaction of Native and Acquired Mechanisms

In most instances the behavior of an animal cannot be classified as depending upon either native or acquired mechanisms. Rather it is necessary to consider the natively determined patterns as furnishing the foundation for learning. As we shall see later in our discussion of the conditional responses, many cases of learning are merely an extension of natively determined patterns. By extending the range of stimuli which will call out a response, learning fits the animal better to the particular environment in which it finds itself.

In other cases the natively determined patterns of behavior are relatively imperfect. The innate mechanisms furnish all of the movements, but it is only after these are coordinated that perfection is reached. It is likely that learning plays an important role in coordinating and perfecting behavior patterns which are determined by inherited equipment. With rather general and incomplete responses to begin with, it is easy for the animal to make certain acquisitions. As the structure of the cat makes it fitted to learn to attack, so the natively determined behavior patterns make it fitted to develop certain other patterned responses. Learning may also influence the direction which the native determinants of behavior will take. It will be recalled (p. 279) that the normal pecking response does not appear in chicks which have been fed by hand in darkness. By this artificial method of feeding the native tendencies have been coordinated into a different feeding behavior pattern. This pattern conflicts with the pecking response which develops under the usual feeding conditions, and the chick can only be caused to peck at grains after tedious training.

The general environmental conditions under which an animal is reared may also modify the character of the later responses of an animal. Rats, for example, tend ordinarily to run along a wall. According to Patrick and Laughlin (1934), this tendency is greatly modified if the rats are raised in an environment without walls. Reared under such conditions the tendency to run into open spaces is greatly strengthened. It is true that repeated exposure to strange rooms increases a rat's exploratory tendencies and consequently overcomes the wall seeking tendency, which is apparently not very strong. Nevertheless, the above experiment demonstrates how even so minor a change in surroundings affects general behavior tendencies.

In the case of the monkey, Foley (1934) found that responses having to do with grooming seem to depend upon contact with other monkeys since a monkey reared in isolation failed to show such responses.

Because learning plays a more important role in the behavior of higher than in the behavior of lower forms, the natively determined behavior patterns of higher forms become much more extensively modified by experience than those of lower forms. It is through such modification that the animal may become adjusted to different environments and escape from the narrow bounds the optimum condition imposes. Higher forms are therefore less dependent upon specific external environmental conditions for survival than are lower forms.

Because of the interaction of acquired and innate factors in behavior it is impossible to classify many behavior patterns. Each type of behavior must be separately investigated. For the proper analysis of a behavior pattern, the underlying mechanisms, whether native or acquired, and the stimuli necessary for setting them off, must be experimentally determined. This is a long and difficult program but a necessary one.

Native Responses as Adaptive Responses

Native behavior patterns tend to terminate themselves by removing a necessary part of the stimulus pattern which sets them off. Feeding behavior, for example, depends on an internal need or drive such as hunger, as well as sensory stimulation from the food. The responses to the food eventually eliminate the need and so terminate the behavior. Sucking responses thus depend upon sensations of hunger as well as stimulation of the mouth parts. When hunger disappears, external stimulation ceases to produce sucking. In a similar manner sex behavior, maternal behavior, nest building, etc., depend on both external and internal stimuli. The *internal* state becomes modified by the response and external conditions alone cease to be adequate. Whether internal conditions direct the nature of the response to external stimulation or merely serve as a source of facilitating stimuli is not always clear. In any case the term *instinct* is ordinarily applied to such behavior patterns.

Protective responses also terminate themselves. Such responses as the scratch reflex, set off by irritation of the skin; the contraction of the pupil in response to light; the withdrawal of the whole or part of the animal from intense stimuli (heat, for example), etc., terminate the response by removing the *external* sources of stimulation. Such responses are usually fairly local and consequently their pattern is more simple. The term *reflex* is often applied to such simple patterns. They differ from *instinctive* responses in that they do not depend upon an internal and temporary condition. The *protective* response is always available and can be repeatedly elicited.

There are also responses which seem to have no adaptive function in the sense that the response has no effect upon any part of the stimulating situation. Such responses as the knee jerk, the extensor thrust (although this may be regarded as protective: pushing-away stimulus), the stepping reflex, etc., are examples of this. Ordinarily these are regarded as complete patterns of response to stimulation and as such they may be nonadaptive. There is no reason why such responses may not exist so long as they do not interfere with the animal's survival. Most responses must, however, be adaptive since an animal with a too limited repertoire of adaptive responses would not perpetuate the species. On the other hand, one should not assume that because a particular response has no effect on the stimulus situation, it is necessarily nonadaptive. It is possible to regard such responses as segments or partial patterns, and when combined with other responses, these segments may have an adaptive function. When one considers that reflexes are an individuated part of the total pattern of behavior, one might expect certain segments to appear in the absence of supplementary segments. One must therefore consider the complete response to stimulation before classifying it according to its "adaptive" value.

Responses which at first have little or no significance for adaptation may also become adaptive when supplemented by learning. Partial patterns may produce tendencies toward the acquisition of certain new acts, and thus become part of an acquired behavior pattern.

Because apparently useless movements may actually become incorporated into responses which make for survival, it is misleading to classify natively determined behavior as either adaptive or nonadaptive. This does not mean that all responses should be regarded as existing for a purpose. Any adaptive function that responses may have is purely accidental. It is because adaptive responses make for survival that they are so frequent. Many responses which may be present and which do not conflict with responses that make for survival may become incorporated with others (through development or learning), and thus also become adaptive.

Suggested Readings

Coghill, G. E. 1929. *Anatomy and the problem of behaviour.* London: Cambridge Univ. Press. Pp. 113.

Holt, E. B. 1931. *Animal drive and the learning process, an essay toward radical empiricism.* New York: Holt. Pp. 307.

Stone, C. P. 1934. *The factor of maturation. Handbook of General Experimental Psychology* (ed. by Murchison). Worcester: Clark Univ. Press. Pp. 352–381.

CHAPTER XIII

THE DIFFERENTIAL REACTION TO STIMULI

Introduction

The number of differential reactions that an animal may make to its environment is limited by the variety of effects that stimulation may have on it. In the higher animal each sense organ responds most readily to a different form of physical energy and because each has distinctive nervous connections, different forms of energy may result in different responses. Since sense organs furnish the sensory data upon which various reactions are dependent, an analysis of their efficiency becomes very essential to an understanding of behavior.

The first problem is to determine whether or not an animal has functional sensory tissues which are especially adapted to respond to specific forms of energy. If sense organs are found, the next problem is to determine the sensitivity of such organs. In higher forms, such as vertebrates, which have highly developed sense organ structures, there is little question that these are functional. The problem then becomes one of determining to what extent each sense organ can respond to variations within one form of energy. Having found, for example, that a rat's eye functions when energy in the form of light strikes it, the problem becomes one of determining what variations in light energy cause different effects upon the visual receptor.

It is not sufficient, however, for the sense organ to respond uniquely to such variations in stimulation. To produce a differential reaction each difference in the function of the sense organ must result in a specific effect in the central nervous system. To continue our example, it is not only necessary that the eye of the rat respond uniquely to, let us say, various wave lengths of light, but each difference in the function must produce its own specific process in the brain in order that the animal respond differently to the various wave lengths. The human eye reacts distinctively to the various wave lengths, and differences in function of some nature are maintained and lead to distinctive experiences which we call hues or colors. The human eye also reacts to various intensities of light resulting in various experiences of brightness; and to various distributions of light resulting in specific experiences of pattern and form. The eye of an infrahuman animal may be deficient or lacking in its ability to produce any of these effects. To the extent that the animal lacks this ability, changes in the lighting of the environment make no difference to it.

To demonstrate that the structures of given sense organs in certain animals are similar to those of man does not establish the fact that such animals are capable of the same distinctions as man. Not only must such structures communicate variations in stimulation to the central nervous system, but the relative effects of these variations must remain the same. For example, the eye of the monkey may respond to changes in the wave length of light as well as to changes in the intensity. To man, varying wave lengths produce marked differences in experience which stand out as dominant over differences in experience produced by intensity variations. The monkey, on the other hand, may have an eye capable of communicating the same distinctions to the brain, but the dominant feature in its experience might be brightness. Hence the difference in the effect of the same kind of stimulation in various animals may be a difference in perception, rather than a divergence in the sensory processes.

Sense organ activities furnish the data for perceptual experiences, but since the perceptual experience represents a certain organization of sensory data, the organization cannot be regarded as equivalent with sense organ activities. Perceptions are determined by the complexity of the nervous system and by the pattern of stimulation. In the absence of sensory data there can be no perceptions, but as circumstances differ, the same sensory data may result in different perceptions. From this it is obvious that the animal's world is by no means readily accessible to man. By limiting the sensory data to changes in one characteristic (such as those produced by changes in the wave length of light) and leaving the intensity and the distribution of light constant, while excluding stimulation of other sense organs as far as possible, one can determine fairly well whether the aspect under consideration has any effect on the animal. Most of the studies on animal discrimination have been of this type. The interest has been one of determining the sensory capacity of the animal, and the problems of perception have been largely overlooked. Recently Fischel (1932) and Klüver (1933) have stressed the perceptual aspects of animal behavior.

Because perception has not been investigated in animals, we leave it with the understanding that it is most important and should not be ignored when interpreting results of sense organ studies. Let us, then, turn our attention to studies which show to what extent sense organs can furnish different sensory data under different conditions of stimulation.

Methods

To determine whether or not different stimuli produce different effects on an organism, some differential response to distinctive stimuli must be isolated. If we wish to know whether two intensities of light produce different effects in the animal's central nervous system we must demonstrate that the light of one intensity corresponds to one reaction and that the other light corresponds to a different reaction. It must be shown in such cases that the difference in reactions is dependent upon a difference in intensity and not upon any other difference between the stimuli. For this reason the attempt is always made to make a pair of stimuli alike in all but one respect.

The different reactions associated with the different stimuli may be of two kinds, (1) natively determined, or (2) acquired.

Natively Determined Differential Reactions to Different Stimuli. Natural or innate differential reactions to different stimuli are used in the study of lower animals in which learning plays a minor role in the repertoire of reactions. Specific reactions to certain aspects of a stimulus situation are often difficult to find and to measure. *Tropisms* or *preference* reactions have therefore been used rather widely. For example, organisms which collect in darkness or which tend to avoid light will aggregate under the less intense of two lights. When the differential reaction to these stimuli disappears, it is evident that the difference in the two lights has become too small to affect the animal. Hence field mice prefer the darker of two compartments so long as these differ sufficiently.

Organic changes such as changes in respiration have been successfully used. Some stimuli increase the respiratory rate and so the reaction can be used to study the effect of different stimulus intensities.

Other characteristic reactions are sometimes discovered by accident and are effectively utilized. Reeves (1919), for example, observed that her fish came to the surface and snapped at bubbles when the aquarium was illuminated with "red" light. She found that when the intensity of "white" light was sufficiently reduced the same reactions occurred. It seemed that this characteristic reaction appeared whenever a light had a low stimulus value for the fish. This reaction was used by Miss Reeves to determine at what intensity different wave lengths of light had the same stimulus value for fish.

Acquired Differential Reactions.—Instead of capitalizing some directly determined reaction one can train an animal to give a certain reaction when one stimulus is presented, and fail to give this reaction or to give some other reaction when a different stimulus is presented. In this manner the experimenter may select a reaction that is very conspicuous and easily measured. The method assumes that the animal is capable of easily acquiring certain reactions to stimuli.

The *method of conditioning* has been extensively used as a means for studying an animal's capacity to discriminate. By this method the animal is trained to react in a certain way to one stimulus situation and not to react to another which is different from the first in all respects or in one particular respect. Suppose salivation is the reaction selected. The experimenter knows that the taste of food calls up this reaction. If now the animal is always fed in connection with one stimulus and is never fed in connection with other stimuli which differ sufficiently from this, the animal soon tends to respond by salivating when this one stimulus is presented and not salivating when other stimuli are presented. Thus a circle may provide stimulus for salivation, whereas an ellipse of equal intensity may not. But if the ellipse is made more and more like a circle, a point will be reached at which the differential reaction (salivation versus no salivation) will disappear. This point may be regarded as the greatest difference between stimuli which cannot be experienced as different. To establish the fact that the discrimination was based on the shape and not upon size and brightness, it must be shown that the differential reaction breaks down only when the difference in shape is eliminated.

The discrimination-box technique requires the animal to learn which of two or more boxes always contains food and to select this one and avoid the others no matter what the relative positions of the boxes may be. If the boxes differ in some characteristic aspect the animal may utilize this aspect to recognize the food box. If, for example, the food box always has a triangle on its front side and the other boxes are alike in all respects except that they have figures which are not triangles, the animal, providing its vision is adequate, has the opportunity of recognizing the food box by means of the triangle.

Yerkes (1907) has developed a discrimination technique which easily controls many variable aspects of various stimuli and which has been widely used in modified forms. In Fig. 64 a typical dis-

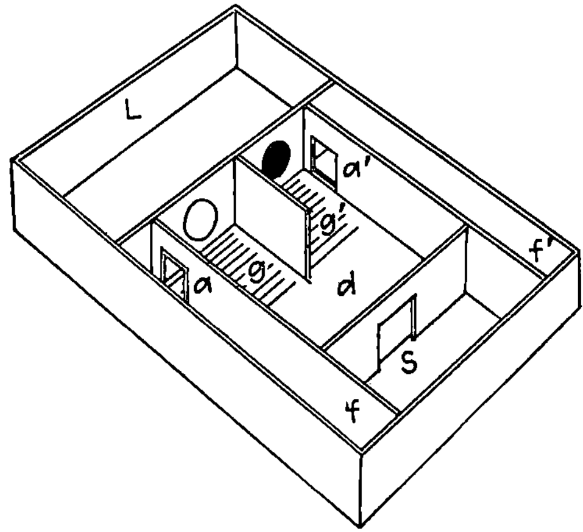

FIG. 64.—A typical discrimination box. The animal is started at *S* and is required to choose between a darkened and a lighted compartment. One of the doors (*a* and *a'*) is unlocked and leads to food at *f* or *f'*. The lighting is controlled by means of apparatus placed at *L*. *(From Lashley,* 1929, *p.* 33. *By permission of the University of Chicago Press.)*

crimination box is shown. The animal starts at *S* and when it arrives at *d*, it must choose between the right or left of two compartments. At the end of each compartment there is a window (or paper figure), the appearance of which characterizes the compartment. Each window may be illuminated from behind and its brightness varied as desired by moving the light source different distances from the window. The window of one compartment may be made to differ from that of the other in either shape, color, or brightness. The animal is required to learn to enter one compartment and to avoid the other. Since the windows are exchanged from time to time the compartment to be entered may be to either the left or the right. In order that the animal will learn to enter the one compartment rather than the other the selection of one window is accompanied with reward (or no punishment) and the selection of the other is accompanied with punishment (or no reward). Reward is administered by permitting the animal to progress through a near-by door (*a* or *a'*) which leads to the outer alley and hence to the nest or to food. Punishment is administered by locking the door which leads to the nest or the food box. In addition the compartments are wired so that the animal may be given an electric shock when it chooses incorrectly.

As the windows are different in one or more respects the animal has the opportunity of learning which compartment leads to reward and which to punishment. If the windows of the compartment differ in color and in no other way, the animal may learn to choose the one of a certain wave length provided that "color" plays a sufficiently important part in determining the animal's perception.

We have indicated that the last two methods require learning on the part of the animal. We shall find later that the conditions for learning vary greatly, depending on the technique employed. At present we are interested in the animal's ability to make discriminations. The learning problems involved will be treated in later chapters.

Discrimination of Wave Lengths of Light

This form of discrimination is highly developed in some animals, whereas in others it is either absent or plays a very minor role. The evolutionary history of "hue discrimination" is very important to a theory of color vision. Comparison of the capacity of various animals is therefore highly desirable. Because many animal forms have been investigated and because the problem has been particularly difficult, studies in the discrimination of wave length of light are not only of value for an understanding of visual reactions but serve also as a means for gaining knowledge of general techniques and methodology.

In Part I we found that certain insects and fishes showed differential reactions to various wave lengths. In no case, however, has the color vision been sufficiently analyzed to make possible a conclusion regarding the kind and extent of wave-length discrimination. The most careful analysis has been made on birds. The methods used in their case will therefore be treated in some detail.

Birds.—Using the Yerkes-Watson color apparatus[1] Lashley (1916) tested the color vision of Game Bantam cocks. At the end of each of two compartments, windows illuminated by spectral lights appeared before the animal. From each compartment a door led to a food box. The chicks were rewarded by finding food if they went to the compartment of one color, and found no food if they went to the other.

One chick was trained to prefer a "red" (650 mμ) to a "green" (520 mμ) and another chick to prefer the above "green" to the "red" light. After 200 trials the preference was well established. Tests were then made to determine whether the reaction was based on a difference in the wave lengths or to a difference in intensity. He found that making either light very bright and the other very dim had no effect, the chicks still preferred their training color. When "white" light was substituted for the "red" so that a choice between "white" and "green" was possible, the chicks trained to avoid "green" went to the "white" and the chicks trained to go to the "green" continued to make the learned response. Similar results were obtained when white light was substituted for green. As "white" light was the brighter member of each of these pairs of stimulus objects, the results can only be explained by assuming the ability to discriminate on the basis of wave length.

The same type of results were obtained when "yellow" (588 mμ) and "blue-green" (500 mμ) lights were used as stimuli. Changing the size, shape, or sharpness of the colored image; or replacing the spectral lights with yellow and blue-green lights produced by color filters,[2] had no effect. These controls show that extraneous differences, which might conceivably be associated with one or the other color, played no part in determining the animal's response.

Lashley then proceeded to determine at what points in the spectrum distinct changes in experience appear for the chick. To

[1] Yerkes and Watson (1911) have described an apparatus by means of which spectral hues may be presented to the animal in an ordinary discrimination box. Their apparatus allowed for careful control of both wave length and physical energy, but was limited in its function in that similar wave lengths could not be presented simultaneously.

[2] Color filters permit only light of certain wave lengths to pass through them. Thus when placed before a source of white light, all but a certain range of wave lengths are excluded. By means of filters monochromatic lights are obtained but these are never as pure as those obtained from the spectrum. They have an advantage, however, in that control of area and intensity is less limited.

the human a slight reduction in wave length in the vicinity of 580 mμ suddenly produces a "yellow." Chicks trained to choose "red" (650 mμ) instead of "green" (520 mμ) continue to choose "red" when "yellow" is substituted for "green." Thus the preference for "red" remains when the other member of the pair is changed. If the "red" is changed, the preference for it should cease when it produces a widely different hue experience. Lashley changed the "red" of the "red-green" pair through the "orange" region of the spectrum without destroying the preference, but when it came in the vicinity of 580 mμ (yellow for man) the preference ceased, and the "green" was chosen as often as the changing color. The chick trained to go to "green" (520 mμ) ceased to prefer it when its wave length was increased above 600 mμ (the region which is orange for man).

Another chick trained to go to "blue-green" (500 mμ) instead of "yellow" (580 mμ) ceased to prefer the "blue-green" when the "yellow" was changed from 565 mμ to 535 mμ. It is between these wave lengths that light changes from yellow to green for man. Although Lashley's experiments are not detailed enough to prove that the chick's color vision is identical with that of man, they do show that it is similar in that the points of greatest change in the quality of spectral rays are alike in location and in number.

Hamilton and Coleman (1933) obtained similar but more detailed results with pigeons. They trained the birds to jump from a stand to one of two perches each of which was located in front of a ground glass screen which could be illuminated by spectral lights. The screens formed doors which were hinged so that when the bird hopped to the correct perch, it gained entry to a dark shelter which served as an incentive. When the bird jumped to the wrong perch it found the door locked, and as the perch was very close to the window the bird lost its balance and fell into a canvas trough below. During the training period the intensities of the lights were changed from trial to trial so that the only consistent association which the birds experienced was that of a certain hue and the open door into the shelter. Training was continued until perfect records were made by the pigeons.

The next step in the procedure was to change the wave length of the "hue" that the birds had been trained to prefer, by small steps (10 mμ at a time), so that its wave length approached that of the avoided "hue." Finally, a point was reached at which the ability to discriminate broke down. For example, a pigeon trained to jump to a hue of 520 mμ (green) and to avoid 700 mμ (red) failed to discriminate when the wave length of the first stimulus was increased in steps until it had a wave length of 620 mμ. At 615 mμ the pigeon still succeeded. Hence the difference between 615 mμ and 700 mμ or 85 mμ was the least difference the bird was able to detect. It may therefore be concluded that when one hue has a wave length of

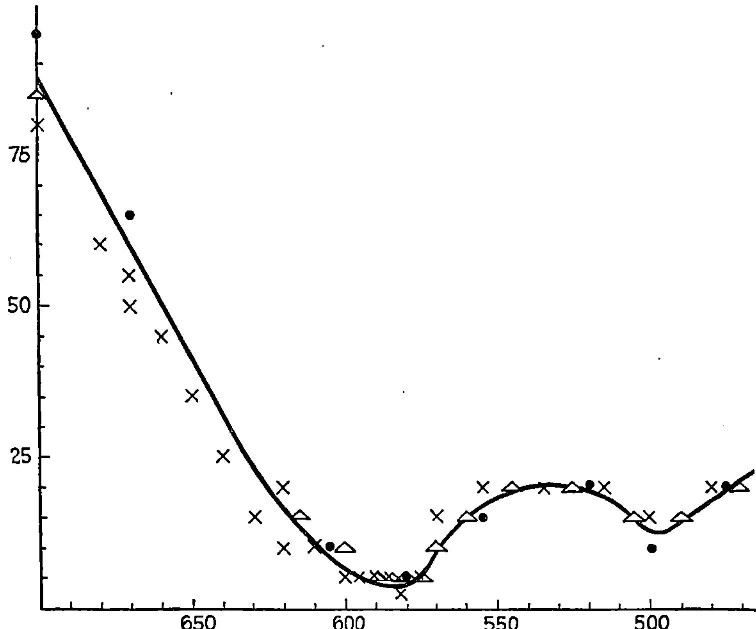

FIG. 65.—Hue discrimination curve of three pigeons indicated by different symbols. The ordinates represent difference thresholds in mμ, the abscissae, wave lengths in mμ. The smooth curve is drawn through the averages. (*From Hamilton and Coleman*, 1933, *p*. 188. *By permission of the Journal of Comparative Psychology*.)

700 mμ, a second hue must have a wave length 85 mμ shorter before it produces a different hue experience.

By measuring the least detectable difference which could be experienced by the pigeons for all points on the spectrum, the points of rapid changes in quality were determined. Figure 65 shows the curve obtained when the least detectable difference is plotted against the various wave lengths. From this curve it can be seen that at two points in the spectrum very small differences in wave length can be detected. At 580 mμ for example, a difference of

5 mμ was detectable. These results show that for the pigeon the spectrum changes rapidly in quality at about 580 mμ (yellow region) and again at about 500 mμ (blue-green region).

By supposing that there are three types of retinal cells sensitive, respectively, to long, medium, and short waves, these results are adequately explained. The points at which small differences are detectable are the points at which one receptor cell (*e.g.*, the one sensitive to long waves) is gradually ceasing to function and the second receptor cell (the one sensitive to medium wave lengths) is gradually functioning more and more. As the different receptor cells produce distinct experiences, the change in experience is very marked at the point of change in function.

The curve of the pigeon is similar to that of man in that there are two low points. It is slightly different in that the points at which the different receptor cells begin to function do not occupy exactly the same position in the spectrum. In the pigeon the middle receptor does not begin to function until a wave length of 620 mμ is reached, but in man it begins at 650 mμ. The human eye is also more sensitive to change in all parts of the spectrum. Thus 20 different wave lengths can be distinguished from one another by the pigeon whereas man can distinguish about eight times as many.

Rodents.—Watson and Watson (1913) attempted to train rats and rabbits to discriminate hues in the Yerkes-Watson color apparatus, but found that whenever a discrimination was set up, it broke down with the introduction of changes in brightness. For example, a rat trained to go to a "yellow" (595 mμ) instead of to darkness continued its preference for "yellow" when a "blue" (478 mμ) of low intensity was substituted for darkness. However, when the intensity of the "blue" was increased, a point was reached at which chance scores were made. In all cases it was found that changes in intensity could be made which would cause a learned preference to break down, indicating that the preference was based upon "brightness" rather than "hue." Tests also indicated that long wave lengths were ineffective for producing experiences of "brightness." The animals reacted to red in much the same way as to darkness. In training an animal to go to "red" instead of "green," it was found that the removal of "green" resulted in chance scores, but the removal of "red" had no effect.

Coleman and Hamilton (1933) and Munn (1932) also failed to train rats to react to wave-length differences. Both used "colored" papers instead of spectral lights. Coleman and Hamilton found that a rat trained to go to "black" instead of "red" changed its preference to "red" when it was made darker and the "black" was changed to "gray." Similarly, changes in the brightness of other pairs resulted in a reversal of the preference, showing that the animals had learned to respond to the darker or to the lighter of the two members of the pair. They also found pairs of colors which produced chance scores. These they called "confusion pairs" since such pairs apparently were alike in brightness for the animals, although they differed greatly in wave length.

Coleman and Hamilton considered the possibility of the animals experiencing hues, but tending to respond to brightness differences because such differences were more apparent. For this reason these experimenters continued training on the "confusion pairs" for a long period, but the animals continued to make chance scores and finally refused to work. To exclude the possibility that the rats may have had the wrong start in that they had learned to use brightness differences during the initial trials of the experiment, a new group of rats was tested on the "confusion pairs." These animals also failed to make any discrimination.

The only positive evidence of color discrimination in rats was obtained by Walton (1933). He used filtered lights which produced relatively impure "colors" and attributed his success partly to his use of larger and more intense patches of color and partly to the fact that he began training his rats with "color" pairs which were equal in intensity to the human eye. He equated the brightness of his pairs in this fashion because he wished to discourage responses on the basis of brightness and considered the possibility that a brightness match for the human eye might be the same for the rat. But Coleman and Hamilton had trained rats on the "confusion pairs" which had been found to be equal in brightness for other rats and found this to be of no advantage.

Walton's results, nevertheless, showed accurate discrimination between the following pairs of colors: "red-green"; "red-blue"; "blue-yellow"; and "red-yellow." Discrimination between the "green-blue" and the "yellow-green" pairs did not reach a high degree of accuracy but was better than chance.

According to Walton, checks were made to exclude responses on the basis of brightness. He states that one or the other member of the pair was made definitely more intense than the remaining member, but does not state how he became certain that he actually had changed the relative brightness of the pair for the rat. If he used his eye to judge changes in brightness, his results show nothing. Since Watson found great differences in brightness sensitivity for various wave lengths, it is necessary to demonstrate that the brightness has actually been changed for the animal. Before one can accept the fact of color discrimination in rats, Walton's results need to be verified. His results are opposed by several other studies and his method is not sufficiently different to suggest that the negative results obtained by other workers was due to inferior methods.

Histological studies which deny color vision to animals (such as rats) because no typical cones are found in their retinas are not convincing. Such conclusions depend upon the assumption that the function of visual cells can be determined from structure. This argument has often been used to prove color blindness in rats.

Walls (1934) has recently demonstrated conclusively that the retina of the albino rat contains cones as well as rods. This does not demonstrate color vision in the animal, but it completely destroys the argument that rats may be color-blind because they have no cones.

Cats and Dogs.—Although Colvin and Burford (1909) found evidence of good color vision in dogs and cats, more carefully controlled experiments have failed to support such conclusions. After carefully controlling brightness and texture of paper, DeVoss and Ganson (1915) conclude that cats are not only color-blind, but have very poor daylight vision. In the case of dogs, Pavlov was unable to establish differential responses to wave length. Smith (1912) tested seven dogs and concluded that some dogs possess a "color sense," but that in their case it is so rudimentary and unstable that it plays no part in the animal's normal existence.

Primates.—It is generally conceded that monkeys have color vision. Watson (1909) tested three monkeys (two *Macacus rhesus* and one *Cebus*) with spectral lights and found that they could distinguish "red" from "green," and "yellow" from "blue" on the basis of wave length. Bierens de Haan (1925) tested a pig-tailed macaque by training it to open one of five colored doors. Behind the door of a certain color, irrespective of its position, a piece of banana was found. All other doors were locked. By this method it was found that "red," "blue," "green," "yellow," and "gray" papers could be distinguished from each other. To control brightness, thirty different "grays" were used from time to time. These were often so nearly alike that they could not be distinguished from each other, but they were never confused with the colors by the monkey.

Klüver (1933) found a squirrel monkey able to discriminate between a "violet-red" and a "yellow." He trained it to choose the "violet-red" and to avoid the "yellow." It was then tested on new pairs of colored stimuli. When either "black" or "white" was presented with "yellow" the monkey never chose the "yellow," and when either "black" or "white" was presented with "violet-red," the latter was chosen 90 per cent of the time. If the animal had been responding on the basis of relative brightness, then the reactions to black and white should have differed because black was always darker, and white always brighter than the color. To avoid consistently "yellow" and to choose consistently "violet-red" when either was present, the animal must either have reacted to color or to the specific brightness values of the colors. To respond to a specific brightness as indicated in the above results, the animal must have learned to avoid the specific brightness experienced from the "violet-red" card. To test this possibility, different pairs of grays varying in brightness were presented. In these tests the animal showed no preference but made a chance score. In other words, some color had to be present before the animal made a preferential response.

Chimpanzees, like monkeys, probably have color vision. Köhler (1918) tested them by mixing colored papers on color wheels which were placed behind the windows of two boxes. Different mixtures of color were made and the ape was taught to choose one mixture in preference to another. The following mixtures were used in one of the tests:

Color *A*	360° "blue"	0° "red"
Color *B*	270° "blue"	90° "red"
Color *C*	200° "blue"	160° "red"
Color *D*	100° "blue"	260° "red"
Color *E*	30° "blue"	330° "red"

The mixtures represent a series of "purples" ranging from "blue" to almost all "red." An ape was then trained to choose color *D* instead of color *B* (*i.e.*, the more red). After this preference had been established, different pairs of the above mixtures were presented (*e.g.*, colors *E* and *C*). The animal showed a marked preference for the mixture containing the greater amount of red. The preference persisted when the brightnesses of the colors mixed were radically changed. This indicates that mixtures of "red" and "blue" form, as in man, a continuous series in that no marked changes in quality appear from such mixtures. Had differences in quality been markedly present, training in colors *B* and *D* would not have resulted in a preference for the more "red" when any of the above colors were presented in pairs.

In a similar manner "yellow" and "red" were mixed and were also found to form a continuous series.

Although these experiments are rather incomplete they suggest that the color vision of the ape is probably similar to that of man, in that the parts of the spectrum forming a continuous series are the same. The method is perhaps not so exact as that of Lashley in which the points of greatest change in the spectrum were determined. The methods of Köhler and Lashley are interesting in that one determines which parts of the spectrum form a continuous series, and the other determines which parts represent change points in the series.

Summary.—Although color vision is highly developed in birds and probably in monkeys and chimpanzees, it plays an unimportant part, if any, in the life of most mammals. That differential reaction to wave length is present in birds (and even in fishes and insects), absent in lower mammals, and reappears in primates is of interest, especially since it reappears in higher animals in much the same form that it is found in birds. Much exacting work, however, still remains to be performed, and the experiments of Coleman and Hamilton should serve as a model.

Discrimination of Form or Pattern

The visual sense organ not only furnishes data which may vary with wave length and intensity, but also data which vary with the distribution of light which constitutes the stimulus pattern. It is our task to determine the relative efficiency of the visual sense organs of various vertebrates in this respect. As in the case of color there are many difficulties to be met. Since the most detailed and careful work has been done on the rat, we shall treat this work first, in order that the difficulties may be appreciated.

Rodents.—Lashley (1912) used an ordinary discrimination box in his first study of vision in rats. At the end of each of two alleys there was a ground-glass window, one in the shape of a circle, the other a square. These were illuminated from behind and were visible to the rat when placed at the starting point. Doors led from these alleys into food chambers. Either one of the doors was always locked and the animal was required to learn which door was unlocked. As the unlocked door always was in the alley with the square window, the animal could learn which alley to choose if it could distinguish the square from the circle.

It was found that after a thousand trials, three rats made an average of 48.3 per cent incorrect responses, which is approximately a chance record. In an attempt to attract attention to the windows Lashley caused the lights to go off and on, but in several hundred trials no improvement was shown.

Other pairs of figures were then used in place of the square and circle. No discrimination was evidenced when one large circle appeared in one window and two small circles (together equal in area to the first) appeared in the other window. The animals did succeed in discriminating a vertically from a horizontally placed rectangle (20 by 30 mm.). They also learned to discriminate between two circles of different diameters but failed when the larger was reduced in intensity so that it gave off as much light as the smaller. When, however, the alleys were lighted so that their illumination did not depend upon the windows, the rats succeeded in discriminating differences in size of circles independent of their intensity. Circles and squares as well as circles and stars different in area, likewise were discriminated, but when their areas were equated the discrimination broke down. To discriminate size the ratio of the areas had to be more than about 4:3. These results seem to be convincing evidence that size, but not form, can be discriminated by the rat.

On the other hand, Fields (1928, 1929) found that rats could discriminate circles from stars and upright from inverted triangles when the animal obtained food by crawling through the figure. However this method introduces numerous uncontrolled factors. Thus Fields found that blind rats also succeeded. When, however, smaller figures were placed above the holes through which the animals were required to crawl (as in Fig. 66), blind rats failed, and normal rats learned. Several control experiments were then introduced to determine whether secondary visual features were involved in the discrimination: (1) Paper figures were used to eliminate the shadows that the illuminated holes might have cast on the floor. (2) A triangle and a rectangle equal in area and base were used as figures, to test the possible effect of a difference in the distance that the center of brightness of upright and inverted triangles might be from the floor. Other control tests were also made. In all cases the discrimination was not affected. These tests indicate that the rats had learned to respond to the figures on the basis of their shapes.

Munn (1929) repeated Fields' experiments with somewhat better control of such factors as olfactory and tactual differences in the

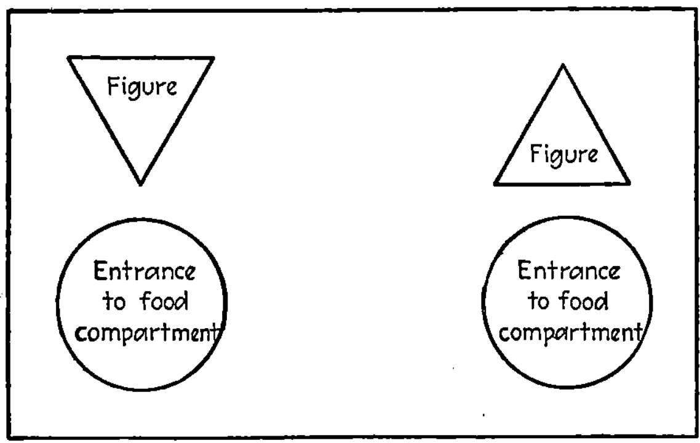

FIG. 66.—A form board used in the discrimination of patterns. (*Modified from Fields*, 1928, *p.* 147.)

holes that the animals crawled through, and the distribution of illumination of the figures as seen from the rat's position when making its choice. Only four out of nine rats learned and these failed when the distribution of the lighting behind the figures was changed. Unfortunately he did not continue training to determine whether or not the discrimination could be re-established. On the basis of his results Munn questioned Fields' conclusions.

Fortunately Lashley (1930*a*) has now established pattern vision in the rat by a method in which the rat jumps from a stand to one of two cards containing geometrical figures. His apparatus is shown in Fig. 67. One card falls over when the rat strikes it, which gives the animal entrance to a platform containing food. The other card is fixed and when it is struck, the rat falls to a net below.

By this method rats have been trained to discriminate the various types of patterns shown in Fig. 68.

Birds.—Form discrimination can be established in the bird relatively easily. Bingham (1922) studied the vision of chicks in great detail and found that they readily learned to distinguish between squares, circles, and triangles when trained in a discrimination box of the type which produced negative results in

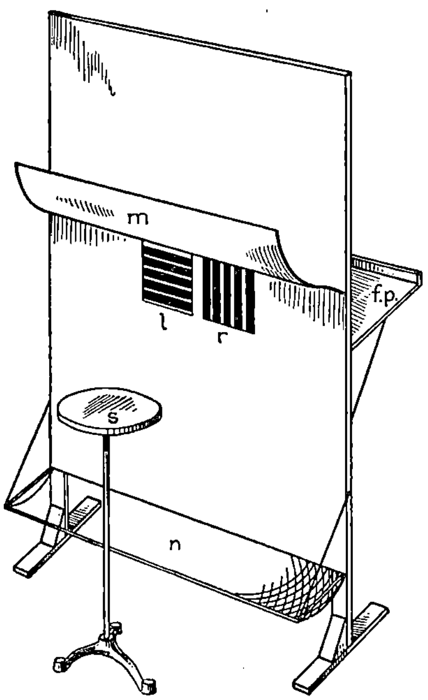

FIG. 67.—The "jumping" apparatus. The animal is placed at S and is required to jump at and knock over one of the exposed cards (l or r). One of the cards is firmly fixed and a response to it causes the rat to fall in net (n); the other falls over when struck, permitting the animal to reach the platform fp where it finds food. *(From Lashley, 1930a, p. 454. By permission of the Journal of Genetic Psychology.)*

the case of the rat. Brightness and size discrimination, however, were found to be more important to the animal and differences in form were utilized only when the other differences were excluded. Further evidence of true form discrimination in chicks was presented by Révész (1924) who found them to be subject to an illusion in size. This is definitely a study of animal perception and suggests that the visual organization of the bird is similar to that of man. He first trained hens to peck food from the smaller of two figures,

After the preference for the smaller figure was well established, he presented them with two figures objectively equal in size but which appear unequal to man. The well-known illusion figures used by him are shown in Fig. 69. It was found that the hens pecked from the figure which appeared smaller to man, the upper one of the two. Munn (1931) has questioned true form discrimination

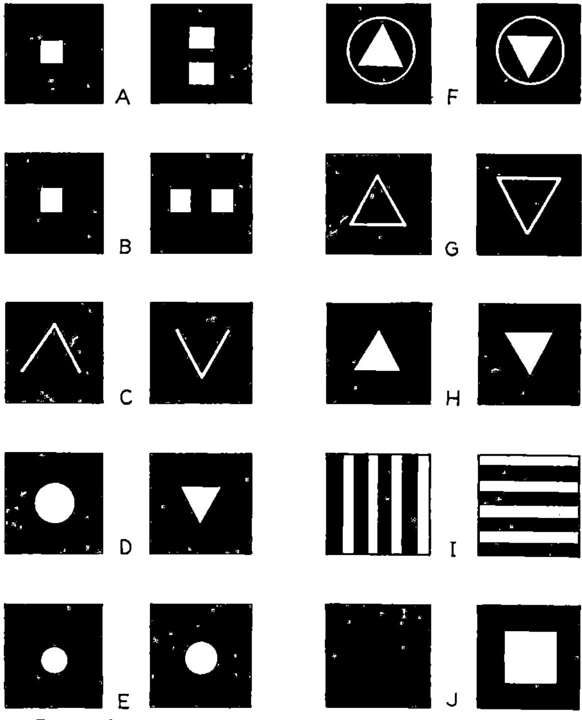

FIG. 68.—Pairs of patterns between which rats have been trained to discriminate. (*From Lashley, 1930a, p. 456. By permission of the Journal of Genetic Psychology.*)

in chicks on the ground that they react to "shape." He claims that the chick's response to such a figure as a triangle is not to a triangle as such but to certain of its characteristics. These may not be the characteristics which make the triangle a "triangle." It is our contention that a response to "triangularity," as understood by man, must be developed by abstraction, and an animal's failure to do so does not prove a lack of discrimination of form but rather indicates that the animal has reacted to some other aspect of the figure. This problem is treated again in Chap. XX.

Dogs.—Pavlov and his students (1927), using the conditioned-response technique, found that dogs could discriminate between circles and ellipses so long as the ratio between the diameters of the latter was greater than 8:9. As the experiments were not sufficiently controlled, this evidence has not been regarded as adequate. Williams (1926), using the discrimination-box technique, failed to establish a differential response to a square and a circle in six different dogs. After 500 trials all of the dogs still made chance scores. Another dog failed to discriminate horizontal from vertical lines.

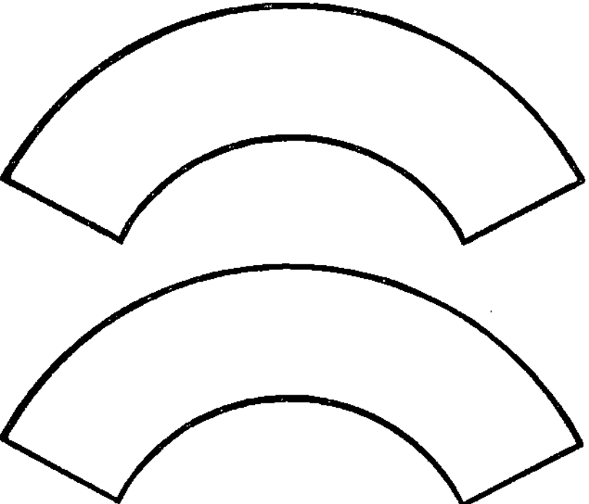

FIG. 69.—Size illusion figures used by Révész (1924). Chicks trained to choose the smaller of two figures choose the upper one of the above two.

On the other hand, Buytendijk (1924) obtained positive evidence of form discrimination by the multiple-choice method. The dog was required to choose a box with a triangle fastened to its front side and to avoid other boxes which had circles, squares, pentagons, or hexagons on their front sides. Karn and Munn (1932) have recently established form vision in the dog by means of a modified form of the Lashley jumping apparatus. As in the case of the other animals, it was not known whether the discrimination was based upon differences in the entire form or upon differences in certain characteristic parts.

Primates.—As might be expected, monkeys are very capable of making discriminations between forms. Bierens de Haan (1927) found it easy to set up a discrimination between a cone and a cube in a pig-tailed monkey (*Nemestrinus nemestrinus*). When these objects were photographed and the pictures used in place of the objects, the animals made chance scores. With training, however, they were able to discriminate between the photographs. (Köhler, 1926, found that apes recognize objects in pictures.) Neet (1933) tested four higher monkeys (*Macacus rhesus*) with a large variety of different forms and found their pattern vision highly developed.

Gellerman (1933) studied form discrimination in the chimpanzee and found that when a preference for a triangle is established the preference persists when the triangle is set at an angle or when it is changed from white to black. In choosing, a subject would frequently trace the outlines of a figure with its finger. The chimpanzee thus seems to make its discrimination on the basis of the total form of the figure. According to Fields (1932), rats must be trained on a triangle placed in many positions before they will recognize it in any position. Reversing black and white also causes the rat difficulty. However, the fact remains that rats can learn to discriminate on the basis of "triangularity." That the rat needs special training and the chimpanzee does not, suggests a difference in perceptual organization. There is no reason for believing that the chimpanzee has true-pattern vision and that the rat has not.

Summary.—In general it may be concluded that pattern vision is rather highly developed in birds and mammals. Differences in ability exist and the relative importance that pattern vision plays undoubtedly varies. It is probable that the perceptual world of some animals is quite different from that of man.

The Discrimination of Tones

The Importance of Careful Controls.—The work of Johnson (1913) on hearing in dogs threw doubt upon the conclusions reached in all previous experiments on pitch discrimination in animals. In a preliminary experiment he trained two blind dogs to place their forefeet on a chair to his left whenever a low tone was sounded, and to mount a low box at his right whenever a high tone was sounded. The animals were rewarded with food each time they responded correctly. One dog required 285 trials, the other 405 trials to learn the above responses to tones produced by tuning forks. Tones produced by other instruments yielded similar results. Under conditions corresponding to those used by other experimenters, Johnson found what appeared to be evidence of pitch discrimination in dogs. When, however, he constructed a discrimination box, and carefully controlled any possible discrimination on the basis of stimuli other than pitch, his results were negative. One of the most important sources of extraneous stimuli were the unconscious movements made by the experimenter (minimal cues). To eliminate the animal's use of such reactions made by the experimenter, Johnson removed himself from the room. The negative results under these new conditions are rather surprising since the responses required of the animal in the discrimination box are much more simple than those required in the first experiment. Failure to react to tones therefore seemed to be due to the removal of extraneous stimuli, rather than to a complication in method which made learning more difficult. The dogs even failed to learn to react differentially to a tone versus silence, indicating actual deafness on the part of the dog. However, this possibility was excluded because three dogs learned to turn left or right when a buzzer to the right or left was sounded. The dogs were thus able to react to noises, but failed to react to tones.

Reactions of Mammals to Tones a Function of the Training Method.—Hunter (1914, 1927), using a similar discrimination box, failed to find evidence of tonal discrimination in rats. Like the dog, they learned to react to a buzzer. When a tone accompanied the buzzer, the reaction was unaffected, but when the buzzer sound was weakened so that it no longer was the dominant sound, the discrimination broke down. Reactions to musical instruments were also learned, but when the source of such a tone was moved some distance away so that the accompanying noises were eliminated, the accuracy fell below that required by Hunter. Thuma (1932) has criticized Hunter's interpretation of results, pointing out that all of Hunter's rats did not fail. One animal made 80 per cent correct responses in 60 trials, another 71 per cent in 90 trials. These scores are better than chance, although they fall below the criterion of 85 per cent correct which Hunter demanded of them.

While difficulty was being encountered in getting animals to react to tones in discrimination boxes, animals were demonstrating a high degree of pitch discrimination by the conditioned-response method. In fact there was evidence that dogs were superior to man in their fineness of discrimination. Studies by students of Pavlov in Russia have been described by Yerkes and Morgulis (1909). In the light of Johnson's criticism, Anrep (1920) repeated the previous work by the conditioned-response method and verified the earlier results. He believed the discrimination-box technique to be unsuitable for measuring sensory processes. Upton (1929) supported this contention by demonstrating that the same guinea pigs which failed to react to different pitches in the discrimination box did so when the conditioned-response method was used.

Horton (1933) compared the tonal sensitivity of guinea pigs to that of humans by the conditioned-response method and found the auditory range of the guinea pig to be 64 to 8,192 d.v. per second, which is similar to that of man. The sensitivity to all tones was, however, greater in man, especially at the extremes of the frequency series. The greatest differences in sensitivity between the guinea pig and man was for high pitches.

That the ears of mammals actually respond when they are stimulated by tones has been demonstrated by Forbes, Miller, and O'Connor (1927) and by Wever and Bray (1930) who found that nerve impulses are set up in the auditory nerve when the ear is exposed to tones. There is therefore little question that tones are effective stimuli. Then why should one method yield positive results and the other not? Muenzinger and Gentry (1931) and Thuma (1932) have partly answered this question by demonstrating that the discrimination-box method can yield positive results. Both of these studies were carefully carried out and cannot be criticized for lack of proper controls. Loudspeakers with tones produced by oscillators were placed at the ends of the alleys. Muenzinger and Gentry used a Y-shaped discrimination box and Thuma a T-shaped box similar to the one used by Hunter. Thuma experienced considerable difficulty in establishing a response. He required some rats to avoid the alley leading in the direction of the sound source, others to go to the right for the tone and to left for silence. He believes that difficulty in localizing pure tones is responsible for the difficulty the animals had in reacting to them. A stimulus must be somewhere before it is reacted to. Muenzinger and Gentry had much less difficulty for reasons which it is difficult to ascertain.

The only study in which the use of the discrimination-box method has yielded a discrimination between tones by rodents is that of Herington and Gundlach (1933). With difficulty they set up a discrimination between tonal frequencies of 500 and 600 d.v. in the case of the guinea pig, but failed to do so with cats. According to them, lower mammals have very poor hearing. The conditioned-response method, they believe, utilizes lower centers in the brain, whereas the discrimination box depends upon higher centers. They suggest that the higher centers are not specialized to receive sensory data in various forms (*i.e.*, afferent impulses in specifically different forms). Hence, although the sense organs respond differently to various pitches, these differences are lost when they reach the higher centers. Habits depending on a difference in the functions of the higher centers therefore fail; whereas those depending on lower centers may be possible. This distinction is of interest, but it must be regarded as highly speculative.

The experiments on hearing in animals, however, leave little doubt that the two methods are not comparable. The conditioned-response method shows that animals at the level of mammals have fine pitch discrimination; the discrimination-box method indicates poor hearing. Perhaps the discrimination method utilizes perceptual differences and although the ear of the animal is adequate, the organization of sensory data is not such as to make pitch play a dominant role. For some reason the conditioned-response method causes the pitch aspect to be dominant. The conditions for learning under the two methods are also markedly different. Since this is a problem in learning it will be treated in detail in Chap. XVI.

The Auditory Acuity of Primates.—Wendt (1934) measured the auditory acuity of five monkeys for tones ranging between 64 and 16,384 d.v. Various intensities of these frequencies were used. The animal was trained to open a drawer containing food when the tone was sounded. The minimum intensity to which animals responded was determined for each frequency. For comparative purposes humans were tested in the same manner. He found monkeys and humans equal in acuity for tones ranging between 64 and 1,024 d.v. Monkeys were slightly inferior for frequencies greater than 2,048 d.v. At frequencies greater than 4,096 d.v. they were markedly inferior, but for frequencies between 8,192 and 16,384 d.v. they were actually superior to man.

Elder (1934) tested the sensitivity of three chimpanzees to tones. The animals were trained to press a telegraph key when a tone was presented through ear phones attached to their heads, but not to respond during silence. A ready signal preceded each test in which a tone was or was not given. The animal was rewarded with food when it responded correctly. The ability to respond to seven pitches ranging in octaves from 128 to 8,192 d.v. was measured by presenting each tone at various intensities. The results obtained show that the ear of the chimpanzee is about as sensitive as that of man; in fact, the measurements show a sensitivity which is even slightly superior to that of the average man. A high degree of sensitivity was found for tones of 8,192 d.v., which is characteristic of the sensitivity curve for children as well as that of the monkey.

The number of frequencies which can be discriminated has not been determined in any of the studies.

Summary.—The experiments on hearing show that mammals in general react to tones and noises, but the experience produced by tones is much less dominant. The method used is a very important factor in determining whether a reaction to a tone will be set up. Primates are found to be especially sensitive to tones. In no case has the number of frequencies which can be discriminated by an animal been determined.

Localization of Auditory Stimuli

Since responses to noises are easily set up in higher vertebrates, these animals can be trained to respond to the differences in the location of such stimuli. Engelmann (1928) repeated some of the earlier work under more carefully controlled conditions, and found dogs, cats, and chickens superior to man in auditory localization. His work on dogs was the most satisfactory. For a noise he used the tapping of an electric bell hammer on metal. The stimulus was placed sometimes behind one, and sometimes behind the other of two identical paper screens. The distance between the screens was varied from time to time in order to determine the fineness of localization. At a given distance from each of the sound sources the dog was required to determine from which of the two screens the sound was coming. When the dog went to the correct screen it received a reward.

The least difference in localization which can be discriminated may be expressed by one half the angle which is formed between lines drawn from the screens to the center of the animal's head (*i.e.*, through the ears of the animal). In Table 20 the smallest auditory angle which yielded correct localization is given for three dogs and three men. It will be seen that the most inferior dog was superior to all of the men in auditory localization.

With screens arranged in a circle about the dog, 16 sound sources could be discriminated when they were equally spaced and each

1.5 meters from the dog. Bandaging either ear reduced the ability to localize and controlled visual cues. In auditory sensitivity the dogs also proved to be superior to man.

Cats and hens were not so easily trained. For this reason more attractive sound sources were used. The noise produced by a mouse in a jar was used for the cat and the peep of chicks was used to test hens (also the cluck of a hen to test chicks). Screens hid the sound sources as in the case of the dog. By this method the cat was found to be superior to the dog, and the hen somewhat inferior. However, comparisons on this basis are not satisfactory since different sounds and different intensities of these were used. Motivation may also have varied greatly.

TABLE 20.—AUDITORY LOCALIZATION IN DOGS AND MEN (FROM ENGELMANN, 1928)

Dog	Auditory angle	Man	Auditory angle
1	2° 9′	1	5° 43′
2	3° 35′	2	5° 43′
3	1° 26′	3	4° 18′

DISCRIMINATION BASED UPON OTHER SENSES

Mammals undoubtedly react to gustatory, olfactory, temperature, and proprioceptive forms of stimulation, but carefully controlled experiments on these senses are so meager that one can conclude little more than the fact that mammals react to such stimulation.

Under ordinary conditions these sense departments seem unimportant for adaptive behavior. Even olfaction, which is often regarded as most important to lower mammals, causes difficulty when attempts are made to set up differential reactions. Liggett (1928) found it difficult to set up a discrimination between two odors in the white rat by means of the discrimination-box method. Swann (1933) encountered similar difficulty, but succeeded in getting reliable results when the rats were required to dig a way through sawdust of a particular odor in order to reach the food. Acuity was not tested.

Tactual and kinesthetic sensitivity have been regarded as especially important in maze learning and have been of interest primarily in this connection. These senses as well as others will again be treated in the analysis of maze learning (Chap. XVII).

Conclusion

The mammal may be regarded as having rather highly developed discrimination capacities, even though data on most senses are incomplete. The experiments reported in this chapter give a general outline of the methods of approach. Obviously, failure to demonstrate a particular kind of discrimination is not proof that this discrimination capacity is absent. The method used is always important and it is difficult to predict what technique will yield best results in a reaction to the specific-stimulus modality which is to be investigated. In all studies having to do with learning it must be assumed that any sensory difference may be used by the animal; but one cannot assume, without demonstration, that some particular sensory difference has been used. Discriminations between objects in an animal's environment may depend upon the differential data coming from a number of senses, or the discrimination may be dependent upon a single aspect of one of the senses. Changes in certain objective conditions of the stimulus situation may cause a shift in the sense organs involved. The transference of responses from the control of one group of sense organs to another is very common. Unlike lower forms, the specificity of reactions to particular forms of stimulation is greatly reduced in mammals.

Suggested Readings

Johnson, H. M. 1913. Audition and habit formation in the dog. *Behav. Mono.*, 2, pp. 78.

Munn, N. L. 1933. *An introduction to animal psychology.* Boston: Houghton Mifflin. Pp. 98–154.

Warden, C. J., Jenkins, T. N., and Warner, L. H. 1934. *Introduction to comparative psychology.* New York: Ronald. Pp. 72–95 and pp. 502–523.

Williams, J. A. 1926. Experiments with form perception and learning in dogs. *Comp. Psychol. Mono.*, 4, pp. 70.

CHAPTER XIV

NEURAL MECHANISMS IN BEHAVIOR

Introduction

The responses an animal makes when a specific sense organ is activated depends upon the nature of the connections between the sense organs and the response mechanism. Connections between sensory and motor mechanisms are made in the central nervous system, and in higher forms the complexity of this intermediate mechanism may be very great. Lower brain structures serve as centers, and by means of such centers nerve impulses from several sense departments may be integrated. Specific cases have been discussed in Part I and serve to illustrate the function of lower brain centers. Since the connections are rather fixed, behavior depending on such mechanisms must necessarily be stereotyped.

In many cases the spinal cord alone is adequate for bridging the sensory and motor mechanisms. This is illustrated by the reflex activities of animals in which the brain has been cut away. Since reflexes of normal animals differ from those of animals in this spinal condition, it must be recognized that the brain also functions in such responses. Because spinal reflexes are exaggerated both in magnitude and in duration, the removal of the influence of the brain from such reflexes may be regarded as a release phenomenon. Lower centers in the brain are involved in the spinal reflexes because they form connections between sensory and motor mechanisms which supplement those made in the spinal cord. These additional connections affect the responses and make possible control from higher centers as well as modification in the response when additional stimulation of other sense organs is involved. With the interaction of several sense departments the complexity becomes very great and it is impossible to trace response mechanisms in detail. This is particularly true in higher forms. The treatment of neural mechanisms in Part I, however, serves to illustrate the role played by lower brain centers in determining the nature of responses to various forms of stimulation.

An important variable in the behavior of all animals is the *intensity* of reactions to various magnitudes of stimulation. Variations in behavior of a quantitative nature cannot be determined by the nature of the connections but depend upon differences in the communication over the same connections. In order to show how differences in intensity of stimulation are transmitted to the response mechanisms, it is necessary to consider the problem of nervous conduction.

The Nervous Impulse and Its Relation to Behavior.—Nervous activity consists of chemical changes which spread along individual nerve fibers as a rapid succession of pulsations. Each pulsation is accompanied by an electrical effect which makes the active region negative with respect to inactive regions. Stimulation initiates such activity; but in no way does it determine the magnitude of the nervous activity, this being dependent upon the available energy in the nerve tissue. The intensity of stimulation, however, determines the frequency of pulsation as well as the number of active fibers. The *frequency* varies with the stimulus intensity because the sense organs as well as the nerves have refractory phases.[1] During recovery from the refractory phase, the threshold of excitation gradually returns to normal, and if the intensity of stimulation is sufficient to pass the heightened threshold, a second impulse is possible at an early stage of recovery, while if the intensity is of lesser degree, a second impulse is not available until further recovery. Frequency ranges between 10 and 100 impulses per second are common. The number of fibers responding increases with stimulus intensity because the thresholds of nerve fibers and receptor cells vary. Intense stimuli activate cells with high as well as with low thresholds, and as a consequence such stimuli activate more fibers than do less intense stimuli.

Upon reaching muscle tissue, impulses produce contraction, the degree of contraction (the number of muscle fibers contracting) depending on the number of nerve fibers conducting and on the frequency at which each responds. By this arrangement intense stimulation leads to intense contraction. When intense stimulation

[1] After the passage of a nerve impulse, the nerve is not subject to further stimulation for a period of a few thousandths of a second. This is known as the *absolute* refractory phase. During the next several thousandths of a second the nerve again responds, but only to more intense stimulation. This period during which the nerve is less susceptible to stimulation is known as the *relative* refractory phase.

leads to the contraction of more body segments than less intense stimulation, it is presumably due to differences in threshold among groups of motor fibers. In Part I we have seen how animals may avoid a certain intensity of stimulation by withdrawing only the part of the body stimulated, but with increased stimulation, move the whole body. The body segments first responding apparently have connections with fibers having lower thresholds; those responding upon increase in stimulus intensity, with fibers having higher thresholds.

Reversals of reactions which result from increased stimulation (such as going toward a dim light and away from darkness, but avoiding bright light and going to darkness) may be explained in a similar manner, but with certain additions. Slight stimulation, let us say, produces a positive response (going toward the stimulus), because it arouses only nervous connections which have a low threshold. These fibers may be connected with muscles which move the animal forward. Intense stimulation may set off fibers having higher thresholds. These may be connected with muscles which make for an avoidance response. But these two sets of muscles would now act in opposition upon intense stimulation, since all fibers would respond. To explain how the function of the first set of muscles (those connected with fibers having low thresholds) is eliminated, it may be assumed that the frequency of the fibers with lower thresholds has greatly increased. We have already shown how stimuli greater than threshold intensity cause fibers to respond during their period of recovery. Nervous impulses set up during the refractory period are known to be of lesser magnitude and hence may be regarded as ineffective in producing muscular contraction.[1] Thus although the nerves connected with the muscles making for a positive response would function during intense stimulation, their function would not lead to contraction. With the dropping out of this set of muscles, the avoidance response would alone remain. Table 21 summarizes the above theory.

This explanation of the variation of function due to changes in stimulus intensity is consistent with the facts of nerve physiology as well as with the evidence on the behavior of lower forms. However, there are many responses which cannot be accounted for with

[1] According to the all-or-nothing law, the intensity of a nervous impulse is dependent upon the condition of the neurone. During a state of partial recovery its condition is such as to produce smaller impulses.

so simple a mechanism. Examples would take the reader too far afield in physiology. Suffice it to say that it is necessary to postulate

TABLE 21.—THE RELATIONSHIP BETWEEN STIMULUS INTENSITY AND THE RESPONSE ELICITED

Stimulus intensity	Group of nerve fibers responding	Frequency of impulses	Magnitude of impulses	Muscles contracted	Response
Low	Low threshold	Low	Maximal	A	Approach
Medium	Low threshold	Medium	Both maximal and sub-maximal	½A	Stop
	High threshold	Low	Some maximal	½B	
High	Low threshold	High	Sub-maximal	None	Withdrawal
	High threshold	Low	Maximal	B	

other principles in order to account for nervous integration in animal forms with highly complex nervous systems. The relationship between excitation and inhibition is a problem which is fundamental to reflex integration, and to account for this interaction the chemical theory has been formulated.

The Chemical Theory of Excitation and Inhibition.—Sherrington's (1925) chemical theory accounts for phenomena in nervous integration. This theory postulates two chemical substances *E* and *I*, which are deposited at the terminals of activated neurones. The deposit is rather temporary since it gradually becomes dissipated. These chemicals have the property of neutralizing each other. An excess of *E* in sufficient quantity leads to the excitement of adjacent neurones. Chemical *I* produces an inhibitory state since it neutralizes any *E* that might be present and which might have resulted in excitation. If it is assumed that different neurones deposit *E* and *I*, then the deposits at any center will be a function of the pattern of stimulation. Since the pattern of response depends upon the condition of the various centers at which deposits are made, the same centers are capable of producing quite different responses at different times.

With the postulation of chemical substances which produce states of excitation and inhibition, it is possible to account for increased activity which arises from continued stimulation, since stimulation of greater duration increases the chemical deposit. Likewise, certain delays in response as well as the continuance of activity after stimulation has ceased (which cannot be accounted for by the properties of nervous conduction alone) become simple problems when one assumes that chemicals must be produced to set up an excitatory state and dissipated before excitation ceases. Many other properties of reflex activity seem to make these or similar postulates quite necessary. The fact that an increase in the duration of a stimulus has the same effect on behavior as an increase in the intensity of the stimulus is easily explained by the chemical theory.

By means of the production of chemical substances, integration becomes a physiological problem as well as an anatomical problem. Bethe (1931, pp. 1210–1214) has actually presented some evidence which supports the postulation that neurones produce chemical substances.

Other Physiological Conditions Important for Integration.—Further evidence which suggests that the physiological condition plays a role in integration is found in the work concerned with the physiological gradients already referred to in Chap. XII and in earlier chapters. It is likely that physiological gradients not only determine the patterns and growth of the nervous system but also the course of nervous activity by changing the relative sensitivity of neurones lying within the gradient field. Stimulation, because it increases the metabolism of the neurones activated, sets up gradients of a temporary nature. Different patterns of stimulation would therefore result in various physiological gradients. There is reason to believe that the spread of nervous activity is affected by such gradients.

The Function of the Brain in Modifiable Behavior.—The chief concern of this chapter is the central mechanism involved in modifiable behavior. Mammals are characterized by a highly developed cortex consisting of millions of cells so interconnected that the structure can be described as superimposed thin layers of cells separated by networks of intermeshed fibers. The cortex is connected directly or indirectly with all other parts of the nervous system. This arrangement makes it quite impossible to trace the structural connections which are involved in a sample of complex behavior. It is quite likely that nearly all of modifiable behavior depends upon changes in this tissue.

Studies of learning in animals in which the cortex has been completely destroyed are rare, although there are many reports of learning in animals in which the cortical destruction was incomplete. Using three dogs as subjects. Poltyrew and Zelony (1930) have reported conditioned withdrawal of the forepaw to the sound of a whistle, but it is not known whether the destruction was complete in any of these cases. Nevertheless the rate of learning was very slow and intense stimuli were required to produce conditioning. Culler and Mettler (1934) found that a completely decorticated dog (the degree of injury was determined at autopsy) was incapable of acquiring a specific response, but its general pattern of response was modified by experience. It must therefore be recognized that lower centers are capable of mediating some modification in behavior, but the degree and specificity of such modification are very limited.

It has been contended that changes in behavior due to learning can be located in specific parts of the cortex. This belief is based largely on the nature of the connections which the cortex has with all other centers. To gain insight into its complex function it is necessary to study the experimental results in some detail. The history of the theories of cortical function shows that two extreme views have appeared from time to time. One extreme view delegates the various psychological faculties or abilities to different parts of the brain. The other regards all faculties as being different functions of the whole cortex. Intermediate positions have maintained localization for sensory and motor functions, with an interaction between these to account for thought processes. In order to evaluate these theories we shall proceed to a study of brain function in the learning of different types of problems.

Cortical Function in Problem-box Learning

The early studies of brain function were rather crude, particularly because behavior was not carefully analyzed. It is through the entrance of psychologists in this field of investigation that refined methods for the measurement of behavior have arisen. Previously the intelligence of an animal after operation was inferred from its general behavior in the laboratory; its vision was tested by observing the response when meat was tossed in the air, and its auditory capacities were determined from general responses to noise.

Franz (1907) trained cats and monkeys to open problem boxes[1] which involved pressing a button or pulling a string. A successful response was rewarded with food. After the animals showed considerable proficiency, they were subjected to brain operations. Following recovery they were again tested to determine any loss in habits. It was found that when both frontal lobes were largely

[1] For a description of problem boxes see Chap. XVI, pp. 358*f*.

destroyed the problem-box habits were lost, but destruction of the frontal lobes of one hemisphere, or in other parts of both hemispheres, had no effect on the habit. Surprisingly enough, those animals which lost the habit by operation were able to relearn it in normal time. Although the number of cases are limited, these results suggest that the frontal lobes contain the changes which have been produced in learning, although no specific part of them seems essential to learning.

Similar, but more detailed, results were obtained by Lashley and Franz (1917) on the rat. The problem-box habit (requiring depression of a platform) was lost only when the frontal portion of the cerebrum was destroyed and in no case was the animal unable to relearn. The retention, however, seemed not to depend on any specific part of the frontal region. That animals with lesions involving as much as one half of the cortex may not be handicapped in learning to open a problem box was strikingly demonstrated by Lashley (1920) when he found that 19 operated rats learned to open a problem box (involving the depression of two platforms) in an average of 79 trials, whereas 10 normals required an average of 143 trials. The destruction involved different cortical areas and in no case were operated animals inferior. In fact the operation apparently facilitated the learning of such a habit. However, Lashley points out that operated rats are less likely to jump over the platforms and therefore more likely to depress them.

Recently Lashley (1935*a*) compared normal and operated rats in their ability to learn to open five different types of problem boxes. Box *A* required them to push a lever; box *B* required them to tear a strip of paper with their teeth or paws; box *C*, to pull a handle toward themselves; box *D*, to pull a chain; and box *E*, to depress a rod and then step on a platform. Boxes *A* and *E* showed no difference in the learning rate of operated and normal animals, but boxes *B*, *C*, and *D* showed the normal animals to be superior. Since different kinds of movement were required to open the boxes, Lashley believes that the inferiority of the operated animals on the three boxes was partly due to a reduction in tendencies to make certain movements. Changes in the repertoire of movements through brain operations may therefore be either an advantage or a disadvantage to learning, depending on the problem. Lashley does not believe that the ability to form associations was impaired. Nor were sensory deficiencies, which may have resulted from the operation responsible, since sense deprivation alone interfered less with the mastery of the problems than did brain operations.

As regards retention, studies with the platform boxes demonstrate a loss depending roughly upon the extent of cortical destruction in the frontal region.[1] Small lesions produced no detectable loss. Thus, cortical destruction, although it does not reduce the ability to learn, nullifies what has been learned about the problem boxes, provided the destruction is confined to the frontal portions, and is extensive enough. Rats which have learned to open a problem box after the destruction of the frontal areas, retain this learning after further destruction of cortical tissue, and no general area has been found which is responsible for such retention. It is probable that all parts of the cortex remaining intact are involved, since the alternative explanation would be that the lower centers have taken over the function.

In the case of the monkey, a simple problem-box habit shows little or no impairment after cortical destruction, but habits involving the manipulation of a combination of several locks in order to open a door showed some loss according to Jacobsen (1931), although his evidence is very limited. Lashley (1924) had previously demonstrated that the destruction of the motor areas in monkeys did not eliminate the problem-box habit, although such destruction produced paralysis from which the animals recovered only gradually. This indicates that such learning produces changes in nervous tissue other than the motor areas.

Cortical Function in Discrimination Habits

The Learning of Simple Discriminations. *Reaction to Light.*—In training rats to go to a lighted as against a darkened compartment, Lashley (1920) found that normal rats required an average of 106.6 trials to learn, whereas operated rats required an average of 80 trials. Less timidity or less susceptibility to distraction may account for the superior score of the operated animals. When normal rats which had learned the habit were subjected to lesions in the posterior part of the brain, the habit was lost, but lesions in other parts of the brain did not destroy the habit. The *learning* of a simple discrimination therefore is not impaired by cortical injury, but its *retention* is impaired if the lesion is confined to a given

[1] The extent of cortical injury is determined at autopsy. The brains are sectioned and reconstructed. The extent of destruction is expressed by the percentage of cortical surface destroyed.

general region. Rats which had learned the habit after destruction of cortex in the posterior part of the brain, still retained the habit after further cortical injury. In their case retention seems independent of any specific structures.

The relationship between the extent of injury in the posterior part of the brain and the degree of loss in retention was found by Lashley (1926) to be very marked, although there was no relationship between the extent of injury and the ability to learn. The correlation between the extent of destruction and the number of trials required for relearning was 0.712, whereas the correlation between the extent of destruction and the number of trials required for original learning was 0.132. This study shows very clearly that the process of learning this visual habit is associated with changes in the posterior part of the brain, the destruction of which results in a loss in retention. Since this loss is proportional to the extent of destruction, it might be supposed that the changes produced by learning are distributed homogeneously over the posterior part of the brain. However, it is also possible that specific structures contain the changes produced by learning and that large lesions are more likely to include these structures than small ones.

Because part of the cortex in the posterior portion of the brain (the black portion of Fig. 71) has connections with subcortical structures related to vision (the lateral geniculate nuclei) Lashley (1935) divided his cases according to the extent to which cortex related to such lower centers was destroyed. He found that a loss in the simple discrimination habit occurred only when such cortex was completely destroyed. As long as some connections with lower centers remained intact, there was little impairment in retention. That the loss in retention was not merely the result of a disturbance in vision is shown by the fact that all rats which suffered a loss in retention were able to relearn the habit in the normal time. Only upon destruction of structures (optic thalami) connected with the retinas of the eyes was the ability to learn the discrimination destroyed. Such destruction, of course, interrupted the visual impulses.

We may therefore conclude that learning a simple visual habit normally depends upon changes in specific visual structures (the striate areas of the cortex which are related to the lateral geniculate nuclei), but that such learning may take place in the absence of such structures.

Reactions to Sound.—Wiley (1932) using a two-choice discrimination box trained rats to turn about in an alley and take the alternative choice whenever a buzzer sounded, but to continue during silence. He then extirpated various cortical areas and found that the retention of the habit was affected when the cortex in the temporal region (area *p* of Fig. 70) was destroyed. Destruction of other regions seemed not to interfere with retention. As in the case of the visual habit, the loss due to temporal lesions was related to the degree of destruction; the correlation coefficient for 49 cases being greater than 0.6. (Whether or not this correlation depends

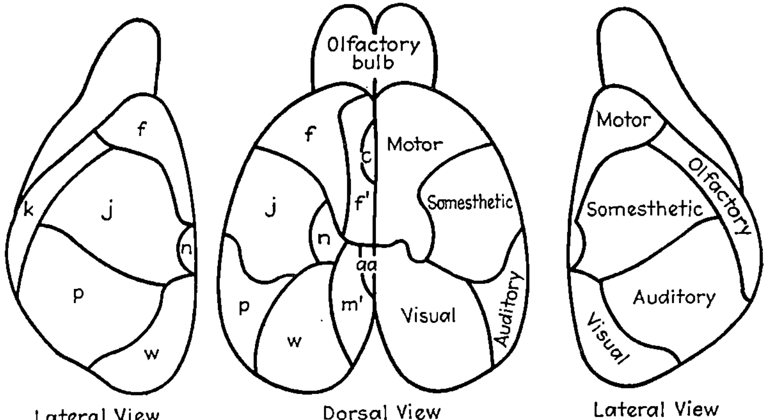

FIG. 70.—Cytoarchitectural areas of Fortuyn as modified by Lashley. The letters on the left half of the diagram indicate cortical regions which are structurely different. The inferred functions of these regions are indicated on the right half of the figures. (*Modified from Lashley*, 1929, *p.* 49.)

upon the fact that large lesions are more likely to include pertinent specific structures has not been determined, but in view of the results on the visual habit it seems likely.) All rats were able to reacquire the habit after their temporal regions had been injured and only two of them required more trials to relearn than any of the rats needed for learning. This corresponds to the results obtained on the visual habit.

Reactions to Smell.—In the case of olfactory discriminations, Swann (1934) found no particular cortical region which, when destroyed, produced a loss in habit. Destruction of as much as 39 per cent of the cortex left the habit intact. These results thus differ from discriminations based upon vision and audition.

This suggests that the olfactory centers are less definitely defined or that such learning takes place without cortical assistance. Since cortex evolves from the olfactory center (*e.g.*, forebrain of fishes), it is possible that the olfactory cortex has remained nonspecific. The other sensory centers investigated are specific in that habits depending upon them for retention are localized; they are nonspecific in that no definite cortical center is necessary for the acquisition of a habit.

The Learning of Complex Discrimination Habits.—When rats were required to discriminate between the brighter of two lights, rather than between one light and darkness, Lashley (1930*b*) found a retardation in *learning* following lesions in the posterior part of the brain. The extent of retardation was found to be related to the extent of injury, the correlation coefficient being 0.58. The same lesions produced no retardation in learning to discriminate light from darkness, verifying previous findings. However, animals which had learned to discriminate light from darkness as well as the brighter of two lights before operation, lost both these habits to about the same extent after the operation, the correlation coefficients for mass versus retention being 0.64 and 0.65, respectively. The similarity of the correlation coefficients suggests that similar structures are responsible for the retention of both habits. Since animals with lesions in the posterior region had more difficulty in acquiring a discrimination between two lights differing in intensity by a 4:1 ratio than between two lights with a 25:1 ratio, it is probable that vision was impaired by the lesions. That vision was impaired is further supported by the fact that fewer operated rats were able to discriminate small differences between lights than normal rats. This impairment was not sufficient to interfere with the discrimination of light against darkness, but seems to have interfered with the discrimination between two lights. However, operated animals able to discriminate any pair of lights required more trials than normals. This would indicate that the loss in ability is, at least in part, due to defective learning.

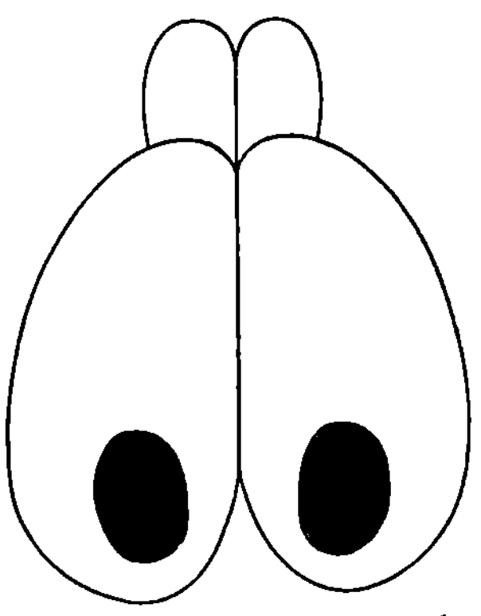

FIG. 71.—Striate area of a rat's brain. The blackened region designates the area which is closely related to vision in the rat. This area varies somewhat for different rats. (*Modified from Lashley*, 1934*b*, *p.* 63.)

In an attempt to determine whether vision is impaired by cortical lesions, Lashley made extensive investigations of the visual mechanisms and pattern vision of the rat. In the study of the neural connections between the retinas, the optic centers, and the cortex, Lashley (1934 *a* and *b*) found a point for point connection between these. The cortical area (striate area) which receives visual impulses is illustrated in Fig. 71. From this arrangement in the conduction system one would expect that lesions in visual cortex should result in scotomas or blind spots in the animal's visual field. In a study of pattern vision in which rats were subjected to cortical lesions

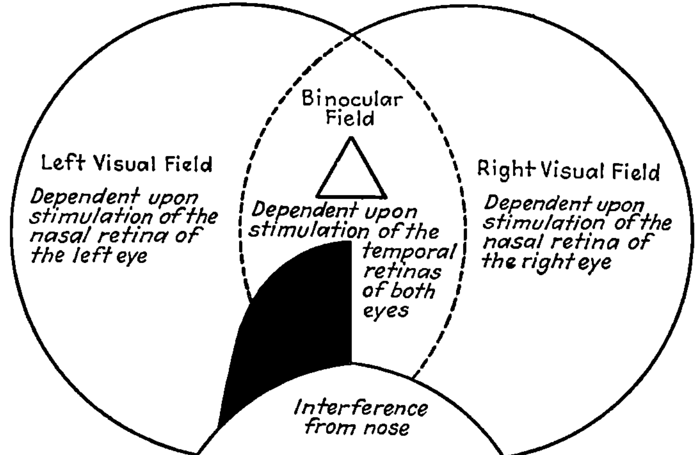

FIG. 72.—Illustration of a rat's visual field when it fixates a triangle 20 cm. distant. The darkened portion indicates a blind spot in the field of vision which would not interfere with the rat's reactions to the triangle. (*Modified from Lashley and Frank*, 1934, *p*. 364.)

in the visual area Lashley and Frank (1934) obtained evidence which indicated that this was actually the case. Blind spots need not necessarily interfere with detail vision as in the case illustrated in Fig. 72. Destruction of the cortical areas which receive impulses from the nasal retinas (resulting from stimuli affecting the lateral part of the visual field) do not disturb the visual habit, but destruction of the areas associated with the temporal retinas (which are involved when the stimuli lie in the median part of the visual field) result in a serious disturbance of the habit. Complete destruction of the striate areas (shown in Fig. 71) of both hemispheres permanently abolishes detail vision.

Thus lesions in the posterior part of the brain may (1) leave the retention of the habit undisturbed; (2) produce a disturbance to which the rat may adjust upon further training; or (3) permanently abolish the animal's reactions to visual patterns. What the nature of the disturbance is depends upon the locus of the lesion. Previously Lashley and Frank (1932) had presented evidence which indicated that the disturbance in reactions to visual patterns following cortical injury was dependent upon the amount of destruction in the posterior part of the brain and was independent of specific structures in the cortex. It was then believed that postoperative disturbances in reactions to visual patterns were primarily due to loss in memory rather than a loss in sensory capacity. At the present time it seems that the reduction in the visual capacity (scotomas, etc.) is primary.

Since the *learning* of a response to a visual pattern produces changes in the brain, it is important to know whether these changes are localized. We have found that the retention of the simple visual habit is localized in the striate areas, and from this it seems likely that the *retention* of a response to a visual pattern as well as the ability to experience a pattern, depend upon these structures. Lashley and Frank (1934) believe this to be true, although their study does not readily lend itself to such an analysis. A study by Kirk (1935) makes such an analysis possible. He destroyed tissue in the striate area of one hemisphere in each of his rats, thereby not eliminating pattern vision in any of them. All of his lesions included the posterior portion of the brain, but one group of lesions was confined to the posterior portion, whereas the other group included tissue lying more forward. He found that the animals with lesions extending forward suffered a greater loss in their retention of a discrimination between an *F* and its mirrored image than did those with lesions limited to the posterior parts. Further, the extent of loss in memory was related to amount of tissue destroyed. These results suggest that for complex discriminations involving pattern vision, *retention* depends upon tissue outside the striate areas as well as upon these areas.

Conclusions.—Present evidence indicates that the cortical changes produced through the learning of complex discrimination habits are rather widespread. Loss in retention is therefore a function of the amount of tissue destroyed. Profound disturbances in such habits may also be produced by small lesions when these interfere with the sensory capacity of the animal. The retention of simple discrimination habits, on the other hand, seems to be more specifically localized. This distinction between simple and complex learning is developed further in the following pages.

The Function of the Cortex in Complex Behavior Situations

When rats with lesions in the cerebral cortex were tested for their ability to learn mazes, Lashley (1929) found them to be decidedly inferior to normal animals. To *learn* a rather simple maze (8 blind alleys) normal rats required an average of 19 trials, whereas the operated rats required an average of 91 trials. The loss in learning ability was not dependent upon lesions in any particular part of the cortex but was primarily a function of the amount of tissue destroyed. This was shown by the correlation between the degree of destruction and the loss in learning ability which was found to be 0.75. Of importance is the fact that no animal, even with 80 per cent of its cortex destroyed, failed to learn at least to some extent.

Rats operated after they had learned the maze, lost the habit, the amount of loss also being a function of the extent of cortical destruction (the correlation coefficient was 0.60) and not of the locus of injury. Unlike the retention of the problem-box habit and the simple visual and auditory discrimination habits, the locus for the *retention* of maze habits was found to be general. Lashley found that a reduction in sensory capacity was responsible for neither the reduction in learning ability nor the loss in retention. Rather the lesions invaded tissue which was responsible for the learning. Simple habits which depend on some particular sense department seem to depend upon somewhat more specific parts of the cortex for their *retention*. Concerning *learning*, it seems that complex habits are affected by lesions whereas simple ones are not. Both the maze and the discrimination of two light intensities (or two patterns) are perhaps more complex learning problems than the other habits described. This matter of complexity will be treated in detail in Chap. XVI.

Loucks (1931) obtained similar results when he required rats to learn a complex alternation habit. Both learning and retention were affected by cortical lesions. Buytendijk (1931) found operated rats equal to normals in setting up a position habit, but when he required that they reverse the habit (*e.g.*, go right rather than left), operated rats were handicapped. They persisted in going to the old food place rather than changing to the new one.

In a test which involved the reorganization of past learning Maier (1932 *a* and *b*) found that small lesions to the cortex reduced this ability and that large lesions (greater than about 25 per cent of the total surface) abolished it entirely. The same lesions which produced failure in tests involving the reorganization of experience had no affect on the formation of simple associations. It seems that reorganization of past learning depends upon a certain amount of intact cortical tissue, but as in the case of learning the maze, no particular part of the cortex is essential.

From the above experiments we may conclude that the acquisition of complex behavior tasks involves the function of the cortex as a whole. Its destruction therefore interferes with such behavior. The learning of simple behavior patterns, on the other hand, depends upon limited cortical tissue, and lesions in general do not retard their acquisition. Since complex tasks are a function of the whole brain, their retention is disturbed by lesions to any part, but simple tasks are only disturbed by lesions when the specific part of the cortex which functioned in learning is destroyed.

Localization versus Mass Action

Histological studies in which nerve fibers have been traced from their source of origin to their points of termination, reveal that the mammalian brain may be divided into projection areas and association areas. Fibers coming from optic centers, for example, radiate to the cortex in the posterior part of the brain, and this cortical area has therefore been regarded as the visual area. On the basis of similar evidence other sensory areas have been isolated. The fibers which are projected from the cortex and which terminate in the cord come from distinct cortical areas which are known as the motor areas. The belief that different cortical areas have specific functions according to their connections is further strengthened by the fact that the histological structure of the cortex is different in the various delimited areas.

Those parts of the cortex having no projection fibers have been known as the association areas. Fibers sweep in bundles between various parts. The association areas are primarily in the frontal and parietal regions of the brain and have been regarded as the centers concerned with adaptive behavior. In Fig. 70 the supposed functions of different cortical areas of the rat's brain are designated. These areas are based upon the histological finding of Fortuyn and have been modified by Lashley (1929) to fit his diagrams of the rat brain. The experiments thus far reported, which deal primarily with the behavior of the rat, fail in many respects to support the histological evidence. On the contrary, the results indicate that the cortical destruction, rather than depriving the animal of some specific ability (such as maze learning ability) depending on the locus of injury, deprives the animal of rather general functions according to the extent of injury. The *mass* of functional tissue rather than its *locus* in the brain seems to be the important factor.

The theory of *mass action* states that the different parts of the cortex do not have specific functions, rather all tissue is responsible for changes resulting from learning. Destruction of some of this tissue reduces the capacity for further changes and results in a loss of what has been learned. Since all parts of the cortex may function alike, the cortical areas have been regarded as *equipotential* in function. (Lashley, 1933). The higher processes which do not depend upon specific sensory capacities best illustrate this theory. Since there is a negligible loss in the *learning* of simple habits after large injuries to the cortex, defenders of this theory may contend that simple habits are not delicate enough tests for detecting the reduction in learning ability. Certainly the theory of localization cannot account for the above fact.

The data on the *retention* of simple sensory habits require a combination of the theories. Destruction of cortical tissue in certain areas produces a specific loss in retention (*e.g.*, loss of problem-box habit) whereas destruction in other areas does not. The function of a particular area, however, even though it plays a specific role in retention, may be taken over by the remainder of the cortex. This *vicarious functioning* of the cortex opposes the theory of specific localization and suggests that most if not all cortical tissue has rather general functions.

The Motor Areas as Specific in Function.—Since electrical stimulation of a given part of the motor areas results in the contraction of a definite group of muscles, the belief that these areas have a direct connection in the spinal cord with motor cells which innervate these muscles, is greatly strengthened. In man the orderly relation between the locus of stimulation and the muscle which is caused to contract is very striking, and at the same time corresponds to what might be expected from histological examination. Dusser de Barenne (1933) has actually located the layer of cortical cells (the layer of large pyramidal cells) responsible for the muscular contraction by destroying certain layers of cells[1] in the motor area of the monkey brain, and then applying electrical stimulation to see what layers must be destroyed to eliminate the function. That paralysis results in man and primates when the motor area is destroyed has long been known. Evidence of this sort encourages a belief in strict localization of cortical function.

Franz (1915) caused even the motor areas to be questioned from the point of view of strict localization, when he demonstrated that they varied considerably in different monkeys and in the two hemispheres of the same monkey. The variation was in the form of individual differences in the relative size and in the locus of the brain areas which, when stimulated electrically, resulted in movement. A little later Ogden and Franz (1917) demonstrated that monkeys which had arms paralyzed by lesions to the motor areas recovered completely under proper treatment (massaging muscles and compelling animals to use paralyzed limbs). This demonstrated that the motor areas although important were not essential to the control of movements.

Further evidence against the theory that the motor cortex has a specific function was obtained by Lashley (1923). He exposed the stimulable area of a monkey's brain, electrically stimulated various parts, and noted the movement elicited. This same test was given on four occasions over a period of one month. On each test day there was a regularity and consistency in the movement produced by stimulation of a given point, but the four test days gave quite different results. Of the 57 points explored, 22 gave a different movement on each test; 26, the same movement on two tests; only 6 points showed consistency on three tests; and no points gave the same results in all tests. The other 3 points gave no responses on any tests. The extent of difference in movements on the various test occasions consisted in opposite movements of the same joint (flexion one time and extension the next time); movements of the different joints; and actual movements of different limbs. On different tests the same movement was often produced by stimulation of rather widely separated areas. Lashley suggests that the movement elicited at any time is largely the function of the temporary physiological condition. Certainly if these results are verified on more animals, the function of the motor cortex must be regarded as rather general, the movements depending for their specificity not upon the cortical structure but upon some general physiological condition.

[1] Destruction of specific cortical layers was accomplished by applying a hot plate to the cortex and leaving it there for different lengths of time. The number of layers destroyed is a function of the time the plate is applied.

One might expect the motor cortex of the rat to be rather non-specific in function since there is no specifically delimited area in which tissue is characteristically homologous to that of primates (Herrick, 1926). Nevertheless, this animal has a stimulable cortex which Lashley (1920) found to produce specific movements.[1] Destruction of these areas does not produce paralysis according to Lashley and Franz (1917). Peterson (1934) found that destruction of the stimulable area for hand movement, resulted in a loss in the functional dominance of the hand which had been controlled by that area. Thus a right-handed rat became left-handed if the cortex controlling movements of the right hand was removed. A change in handedness suggests that the operation must have reduced the efficiency of the previously preferred hand, although actual paralysis may not have resulted.

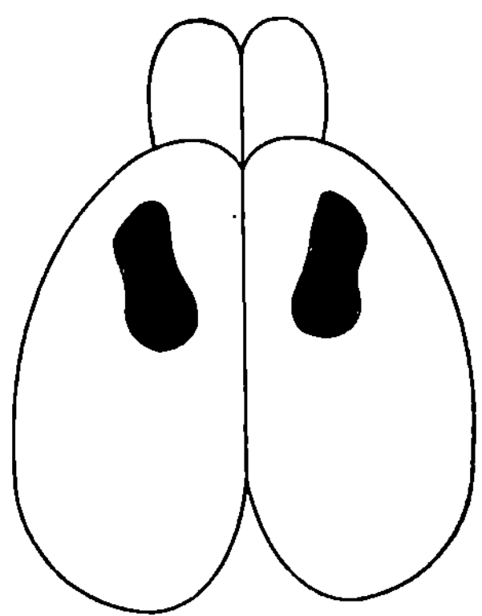

FIG. 73.—Cortical area concerned with coordinated walking. Rats having locomotor difficulties as a result of cortical injuries are found to have lesions primarily confined to the region indicated in black. (*Modified from Maier*, 1935.)

Maier (1935) found that the destruction of certain areas resulted in incoordinated walking which was observable when the rats ran on narrow pathways. Animals having a common area destroyed which is designated in Fig. 73, showed a marked handicap. Groups of animals with lesser destruction of this area showed less incoordination. The degree of incoordination thus seemed to depend on the degree of destruction in this area. Further, surrounding areas contributed to the handicap. For example, an animal with a certain degree of destruction in this area, but little destruction in the surrounding regions, was less handicapped than one with the same destruction in the critical area, but with considerable destruction outside. Whether incoordinated walking is a sensory or motor defect, or a combination of these cannot be determined at present. The isolated region falls partly in Fortuyn's motor area and partly in his somesthetic area, which suggests that the coordination of sensory and motor processes is involved.

Sensory Areas as Specific in Function.—The studies of Lashley (1934*a* and *b*) and Lashley and Frank (1934) on the visual areas

[1] The motor area of the rat is distributed rather generally over the frontal two thirds of the cortex. The more frontal points produce movements in muscles of the fore part of the body, the points most posterior produce movements of the hind part. The locus of points varies widely in different rats, but the general pattern is rather constant.

already referred to demonstrate a cortical area, each point of which has a specific functional and structural connection with a different part of the retina. This cortical area (Fig. 71) has at least one function, that of receiving visual impulses arising from stimulation of the retina, which is specific in nature. To what extent other sensory projection areas are specific has not been determined in the rat. In higher mammals there is reason to believe that most sense organs have specific cortical areas. It does not follow, however, that this specificity carries over to learning and thus gives rise to specific association centers.

The Limitations of Mass Action

Mass action in the extreme sense would make behavior dependent purely upon the mass of functional cortical tissue. Studies already reported delimit this generalization in that the learning of simple habits is not affected by lesions and in that the retention of such habits is localized. When retention depends upon delimited regions, these regions differ *qualitatively* from others in their potentiality for function, even though other regions may take over the function. There is also some evidence which suggests that there may be *quantitative* differences in the potentiality. Maier (1932*b*) found that cortical lesions in the posterior part of the brain had the same effect on the "higher processes" as lesions in the anterior part of the brain, but the lesion necessary to produce the same effect had to be greater in the former case. Further evidence, however, is required to prove the point.

Evidence which shows that the degree of deterioration is dependent upon the shape of the lesion has been presented by Maier (1934). He found that when the extent and locus of the injury remained constant, elongated lesions were much less destructive than round ones. The relationship between the ratio of length to width of lesion and efficient performance was linear as shown in Fig. 74. (The length of the lesion was measured in the anterior-posterior direction, the width of the lesion in the lateral direction.) This effect of the lesion pattern on behavior may be due to either a difference in dynamics which results from the shape of the lesion, or to a difference in the manner in which differently shaped lesions interrupt the association fibers. Long lesions would interrupt more association fibers running in the lateral direction than round lesions which would interrupt relatively more longitudinal fibers. In any case the concept of "mass action" must be modified by these findings.

Since most of the evidence for mass action has been obtained from studies of the behavior of the rat it can be argued that the theory does not hold for corresponding behavior in higher mammals. Primates have more definitely defined cortical areas and generalizations based on the rat might be irrelevant. However, the review of the few studies which dealt with higher mammals suggests that the generalization holds for them. Lashley (1929) carefully examined the clinical literature in which attempts were made to isolate

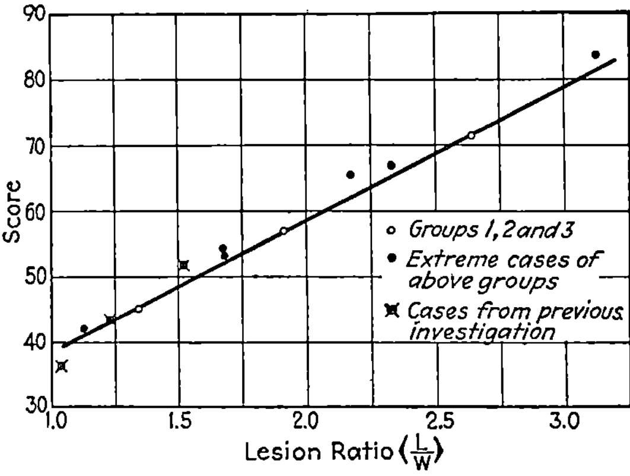

FIG. 74.—Curve showing the effect of the shape of a lesion on the ability to solve problems. As the ratios between the length to the width of lesions increase. performance tends to increase proportionally. The extent of the lesions was approximately the same for groups of animals having differently shaped lesions. (*From Maier,* 1934, *p.* 426. *By permission of the Journal of Comparative Neurology.*)

the effects of accidental brain injuries in man and found no evidence which refuted the theory. Little of a positive nature can be concluded from the clinical literature because adequate tests are seldom made both before and after injury. The brain is seldom obtained after the death of a patient and studied in its relationship to the behavior defects. General observations of behavior over a short period of time, and approximations as to the extent and locus of the lesions, are not satisfactory. The results, generally speaking, are most confusing and contradictory. At present many clinical workers find the theory of mass action as helpful as the theory of localization.

The studies of Lashley on vision further lessen the probability of a fundamental distinction between the functions of brains of rats and higher mammals. Since the rat's visual area is as specifically localized as that of higher forms, distinctions between mammalian groups on the basis of differences in localization are rather futile.

In the light of his studies on the rat Lashley (1934*b*) suggests that the orderly projection of fibers from the retinas to the cortex is essential to the "establishment in the cortex of a pattern of excitation corresponding in its spacial properties to the spacial pattern of excitation in the retina" (p. 72). By this means, certain relationships in the visual field can be transmitted to the brain. Strict localization in the cortex, therefore, may be regarded not as indicating localized centers which control behavior, but rather as a means for furnishing a basis for responses to visual space relationships. But once the relationship is set up in the brain, the responses resulting are largely a matter of generalized cortical function. Different spacial relations thus furnish a basis for different cortical integrations. In this manner the facts supporting both theories of localization and mass action may be harmonized.

The only behavior evidence which suggests that the brains of rats and higher mammals have different degrees of localization in function is a preliminary report by Jacobsen and Wolfe (1934). They found that lesions in the lateral parts of the frontal lobes of the brain of the monkey and the chimpanzee produced specific losses in the ability to solve complex problems. Lesions to other parts produced disturbances of a different nature.

Until more evidence is available, the universality of the theory of mass action cannot be established. No matter what the outcome, "mass action" is an important characteristic of brain activity which must be incorporated in any theory of cortical function. Strict localization for all functions is undoubtedly out of the question. Differences in the brain structure of lower mammals and primates are undoubtedly factors and primates perhaps have a more differentiated cortex. Just how this differentiation will affect the theory of cortical function cannot be predicted. In any case no theory of learning which implies a change in specific neural structure can be accepted as consistent with the function of the nervous system. Rather one must seek dynamic principles to account for plastic behavior.

Suggested Readings

Freeman, G. L. 1934. *Introduction to physiological psychology.* Pp. 26–43, and 222–243. New York: Ronald.

Fulton, J. F. 1926. *Muscular contraction and the reflex control of movement.* Pp. 336–363. Baltimore: Williams and Wilkins.

Herrick, C. J. 1926. *Brains of rats and men.* Chicago: Univ. Chicago Press. Pp. 382.

Lashley, K. S. 1929. *Brain mechanisms and intelligence.* Chicago: Univ. Chicago Press. Pp. 186.

PART III

THE NATURE OF MODIFICATION IN BEHAVIOR: CLASS MAMMALIA[1]

The susceptibility to modification of behavior through previous exposure to a situation gives the animal a new avenue along which to develop. Since this development is a matter of past experience it becomes a means for adjustment to surroundings. Because of it adaptation may appear in the individual as well as in the species. While adaptive behavior patterns survive in the species through natural selection, in the individual they survive through the selective process of modification by experience.

The process of modification by experience is not a simple one. It is one thing to demonstrate the presence of modification in behavior, but quite a different thing to analyze the nature and limitations of such modification. Lower animal forms offer evidence of modification, but higher forms, because they have great capacity for modifiability, offer evidence which is useful for determining its nature. For this reason Part III concerns itself primarily with the study of mammalian learning behavior.

The nature of modification by experience expresses itself differently in various situations and in various animals; and these differences are both qualitative and quantitative in nature. For this reason it is important to study the behavior of one animal form in a variety of situations; and that of a variety of animals in the same situation. Comparative studies yield data which are pertinent for an understanding of various animals as well as the process of modification in general. Variations in technique which are guided by

[1] Class *Mammalia* includes homoiothermic (warm-blooded) vertebrate animals which suckle the young, are covered with hair at some stage, and possess a diaphragm. Among the important orders of mammals are: *Marsupialia* (young carried in pouch), including opossum, kangaroo; *Carnivora* (clawed, flesh-eating), including dog, cat, raccoon; *Rodentia* (gnawing mammals), including squirrel, rat, porcupine; and *Artiodactyla* (hoofed, with even number of toes), including pig, deer, domestic cattle; *Primates* (possessing "nails," opposable first digits, large brain), including monkey, chimpanzee, and man.

theoretical concepts, and which are more than the experimental controls demanded by scientific method are most essential. In many cases theory does not demand that a particular animal form be studied or that an established technique be used. Too often old experiments are repeated with the substitution of a different animal when comparison of results on different animals has no point. In many cases experiments are performed merely for the purpose of measuring something, and with no idea as to what has been measured or why it has been measured. If experimenters would ask themselves "What difference does it make if the results turn out one way or another?" much worthless work would be avoided. No one can justify the attitude that some day someone will find all facts useful to some theory. The problem investigated, the methods used, and variations in procedure which are carried through must grow out of the requirements of a theory. New concepts produce great strides in our knowledge of adaptive behavior particularly because they require new and specific procedures.

The difficulty of our problem is increased by the fact that the behavior of certain animals is so easily modified that it ceases to be a constant specimen for any experiment. Changes in the animal taking place during the process of experimentation must therefore be considered and dealt with.

Modifiability in behavior perhaps depends upon several distinct processes. Its study is often complicated by an interaction between these since tests seldom measure a single process in isolation. The following analysis represents an attempt to isolate the separate functions which may be present and to organize the evidence around them.

CHAPTER XV

INTRODUCTION TO THE PROBLEM OF LEARNING

In our treatment of lower forms, it was found that the behavior of very primitive organisms can be modified by experience.[1] Often the modification is temporary in nature, but in many cases it is permanent. This modification in behavior is expressed by the elimination of certain movements because of previous stimulation, and by the dominance of a particular combination of movements. In the earthworm, past experiences have led to the establishment of a certain pattern of movements which free the animal from a simple maze so that these movements dominate when the animal is placed in the maze, whereas others drop out. In order specifically to define learning, and to isolate the necessary conditions under which it takes place, it seems advisable to examine an animal form in which a large number of simple but permanent modifications in behavior have been produced by previous experience. The experiments best suited for this purpose are those performed on dogs by Pavlov and his students (1927).

THE CONDITIONED REFLEX

Pavlov's technique involves the study of the salivary response to various sorts of stimulus situations. Without previous experience, dogs are equipped with the ability to salivate when meat is placed in their mouths. We are not concerned, at present, with the origin and development of this ability. Because it is present without subjecting the animal to any special conditions, Pavlov believes

[1] The term *experience* as used here and on following pages refers to the effect of stimulation on the organism. Consciousness may or may not accompany such stimulation, but this cannot be known and really makes no difference. We have therefore used the term in its broad sense and have not used it as a reference to consciousness. To limit the meaning of the term *experience* to a conscious state is likely to imply that consciousness is an important determiner of behavior and therefore worthy of a terminology which refers to it. The point of view of this treatment opposes any such assumptions.

this response to be innate, and calls this response to food in the mouth an *unconditioned reflex.* The physical object which sets off the response is known as the *unconditioned stimulus.* The flow of saliva is a particularly desirable response to investigate because its intensity can be easily measured. By means of a minor operation the saliva can be led from the dog's mouth and caught in calibrated tubes or measured by electrical recording instruments. In all cases Pavlov indicates the intensity of the response by stating the number of drops of saliva produced.

Using the unconditioned reflex as a starting point, it is possible to investigate in what manner the animal may be modified by past experience. If, for example, a bell is sounded, the untrained dog responds by pricking up his ears (an unconditioned response to an auditory stimulus), but if the bell is repeatedly sounded just before meat powder is placed in the dog's mouth a new condition is set up. Gradually the sound of the bell produces less of the response that it formerly evoked and more of the response that the meat powder evoked. After half a dozen repetitions of the sounding of the bell, followed by feeding, the salivation actually begins before the food is presented. In fact, the sound of the bell alone eventually produces approximately as much saliva as the taste of the food. In other words, the bell has ceased to be the unconditioned stimulus for pricking up the ears and has become the condition for producing the salivary response. This new response to the sound of the bell is known as a *conditioned reflex.* An old reflex has been so modified that it can be elicited under a new condition. The new stimulus is known as the *conditioned stimulus.*

In order to appreciate fully the fundamental importance of the conditioned reflexes, let us examine a few illustrations. A dog ordinarily salivates when it comes within sight of food. This response to the sight of food has been acquired through the ordinary past experiences which dogs encounter. When a dog is fed, it ordinarily sees the food before it tastes it. Like the sound of the bell, the sight of food becomes the conditioned stimulus for the salivary response.

During investigations on the effects of drugs in Pavlov's laboratory, it was found that preparing the dog for an injection of morphine produced the same effects as the actual injection. The unconditioned response to morphine is nausea and vomiting followed by profound sleep. As the preparations for the injection always preceded the actual injection, this preparation actually became the conditioned stimulus for these same responses.

Painful stimuli, such as electric shocks, which normally produce avoidance reactions such as struggling (unconditioned reflexes) can be converted into conditioned stimuli which produce positive seeking reactions. By following the application of a painful stimulus with a pleasant stimulus, such as food, the painful stimulus, instead of producing its former reaction, produces such reactions as salivation and a smacking of the lips. From all appearances the unpleasant stimulus has been converted into a pleasant stimulus.

Stimuli Which May Be Conditioned.—Any stimulus which can be experienced by an animal can be converted into a conditioned stimulus provided the animal has a nervous system which is highly enough developed. The dog, for example, can be made to salivate (or produce some other specific reaction) in response to stimuli applied to any sense department and even to different intensities or qualities of the stimuli. Thus a bell of particular pitch, or a visual pattern of specific geometrical design may be made to produce salivation. Even the experience of certain time intervals may be conditioned. If a dog is presented with food every 15 min., it will soon salivate at the end of such intervals even when food is not present. Dogs have also acquired conditioned reflexes to the following kinds of stimuli: a certain frequency of beats of a metronome; the beginning of a stimulus, such as a noise; the termination of a stimulus; the last half of a stimulation period of considerable duration; and a particular temporal pattern of stimuli, such as a series of tones.

Conditions Necessary for Establishing a Conditioned Reflex.—In order that a conditioned reflex be established, it is necessary that the conditioned and the unconditioned stimuli be active at approximately the same time. The reflex is most efficiently established when presentation of the conditioned stimulus (*e.g.*, the sound of the bell) precedes the unconditioned stimulus (*e.g.*, the taste of food) by the fraction of a second. It is least efficiently established when this order of presentation is reversed.[1] Generally speaking, the

[1] Wolfle (1930) found in conditioning the human subject that an interval of ½ sec. between the conditioned and the unconditioned stimulus yielded optimal results. Variations in time in either direction were less effective. When the order of stimulus presentation was reversed, conditioning was less frequently produced apparently because of the deviation from the optimal interval. Pre-

two stimuli must always be presented in temporal contiguity with each other. The interval between them must not be too great otherwise the contiguity is lost. An interval of several seconds causes considerable difficulty.

As each of the stimuli may produce unconditioned reflexes of its own, the response which eventually becomes conditioned may be regarded as the more basic. In our illustrations, the response to the bell is therefore biologically less important than the response to food in the mouth. The relative intensities of the stimuli used are also an important factor in determining which responses will become conditioned and how quickly they will be conditioned.

Not only are certain objective conditions required, but the condition of the animal is also important. Pavlov points out that the animal must be in good health and alert during the learning period. If it is in poor health, internal stimuli are present, and these may interfere with the dominance of the conditioning stimuli. To say that the animal must be alert or attentive means that the conditioned stimuli must stand out or dominate in the animal's experience.[1] To keep the animal alert and attentive, extraneous or distracting stimuli must be excluded.

The above mentioned objective and organic conditions are necessary for the establishment of the conditioned reflex. Other conditions vary the speed or ease with which these reflexes are established. We have already pointed out that varying the contiguity of the conditioned and unconditioned stimuli varies the ease with which an animal may be conditioned. The rate of acquisition also varies from animal to animal and different stimuli vary in their effectiveness for the same animal. Invariably, the first conditioned reflex is more difficult to establish in an animal than are later ones. Perhaps this is due to the animal's learning to become attentive and alert in the experimental situation.

Experimental Extinction.—Once a conditioned reflex has been established, it remains for some time but gradually wanes. If, however, the conditioned stimulus is repeatedly given without being followed by the unconditioned stimulus (*i.e.*, without being reinforced) the response soon disappears. Thus sounding the bell repeatedly without the accompaniment of food soon results in no

---

senting the unconditioned stimulus first reduces the effectiveness of the conditioned stimulus just as does too great an interval.

[1] See footnote on p. 337.

salivation. The destruction of an acquired response in this manner is known as *experimental extinction.* The phenomenon is but temporary in nature and a response which has been extinguished on one day may reappear in full strength the following day. Extinction seems to be a form of negative learning, *i.e.*, the animal learns not to respond. Like the formation of the conditioned response, its destruction is achieved more quickly when the animal has experienced extinction more frequently. It is likely that a negative response has been conditioned and replaces the original conditioned response.

The Secondary Conditioned Reflex.—We have seen that the conditioned reflex is essentially an old response to a new stimulus. In order to modify the animal it is necessary to begin with a response of which the animal is already capable. If a conditioned reflex has already been established, it is possible to begin with this reflex instead of the unconditioned reflex, and condition this response to a new stimulus. In this case a conditioned reflex would be conditioned. Suppose a dog has been trained to salivate at the sound of a bell. If now a light is flashed just before the bell is sounded, the flash of light should, in its own right, soon call out the response of salivation. However, by the time the response to light is acquired, the bell will have been presented several times without reinforcement (food) and as a consequence the salivary response to the bell extinguished. To avoid this, it is necessary to reinforce the response to the bell from time to time. Thus if the bell-food combination and the light-bell combination are presented alternately, the difficulty is overcome. The result is a conditioned reflex of the second order, or a *secondary conditioned reflex.*

In a similar manner, conditioned reflexes of the third order have been successfully established in some cases, but no one has succeeded in establishing conditioned reflexes of the fourth order. Apparently experimental extinction sets in before all of the necessary pairs of combinations can be renewed.

Generalized and Specific Conditioned Reflexes.—In the early stages of the acquisition of a conditioned reflex the stimuli which call out the response are of a rather general nature. Thus if the sound of a bell is used as the conditioned stimulus, the dog does not learn to salivate to the sound of that particular bell, but rather to the sounds of all bells, or even to auditory stimuli of any sort. In order to make the response specific for some particular characteristic of the stimulus it is necessary to present the animal with a variety of stimuli at different times, and to reinforce only those which have that characteristic. Thus if food always accompanies the sound of a bell of a certain pitch, but never accompanies the sounds of bells of different pitches, a dog soon salivates only to this particular pitch. This procedure results in some specific aspect of the stimulus situation becoming the positive stimulus, all other aspects of the situation becoming ineffective as stimuli for salivation. The development of specific responses to specific stimuli through learning parallels the

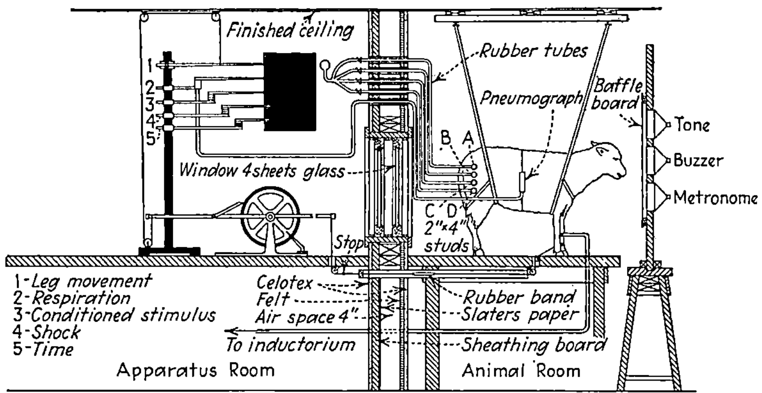

FIG. 75.—Apparatus for studying the conditioned foreleg flexion in the sheep. Electric shock is used as the unconditioned stimulus. The apparatus allows the presentation of a large variety of conditioned stimuli as well as accurate measurement of the response. (*Redrawn from Liddell, James, and Anderson,* 1934, *p.* 89. *By permission of the authors.*)

development of specificity through maturation already treated on page 277.

Conditioning as a Method for the Investigation of Learning. Since the conclusions thus far presented have been based upon a specific response (salivation) and a specific animal (the dog), it might be supposed that the principles discussed may not hold for other responses and other animals. However, a great deal of work has been done which eliminates such a possibility. In Fig. 75 a modern laboratory for investigating conditioning is illustrated. Liddell and Anderson (1931) investigated the defense reaction (withdrawal by flexion of foreleg) using electric shock to the leg as the unconditioned stimulus. The rabbit, sheep, goat, and pig were used as subjects. The results are in harmony with those obtained from the study of the salivary response in the dog. As might be expected, the speed at which various animals form the response varies, being most rapid for the pig and least for the rabbit. Krasnogorski (1926) used the salivary response (the intensity of which was measured by the frequency of swallowing) as well as other food-taking reactions in his investigations with children, but a wide range of other forms of responses have been used in investigations with humans (see Hull, 1934).

It must not be assumed that all responses can be conditioned. Some offer difficulty, but no generalization can be made regarding the class of responses most easily conditioned, since examples of the conditioning of skeletal and visceral muscle responses as well as responses of glands can be found in the literature. Animals of the same species, particularly among human beings, also show great variation in their susceptibility to the conditioning of certain responses.

In the present chapter our interest in the conditioned response has been primarily one of developing a background for a general treatment of learning. We will now pursue this problem, but will again return to the conditioned response in the following chapter and discuss it in relationship to other forms of learning.

Definition of Learning

The varied experiments on the conditioned response have made it possible for us to isolate the conditions under which an animal may be modified by experience. From these conditions it should be possible for us to define learning in a restricted sense. Since there are many varied situations in which animals modify their behavior and since many animals have abilities other than the ability to learn, a specific definition of this ability is essential to any detailed analysis of animal capacities.

Let us therefore define learning ability as *the ability to combine (or associate) two or more contiguous experiences in cases in which the contiguity is determined by the environment.* In the conditioning experiments two stimuli (*e.g.*, bell and meat in the mouth) are presented in contiguity. Each of these produces a brain reaction (an experience) in the animal. But the contiguity of the stimuli does not necessarily insure a contiguity in experience. To insure this, the animal must be alert and distracting stimuli excluded. In other words, the two stimuli must produce two simultaneous centers of brain activity which dominate over other centers of activity. The dominance of these simultaneous centers of activity constitutes the contiguity in experience.

That past experience is required for the formation of such combinations is implied. The two stimuli must be experienced together in order to become combined, hence at least one repetition is necessary. Reactions to situations which occur without previous repetitions are therefore not learned reactions.

The degree of learning, all other conditions remaining equal, depends on the number of repetitions, and every occasion on which the two experiences are not in contiguity weakens their combination, since on such occasions they are in contiguity with different stimuli. For example, if the bell is sometimes experienced in contiguity with food and not at other times, the degree of learning is less than if they had been experienced together without exception. The formation of responses to specific aspects of a stimulus is based on this fact. The response to the general stimulus becomes weakened because the contiguity is not always present, whereas the specific characteristic is common to all stimuli which are reinforced, and the response to this characteristic becomes strengthened and finally dominates.

The process of learning may therefore be regarded as a development of responses to limited aspects of the environment. At first many different stimuli are equal in the sense that they call forth the same responses. As training progresses, the animal responds more and more specifically. In other words, further training makes the animal increasingly adaptive and efficient because the responses become linked with specific aspects of the environment. The animal's repertoire of differential responses is thus limited only by its discriminative capacity.

Problems in Learning

Performance an Index of Learning.—One obtains evidence of learning by observing an animal's performance; one does not know of an ability unless it can be brought to expression. In the conditioned reflex, the act learned is simple and can be elicited at any time by means of the unconditioned stimulus. In training an animal to traverse a maze, however, one is dependent upon a goal at the end of the maze for bringing about a certain performance. To cause an animal to perform to its best ability, or to perform at all, is often a perplexing problem.

Because evidence for learning depends upon performance, learning in the objective sense must be regarded as an acquired response to some form of stimulation. The problem for the psychologist is to determine the stimulus which is effective in eliciting the acquired response, as well as to determine the nature of the response which is elicited. In the conditioned reflex a given response is merely elicited by a new stimulus. In other cases of learning, one of several responses becomes the dominant response to a stimulus. The response which dominates acquires its dominance by satisfying a need, and is not a response which has been transferred from one sense department to another. In many cases the response which is acquired is only part of the animal's pattern of behavior. This requires an analysis of the behavior in order to determine what part of the pattern has been acquired.

Analysis of the Behavior in Learning Situations.—Just exactly what responses have been acquired in any particular learning situation become apparent as soon as a certain consistency appears in the pattern of behavior. When placed in the situation repeatedly, an animal tends to repeat certain acts. The specific acts which it repeats whenever placed in this particular situation become a function of the situation and are what the animal has learned or acquired in that situation.

For example, let us say that a rat must turn to the right at a certain junction of a maze in order to obtain food. If we know that the animal needs food, we may expect it to perform if it is able. Let us say that the rat turns to the right and receives food. Suppose it turns to the right each time it is confronted with that junction. We can then say that the turning to the right has been learned, because this response to the same situation occurs repeatedly. However, we find that the animal's acts are not exactly alike each time it makes the trip. Sometimes it runs rapidly, sometimes slowly; sometimes it stops along the way, sometimes it does not; sometimes it makes the turn in a different manner than at other times. In other words, the same succession of acts is not reproduced on each trial. This being the case, we can say that the rat has *not* learned a succession of acts which involve the same kind of movements. But it has learned to turn to the right for food when placed in a particular situation. The turning to the right, not the manner in which it turns to the right, is thus the pattern which it has acquired. If this is the behavior pattern in which we were interested when planning the experiment, we say that the animal has succeeded in mastering the problem.

Analysis of the Stimulus Situation.—Since a learned response is elicited by a stimulus, it is necessary to determine just what aspect of the stimulus is effective. A dog may sit up when commanded. Does it do so in response to any voice, or to its master's voice? Must the command be given at a certain pitch or intensity? Is the sight of the person giving the command required? In giving the command the dog's master produces a complex of auditory and visual stimuli. Which of these produces the response? It is possible that a learned act is a response to stimulation of some general sense department, but it is also possible that several sense departments together are responsible. Certain attributes of the stimulation of one sense department may also be sufficient. Thus a certain pitch or intensity or duration or quality of a sound alone may be responsible. Animals can be trained to respond to any of these attributes of a stimulus but in learning situations it often becomes a difficult problem to determine which attribute, or what combination of attributes, forms the basis of a learned response. The nature of the sensory processes used by the animal in running a maze is still a warmly debated issue despite the fact that this has been a favorite problem for investigation.

The Rate of Learning a Function of the Animal.—The ability which an animal demonstrates in modifying its behavior is largely dependent upon the plasticity of its nervous system. For this reason individual differences in learning ability among animals of the same species become apparent. Among animals of different species this variation in the rate of learning will be much more definite because of the wider difference in neural structure. Comparisons of learning ability between animals, and correlations of differences in learning ability with differences in neural structure will therefore add to an understanding of learning. Diversity in temperament as well as in physical structure must also be considered since these greatly affect the performance of an animal.

The Rate of Learning a Function of the Test Situation.—The behavior of animals has been studied in a variety of experimental situations. These situations are the instruments of the psychologist and give him a set of measurements. If he uses learning instruments, we might expect that he is measuring learning. If this were the case, all learning instruments would give similar results, and any differences which appeared could be attributed to differences in the sensitivity of the instruments. However, it is found that the measurements are inconsistent with each other and do not agree even in their qualitative aspects. For this reason it is necessary to analyze the various learning situations and see whether or not it is possible to make the inconsistencies understandable. It seems likely that the various learning instruments presuppose a too general concept of learning ability. If we limit its meaning (as we have done in our definition) and suppose that other abilities may also be involved, then it is reasonable to account for the inconsistencies by saying that *different learning instruments measure different combinations of abilities*, although learning ability may be a part of the combination in each case.

With this possibility in mind, we will set for ourselves the task of analyzing various learning situations in order to determine first, the actual aspect of the learning situation which is linked with a certain response; second, the extent to which learning is demanded by the situation; and third, the possible pattern of abilities which may contribute to the measurements. The two following chapters are a presentation of such an analysis.

Suggested Readings

Hull, C. L. 1934. The factor of the conditioned reflex. *Handbook of General Experimental Psychology* (ed. by Murchison). Pp. 382–455. Worcester: Clark Univ. Press.

Liddell, H. S. 1934. The conditioned reflex. *Comparative Psychology* (ed. by Moss). Pp. 247–296. New York: Prentice-Hall.

Tolman, E. C. 1932. *Purposive behavior in animals and men.* Pp. 3–23, New York: Appleton-Century.

CHAPTER XVI

THE SPECIFIC PATTERNS LEARNED IN VARIOUS LEARNING SITUATIONS

The Position Habit

One of the simplest forms of learning is the acquisition of the position habit. In mammals it is established in a few trials. The animal is merely required to learn to go one place rather than another. Usually a T maze such as was used to detect learning in lower organisms (see Fig. 26) is used. The animal is placed at the starting point and must choose between two alternate alleys. To cause the animal to perform it is rewarded (*e.g.*, with food) when it takes the correct alley, and punished (with an electric shock or no food) when it takes the other alley. In order to learn which alley contains food the animal must be able to discriminate between the two alternate choices.

Since these differ in position, each alley will produce specific visual effects (*e.g.*, one will stimulate the right part of the visual field), and the turn into each alley will give rise to specific muscular (kinesthetic) sensations. Learning consists in associating the experience of responding to a specific part of the situation with the experience of food. Only one association must be formed. Contiguity of these two experiences is assured because the experience of a certain response to the junction situation presents the animal with the experience of food. The only possible difficulty is the establishment of a dominance of these contiguous experiences. This difficulty is not very great in the ordinary position habit because a response to any differential experience of the two alleys which happens to be dominant will be adequate. Some visual and kinesthetic differences are always present and in addition olfactory, auditory (sound reflections), and tactual differences may be present. All or any of these are adequate. Since animals which have learned the habit continue to make the response after different sensory features are excluded, it is probable that kinesthesis plays an important part; the turn at the junction being in response to the experience of running up the starting alley to the junction.

However, one cannot conclude that the position habit is merely a response based upon kinesthesis, otherwise the animal would make the same turn every time it went through a certain pattern of movements. The animal only makes the turn when it is placed in the test situation. This situation must therefore furnish stimuli which supplement kinesthesis. These stimuli are most likely tactual and visual in nature. By means of them the animal responds to the starting place and the junction; they need not determine the nature of the response at the junction. Since it is the differential response at the junction which one wishes the animal to learn, one may regard this response as the one which has been acquired. The supplementary responses which accompany it are present in all learning situations and may be disregarded in making comparisons. They must not, however, be regarded as nonexistent.

Discrimination Habits

The Process of Association.—In the chapter on sensory discrimination the emphasis was upon what the animal could discriminate. We are now concerned with these habits from the point of view of learning. Discrimination habits, as far as the actual process of learning is concerned, are very simple and specific. They are simple because but one association is required; specific because the association pattern which is to be formed is determined (or at least should be) by the experimenter. They are often difficult for an animal to learn because the particular sensory aspect of the stimulus (the sensory cue) selected by the experimenter is not readily reacted to by the animal.

In order that an animal be able to go to light for food (see Fig. 64), it must, of course, be able to experience the difference between light and no light; but assuming this sensory discrimination, it must in addition associate a response to the lighting situation with food. Food may be experienced in contiguity with going to the lighted compartment, but position differences, odor differences, and other peculiarities are also present, and may be reacted to and experienced in contiguity with the food. The experiment must therefore be continued until the reaction to the desired peculiarity in the situation (*i.e.*, the aspect of the stimulus which the experimenter has selected) becomes associated with the food. A difficulty in the formation of the habit thus becomes one of isolating the "correct" sensory process. Usually the animal will succeed, because in the course of many trials only the "correct" aspect of the stimulus situation (*e.g.*, lighting difference) is present with the food on all occasions. Other sensory aspects (olfactory, kinesthetic, etc.) vary from trial to trial, so that no particular association tends to be formed. These become associated with food on one trial and with no food on another. Hence it is theoretically possible that the animal will eventually experience a response to the desired sensory process in conjunction with the experience of food.

The Process of Selection.—Krechevsky (1932 *a* and *b*) has very nicely demonstrated that the rat first responds to one sensory aspect in relation to food, then to another, and so on. One of his methods involved the use of a multiple discrimination box in which the rat was required to choose between a succession of right and left turns. There were obvious position and lighting differences at each junction. The problem was, it happens, insoluble because neither the turning to the right, to the left, to the light, nor away from the light, was correct at all junctions. The responses of the rats showed, in turn, definite responses to the several sensory aspects of the situation. A rat might respond by going to the right at each junction, then by going always to the left, next by a succession of alternations, and on a different occasion the responses toward or away from the light might dominate the behavior. Different rats responded to these possibilities and combinations of them in different orders, but all tended to respond in some systematic manner at any particular stage of learning. The choices were not according to chance.

Thus it is obvious that the difficulty involved in the formation of discrimination habits is not one of combining certain elements into a pattern of behavior, but is rather one of selection. The difficulty encountered in selection varies for different sensory processes, as well as for the same sensory processes in different test situations. A differential response to tones is much more difficult to establish than one to lights, even when the techniques used seem almost identical. The response to the same visual patterns is easily established in rats by the use of Lashley's jumping technique, but is difficult to establish by the use of a discrimination box. Lashley (1930*a*) has shown that the jumping technique requires only from about 3 to 10 per cent as many trials for the establishment of certain responses to visual pattern as do other techniques. In fact, some techniques fail to cause the establishment of differential responses to visual patterns.

The reason the acquisition of some discrimination habits is more difficult than others may be because some sensory aspects of a stimulus are more evident to the animal than others. A tonal stimulus is difficult to localize in space and consequently it cannot become something definite to which to react.[1] Noises are more easily localized and the results of experiments show that a response to noise is relatively easily established. Responses to visual stimuli are more easily established than to auditory stimuli, apparently because visual stimuli always occupy a definite position in space.

The fundamental importance of the degree of the directness of a response to the stimulus is further illustrated when the different techniques utilized in setting up discrimination habits are analyzed. In the jumping technique a direct reaction to the stimulus pattern is required; in the discrimination-box method, a response to the junction is demanded, but the visual pattern which serves as the basis for the discrimination is somewhat removed and is less likely to be reacted to at the time a choice is required. For this reason the problem of selection is much less complex in the former than in the latter method.

A high degree of intelligence[2] is unlikely to be an asset in learning such habits because the isolation of the sensory process to which the animal reacts is largely a matter of trying out reactions to different sensory aspects. There is no more reason why food should be associated with one factor in the situation than with another. Only continued and varied activity will eventually cause a response to a particular characteristic of the situation to be experienced with food. This an unintelligent animal can do as well as an intelligent one. Once the response to this characteristic is experienced in conjunction with the food, the establishment of a preference for this reaction is a simple process.

The Response to Relative Differences

Brightness Discrimination.—If two windows of a discrimination box differ in brightness, an animal can readily be trained to prefer

[1] The importance of localizing cues is pointed out by Thuma (1932).

[2] Intelligence may here be regarded to mean plasticity or adaptability to modification in behavior.

either one of them regardless of their relative positions. Suppose the animal has been trained to choose the compartment with the brighter window. We may now raise two fundamentally different types of questions: (1) Has the animal learned to go to the brighter window, or has it learned to avoid the less bright one, or has it learned both to avoid the one and choose the other? (2) Has the animal learned to respond to the brightnesses of these windows as particular stimuli, or has it learned to respond to the *brighter* of the two windows, irrespective of their absolute intensity, and thus to respond to the situation as a pattern?

Several experiments have been performed to answer these questions in the case of a variety of animals. Köhler (1918) experimented with chickens, a chimpanzee, and a child. His procedure in the case of chickens was as follows: Four chickens were trained to peck grains from a gray card of a certain brightness rather than from a neighboring gray card of a different brightness. The relative position of the cards was, of course, changed for different trials in order to prevent the formation of position and alternation habits. The preference for pecking from one gray card rather than another was established in two different ways. (1) The grains on the gray card to be avoided were covered by a piece of glass, whereas the grains on the other gray were obtainable. The chickens soon learned to avoid the card covered with glass and continued to avoid it when the glass covering was no longer employed, showing that the preference was based on the cards rather than the glass. (2) When the chicken approached the one gray, it was always chased away; when it approached the other gray, it was allowed to eat. Soon it became unnecessary to chase the chickens away.

By using a series of grays varying in brightness from black to white and numbered from 1 to 49, respectively (Zimmermann series), numerous pairs of grays of known relative brightnesses could be presented to the chickens. For the training series, Köhler used cards numbered 5 and 30. Two chicks were trained to prefer the lighter gray (card 5) and two chicks the darker gray (card 30).

After the preference had been established (requiring several hundred trials) the chicks were given some critical tests. The two chicks which were trained to go to card 5 (the positive stimulus) were presented with cards 1 and 5. The two chicks trained to go to card 30 (the positive stimulus in their case) were presented with cards 30 and 49.

The results of the tests showed that the new card was preferred 69.4 per cent of the time in a total of 85 choices. The card which had been the positive stimulus in the training period was avoided in the test period. Hence the training period was not such as to establish a preference for a particular card. However, if the chicks trained positively to card 5 actually learned to go to the lighter card, and the chicks trained positively to card 30 learned to go to the darker card, then the results are as one would expect, since card 1 was lighter than card 5, and card 49 was darker than card 30.

These results can also be explained by postulating a natural tendency to choose a new card. To exclude this alternative explanation, a second test was given in which the card which had been negative in the training period was presented to the animal together with a new card. On the basis of a response to relative brightness the new card was incorrect in all cases. In 78 test choices the four chicks chose the negative stimulus card of the training period 66.7 per cent of the time. Again the chicks seemed to respond according to what one would expect if the training period set up a response to a relationship, and contrary to what one would expect if chicks had a preference for new cards.

Although these results show a marked reversal in preference for certain cards used in the training and test periods, the reversal is not complete. Apparently the absolute appearance of the cards of the training period is a factor. If a chick is given tests of the above mentioned type directly after a number of training trials, the memory of the appearance of the particular training cards should be stronger, hence in immediately following tests the presence of new and strange cards should be more likely to be noticed. This naturally should interfere with a response to the relative properties of the stimulus pattern. Tests under these conditions showed a less definite response to the relationship. Thus we have evidence which indicates that although the response to the properties of the two cards as a pattern is primary in the learning, the absolute characteristics of the cards also play a part, at least in immediately succeeding tests.

Köhler tested a chimpanzee in a similar manner. Two small boxes with gray cards fitted into place on the front sides of the boxes were placed before her cage. Food was to be found in one of them. The animal indicated her choice with a pole. (At first the animal reached the box with the pole. Then the experimenter helped the chimpanzee draw in her choice in order to save time. Finally the chimpanzee learned to indicate her choice by merely laying the pole on the desired box.) Cards 5 (the lighter) and 24 (the darker) were used in the training period. Card 5 was made the positive stimulus-card. In the last 75 of 232 trials only one error was made.

The results of the tests are similar to those of the chicks. The scores made by this animal in the different test periods are given in Table 22. The original pair of cards learned was presented from time to time to test whether the learning was modified by the test periods.

TABLE 22.—SCORES MADE BY KÖHLER'S CHIMPANZEE SHOWING RESPONSES TO A RELATIONSHIP

Test cards		No. of preferences for	
		Lighter gray	Darker gray
Critical pair	24–49	10	0
Learning pair	5–24	10	0
Critical pair	1–5	3	2
Learning pair	5–24	38	2
Critical pair	24–49	9	1
Learning pair	5–24	10	0
Critical pair	1–5	10	0

A three-year-old boy was tested in much the same way as the chimpanzee. He was merely asked to choose one of the boxes. From his fifteenth to his forty-fifth trial he made no more errors, always choosing the box with card 5 attached (and receiving his reward) in preference to the box with card 24 attached.

The results of the test series are shown in Table 23. These results show an even more marked response to the relative properties of the stimulus situation than do the results of the chimpanzee and the chickens.

TABLE 23.—SCORES MADE BY A CHILD SHOWING RESPONSES TO A RELATIONSHIP

Test cards		No. of preferences for	
		Lighter gray	Darker gray
Critical pair	1–5	10	0
Learning pair	5–24	10	0
Critical pair	24–49	10	0

Similar evidence has been obtained in the study of other animals. Washburn and Abbott (1912) vaguely suggested the possibility in their study of vision in rabbits. Helson (1927) found such evidence in the behavior of white rats, and Moody (1929) in his study of mice. The evidence indicates that the response to a relationship is an elementary response rather than one requiring intelligence. This does not mean that a relationship is experienced, but merely that the response is based upon the properties of the stimulus situation as a whole, rather than on the specific properties of one or the other of the stimulus objects.

Size Discrimination.—Gulliksen (1932) studied the response to the relative size of disks and made some quantitative measurements of the tendency to respond to relationships under different conditions. He used Lashley's "jumping apparatus" (see Fig. 67) in order to insure the rapid learning of the habit of choosing between a 9-cm. and a 6-cm. circle. After a discrimination had been established, the animals were confronted with new pairs of circles in order to determine whether or not they would respond on the basis of relative or absolute size.

Gulliksen's results corroborate Köhler's conclusions and show that the response to certain relative properties of the stimulus situation is more natural for the rat than are responses to absolute properties when it is placed in a situation which involves quantitative discriminations. However, Gulliksen's results also show that the response to individual quantities is a factor, although a minor one. It was never important enough to dominate over the response to the relationship, but it did lower the degree of transfer from the cards used in the training series to the cards used in the test series. The substitution of a new card for either the negative or the positive card seemed to effectively lower the score, whereas the substitution of two new cards had little or no effect on the response to the relationship.

The Equivalence of Stimuli.—The experiments of Klüver (1933) yield further information regarding the properties of the stimulus situation which are responsible for eliciting the response in discrimination situations. He trained monkeys to choose one of a pair of stimulus objects. When the training was completed, tests were given in which these objects were modified in various aspects. If the modified stimulus objects produced a preferential response to one of the objects, the modified pair of stimulus objects were regarded as equivalent to the pair used in training. If, however, the training broke down and no preference was shown, the pairs of objects were regarded as nonequivalent. In the discrimination of weights the animals were trained to choose between a box weighing 450 gm.

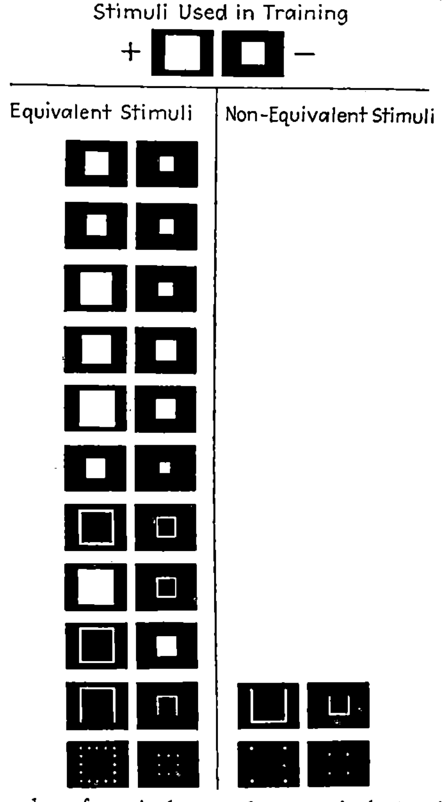

FIG. 76.—Examples of equivalent and nonequivalent stimuli. Monkeys were trained to choose the right member of the above pair of stimuli. This training set up a preference for one member of each pair of stimuli labeled equivalent, but not for those labeled nonequivalent. *(From Klüver, 1933, p. 175. By permission of the University of Chicago Press.)*

and one weighing 150 gm. Strings were attached to the boxes and the monkeys soon learned to pull in the 450-gm. box which always contained the food. A trained monkey would pull first at one string, then at the other, sometimes making several comparisons, and then pull in the correct box. Critical tests were then introduced. If the weights of the boxes were modified, but remained alike in all other respects, a preference for the heavier box was shown. Changing the color of the front sides of the boxes, the amount of friction between the boxes and the floor, or the distance which the boxes had to be pulled, failed to affect the preference for the heavier box. Even the addition of other stimuli, such as placing buzzers in the boxes, failed to disrupt the preference. When, however, the weights of the boxes became approximately equal, a preference for one of the objects disappeared.

With two weights attached to each string (the box and an additional weight attached to the middle of the string), and with slack string between the weights, the choice was made in terms of the first weight rather than the weight of the box. Thus if for one alternative the first weight was 450 gm. and the attached box weighed 150 gm., and for the other alternative the corresponding weights were 150 gm. and 450 gm., a preference was shown for the first alternative. The monkeys thus responded to the difference in tension between the strings which first became noticeable when comparison was made, irrespective of whether or not this tension depended upon the weight of the boxes. This was done despite the fact that the monkeys watched the movement of the boxes carefully and always reached for the boxes to obtain food.

In a similar manner Klüver tested the responses to objects which differed in their visual appearances and determined the aspects of the stimulus situation upon which reactions to them depended. Figure 76 illustrates some of the equivalent and nonequivalent stimuli. It is obvious that great changes can be made in the visual patterns without affecting the preferential response set up during training. Certainly the animal's response is not dependent upon all of the properties of the situation, but rather upon certain differential aspects. The differential aspects are not the property of either of the stimulus objects, but are a product of the two.

Concluding Remarks.—The foregoing experiments demonstrate that in discrimination situations the animal does not react to the specific properties of either of two stimulus objects, but rather it responds primarily to certain relative features of both stimuli. This means that the response depends upon certain aspects of the whole situation. If the stimuli are changed too radically, the situation ceases to be comparable with the learning situation; consequently the training does not carry over to the changed situation even when the same relative features are present. However, many features of the situation may be radically modified without affecting the response. Only changes which eliminate the aspects of the situation upon which the response is based or which destroy the aspects that characterize the situation for the animal disrupt the behavior.

A response, therefore, depends upon certain features of the whole stimulus situation. Two cards are not responded to as separate stimuli but as parts of this total situation. Certain differences between the cards form the basis for the preferential response, whereas other features characterize the situation. For example, a difference in brightness between two lights may produce a differential response in the discrimination box, but the same difference in brightness in different parts of the room might not produce such a response. Being placed in a discrimination box and on the floor of a room are so different that they fail to produce a background for the same responses.

The Solving of Problem Boxes

Introduction.—Some of the first controlled experimental work adapted to a psychological investigation of animals involved the use of the puzzle box or the problem box. The animal is placed inside (or outside) a box and food is to be found outside (or inside). The problem for the animal is to get to the food. This can only be achieved by the opening of a door which is held in place by some form of latch. The latch may be released in the various problem boxes in different ways, the method of release being such as to fall within the scope of the animal's ability. Thus when the animal steps on a platform, pulls a string, or pushes a lever, the door springs open. The animal must learn just what to do in order to get access to the food.

Thorndike (1898) was a pioneer in the investigation of learning in animals. He tested dogs, cats, and chickens in their ability to escape from various types of puzzle boxes. In Fig. 77 a typical problem box is illustrated. As the behavior of his cats is characteristic of that of the other animals, we shall confine our discussion to them.

A hungry cat was placed inside a cage, the door of which opened when a string inside was pulled. Food was placed just outside the cage. The behavior at the outset was that of a "mad scramble." The cat bit and clawed at all parts of the cage. For no reason in particular the claw caught in the string and the door opened. By reconfining the cat in the cage many times, Thorndike observed that the time required for opening the door gradually became less and less and that the cat began to confine its activity more and more to the vicinity of the string. Finally it succeeded in releasing itself by making only such movements as were necessary for its success. From many results of this sort, Thorndike concluded that success is a matter of "trial-and-error" learning. Intelligence and understanding are not involved. Rather the animal makes many random movements, some of which fail to be repeated because they do not

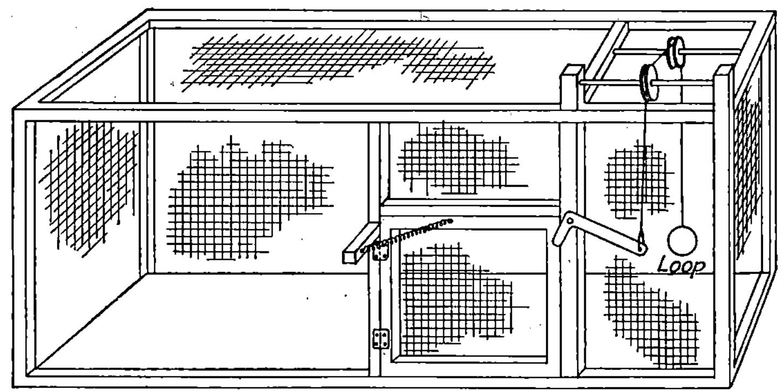

FIG. 77.—A typical problem box. The animal is confined inside the cage and is required to learn to pull the string which opens the door and makes the food accessible.

"satisfy" and so drop out, others of which do "satisfy" (result in food) and so are retained and become the learned act.

Analysis of Learning the Problem-box Habit.—In the light of our definition of learning we might say that the animal is required to associate a certain reaction to the string (in response to some aspect of the situation) with the experiencing of food. The experience of giving a certain reaction to some aspect of the stimulus situation and the experience of food are likely to be contiguous since the second experience must always be preceded by the first. However, the contiguity is not inevitable because the animal may not leave the problem box and obtain food for some time after making the response. The animal reacts to many aspects of the situation in a variety of ways, but one particular reaction to the situation must become preferred because of its connection with food. In other words, a particular part of the mass of behavior must become isolated and connected with the obtaining of food. This is a process of selection and is complicated by the fact that many reactions to a large number of stimuli are possible in the situation. In our analysis of the learning of discriminations we found the number of responses and the number of differential stimuli to be much more limited. The actual learning required by the problem box, however, is very limited, since only one association must be formed.

The difficulty encountered in the acquisition of the problem-box habit may therefore be regarded as primarily due to the difficulty arising from the selective process. The complexity of the selective process depends largely upon the nature of the response which the problem demands. Some responses are readily elicited, others seldom occur. Sometimes the required act is part of a larger pattern of behavior and never becomes an isolated response to some aspect of the situation.

The *sensory* and *manual* equipment of the animal is also important in determining what responses are likely to appear and to become isolated. A simple association may be formed by the cat in one or two trials, whereas an adequate solution to the puzzle box may require twenty-four or more trials. The situation is similar to that of the discrimination problem, except that instead of a few general responses to a limited number of aspects of the stimulus, one of which must be associated with food; there is a complex pattern of behavior called out by the situation, one small part of which must be associated with the desired food experience. The isolation of some particular movement depends on the fineness of perception as well as on the manual dexterity of the animal. Animals with poor vision and lacking in ability to manipulate the fore limbs would find it difficult to isolate from all their activity such a movement as the pulling of a string, even if their random activity happened to include such an act. Yet, such animals might readily succeed in learning to open the door by stepping on a platform.

An Evaluation of the Interpretations of Thorndike and Adams. Thorndike naturally stressed the factors in the animal's behavior which indicated that cats did not figure out how to open the latches, but learned to open them through random trial-and-error activity. Later writers have opposed this extreme position and believe that random activity is not the exact description of the animal's behavior. Adams (1929) has recently repeated Thorndike's work under more favorable circumstances and finds that the behavior of cats in the puzzle box is not so random and haphazard as Thorndike describes. The cats used by Adams were much better adapted to the situation and did not show the excitement that Thorndike's cats displayed. He does not believe that learning to open the door was the mere elimination of "useless" movements. Rather, he believes that higher mental processes are displayed, but tells us very little of their nature. This is probably because the technique is inadequate for demonstrating the presence of higher processes. The fact remains that the animal must learn to open the latches and the particular act must be experienced in connection with the food. There is no *a priori* reason for a cat knowing that pulling a string opens a door any more than there is for a man knowing, without having had the necessary past experiences, that pushing a button turns out the lights. He might discover that the two events were related, but in order to make the discovery the two events would have to be experienced in contiguity. If he made many movements at the same time, only one of which depressed the button, it would require many trials before the necessary act was isolated.

The Problem Box as an Instrument for Measuring Learning.—The problem box has been a rather popular technique for testing the relative intelligence of different animals. Different types of boxes have been designed so as to be adapted to different animals, and combinations of two or three latches have been placed on the same door to add complexity. However, comparisons between animals in these tasks have not been very successful, since the designs must vary to meet the differences in sensory and motor equipment among the various animals. In order partly to eliminate the above difficulty, Jenkins (1927) designed a standard problem box having release plates which can be operated with ease by animals differing widely in manual dexterity. The difficulty of the problem can also be varied by combining two or more release plates.

Another difficulty with the problem-box method arises from the fact that the first appearance of the correct response is largely a matter of chance. For this reason differences in "level" of activity and nature of characteristic movements among animals become an important factor in their success. Animals which are naturally active and which vary their activity will achieve the first success in less time. Since the amount and kind of activity are not dependent upon speed of learning, but vary for other reasons, the learning ability of some animals becomes obscured.

If we regard the problem box as a test involving primarily the selection of a particular response, it ceases to be a measure of learning. The process of selection depends first upon the chance occurrence of a certain movement and secondly upon the making of this movement in response to some aspect of the situation. A complex situation which involves selection largely masks any abilities that might be present and it is therefore difficult to know just what has been measured. In other words, the test is not an analysis of any particular ability but gives us an end score which is the product of chance factors, level of activity, ability to observe, sensory and motor equipment, and learning ability. In many cases an end score of this sort is desirable. Because the end score obtained by the use of the problem box is fairly reliable, it is a useful test so long as one does not care to know what specific abilities have been tested.

Typical Results Obtained in the Study of Various Animals. Porter (1904) studied the sparrow's ability to enter a cage made of wire mesh by pulling a string which released a door. The problem was simplified by the fact that the string was suspended near the door. This fact tended to confine all activity in the vicinity of the door and thus make the successful act more likely to occur. Porter had previously trained the bird to enter the door and so localized the bird's activity at the outset of the problem. The learning curves of sparrows in this problem were comparable to that of the higher forms. As in the case of cats, the great improvement was after the first successful trial. The general locality of the successful movement (pecking in the vicinity of the string) was most quickly isolated. After this the finer details of the movement had to be experienced, and before the correct pattern of movements occurred, a small amount of excess time was consumed. The final perfection was therefore rather gradual as far as time measurements were concerned.

Small (1899) in his study of rats, used a box into which a rat could gain entrance by digging through sawdust. The character of the results may be indicated by presenting the results of a typical rat. On the first trial one of the rats succeeded only after 1½ hr. On the second trial, given the following day, only 8 min. were required. Of special interest is the fact that the rat did not begin digging for 4 min., but once it began, it continued until it had gained entrance. On the third trial the time dropped to 2½ min. In this type of puzzle box the digging and the place to dig are easily isolated by the rat. The first trial isolates digging in the vicinity of the box; on the second trial the place for digging is further isolated. The very rapid drop in the learning curve is accounted for in this manner. The learning curve of the rat in this case is as rapid as is found in higher forms, and for this reason it is rather futile to attempt to demonstrate the superiority of apes over rats by pointing to sudden drops in the ape's learning curves.

Small also used a box which could be entered by the rat when it tore away a sheet of paper. In this experiment the rats experienced several days of complete failure, but finally after one accidental success the problem was easily mastered. This puzzle box required a response which was unlikely to occur in the rat's routine of acts. In this sense the problem might be regarded as difficult. As far as selection was concerned, this problem offered very little difficulty.

Yoakum's (1909) squirrels learned a sawdust box and a latchbox with ease. Sackett (1913) taught porcupines to open puzzle boxes requiring the use of levers, the raising of a hook, the removal of a plug, the turning of a button, and combinations of these. Davis (1907) in his study of raccoons, and Kinnaman (1902) in his study of monkeys, used similar puzzle boxes. The results of the three studies each show a rapid fall in the time score after the first success, but not so rapid as that obtained by Small. This is to be expected when we remember that the successful act required of the higher forms was very detailed and specific. Cole's (1907) raccoons opened boxes with combinations of as many as seven previously mastered locks, and with nearly as much success as Kinnaman's monkeys. On the basis of his results he believes that raccoons stand somewhere between monkeys and cats in ability. It is perhaps necessary to add that we must here understand "ability" to mean the specific ability to open problem boxes of the type used.

The aspect of the situation which calls out the correct response has not been carefully determined at the present time. The above studies do show that the response is specific to part of the situation. The animal learns to push a lever, for example, and this is accomplished in various ways. One foot may be used at one time, the other a second time, and the head the third time. After a few successes the animal applies its activity to the lever rather than to a locality. This is especially clear in Cole's work. The animal learns finally to push a lever, not to make a specific movement at a certain part of the cage.

THE LEARNING OF MAZES

Introduction.—The historic hedge maze in the gardens of Hampton Court Palace was adapted by Small (1901) for the purpose of investigating the animal mind. He built miniature mazes of both wood and wire mesh suitable for rats and in accordance with the plan of the Hampton Court maze, modifying only the relative dimensions in such a way that the outside form was a rectangle rather than a trapezoid. This first animal maze contained both "blind alleys,"

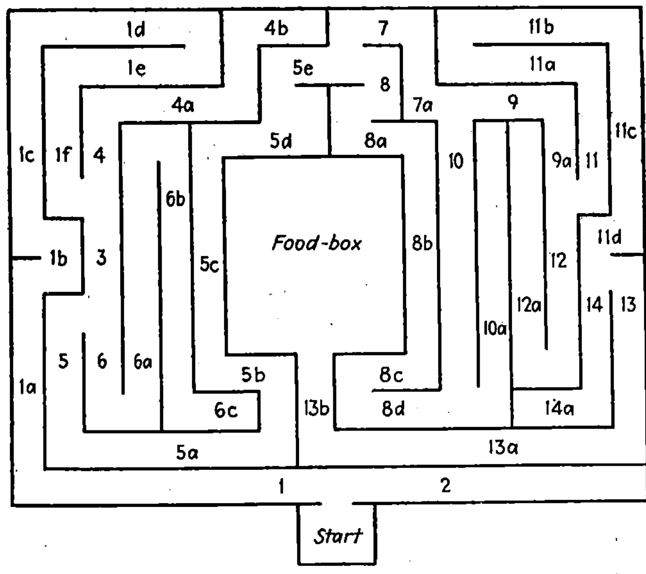

FIG. 78.—The Hampton Court maze. The animal is required to learn the route from the starting point to the food. (*Modified from Small, 1901, p. 207.*)

and long and short alternatives. It was rather complex when compared with many of the mazes that have since been used. Figure 78 is a reproduction of Small's rat maze. Ordinarily an animal is required to learn the route from one part of the maze (the starting point) to some other part (the food box). To reach the food in the most efficient manner, blind alleys must be avoided. If the animal is active and hungry, it continues to wander about investigating the strange territory or seeking an outlet. If the maze is not too large and complex, the food will be found before the animal ceases activity; if it is too large and complex, the animal may not reach the food unless placed in the maze on several occasions. Having found the food, the animal soon learns that the maze situation means food. Learning to run in the maze situation for a reward, however, is not learning the maze pattern. It is the latter that investigators are usually interested in, and for this reason learning to run in the maze situation is a prerequisite to the actual maze learning.

In order to investigate the learning of the maze, different investigators use different techniques to establish the prerequisite food association with the maze. (1) Some feed the animal in the food compartment for several days before beginning the actual experiment. This method establishes the food association merely in part. (2) Others do not take records on the animal's first trip through the maze, but regard it as a trial in which the rat forms the food association and learns nothing else. The obvious objection to this method is that we have no reason for assuming that the food association is formed in but one trip, or that nothing additional is learned. (3) The best technique is perhaps that of permitting the animal to learn a practice maze. The objection to this method is that successive maze patterns may have certain things in common; hence what is learned in one situation may be transferred to the next. Such previous learning might also interfere with the learning of the new maze since some parts of it which are similar to previously learned mazes might require a different reaction. However, successive mazes may be planned to check and measure such possibilities.

In any case, before the study of maze learning is actually begun, the animal should have formed the food association. In addition it should be well tamed, and made perfectly at home in the experimental situation. The last of the three above mentioned methods insures this preliminary training to the greatest degree, the second method fails to the greatest degree.

Comparison of the Maze with the Discrimination Box.—If the animal has been given the necessary preliminary training and the foregoing objections eliminated, when introduced into a particular maze it must learn its way to food by the most economical route. Subject to the "hunger drive," it tends to take only those paths which progress toward the food. For this reason blind alleys are avoided. Because of these conditions the animal's performance in the maze will be an indication of its learning.

Let us now turn to the nature of the learning situation which the maze presents. At each junction in the maze the animal is presented with two or more alternatives. By trial the animal must learn which of the alternative turns lead forward and which do not. Each junction thus presents the animal with a discrimination. In the discrimination box, however, we are interested in teaching the animal to respond to some particular sensory feature, whereas in the maze it may respond to any fixed characteristic that it wishes. A further difference is that in the maze, the position (or the direction of the turn) of the correct choice remains unchanged so that a right- or a left-position habit may be formed at each junction. (In the discrimination habit, the position of the choice is a variable.) But the maze greatly complicates the simple-position habit because it is actually comparable to a succession of position habits. Conceivably this might lead to confusion. A further difficulty also arises in the event that the animal reverses its direction and returns toward the starting point. Such reversals often occur in the early stages of learning when the animal apparently becomes confused. Because of the conflicting experiences that the use of a position habit might bring about, it is only reasonable to believe that such habits would soon be discontinued.

Since the maze contains a succession of turns, the "rightness" or "leftness" of a turn may be determined in preceding alleys. Thus the turn at one corner might conceivably be the stimulus for a turn to be made some distance ahead. The characteristic experiences coming from such turns involve among others, those arising from stimulation of the kinesthetic sense organs. Hence this sense may play a most important role in maze performance. That it is important can easily be demonstrated by changing the length of an alley way. Rats will run into the wall which has been introduced when an alley is shortened in a well-learned maze (Carr and Watson, 1908).

It is therefore likely that when the maze has been learned to perfection, the animal does not make a series of isolated choices but rather runs it as a pattern. Nevertheless, other characteristic differences, especially at the junctions, are most important, both in aiding to establish the kinesthetic pattern and in supplementing it.

During the learning of the maze it might be supposed that the animal either learns which way to go, or which way not to go. This is similar to the problem raised in our discussion of the learning of relationships. In that case, it was found that the animal learned neither to choose one card nor to avoid the other, but rather its choice was a response to some difference in the stimulus situation.

It is therefore most likely that at any particular junction, the animal finally learns to make a response to the situation. The situation in this case not only includes the presence of a right and left going path, but any characteristic experiences the animal may have at the junction as well as any characteristic experiences (primarily kinesthetic) it may have in previous parts of the maze. All of these factors determine the nature of the animal's experience when it arrives at a junction in the maze.

The Value of the Maze as an Instrument for the Study of Learning. Since the maze has been used both as a test of learning and as an instrument for analyzing learning, it is undoubtedly the most important instrument which has thus far been devised for animal studies. It has the following virtues, all of which other learning situations do not have:

1. The past experience of the animal need have no bearing on the study. A maze built in the laboratory involves a pattern which the animal has never before learned. We therefore know that any learning which the animal displays has been acquired under the experimenter's supervision.

2. The process of learning rather than the process of selection characterizes the acquisition of the maze habit. Many associations must be formed and this is a process of learning. The process of selection is minimized because the desired response is reduced to two or three, turning right or left, or going straight ahead. Other responses such as scratching and biting are eliminated because the animal is previously trained to run in the maze situation. The sensory differences at any choice point are usually obvious, since several differences are present which are adequate for initiating the response.

3. The degree of complexity, both in the arrangement and in the number of elements, can be readily varied.

4. Objective measurements of the animal's behavior can be easily and accurately made.

5. The maze situation is natural for most animal forms. All higher animals learn their way about in nature, and all have the ability of locomotion to a degree satisfactory for the purposes of investigation. Their sensory equipment may differ, but in each case the sensory factors involved can be successfully isolated.

Because of the latter virtues, the maze becomes a most useful tool for a comparative study of learning ability. Warden (1929) has devised a standard maze which he suggests should be used by different investigators to facilitate comparisons of results. His maze, however, has not been widely used, partly because most interest has centered about the analysis of learning. Until more is known of the nature of learning, comparative studies contribute little of scientific value.

The analysis of learning has been and still is the chief concern of most of the investigators. The studies are numerous and varied and will be treated in the following chapters.

Learning Situations as Measures of Learning

The foregoing analyses of different learning situations clearly show that they present an animal with tasks which require varying degrees of different abilities. Selection and association are dissimilar processes and the learning situations discussed above involve them in different proportions. An animal's success in one type of problem would therefore not insure the same degree of success in others.

Commins, McNemar, and Stone (1932) have given us pertinent experimental measurements. Table 24 gives the average intercorrelation which they find for various learning situations. Inspection of the table shows that the intercorrelation for different learning situations is about zero, but the intercorrelation for different types of the same class of learning situations is greater than 0.5. Even very different mazes, such as the enclosed and elevated mazes give an intercorrelation of over 0.5. These types of mazes differ in

Table 24.—Intercorrelations of Scores Made by Rats When Tested in Different Learning Situations (Data Taken from Commins, McNemar, and Stone, 1932)

Testing situation	Light discrimination box	T maze	Elevated maze 2
Platform problem box	+0.02	−0.02	......
Light discrimination box	......	0.00 & −0.10	−0.02
Elevated maze 1	+0.01	+0.59	+0.57
Elevated maze 2	−0.02	+0.51	......

the sensory processes which they arouse, but such differences apparently are not pertinent.

As the reliability of each of the testing instruments was found to be very high, these results take on significance. They substantiate the foregoing analysis and clearly demonstrate the need for careful investigations of the abilities shown by animals when performing in various test situations. In order to get at the process of learning, learning ability as such must be isolated. Thus far we have primarily measurements of performance in various test situations, with but little knowledge of the processes and abilities which go to make up the composite score.

A Comparison of the Conditioned Response with Other Forms of Learning

As stated in Chap. XV, the conditioned response is characterized by the association of a given reaction with a new or neutral stimulus. The response to be acquired as well as the stimulus situation are therefore under the control of the experimenter. When the two stimuli are presented in succession, the response necessarily follows, and as a consequence is in temporal contiguity with them. These three contiguous events set up activity in three brain centers which undoubtedly overlap in time of activity and thus constitute contiguity in experience. Before the acquisition of the conditioned reaction, the unconditioned stimulus arouses the motor center which brings about the response. After the acquisition of the conditioned reaction the conditioned stimulus is capable of arousing the motor center because of some change which has taken place in the nervous system. It is possible that the change has permitted the conditioned stimulus to arouse the motor center directly. It is also possible that the conditioned stimulus arouses the sensory center for the unconditioned stimulus which in turn arouses the motor center. Which of these alternatives better describes the modification in the nervous system has not been definitely established. That the process is one of indirect arousal of the motor center is favored by the following facts: (1) the time elapsing between the presentation of the stimulus and the appearance of the response (the latent period) is longer for the conditioned than for the unconditioned response (indicating that the former is less direct); (2) usually the conditioned response is less intense than the unconditioned (indicating incomplete arousal of the center for the unconditioned stimulus); (3) the conditioned response shows signs of being an anticipatory reaction to the unconditioned stimulus (salivation in anticipation of food); (4) the appearance of experimental extinction when the conditioned stimulus is repeatedly presented without the unconditioned stimulus (indicating that the arousal of the response depends upon the activity of the sensory brain center for the unconditioned stimulus). During the process of experimental extinction the brain center of the conditioned stimulus as well as the motor center are obviously involved. If only these two centers were necessary, the response should remain intact, which it does not.

Other learning situations differ from the conditioned response in that the response is not in the control of the experimenter, but must grow out of the situation. In the acquisition of the position habit the animal may respond by turning right or left at the choice, or by lying down or scratching itself, or by doing any number of other things. Of these reactions, one *leads* to food, but the food does not *produce* the reaction as in the conditioned response. The ordinary learning situation thus requires the selection of a response. If the animal were fed only when it salivated to the sound of bell, it would have considerable difficulty in acquiring the response which is so easily set up by conditioning. We have already shown how the discrimination habits and the problem-box habit involve a complex selection process.

In general it may be said that the ordinary learning situations produce a preference for particular reactions to certain stimuli because reward is given in contiguity with them. Since the reaction to the stimulus takes place before the reward is given, it seems that the process of learning is the establishment of some connection between the experience of the reaction to a certain stimulus and the experience of food. This connection causes a certain kind of behavior to dominate in the situation. In contrast to this, the conditioned-response situation has the response predetermined and the establishment of a preference is unnecessary. Reward or punishment, used in other learning situations to produce a preference for or an avoidance of a certain response to some aspect of the stimulus, is not needed. Through training, the sound of a bell may result in the contraction of the pupil of the eye (if light is used as the unconditioned stimulus) as well as salivation. Food in the conditioned-response situation is therefore not a reward. The pupillary conditioned reaction appears to be a useless and senseless response as compared with the apparently sensible behavior of the animal in other learning situations. The apparently sensible behavior which the trained animal displays in the latter situations may be regarded as arising from the fact that a preference for certain kinds of behavior has been established. Because the animal receives a reward for its behavior, it appears to react in a sensible manner.

Because the conditioned-response method controls the response and excludes many unnecessary forms of stimulation, positive evidence for learning has been obtained in certain instances when other methods failed. It is difficult for an animal to respond to a pure tone in a certain manner when the desired response cannot be controlled by the experimenter. This may well explain why the discrimination method has failed to demonstrate pitch discrimination in animals.

It is also probable that the "jumping technique" yields results more quickly than the discrimination box, because in the former case the response is directed at the stimulus object from the outset. Furthermore, the reward in the jumping apparatus is always in the same place and the animal obtains it immediately after responding. In the ordinary discrimination box the food is sometimes in the right alley and sometimes in the left one (see Fig. 64), and the animal must go some distance after responding before it obtains food. As a result the response to the alley rather than the response to the stimulus cards dominates the animal's behavior.

Experimental Neurosis and Its Relation to the Learning Situation

An interesting and most important discovery in connection with studies on conditioning is the appearance of pathological neurosis under certain conditions. If, for example, a dog is conditioned to salivate in response to a circle and not to salivate in response to an ellipse, this differentiation can take place as long as the animal is capable of making the discrimination. If, however, the ellipse is modified so that its two diameters are nearly equal, a point will be reached when the ellipse is both a stimulus for salivation and for no salivation. Pavlov (1927) found the critical ellipse to be one with its diameters in the ratio 8:9. Such a stimulus produces a conflict between two opposing responses and the result is often a clear case of neurosis. In some dogs excitement dominates, making them nervous, noisy, and difficult to handle. In other dogs the effect is one more like melancholia; the animals become unresponsive, passive, and sleep a good share of the time. In both types of neurosis the effects of training disappear, and the animals are unable to make any of the differentiations they had previously mastered. The animal must be retrained with large differences between the stimuli, and as soon as the discrimination is again established, the pathological behavior disappears. A state of neurosis can be produced and cured any number of times by this procedure.

Similar results have been obtained by Liddell and Bayne (1927) on sheep, and by Bajandurow (1932) on birds.

Cases of neurosis have not been obtained from the use of the discrimination-box method. This indicates that a fundamental difference between the two methods exists. The analysis which we have already given may well be applied in this connection. Since we have regarded the responses acquired in conditioning to be determined by the procedure, two opposing responses (salivation and no salivation) to the different stimuli which can be discriminated have been made available. As long as discrimination is possible, one of these responses can be made. When, however, discrimination fails, the stimulus becomes adequate for both responses. This condition necessarily sets up a conflict which cannot be avoided since there are only two possible reactions to the situation; and these are in opposition to each other and are called out at the same time. As a result there is a general disruption of behavior.

Opposing responses conflict only when alternatives are not available and when the stimulus calls them both out at the same time. The discrimination box allows for alternative responses since the stimulus does not call out the responses but merely causes one of many to be preferred. When two stimuli which cannot be discriminated are presented to the animal, a preferential reaction to the learned aspect of the situation is impossible. The animal may fail to react, it may wash itself, or it may choose either one of the compartments. Whatever it does is a preferential reaction to some other aspect of the situation, and what this reaction will be cannot be predicted because no reaction is forced by the situation.

Since the above has been written, Liddell, James, and Anderson (1934) in their study of sheep in various learning situations have made a strikingly similar contrast in the behavior of animals confronted with difficult discriminations. In discussing nervous strain they state the following:

No sheep ever showed evidences of nervous strain during maze learning experiments even when it could not solve the problem presented. For example, it could not form the double alternation habit. However, it was willing to enter the maze each day after more than a year of failure. The experimental neurosis developed in the sheep only as the result of training in the conditioned reflex laboratory.

Now, the essential difference between the maze situation and the situation in the conditioned reflex laboratory is that in the maze the animal can resort to evasive behavior. If it cannot do what is required, it can do something else. Procrastination and stereotyped responses were commonly observed during maze learning experiments with the sheep. *In the conditioned reflex laboratory evasive behavior is severely curtailed.* Through training, spontaneous activity is almost completely suppressed and the animal remains quietly on the table because of *self-imposed* restraint. In the sheep this habitual restraint *of itself* need have no serious consequences, for in no case have we observed either the neurosis or its premonitory signs to develop as a consequence of the preliminary training or of simple positive conditioning. *It is only when we introduce negative conditioned stimuli into the training routine that appreciable nervous strain develops. Restraint is then piled upon restraint* (pp. 77–78).

The Process of Learning Compared with the Process of Development

Coghill (1930*b*) has pointed out the similarity between the development of specificity of responses to stimuli in native and acquired behavior. We have already discussed (Chap. XII) the restriction of the sensitive zones as well as the individuation of specific responses which take place as maturation progresses. In learning situations the animal is often required to make a specific response to some specific aspect of the stimulus. For example, in the puzzle-box situation the animal's general activity must be replaced by certain definite responses, and the stimulus to which the animal responds must become some aspect of the situation rather than the situation in general. As in the process of maturation, the learned pattern of activity becomes more restricted as some specific activity acquires dominance over the generalized reaction. The effective stimulus for the acquired responses, like the responses in maturation, gradually narrow down as reward consistently appears in contiguity with a response to some specific aspect of the situation. Responses to other aspects appear in contiguity both with and without reward. The restriction of the stimulus situation is most pronounced in discrimination studies.

Trained animals attack new situations by immediately responding to limited aspects of the situation. As a consequence such animals master new situations more quickly than do inexperienced animals. Nevertheless, it may be said that the first phases of learning are characterized by responses to the total situation. Pavlov, as we have seen, points out how the conditioned reaction begins as a response to a general situation and how, through further learning, the reaction is made specific to some aspect of the stimulus.

Development, which is a process of both maturation and learning, shows outward manifestations of both these processes in a similar manner. Consequently the observation of a pattern of behavior during development gives no clue as to which process is responsible for certain changes which may appear from time to time. As we have seen in Chap. XII, changes due to maturation and to learning are most difficult to differentiate.

Suggested Readings

Adams, D. K. 1929. Experimental studies of adaptive behavior in cats. *Comp. Psychol. Mono.*, **6**, pp. 168.

Heron, W. T. 1934. *Learning.* Comparative psychology (ed. by Moss). Pp. 297–333. New York: Prentice-Hall.

Klüver, H. 1933. *Behavior mechanisms in monkeys.* Chicago: Univ. Chicago Press. Pp. 387.

Liddell, H. S., James, W. T., and Anderson, O. D. 1934. The comparative physiology of the conditioned motor reflex. *Comp. Psychol. Mono.*, **11**, pp. 89.

CHAPTER XVII

THE ANALYSIS OF MAZE LEARNING

The Sensory Control of the Maze Habit

One of the first problems which arises in connection with maze learning is that of determining what differential sensory experiences an animal uses when it runs a maze. This problem immediately suggests a variety of experiments. For example, one may change one sensory condition, leaving all others the same, and observe whether such a change alters (*a*) the performance of an animal in a previously well-learned maze, or (*b*) the learning rate of a new maze. The sensory condition may be altered in two different ways. In the first place, one may remove, alter, or exaggerate the stimulus characteristics which give rise to the sensory condition in question. In the second place, one may alter the animal by operation, and so make it insensitive in the sensory mode that is to be tested.

There are three possible effects that such changes may have on the behavior of the animal and each of the possibilities has different consequences: (1) If each of the two types of change has no effect, we know that the particular sensory condition in question is not essential to the performance, but we cannot conclude that it is not an adequate sensory condition for learning. It might well become the basis for learning in the absence of other sensory conditions. (2) If the changes make the learning of the maze impossible, we know that the removed sensory condition is essential. (3) If the modified condition merely makes the performance less perfect, or affects only some of the animals tested, no satisfactory conclusions can be drawn. Poor performance may be the result of distraction or of additional difficulty introduced by the modified conditions. If the method of extirpation of sense organs has been used, it is difficult to ascertain whether the operation has merely removed a sensory modality or whether the animal's physical condition has been altered in some other essential way.

From this discussion it is evident that our problem is a rather difficult one. One must therefore interpret with care the experimental results which follow.

Experiments Which Stress the Fundamental Importance of Kinesthesis.—When Small (1901) used the Hampton Court maze (see Fig. 78) to study the mental processes of the rat, he became convinced that running a maze was primarily a kinesthetic habit. By using sawdust on the floor of the maze and changing this from time to time, he convinced himself that rats did not smell their way through the maze. He also modified the visual situation by (1) changing the position of the light above the maze, and (2) introducing red posts at the junctions of the maze. These modifications seemed to have little or no effect upon the rat's performance. He also tested blind rats and rats with vibrissae removed, and obtained no substantial differences in performance between these and normal rats. Since only kinesthesis and tactual sensations were present in all cases, he concluded that they furnished the essential data for the necessary discriminations.

Watson (1907) reported a detailed investigation on the sensory control of the rat in the Hampton Court maze. He established certain norms for the learning of the maze by rats under ordinary conditions and compared these norms with the records made by rats tested under various experimental conditions. He found that darkening the room, blinding the rat, destruction of the olfactory bulbs, destruction of the middle-ear bones (producing partial deafness), removal of vibrissae, anesthetization of the soles of the feet or noses, and the introduction of air currents and temperature differences in the maze had no final detrimental effect on the rat's performance.

A rat which was blind, anosmic, and without vibrissae finally, after considerable difficulty, learned to run the maze very accurately.

The above results indicate that maze learning can be quite independent of visual, auditory, olfactory, and tactual sensations, but other of his results seem to be in disagreement. Thus rotating the maze 180 deg. confused normal, anosmic, and partially deaf animals, but not blind animals. However, rotation through 90 deg. also slightly confused blind rats. As rotation changes the visual and perhaps the auditory environment, it seems that these results indicate that vision, in some way, is functional in maze running.

Taking the results as a whole, it is rather difficult to draw any specific conclusions. Watson realized the danger in dogmatic statements about the sensory modes which were used by the rat, and claimed only to have demonstrated which sensations were *not essential* to maze learning. He believed, however, that the maze habit was kinesthetically controlled and that kinesthesis was perhaps coupled with certain organic and static sensations. Since, however, all extraorganic sensations were never excluded at one and the same time in any rat, their contribution to maze learning can merely be regarded as a possibility. Maze junctions differ in many ways, and unless all differences are excluded at once, with the breakdown of the habit as a consequence, it is difficult to decide on the importance of any one of them.

Positive evidence of the use of kinesthesis in maze running was obtained by Carr and Watson (1908). They found that if the alleys of a maze which had already been learned by rats were either lengthened or shortened, the rats were greatly disturbed. In the case of shortened alleys, the rats often ran head long into the end walls, and in the case of lengthened alleys they tended to make their turns at the points where the junctions had previously been. These results seem to indicate definitely that the rats were running the modified form of the maze on the basis of the kinesthetic sense.

Experiments Which Indicate that Kinesthesis Is Supplemented by Other Senses.—Bogardus and Henke (1911) took records of the number of times blind rats and rats without vibrissae made contacts with their noses at junctions during maze learning. They found that in new mazes and in slightly altered familiar mazes, the number of contacts paralleled the number of errors. They believed that the number of contacts were, therefore, sources of sensory data to which the rat resorted when it became confused. As contacts were not present during perfect performance, kinesthesis seemed adequate. They concluded that the maze habit depends on tactual sensations during the process of learning, but that the sensory control is gradually transferred to kinesthesis.

Vincent (1912, 1915 *a*, *b*, and *c*) found that by exaggerating certain sensory factors the total number of errors made in learning the maze could be modified. Thus differentiating the true paths from blind alleys by making one black and the other white, or by laying an olfactory trail (*e.g.*, cream cheese rubbed on the floor) on one set of paths (*e.g.*, the true paths) and not the other (*e.g.*, the blind alleys), tended to reduce the total number of errors produced during learning. The removal of the side walls of the maze (thus forming an elevated rather than an enclosed maze) produced a slight saving in learning. This saving, Vincent concluded, is attributable to the fact that the open maze requires more tactual control, and so the tactual sensations become exaggerated. Thus the exaggeration of any sense department for which the true paths and the blinds are different seems to be an aid in maze learning.

She also found that blind rats and rats lacking tactual sensitivity in the nose (fifth cranial nerve cut) made about twice as many errors as normal rats in the open maze. Rats without vibrissae made scores about equal to normal rats, but blind rats without vibrissae were the most handicapped.

These results thus furnish further evidence of the function of sensory processes other than the kinesthetic operating in maze performance, but they still furnish no evidence bearing upon the essential importance of these. Vincent favors the view that kinesthesis is fundamental and that other sensations function primarily in the early stages of learning.

Carr (1917*b*) summarized the evidence obtained in previous studies by stating that the rat learns the maze primarily in terms of touch and kinesthesis, but that touch gradually drops out as the maze becomes mastered. He confined his investigations (1917 *b*, *c*, and *d*) to a study of the effect of changes in the environment outside the maze upon the maze performance of normal, blind, and anosmic groups of rats. These changes involved the position of the experimenter when placing the rat in the maze; covering and uncovering the maze; rotating the maze so as to change the points of reference in the room; and changes in lighting both inside and outside the maze. The groups of animals showed no marked differences in performance, but the results indicate that visual changes affected blind rats the least. Cleaning the maze affected blind rats the most and anosmic rats the least.

Carr found further that learning efficiency was reduced in normal rats when the maze was rotated each day. His results also showed, contrary to previous studies, that blind rats were less efficient in maze learning than normal rats. He attributes this inferiority, not to loss of important visual experiences, but to either the probable loss of certain tonic effects which visual sensations may exert, or the possible injurious effect of the operation. Carr points out that instead of being useful, vision is often a handicap because certain visual changes might distract the animals.

All of Carr's rats showed marked individual differences in their reactions to the changes he introduced. This suggests that none of his changes involved a fundamental sense department, but rather that all sensory modes may have played some part. Certainly he presented no conclusive evidence to the contrary.

The most recent defender of the fundamental importance of kinesthesis is Dennis (1929). He used a simple maze with wide alleys and found that vibrissaeless rats could not perfect a maze habit without resorting to contact with the walls. He regards contact and vision to be the senses which are necessary to elicit the turn, but kinesthesis to be the sense which controls the direction of the turn.

Experiments Which Question the Importance of Kinesthesis.—In the experiments thus far reported, the fundamental importance of kinesthesis has been largely inferred, although some convincing positive evidence for it was found. If kinesthesis is important, a marked modification of muscular sensations should greatly disturb maze performance. In an experiment by Lashley and McCarthy (1926) rats which had previously learned a maze were retested after cerebellar injuries. Such injuries destroyed the rats' equilibrium and coordination. As a consequence the behavior was greatly modified. Some of the rats literally rolled their way through the maze. Nevertheless, the route through the maze was perfectly retained. Even rats which were blinded in addition to such injury, and were thus unable to use visual reflexes, made no entrances into blind alleys on the retest.

Similar negative results were obtained by Lashley and Ball (1929) and by Ingebritsen (1932) with injuries to the spinal cord of the rats. Kinesthetic, organic, and tactual sensations from regions below the neck reach the brain by way of the cord; yet severing any group of such conduction paths neither destroyed maze retention nor affected the learning ability of the rats.

Hunter (1929) argues that if the maze is learned on a purely kinesthetic basis, a rat should be unable to learn a maze in which it must make two right and two left turns in sequence, if all other sensory differences in the maze are eliminated. In such a maze the turn to the left is followed by another turn to the left, but this second turn to the left is followed by a turn to the right. Since each left turn must produce kinesthetic effects which are alike, it is difficult to understand how two like forms of stimulation can sometimes produce a response to the left and sometimes a response to the right. It could, of course, be that the two responses to the left together produce the stimulation for the right turn, but Hunter regards the rat as too simple a creature for such a complex process.

Because a maze with three pairs of right and left turns (*rrllrr*) was learned by three of his six rats, he suspected that some other sensory factor must have supplemented kinesthesis, and so produced pairs of sensory effects which were actually different. As his rats were blind and without vibrissae, and as the elevated poles which constituted the pathways of the maze were well machined and carefully

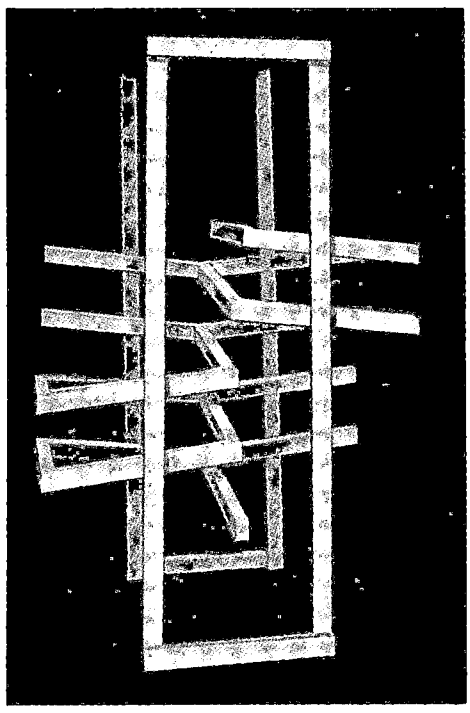

FIG. 79.—The tridimensional maze. (*From Hunter, 1929, p. 518. By permission of the Journal of Genetic Psychology.*)

washed each day, he believed that a constant noise from one side of the room caused some difference in experience for the rats when making the two turns to the same side. To eliminate this possibility he built a tridimensional maze. (See Fig. 79.) In this maze, progress, instead of being forward, was upward, each leg of the maze being slightly on the incline. Only four junctions were present (*rrll*). Of 23 normal rats, only 6 succeeded in making one perfect run. This procedure made the problem more difficult, but the disturbing fact was that some of the rats still learned the maze. On repeating the experiment with the order of the turns changed (*llrr*) 10 out of 11 rats learned the maze. Because these rats had to relearn the maze after being blinded, Hunter concluded that vision must have been important.

The temporal maze was the only one with which Hunter was able to obtain perfect runs which were so scarce that they might have been due to chance. A diagram of this maze is shown in Fig. 80. Rats started at *S* were required to run in a circuit to the right (*S-A-B-C-S*) twice in succession, and then make a circuit to the left (*S-A-D-E-S*) twice in succession. In this maze Hunter believed he

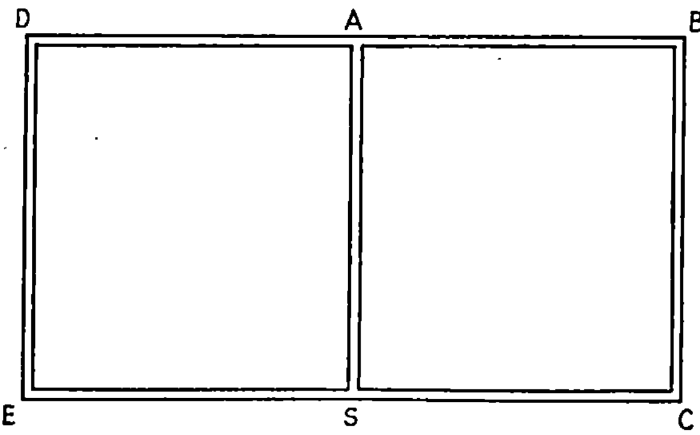

FIG. 80.—Diagram of the temporal maze. The pathways consist of elevated poles. The animal is required to run two right circuits (*S-A-B-C-S*) and two left circuits (*S-A-D-E-S*) in succession after which it is rewarded. During the training the direction of the animal's progress is controlled by blocking off the incorrect route. (*Modified from Hunter*, 1929, *p*. 527.)

had a perfect case in which the animal must make a turn in one direction after a certain pattern of kinesthetic stimulation, and then make a turn in the opposite direction after the same sensory effect. (A circuit to the right should always produce the same kinesthetic stimulation.)

However, if it is supposed that the rat's sensory discrimination is so keen that it is aware of the fact that it is repeating its route in this maze, and that it was not repeating its route in spatial mazes, then the rat may refuse to make an unnecessary circuit when it comes to the proper junction for the same reason that it learns to avoid blind alleys. Since point *A* is the same for each circuit a turn to the right is equivalent to entering a blind alley. In that event failure would not be due to a lack of sensory discrimination, but rather it would be due to its very marked presence. And this is exactly what seems to be the case. Shepard (1931) has demonstrated that the rat refuses to run in a circuit. If a door opening to alley *F* of Fig. 81 is left closed until the rat has passed it and has completed the circuit *B-C-D-E*, the rat soon refuses to make the unnecessary trip, but waits in front of, or scratches at, the closed door which it found open on previous occasions after making the circuit. Control experiments eliminated the possibility that the door was located by any characteristic markings, rather, the rat

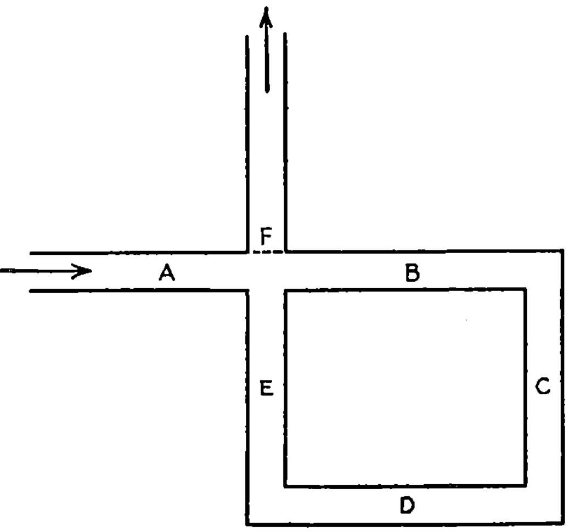

FIG. 81.—Diagram of circuit rats refuse to run. As the rat approaches *A* the opening into alley *F* is closed until the rat has progressed beyond alley *B*. On arriving at *E* it continues to *F*. After several repetitions, the rat refuses to run *A-B-C-D-E-F*, but attempts *A-F*. (*Courtesy of J. F. Shepard.*)

recognized the junction. Thus, despite the fact that alley *F* was at first entered from *E*, this alley, although closed, was soon reacted to by the rat when it emerged from *A*.

Experiments Which Attempt to Exclude the Function of All Sensory Processes Other than Kinesthesis.—After many years of work in the analysis of learning of various types of mazes, Shepard built a "unit" maze, a typical form of which is shown in Fig. 82. (The points indicated by *X* represent doors which join adjacent units.) This maze pattern consists of a succession of identical maze units, from each of which the rat can pass into the next by making a characteristic turn at the junction (*e.g.*, always taking the alley to the left). However, one of the turns is an exception to this rule (the last unit in which the turn is to the right). In order to pass through all units without error, this exceptional junction must be located and responded to differently. If the units are actually alike, the exceptional unit can be located only kinesthetically. That is, the rat must experience a certain rhythm, at the end of which it makes the exceptional response. Recognizing the exceptional unit in this way would correspond to the way we recognize the last step of a familiar stairway in darkness.

To the surprise of the experimenter, rats not only learned these mazes, but they learned them with ease. At first the characteristic general response to the junctions was learned and was even applied to new parts of the maze. Next the exceptional unit was located and the appropriate turn made. Increasing the number of units to as many as 29 did not produce failures. This demonstrates either a remarkable kinesthetic sense or the presence of some unknown uncontrolled sensory difference. Tests involving changes in the visual and olfactory situation proved negative. The maze was torn down, the wall sections interchanged, and the maze rebuilt in order to test whether the rats were using any characteristic difference in the walls. Again the results were negative.

A test of the presence of a kinesthetic rhythm was then made (Shepard, 1929). As the units were all alike in construction, any unit could be made the starting point by merely closing the entrance (indicated by *X* in Fig. 82) to that unit. The different possible starting points are indicated in the diagram of the maze. Since different starting points vary in their distance from the exceptional unit, a kinesthetic rhythm would be useless for locating the exceptional unit. It was found that no matter from which unit the rats were started, the exceptional unit was successfully located. This meant that despite the various tests to the contrary, some local sensory discrimination, other than that based on a kinesthetic pattern, was present.

Changes in the composition of the floor were then introduced; and as a consequence errors resulted. The floor structure, beneath the linoleum sections on which the rats were running, had been supplying them with different sensory effects in the various units. As the rats had no immediate contact with this subfloor, Shepard believed the differentiation to be auditory in nature. Different parts of a flat surface vary in their vibration rates, and the patter of the rats' feet seemingly resonated differently on different parts of the floor. In any case this differential experience from the floor very nicely explains the difficulties and inconsistencies which arose in the previous experiments.

Having located this sensory difference, the next step is to eliminate it and force the rats to rely on kinesthesis alone. If they can still learn the maze effectively, the results will not be conclusive, because the learning may be due either to kinesthesis, or to some other unexpected discrimination. If, however, the elimination of the sensory differences from the floor breaks down the learning in some

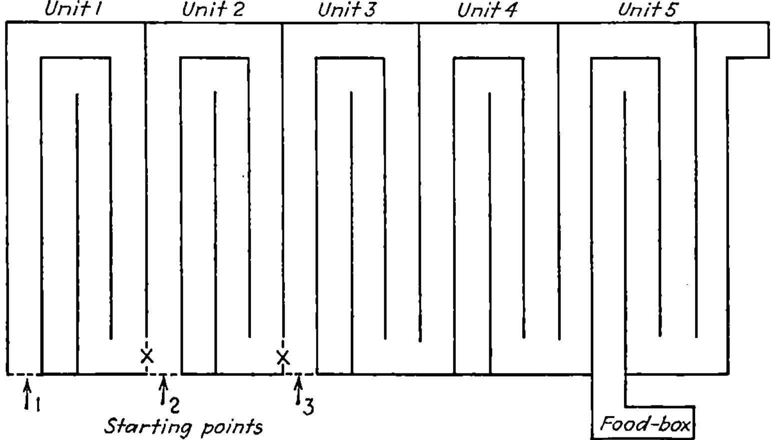

FIG. 82.—The unit maze used by Curtis. The arrows indicate different starting points which may be used. The sections marked *X* are removable. They are inserted only when the latter part of the maze is used. *(Courtesy of Q. F. Curtis.)*

detectable fashion, or makes the learning of long unit mazes impossible, the results may be given a definite interpretation.

The Elimination of the Floor Cue.—The task of constructing a maze in which the sensory differences arising from the maze floor were to be eliminated was undertaken by Curtis (1931). In order to eliminate the vibrations from the maze floor, he covered the concrete floor of the building with sand to a depth of 1½ in. The sand was then covered with black oilcloth. Upon this floor he built the unit maze shown in Fig. 82. Lights were equally spaced in each unit so as to eliminate any visual differences either inside or outside the maze. The last three units of the five-unit maze were used during the learning period, the last unit having the exceptional junction. Thus at the junctions the rat had to turn left at the first two units and right at the third.

Ten rats which had mastered the five-unit maze in an average of 5.3 trials, now required an average of 72.25 trials in order to learn three units. Not only were more trials required before three consecutive errorless runs were made, but the perfect runs were made with much more hesitancy. Further, two of the rats failed to reach the criterion of learning in more than 100 trials. As the possible number of errors is small, the criterion for learning is not a difficult one. (In more complex mazes the criterion of learning is often 10 consecutive errorless runs.) These results therefore show a marked increase in the difficulty of maze learning because of the change in the floor.

After the maze had been learned, a series of tests were made in order to determine how the rats had learned it. If the units are alike for the rat, the starting point and the exceptional junction are characterized only by being in the first and third units, respectively. If the units are different in sound effects, visual appearance, smell, or touch, then the exceptional unit occupies a certain fixed position in the maze series. Therefore, when the rat is introduced into the maze at various points, it should make the exceptional response at the third unit from the starting place, provided it has learned the maze kinesthetically; but it should make the exceptional response at the last unit if it has learned the maze in terms of local sensory differences.

The results of a series of tests of this sort showed that the exceptional response was made in the third unit as often as it had been in the learning runs. The rats showed no signs of strangeness when started in the various units, and behaved as if there had been no changes. The responses, therefore, were not made to a particular part of the maze, but rather to a *particular part of a kinesthetic pattern.*

The Elimination of Kinesthesis.—In a later study (unpublished), Curtis trained rats on a similar maze, but from the outset eliminated any use of kinesthesis. The position of the food box was always the same, but the starting point varied, so that the rat was required to run 2, 3, or 4 units before reaching the food. Mastery of this problem required the use of some local sensory difference in locating the food unit which contained the exceptional turn. Most of the rats trained in this manner learned the maze to a point better than chance, but not to perfection. Various changes in the floor produced only temporary disturbances. Since kinesthesis was eliminated, some undiscovered sensory differences must have been available to the animals, although these were apparently somewhat unreliable since the number of perfect runs was limited. In the first study of Curtis these sensory differences were not used by the animals. This is not surprising since kinesthesis was available to them.

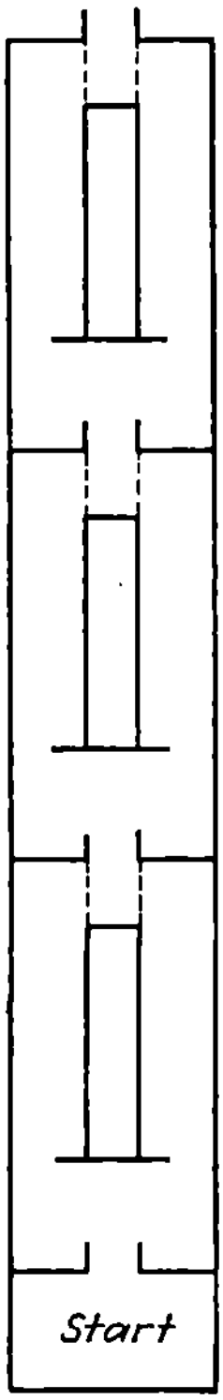

Start

FIG. 83.—Walton's Sectional maze. Any number of units may be placed end to end and either the right or left route through a section may be blocked by inserting a board at any of the points indicated by broken lines. (*Modified from Walton, 1930, p. 55.*)

Anticipatory Behavior and Its Relation to Kinesthetic Sensitivity. That kinesthesis is used by animals in maze running, but that it must be supplemented by other sensory differences in order to obtain high efficiency in the maze, is further demonstrated by an experiment of Spragg (1933). He used an eight-unit maze, three sections of which are illustrated in Fig. 83. The various local sensory differences were eliminated by interchanging the units from time to time. This forced his rats to use kinesthesis. In this experiment the rat was required to make seven right turns followed by a left turn. The exceptional turn was thus characterized by being the eighth junction. After 100 trials the maze was still unlearned, each rat averaging 1 error per trip in the last 25 runs. The majority of the errors were made at the seventh unit. The rats tended to turn left in this unit in anticipation of the last exceptional unit. Apparently the seventh unit could not be discriminated from the eighth unit on the basis of a temporal kinesthetic pattern.

In Table 25, Spragg's data obtained from four rats have been analyzed to bring out this point. The errors made by the four rats at each unit have been grouped for each successive 25 runs. From the table it can be seen that during the first 25 trials the errors were concentrated in the last four units, during the next 25 trials they are limited largely to the last three units with a concentration at the seventh. This concentration at the seventh unit increases during the third group of 25 trials and in the last 25 trials the errors are primarily at the seventh unit. The rat has apparently learned that an exceptional turn must be made, but its ability to form the necessary discrimination is not adequate. At the outset the rat learns to make right turns. This is shown by the fact that errors are made largely in the eighth unit during the first 25 trials. Next it learns to make an exceptional turn, but the limitations of kinesthesis prevent it from making this turn at the proper unit.[1]

TABLE 25.—ANALYSIS OF RESPONSES BASED ON KINESTHESIS (DATA OBTAINED BY SPRAGG, 1933)

Unit	1	2	3	4	5	6	7	8	Total
Response required (right or left turn)	*r*	*r*	*r*	*r*	*r*	*r*	*r*	*l*	
Errors in 1st 25 trials	21	11	9	9	15	29	33	48	175
Errors in 2d 25 trials	14	4	7	4	5	33	46	31	144
Errors in 3d 25 trials	5	3	6	5	11	26	72	18	146
Errors in 4th 25 trials	3	2	2	3	2	14	67	7	100
Total errors	43	20	24	21	33	102	218	104	565

In a later study Spragg (1934) used a unit maze in which rats could go either right or left in the first seven units, but were forced to go left in the last unit. As learning progressed, left turns at all units were found to increase, but the increase was more marked in the last few units. When the last turn was changed to a right turn, the number of left turns at each junction decreased with trials, and the decrease was most marked in the last half of the maze. In this experiment fine discrimination was not required since turns to either side were possible in the first seven units. This explains why the "anticipatory" response was spread over the last half of the maze rather than concentrated at the seventh unit.

Conclusions.—From the foregoing experiments one may conclude that when the rat is deprived of all other sensory differences it can learn a maze on the basis of kinesthesis; but the learning is more difficult and as a result more trials are required or perfection is not reached. Because of the limited efficacy of this sense we cannot

[1] Spragg regards the error at the seventh unit as due to anticipation. This seems unlikely since anticipation is also present in other mazes. Rather, the inability to make the necessary discriminations causes the rat's anticipation to become evidenced.

regard the learning of complex mazes as a purely kinesthetic-motor habit. Other sensory differences will be used by the rat whenever available, and these other differences are not merely supplementary, but form a real basis for maze learning.

By constructing unit mazes of the above types and introducing various sensory differences, one at a time, we can determine just what sensory processes the rat is capable of using. We have, therefore, just arrived at a point in our knowledge where a constructive study of the sensory control of the maze may begin. The overemphasis on kinesthesis must be abandoned and much of the earlier work charged to experience. It is most likely that all sensory differences, when above the rat's threshold of discrimination, will be utilized. Which sensory differences are most effectively utilized, which preferred, and other questions of this sort must be answered by future work.

This discussion has entirely centered about the rat. It is a typical mammal and generalizations having to do with its sensory control are very probably applicable to other animals. Until thorough work with other forms is forthcoming we must rest content. Animals with finer sensory discrimination may show quantitative differences. Those with more intelligence may make different uses of their discrimination ability. But the fact that maze learning is not a kinesthetic-motor habit unless all other sensory differences have been eliminated, has been demonstrated. How great a kinesthetic temporal pattern an animal can master may vary greatly among the mammals, and the extent of this pattern may be a function of intelligence. However, even in the rat the responses are often dependent upon complex temporal stimulus patterns and need not be a simple response to a preceding stimulus as behaviorists have contended.

Quantitative Differences in Mazes and Their Effects on Learning

Introduction.—When we place a hungry animal in the maze we assume that it will go through the maze as efficiently as it is able. Accordingly, if the animal is properly motivated but nevertheless continues to make errors, we conclude that the errors are an expression of incomplete learning. Persistence in errors is consequently regarded as due to inability to distinguish (at that particular stage of learning) the efficient from the inefficient route. That a rat will choose, as far as it is capable, a line of activity which most effectively removes its hunger is assumed whenever we test an animal. When it does not behave in this fashion we are satisfied that it cannot distinguish between the efficient and the inefficient manner of satisfying its need.

Nevertheless, the literature contains several studies which claim to demonstrate that the rat will prefer the course of least action. Gengerelli (1930*a*) and Tsai (1932) have shown experimentally that excessive effort is eliminated by animals during the course of learning and have formulated the principle of *minimum effort.* However, the experiments really demonstrate nothing more than the rat doing what the experimenters regard as most efficient. The study of the animal's preference for one of several ways of reaching a goal, therefore, becomes a study of discrimination ability. If a rat is to learn a maze, it must discriminate between alternative routes. According to DeCamp (1920) two alternative routes must differ in length by at least 1/10. Yoshioka (1929) performed a careful experiment to determine whether the ability to discriminate the spatial difference between a short and a long route satisfied Weber's law. He found that the absolute lengths of the routes were unimportant. As long as the ratio between the long and the short routes was approximately 1.14 : 1 or greater, the discrimination was made.

When a maze contains blind alleys, the true path is not only the shorter route, but it is also the only route which does not lead to a block and subsequent retracing out of the blinds. It therefore becomes necessary in our analysis to determine whether the experience of the block, the experience of retracing, or the experience of greater distance traversed is the cause for the elimination of blind alleys. The solution of these and other problems depends upon a careful study of the behavior of animals in a great variety of mazes differing in both quantitative and qualitative aspects.

The Elimination of Short and Long Blind Alleys.—Peterson (1917*a*) compared the relative ease with which rats eliminated 22-in. and 9-in. blind alleys, and found that the shorter ones were eliminated first. In both types of alleys there was a period of partial entrances which preceded complete elimination.

White and Tolman (1923) using somewhat longer alleys and in addition having a right-angle elbow in the alleys, obtained just opposite results. They found that long alleys were eliminated more readily than short and explained Peterson's results by calling attention to the possibility that his rats could see the ends of the blinds. In such case one would expect the short to be eliminated before the long blinds.

Shepard (unpublished study) has studied the elimination of short and long blinds by rats experienced in maze running. His long blinds were sometimes 30 ft. long and contained many turns. The short blinds were very much shorter than the long ones and had but a few turns. Thus the inconvenience caused by retracing in long blinds was much greater than that caused by the short. The results, however, show but a slight difference—the shorter blinds, on the whole, being eliminated first. In all cases the blinds were gradually eliminated.

From the above results it seems that the length of a blind alley plays only a minor role as far as difficulty of maze learning is concerned. This suggests that a blind alley is primarily eliminated because it blocks progress rather than because it increases the length of the route to food. Since the blinds are gradually eliminated it seems that the entrance into a blind soon causes the animal to anticipate[1] the closed end. More learning means that the end is anticipated sooner. The blind is completely avoided when the end is anticipated at the junction (*i.e.*, when the sensory control for the response has been transferred to the junction). The efficiency of the anticipation seems to depend more upon the character of the blind alley than upon the distance from the end of the blind because on the whole the short blinds are not eliminated with markedly greater readiness than the long ones.

The Learning of Mazes Having Different Numbers of Blind Alleys. The maze may also be quantitatively varied by changing the number of blind alleys. Warden and Hamilton (1929) studied the relative ease with which various groups of untrained rats each learned a simple maze having either 2, 4, 6, 8, or 10 blind alleys. They found no reliable difference in score between the groups in learning the different mazes. This would seem to indicate that all of the mazes studied were so simple that the learning in each case was negligible. As many as 15 trials were required before a group of rats reached a criterion of four perfect runs in five, but this score seems little more

[1] To anticipate merely means that the response originally made to the end of the blind is made to some other part of the blind alley. This is to be expected because other parts of the blind alley are in contiguity with the end of the alley and this contiguity is the condition for learning.

than what is required by the rat before it completely adapts to the maze situation.

TABLE 26.—THE EFFECT OF LENGTH OF MAZE ON THE LEARNING OF NORMAL AND PARTIALLY DECORTICATED RATS. (LASHLEY AND WILEY, 1933)

Size of maze	Normal rats		Partially decorticated rats	
	Trials	Errors	Trials	Errors
4 culs-de-sac	24.3	22.0	77.7	246.7
8 culs-de-sac	33.8	66.7	103.8	921.1
12 culs-de-sac	42.1	85.7	114.6	1039.1
16 culs-de-sac	61.3	121.5	109.0	1382.9

Lashley and Wiley (1933) compared the learning of 4, 8, 12, and 16 cul-de-sac mazes by normal and partially decorticated groups of rats. The results are shown in Table 26. To learn a series of mazes having different numbers of blind alleys, both groups required more trials and made more errors when the number of blind alleys was increased. But the increase in the number of trials and errors was no greater for the operated than for the normal animals. As the mazes increased in complexity in the ratio 1:2:3:4, the error score of normal rats increased in the ratio 1:3:4:5, and the error score of operated rats increased in the ratio 1:3.7:4.2:5.6. This indicates that the relative difficulty in learning does not increase with the number of blind alleys. If the difficulty or qualitative complexity of the problem increased out of proportion to the number of blind alleys, the rats with inferior ability should be at a relatively greater disadvantage with an increase in the number of blinds. Problems of increasing complexity in all other situations cause the difference in score between normal and partially decorticated rats to increase.[1]

[1] The method of scoring partly explains the increase in errors and trials of both groups of rats as the number of blinds was increased. A maze was considered learned when ten consecutive errorless runs were made. But the possibility of errors in large mazes is greater than in small mazes. Thus in the 12-cul-de-sac maze a total of 120 errors had to be avoided before the criterion was reached. In the 4-cul-de-sac maze the possible number of errors in 10 trials was only 40. When a certain number of perfect runs is regarded as evidence of learning, the criterion for learning is actually much higher for large than for small mazes.

We may, therefore, conclude that increasing the number of blinds may make a maze more difficult, but it does not make the maze qualitatively more complex. In fact, the longer mazes are relatively easier to learn, according to Ballachey (1934) who found that errors per unit of maze were less for long than for short mazes. As his mazes consisted of alternate right and left turns, this reduction in relative complexity may be attributed, at least in part, to the setting up of alternation habit which is as effective for long as for short mazes. Studies using mazes with more junctions and having patterns that are not repeated are required to satisfactorily solve this problem.

Qualitative Differences in Mazes and Their Effects on Learning

Maze patterns may be qualitatively different in the following three ways: (1) the type of junction used; (2) the patterns of true path; and (3) the kind of blind alley employed.

The Effect of Different Types of Junctions on Maze Learning.— Hubbert and Lashley (1917) found that in their use of junctions of the type shown in Fig. 84*A*, rats, in their first trip through the maze, went through the opening to *b* or *c* three times as often as they went past it and into *a*. They also went to *b* more often than to *c*. On the other hand, Dashiell (1920*a*) found that in mazes with junctions such as shown in Fig. 84*B*, rats, when in a maze for the first time, had a tendency to go forward rather than to turn. The ratio of the number of times that an alley opening to the side was passed to the number of times it was entered was 5 to 3. If the forward leading alley directed the rat into a blind, the animal, on emerging, continued correctly 3 times in 5. If the alley leading off to the side took the rat into a blind, the rat, on emerging, continued correctly 7 times in 9. By going straight ahead on emerging from the blind in the first case the rat was started toward the starting point, but in the second case it had to turn at the junction, and in this case it was less likely to retrace in the direction of the starting point.

Both studies indicate that the nature of the junction and the direction of approach to the junction partially determine a rat's response, but Dashiell's study indicates a "forward going" tendency, and Hubbert and Lashley's a tendency to enter an opening. As the mazes used in these studies are very different in pattern it is difficult to determine whether or not these results are contradictory. In both studies the rats were inexperienced, and the mechanics of running and lack of attention on the part of the inexperienced rats may have had much to do with their entrances. Such tendencies are very probably not so markedly present in experienced rats.

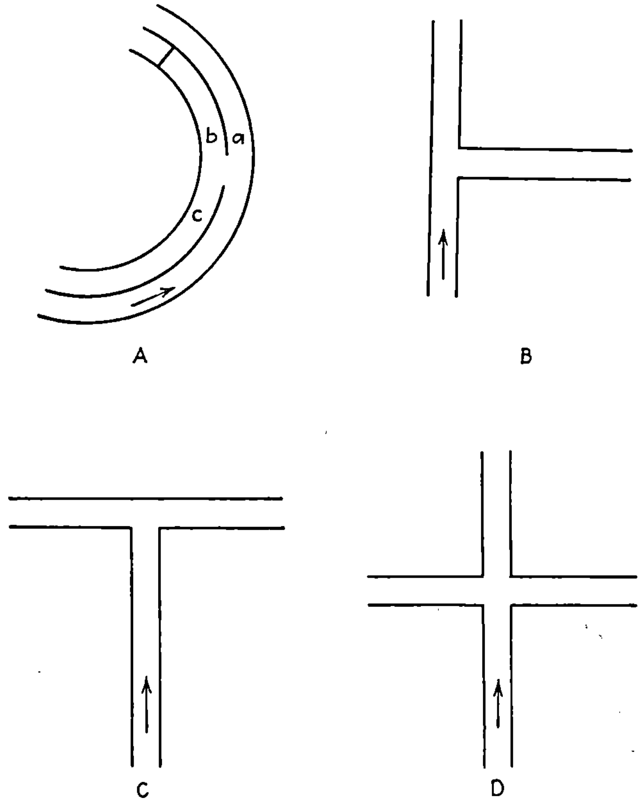

FIG. 84.—Types of maze junctions. The arrow indicates the direction of approach to the junction. (See text.)

There are also individual differences in rats as regards the preference for right or left turns at junctions such as in Fig. 84*C*. Yoshioka (1928) found that some rats prefer right turns, some left turns, and this preference seems in part to be due to the shape of the skull bones. None of the natural tendencies of the animal seems to be of major importance.

Junctions of the type shown in Fig. 84*D* are the most difficult, but this is probably due to the fact that such junctions present the animal with more opportunities for errors.

The Effect of the Maze Pattern upon Behavior in the Maze. *The Centrifugal Swing.*—The conditions arising from the nature of the maze pattern are of even greater importance in determining maze behavior. We have already referred to what Schneirla (1929) called the *centrifugal swing* in our analysis of the behavior of ants (pp. 160*f*.). This tendency which is set up by the nature of the turn

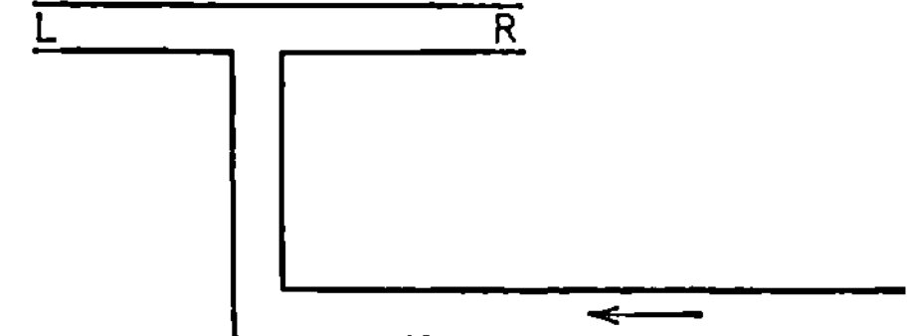

FIG. 85.—A maze alley which determines the animal's choice at the junction. The arrow indicates the direction in which the animal is traveling. When it reaches the junction it tends to turn toward *L*.

in the true path, is also present in the case of rats. Dashiell and Bayroff (1931) observed that if rats are run through an alley such as in Fig. 85, there is a tendency for them to turn left at the junction. This the authors regard as a tendency for rats to continue in the original direction and is called the *forward going* tendency. The elbow turn in the true path is a temporary obstruction which interferes, for the time being, with the direction of progress. According to Schneirla (1933*b*) however, the rat's momentum, which is a contributing factor to the centrifugal swing, is responsible for the above behavior. The elbow in the true path forces the animal to the outer wall, and this causes the turn toward that wall, which in this case is to the left. To test the two alternative interpretations Ballachey and Krechevsky (1932) devised a maze in which the centrifugal swing would cause the rats to turn in a direction opposite to that in which the forward going tendency would cause them to turn. A section of their maze is shown in Fig. 86. Their results show that the predominance of turns is to the side expected if we assume the principle of centrifugal swing.

The validity of the centrifugal-swing concept in the maze behavior of rats is further strengthened by the results of Ballachey and Buel (1934). They found that the path traversed by a rat in an alley containing elbow turns was toward the outer wall.

From the foregoing analysis it seems that the length of a straight alley determines the rats' momentum, and the direction of the elbow turn brings the centrifugal force into play. The operation

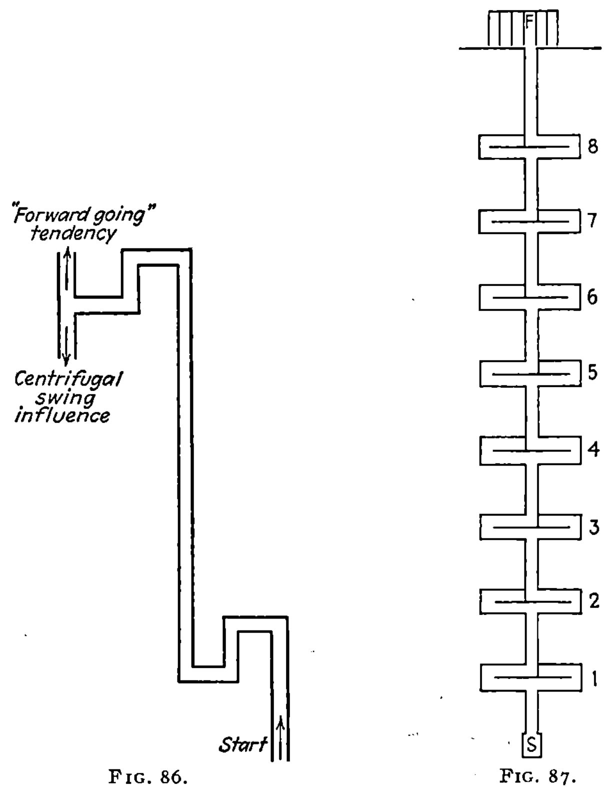

FIG. 86. FIG. 87.

FIG. 86.—Maze plan in which the "forward going tendency" and the influence of the "centrifugal swing" are opposed. The nature of the choice made at the junction indicates whether the animal is influenced by the direction assumed in the long alley; or whether the turns preceding the junction force the animal toward the wall and into the alley labeled "centrifugal swing." (*Modified from Ballachey and Krechevsky*, 1932, *p.* 88.)

FIG. 87.—Plan of linear maze. The location of the food cannot influence the direction of an animal's choices in this maze since none of the choices point in the direction of the food. (*Modified from Buel*, 1934, *p.* 186.)

of these two factors, represented by centrifugal swing, depends upon the pattern of the alleys preceding the junction and greatly determines the choice the animal will make at the junction.

The Direction of the Food Turn.—Another tendency which has been regarded as important is the preference for turns lying in the direction of the food box. Dashiell (1930) has found that if the final choice which leads to food is to the right rather than to the left, rats will tend to turn right in earlier parts of the maze whenever they have a choice between right and left turns. Dashiell regards this as evidence that the rat has a general sense of direction. This sense causes the rat to make characteristic errors. That this tendency to turn in the direction of food is due to some general orientation toward the food may, however, be questioned.

Maier (1929*a*, Experiment 9) has shown that if rats experience the position of food with reference to the starting point they will not turn more often in the direction of the food than away from it, when running over an unfamiliar pathway. If the rat uses a general-direction sense when running a maze, familiarity of the pathways should not be a prerequisite. As a matter of fact, Yoshioka (1933) has found that in a maze in which the pathways form a diamond, rats do not tend to run from opposite points of the diamond along the side that corresponds to the side where the food is placed. Food to the right of the diamond does not cause the rats to run around the right side of the diamond.

Dashiell's results can be equally well explained by assuming a tendency on the part of the rat to anticipate the turn toward the food. As we shall see later in the chapter, the maze is learned in a backward order. If the rat learns the turn into the food box before it learns which way to turn at other junctions, it is reasonable to suppose that it will use the learned response at earlier junctions.

An experiment of Buel (1934) demonstrates this point very nicely. Buel used a maze in which rats had to make either a right or left turn at each junction. The last turn led to a food box, the position of which was straight ahead, as shown in Fig. 87. In analyzing the errors it was found that when the food turn was to the right (as in Section 8 of this figure) most of the errors were due to a tendency to turn right, and when the food turn was left, most of the errors were due to a tendency to go left. As the actual position of the food was the same in all cases, direction orientation could not have determined the turns. Rather the response at the last junction tended to be applied to other junctions.

The Kind of Blind Alley Used and Its Effect on Behavior.—The third way in which the maze pattern may be changed is the use of various kinds of blind alleys. Shepard (unpublished studies) has extensively used mazes which may be characterized as having "dead-end" blinds, "circle" blinds, and "long-short" pathways. Examples of parts of such mazes are given in Fig. 88.

In the *dead-end blind* the animal comes into an alley which has a closed end and it must turn about and retrace its steps in order to continue. In the *circle blind* there is no closed end. After entering such a blind, the animal may continue in a circle, or it may retrace a few units and get out of the blind. (From the figure it can be seen that the rat may enter *A* and so arrive at *B* or *D*. It can now run a circle in either direction or go back to *A*.) In order for

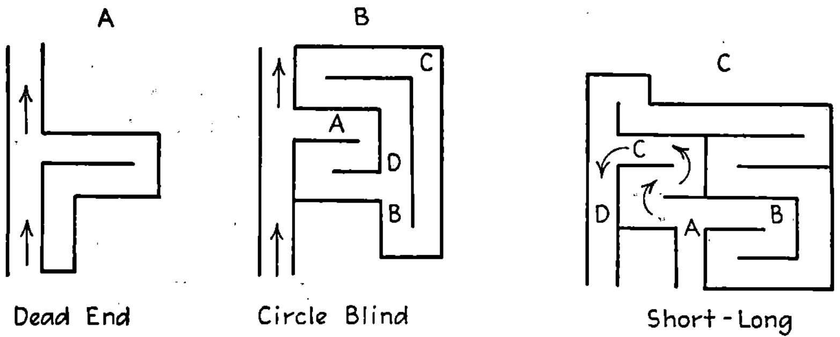

FIG. 88.—Types of blind alleys.

the animal to learn that it is not making progress, it must either experience the fact that it is running over the same area, or it must in some general way experience a lack of progress. In the *long-short* type of maze there are no blinds in the strict sense. Two alternative routes, one long, the other short, each lead to the same point. For the sake of rapid progress the long route must be avoided. The long routes may be regarded as blinds in the sense that entrance into them increases the distance to the food. In addition to having a long route, such mazes also make it possible for the animal to run a circle. The rat may, for example, take the longer route from one junction to the next, and then instead of continuing to a new part of the maze, may enter the short route and get back to the junction it has just left. (From the figure it can be seen that the rat may go from *A* to *B* and around and then choose *C* rather than *D*, and so again arrive at *A*.)

The relative difficulty of these various types of mazes has not been exactly determined, but the dead-end type of maze is markedly less difficult to learn than either of the two other types which are about equal in difficulty. In the long-short type of maze the rats tend to run in a circle, and it is this running in a circle which they revolt against and therefore tend to avoid. Whether the dead-end blinds are less difficult because of the closed end which blocks the animal's progress, or because the opportunities for errors and confusion are less in such mazes cannot be said. The indication is that the end wall furthers learning because the place causing the trouble is easily located and for this reason reduces the confusion. When the place causing the difficulty is not definite, the task of selection is greatly complicated.

The Serial Position of Blinds in the Maze and Its Effect on Learning

Introduction.—The ordinary maze contains a number of blind alleys. The question which naturally arises from this condition is whether there is a difference in the elimination of blind alleys due merely to their serial position. It is possible that the maze habit is (1) built up from the starting point and learned primarily in the forward direction; (2) established with reference to the food box and learned primarily in the backward direction; or (3) acquired without reference to either end and learned equally well in all parts. Thus if the serial position of a blind alley is a factor in its elimination, different parts of the maze should be mastered at different rates.

The study of the order of the elimination of blinds has several difficulties. As we have already seen, there are certain factors which make different parts of the maze of unequal difficulty. Unless these factors are taken into consideration and eliminated as much as possible, the data on the order of elimination of blinds will not solve the problem. A second type of difficulty is that resulting from an animal's retracing parts of the maze. During the early stages of learning the rat gets lost and may find itself back at the starting point on many occasions. Because of this, the early part of the maze is traversed in both directions more often than the later parts. An animal, therefore, has more opportunities for making errors in the first than in the last part of the maze, and may also become confused because of traversing it in both directions. If we count the number of entrances into the various blinds, we shall not have a true measure of their relative difficulty because the early stages of the learning activity are primarily limited to the first part of the maze. The number of trials required before the various blinds are eliminated is also an unsatisfactory measure because each trial (a trip from the starting point of the food box) may include much more contact with the first than with the last part of the maze. This additional contact in the first part of the maze should be of aid in learning it. If retracing in the maze were eliminated (*e.g.*, by a series of doors which would close behind the animal as it progressed) a more reliable index of the order of elimination of blinds could be obtained.

Results from Mazes with Fixed Patterns.—Carr (1917*a*) made a careful study of his own and other data obtained from rats while learning similar mazes, and found that the distribution of errors was a function of the spatial order of blinds, sometimes with reference to the food box, and sometimes with reference to the starting point. He also found that, in all cases, the blinds which were least attractive (*i.e.*, the least frequently entered) were the first eliminated.

Warden (1923*a*) studied the order of elimination of blinds by counting the number of trials required by rats before a blind was no longer entered. He interprets his results as showing no indication of a particular order of elimination. He verified Carr's findings regarding the ease of elimination of blinds infrequently entered but, unlike Carr, regards it as the basis for the irregular elimination of blinds.

If, however, we analyze Warden's data in a different way, the relation between the elimination of blinds and their serial position in the maze is rather striking. By dividing the nine blinds of his maze into groups of three, we can partially average out any individual differences in difficulty among the various blinds. When we do this, we find that the average number of trials required to eliminate the first three blinds is 29.9, the middle three blinds, 32.5, and the last three blinds, 13.6. These figures show a strikingly small number of trials required for the elimination of blinds in the vicinity of the food box. The first part of the maze seems slightly easier to learn than the middle portion. If we take the individual records of the 35 rats we find that 94.3 per cent of them made their lowest score on the three blinds nearest the food box. Only 45.7

per cent of them made lower scores on the first part than on the middle part of the maze. Considering Warden's results in this manner, there seems to be good evidence for the easy elimination of blinds near the food box. Blinds farther away are more difficult, but all of these seem to be eliminated about equally well.

By using a maze which consisted of a succession of V junctions in which one of the paths led to a blind and the other to another junction of the same type, Warden and Cummings (1929) attempted to eliminate the individual differences among the various junctions. In this maze the blinds differed only in their proximity to the food box. Their results show that the blinds in the second half of the maze were eliminated before those in the first half, but the order of elimination was not perfectly backward. For the 10-cul-de-sac maze the order of elimination from first to last was alleys number 10, 8, 4, 6, 5, 3, 7, 9, 2, 1. Because the order is not exactly backward in a maze having blind alleys and junctions all alike, the authors believe that they have evidence against the backward order of maze learning. The fact that there is a higher average error score in the first than in the last half of the maze is explained as being due to retracing; the retracing being primarily confined to the first part of the maze.

The irregular order of elimination shown above becomes clear, however, when we consider the character of the last turn in a maze and its influence on maze performance. The maze used consisted of a succession of right and left turns. The odd-numbered blinds turned to the left as did the food box, whereas the even-numbered blinds turned to the right, opposite to the food-box turn. If the turn into the food box is anticipated, then the odd-numbered blinds should cause difficulty. This is exactly what the above results show. Blinds 7 and 9, which are near the food box, were eliminated relatively late, whereas blind 8 is eliminated early. The individual differences in the junctions were therefore not eliminated as the experimenters believed. When corrected for anticipation, however, the order is backward. (See Buel, 1934.)

Results from Mazes with Variable Patterns.—Borovski (1927) adopted a rather interesting technique to test the backward order of learning the maze. He used a box (see Fig. 89) divided into two halves. Each half presented the rats with four forward-leading paths (alleys 1, 2, 3, 4 and *A*, *B*, *C*, *D*). Each of these alleys had removable ends, but only one was open at any time; the others served as blind alleys. There were also a number of permanent blinds. The two halves were identical except for the fact that the food box was at the end of the second half in a position corresponding to the end of the first half. If learning is backward from the food box, a change of the pattern in the second half should interfere with the establishment of the habit more than a change of the pattern in the first half. To test this, 83 rats were divided into three groups. One group learned the route to food without any modifications. A second group learned the maze with the first part of the maze changed (a different path open) daily. This group was therefore prevented from learning a specific route through the first half of the maze. The third group was treated as the second, except that the changes from day to day were made in the second half of the maze instead of the first.

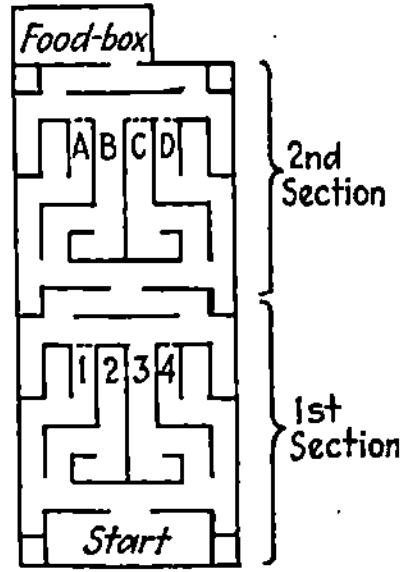

FIG. 89.—Plan of Borovski's maze. The two halves of the maze are alike in pattern. The route through each half may be altered by removing or inserting sections in alleys 1, 2, 3, 4, and *A, B, C, D*. (*Modified from Borovski, 1927, p. 493.*)

The results may be summarized as follows: (1) the first group of rats learned the second half of the maze before the first part; (2) after the first group learned the maze, changes in the second half of the maze caused more confusion than changes in the first half; and (3) the third group of rats made many more errors during the learning period than did the second group. The results show that the learning is largely built up backward from the food box. However it must be noted that the first part of the maze was learned even though the second part was changed.

The Goal-gradient Hypothesis and the Serial Position of Blind Alleys.—Hull (1932) approaches the problem from a study of motivation. He postulates an excitatory gradient increasing in intensity from the beginning of the maze to the food box. Hull was led to this conclusion by his observation that rats ran more rapidly in the latter part than in the first part of the maze. From this postulation of a goal gradient it follows that deviations from the true path are more critical and effective when the gradient is high. We should therefore expect the part of the maze near the food box to be learned first. We should also expect the gradient to be higher at the beginning of a short maze than at the beginning of a long one. If a maze contains both a short and a long route to food, the short route should be learned first. As the gradient differences would exist at the junctions, the true path being nearer the goal, should be preferred. But if this line of reasoning is correct, all blinds should be eliminated at the junctions. Partial entrances into blinds are, however, very common before they are completely dropped out. Further, long blinds, because they carry the rat farther from the food than do the short, should be eliminated more quickly than short blinds. This also seems to be contrary to the actual facts.

Assuming the importance of the goal-gradient hypothesis and the tendency of rats to turn in the general direction of the food location, Spence (1932) examined some maze data obtained by Tolman and Honzik on a 14-junction maze in which all junctions were alike in that each presented the rat with a choice between a right-going and a left-going path. By assigning certain relative values to the pathways at each junction, the value depending on the nearness of the pathway to the goal and on the direction of the pathway with reference to the position of the goal, he arranged the blinds in their theoretical order of difficulty. On comparing this theoretical order of difficulty with the actual order of elimination Spence found a high degree of correspondence (a correlation of 0.9 or more). He concluded that the difficulty of blind-alley elimination depends on (1) the distance of the blind from the goal; (2) the direction in which the blind leads with reference to the position of the food box; and (3) the tendency of the rat to anticipate the goal reaction.

Spence and Shipley (1934) found the order of blind-alley elimination to be backward during the first few trials, but as learning progressed, the order of elimination changed to a forward one. The forward order appears attributable to a tendency to anticipate the final turn in the maze, since most of the errors were made in blinds which turned from the true path in the same direction as the final turn which led to food. Thus the forward order of blind-alley elimination was not due to a more rapid elimination of the blinds in the first part of the maze but to an increase in errors in the latter part of the maze. If the goal-gradient hypothesis is applied to the results of this experiment, it would have to be used to explain the increased intensity of the anticipatory tendency and the consequent increase in errors in the latter part of the maze. This being the case, it is difficult to understand how the same goal gradient can, at the same time, account for the backward elimination of blinds.

It seems to the present authors that the results can be satisfactorily explained by assuming (1) the establishment of a food association in the first few trials, and (2) the inadequacy of the kinesthetic process for making differentiations between the similar junctions. As a consequence, the learned goal reaction is made at the wrong junction. This accounts for the persistent errors in the latter part of the maze, on the basis of inadequate sensory discrimination and not on the basis of motivation.

If this is the case, one might assume that anticipatory reactions would be independent of the position of the goal. A study by Spragg (1934) actually bears this out. He used a unit maze in which the turns required were *rrrrlrrr*. In this experiment a left turn is the exceptional turn and it does not lead to food. Nevertheless, errors were concentrated at the junction preceding the exceptional turn as is shown in Table 27. However, the concentration of errors was not so great as when the exceptional turn was the last and led to food. This is to be expected since the problem required is a discrimination between running four and five units, rather than seven and eight. As the difference between four and five is relatively

TABLE 27.—ANTICIPATORY RESPONSES TO THE EXCEPTIONAL TURN IN A UNIT MAZE (FROM SPRAGG, 1934)

Unit	1	2	3	4	5	6	7	8	Total
Response required	*r*	*r*	*r*	*r*	*l*	*r*	*r*	*r*	
Errors in 1st 25 trials	27	26	18	57	41	8	8	14	199
Errors in 2nd 25 trials	10	13	30	45	48	6	1	5	158
Total errors	37	39	48	102	89	14	9	19	357

greater than the difference between seven and eight, fewer errors would be expected. Furthermore, a right turn followed the exceptional left turn, hence a discrimination between the fifth and sixth units was also required. This should cause errors at the fifth unit and that is actually what Spragg found.

Thus by assuming that the exceptional turns are the first to be learned and that they must be differentiated from the others, the results of the serial order of blind-alley elimination can be explained without recourse to goal reactions as such. Goal reactions are likely to appear important, not because they are goal reactions, but because *the turn into the food alley is an exceptional one*, being characteristic in the sense that *this turn leads to something different from all the other turns*. The goal-gradient hypothesis assumes that the speed of learning is a function of the motivating situation, whereas the above explanation makes no such assumption.

Conclusions

From the foregoing analysis of maze learning we may regard the maze habit as the establishment of a unified pattern of responses which depend upon a variety of stimuli. The rat must set up preferences for alternate alleys and in order to do this, various points in the maze must differ for it. Traversing a learned maze is not merely the running of a rhythm of movements, because the rat can, when placed in any part of the maze, find its way successfully to the next point. We may greatly interfere with the animal's rhythm of running without causing it to become lost. As long as various parts of the maze can be *differentiated* by the rat, it is capable of responding correctly to any particular part of the maze pattern without first experiencing the preceding parts of the maze. As such, the maze pattern is not an indivisible unity. The kinesthetic sensations alone are not a sufficient sensory basis for the learning of large mazes, since the differential experiences between them are not reliable and consistent enough to be useful. The junctions in the maze usually are the critical points, and it is there that characteristic differences must be found by the animal.

To make a correct run an animal must not only integrate certain parts of the maze, but it must first *select* that which must be integrated. The problem of selection is increased in difficulty by increasing the number of items. This is what happens when we increase the number of junctions or complicate them by adding to the number of paths leading from them. Selection is also made more difficult by using blinds which are difficult to isolate in that they have no definite point of termination.

The actual *learning* part of maze behavior is one of *integration*. Certain factors increase the difficulty of this process. The number of units to be integrated is, of course, one of the factors. Other factors are less obvious.

The swings or turns in the maze in relation to the length of alleys influence the mechanics of running and cause certain alleys to be more obvious and more easily entered than others. This causes difficulty because it introduces certain natural tendencies which must be overcome in learning. Junctions which are preceded by the centrifugal swing, therefore, not only require the formation of the proper association, but also the inhibition of certain mechanical tendencies in running. Hence this factor increases or decreases the difficulty of the formation of the proper associations, depending on whether the mechanics of the swing favors the entrance into the blind or into the true path.

Another factor which must be considered in maze learning is the *interference of partial learning with* a more *complete learning*. The experience of receiving food in a certain place is an important event, and the characteristic experience which accompanies it is the last choice which leads to the food. The contiguity of these two experiences results in their combination. As soon as this association is formed, the response to food tends to be repeated at the earliest opportunity. The animal therefore applies the characteristic turn which it has learned to other junctions in the maze and tends strongly to enter certain blinds. Not until the rat makes further distinctions between the last turn to food and other turns in the maze can it overcome this tendency to apply an act already learned to wrong parts of the maze. After the animal has learned to apply the characteristic response to the correct junction, it learns to associate the preceding contiguous experience with this junction. In this manner the maze responses are built up backward.

In this connection it is perhaps important to recognize that the mammal is not merely a learning creature, but is also capable of *reorganizing* its experiences. If a rat has explored a region, such as a simple maze, and is then fed at any point in the maze, it is capable of going directly to the food when placed at any other point in the maze. (This problem will be treated at length in Chap. XX.) In such a case the associations have not been built up gradually, rather the path to food has been simultaneously integrated. If a rat is capable of reorganizing and integrating such random experiences as are obtained during free exploration, it must have an ability which will be useful in building up the maze pattern. It seems that this ability to reorganize and to spontaneously integrate experiences plays an important part in the integration of the maze habit.

When we realize the number of qualitative factors which are at play in the process of learning and consider that abilities other than learning may also be involved during the process of building up the maze pattern, it becomes clear why the behavior of animals in the maze had been so difficult to analyze. The animal's responses may not only be the resultant of an involved learning process, but of other processes as well. This being the case, the maze is not a measure of pure learning ability. It is perhaps more accurately described when termed an *intelligence* test—with intelligence regarded as a composite of these abilities which are concerned with adaptation through experience.

Suggested Readings

Buel, J. 1935. Differential errors in animal mazes. *Psychol. Bull.*, **32**, 67–99.

Dashiell, J. F. 1930. Direction orientation in maze running by the white rat. *Comp. Psychol. Mono.*, **7**, pp. 72.

Hull, C. L. 1934. The concept of habit-family hierarchy and maze learning: Part I and Part II. *Psychol. Rev.*, pp. 33–52 and pp. 134–152.

Munn, N. L. 1933. *An introduction to animal psychology.* Pp. 155–205. Boston: Houghton Mifflin.

Watson, J. B. 1907. Kinaesthesis and organic sensations: their rôle in the reactions of the white rat to the maze. *Psychol. Mono.*, **8**, pp. 100.

CHAPTER XVIII

FACTORS DETERMINING PERFORMANCE IN LEARNING SITUATIONS

INTRODUCTION

As has been pointed out, the degree of learning can be detected only through a study of an animal's performance. This does not mean, however, that performance is a direct measure of learning. Errors in a maze may not always be due to lack of learning but may arise from poor motivation, inattention, and chance. Until these are eliminated, performance is not a reliable index of learning.

Different motives often greatly alter an animals behavior. For example, hunger may result in a preference for the true path in a maze, whereas curiosity may result in a preference for blind alleys. Because curiosity elicits exploratory behavior, changes introduced into the maze situation are likely to alter performance. Maier (1929*a*) has shown that when new paths are opened in a maze, rats invariably take the new path in preference to the learned true path.

In discrimination experiments an incorrect choice is often accompanied by punishment (*i.e.*, mild electric shock), whereas a correct choice is accompanied by reward (food). The use of such a technique might be expected to improve performance because it reduces responses based on curiosity. The use of punishment, however, is sometimes unsuccessful. During the early stages of learning an electric shock may set up an emotional condition and so interfere with normal behavior. Or the animal may associate maze running with pain and so refrain from all activity. An electric shock must be so adjusted that it irritates but is not painful.[1] This is somewhat difficult because the effective intensity of an electric shock in acting upon the tissues depends on the nature of the contact of the animal's foot with the electric grill and the amount of moisture in the animal's

[1] The methods for the administration of electric shocks in animal training have been carefully investigated by Dunlap (1931); Dunlap, Gentry, and Zeigler (1931); Muenzinger and Walz (1932); and Muenzinger and Mize (1933).

foot (skin resistance to current), as well as on the voltage. If a shock is used, it is well to introduce it after several choices have been made so that the making of a choice does not become associated with punishment.

There is a prevalent opinion that in order for an animal to learn a maze or to form a problem-box habit a reward of some kind is necessary. It is true that an animal will not make a response of a certain type unless it is motivated, but failure to make such a response does not demonstrate inability to learn. An experiment by Lashley (1918) shows that learning does take place in the absence of a reward. He permitted a group of rats to explore freely a simple maze for 20 min. After the period of exploration the rats were required to learn the maze in the usual manner. Running 10 trials per day these rats required an average of 29.4 trials before they made 10 consecutive errorless runs. Without the period of exploration another group of rats required 51.7 trials before reaching the same criterion. Maier (1932*d*) has shown that a simple maze may be entirely mastered through random exploration. The introduction of a motive therefore merely causes the animal to demonstrate its learning.

Because performance varies for reasons other than the degree of learning, it is necessary carefully to distinguish between learning and performance. The present chapter is therefore concerned with an analysis of the factors which affect performance but do not affect learning.

Motivation

Changes in Motivation Produce Changes in Performance.—Tolman (1932, Chaps. III and IV) emphasizes the importance of the relationship of various goals to the need of the animal and uses experimental findings to show that when a goal satisfies the need of the animal its performance is better in an experimental situation than when the goal does not satisfy the need. He inclines toward the interpretation that learning is actually facilitated by a satisfying goal. We shall, however, discuss the experiments on motivation from the point of view of *performance* rather than of *learning* and attempt to show that the available evidence demonstrates only that performance is affected by various motivating conditions.

Blodgett (1929) divided his rats into three groups: (1) a control group which received food at the end of the maze after each day's run; (2) a group which received no food at the end of the maze until the seventh day; and (3) a group which received no food in the maze until the third day of running. Each of the rats made one trip per day.

The results clearly demonstrate that the performance of the rats not receiving food in the maze was decidedly inferior to that of the rats receiving food, but two days after the reward was introduced for the second and third groups the performance of these animals equaled that of the control animals. Thus the introduction of

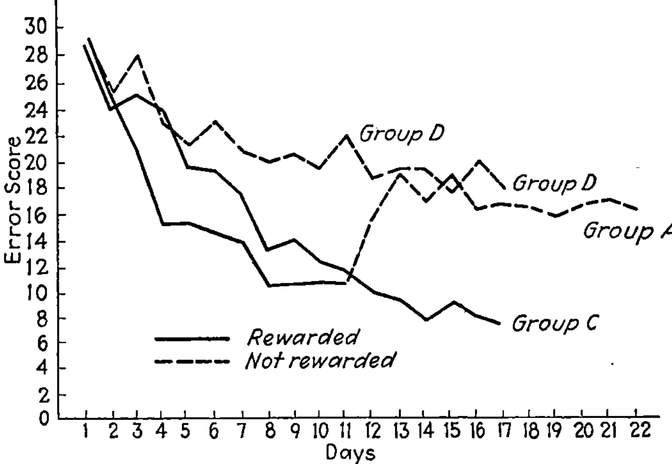

FIG. 90.—Curve showing the effect of the removal of reward on performance. The solid line and broken line curves (*C* and *D*) show the performance of the control groups of rats which were rewarded and not rewarded, respectively, during the entire test period. Curve *A* shows the performance of a group of rats which was rewarded until the eleventh day, and thereafter not rewarded. (*Modified from Tolman and Honzik,* 1930*a, p.* 262.)

the reward was essential to produce satisfactory maze performance. Elliott (1929*a*) obtained similar results. He found that thirsty rats performed best when they received water at the end of the maze, and hungry rats performed best when they received food.

In a more elaborate experiment of the same type, Tolman and Honzik (1930*a*) compared the performance of four groups of rats. All groups were required to learn the same maze and each rat made one trip through the maze each day. The groups were alike in every respect except for the motivating conditions. Thus Group *A* received food at the end of the maze for the first 10 days, thereafter it received no reward at the end of the maze; Group *B*, however, received no reward during the first 10 days, but from the eleventh day on it was rewarded; Group *C* received reward during the whole period; and Group *D* received no reward during the whole period. The error scores made by these groups are shown in Figs. 90 and 91. The solid line represents the period during which reward was given, the broken line the period during which no reward was given. Since the motivating conditions remained constant for Groups *C* and *D*, they serve as control groups. The records made by these are compared with Group *A* in Fig. 90.

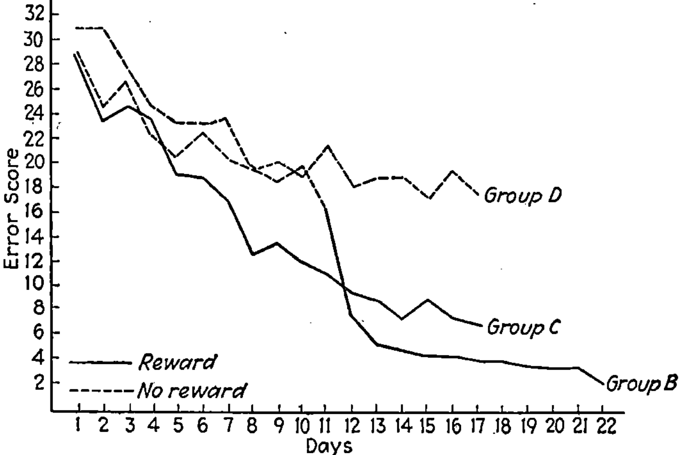

FIG. 91.—Curve showing the effect of the introduction of reward on performance. Rats in group *B* were rewarded on the eleventh day and thereafter, as indicated by the broken line which changes to a solid line. (*Modified from Tolman and Honzik*, 1930*a*, *p*. 267.)

As one might expect, Group *C* showed an improvement in performance from day to day, whereas Group *D* showed very little improvement. The motivating conditions for Groups *A* and *C* were alike during the first 10 days, and the performance of these groups during this period was strikingly similar. After the eleventh day the conditions of Group *A* became like those of Group *D* and as a consequence the performance of Group *A* became as poor as that of Group *D*. From Fig. 91 we see that the conditions under which Groups *B* and *D* performed were alike for the first 10 days and both of these groups showed little if any improvement during this period. After the eleventh day Group *B* performed under the same conditions as Group *C*, and the performance of Group *B* not only became as good as that of Group *C* but actually improved upon it. It would seem that during the period Group *B* was not rewarded it actually learned more than Group *C* which was rewarded. If we consider that during the nonreward period Group *B* was making a very poor performance and consequently was spending more time in the maze than Group *C*, which was performing very well, it is not altogether surprising that Group *B* should have learned more. It is reasonable to suppose that a rat learns more from wandering about the maze and finally arriving at the end than it does from hastily running through as best it can.

From the foregoing experiments and from others of similar nature, it may be concluded that an animal performs best, that is, it runs to the goal most directly, when the goal is desired or needed. When the goal is not desired, then there is no satisfaction at the end of the maze, unless it is the satisfaction of release. For this reason the rat does not tend to run directly to the food box, but rather runs to other parts as well and so makes errors. If we assume that the rat is a very simple mechanism and responds only to its immediate surroundings, then the animal should respond in the same way in the maze, no matter whether a desired goal is at the end of the maze or not. Because the performance of the rat is a function of the goal, its behavior in a situation cannot be explained in terms of its immediate source of stimulation. When the rat runs the maze, it is not merely responding to different parts of the maze, but to memories of food as well. For this reason we may say that the animal anticipates food at the end of the maze.

The Anticipation of the Reward.—There is experimental evidence which shows that animals respond not only to a memory of a reward, but to certain properties of the reward. Elliott (1929*b*) ran a group of thirsty rats in a maze (one trial each day) and gave them water for a reward. On the tenth day the drive or need was changed to hunger and the reward was changed to food. On this day the rats did much more poorly than usual. When they arrived at the end of the maze they found not water, as on previous occasions, but food, which satisfied their present need. The following day they were again hungry, but their performance was as good as that of a control group which was not subjected to the changes in running conditions. Thus the nature of a goal which could only have been present in memory played an important part in an animal's behavior. This memory phenomenon corresponds to what is called anticipation in human beings.

In order to demonstrate this phenomenon in goats, Fischel (1930) placed before them two boxes with lids which could easily be raised by the goats' noses. One box contained bread, the other grain. Of the two kinds of food the bread was preferred, in the sense that it was always eaten first when there was a choice between them. The animals soon learned that one of the boxes contained bread, but they could not detect in which box it was without first opening the lid. It was found that whenever the box containing the grain was opened first, the goats did not trouble to eat it, rather they merely turned their heads to the side and opened the other box containing the bread. For the goats actually to ignore food of a particular kind they must have had a memory of the other food, or we might conveniently say they must have expected bread in the other box. Ordinarily when goats find food, they eat it as they find it.

Tinklepaugh (1928) found that monkeys definitely manifested surprise when a desired food (banana) which they had seen placed under a container was not present when they turned over the container. While unobserved, the experimenter had substituted lettuce for the banana. Although the monkey ordinarily accepted lettuce, it refused to eat it when found under such circumstances.

A Comparison of the Effectiveness of Various Incentives.—If performance in learning situations varies for different incentives, the maze becomes an instrument for measuring the relative desirability of different incentives when the animal is in different physiological states.

Szymanski (1918) lists a number of incentives which motivate animals sufficiently to produce accurate performance in learning situations. Thus he finds that rats perform perfectly in a maze for food only when they are hungry. Return to litter is a sufficient motive for some mother rats, but not for others. In all cases when nursing ceases to be a need on the part of a female, the return to litter ceases to be a motive. The return to the home cage was found to be an unsatisfactory motive.

Many experiments have been designed to measure the relative effectiveness of various incentives. Stone and Sturman-Huble (1927) found food and sex to be about equal incentives for maze performance. Simmons' (1924) experiments indicate that bread and milk, a female in heat (for male rats), sunflower seed, and return to litter (for female rats) are decidedly more effective incentives for rats in running a maze than escape from the maze or return to the home cage. Their effectiveness is approximately in the order listed, and the difference between them is greater for time than for errors. She also found that a combination of rewards such as "bread and milk" plus "return to the home cage" was a more effective incentive than was either of them alone. On the other hand, MacGillivray and Stone (1931) found that a combination of food and escape from water was no more effective than was food alone for determining performance in the discrimination box. It might be that the rat performs to its maximum capacity for food. Foods varying greatly in preference, however, produce only a slight difference in performance, according to the results of Maslow and Groshong (1934) in their study with monkeys.

Grindley (1929) found that the amount of reward was a factor determining the maze performance of chicks. When the reward was increased sixfold, the number of trials required to reach a certain standard of performance decreased 25 per cent.

Hamilton (1929) measured the effects of certain intervals of delay in feeding after the rat arrived at the food box. She found that an interval of 1 min. approximately doubled the errors, trials, and time. Delays of 3, 5, and 7 min. did not further affect the performance except for a slight increase in the running time. Whether or not the attractiveness of the reward is actually changed by making it more and more remote cannot be stated. It seems that the longer delays should have further lowered the level of performance, unless the rat's sense of time is poorly developed. There is also the possibility that delayed feeding actually increases the difficulty of the problem. The rat must learn to associate food with running the maze and if a delay is present, this association is more difficult to make. The formation of the conditioned response is also greatly hindered when the interval between the two stimuli is greater than a few seconds.

Wood (1933) obtained similar results with chicks learning a discrimination habit, but found that increasing the delay up to 3 min. continued to cause an increase in the trials and errors required to reach the criterion of learning. Delayed punishments seemed to have no effect on performance.

The Relative Effectiveness of Punishment and Reward.—Since both punishment and reward are effective for controlling an animal's performance, it becomes of interest to determine which is the more effective. Yerkes (1907) argues that food is unsatisfactory as an incentive because the degree of hunger cannot be controlled, and utter hunger may interfere with performance.

Hoge and Stocking (1912) found that in discrimination experiments with rats it was more effective to accompany incorrect choices with punishment (an electric shock) than to accompany correct choices with reward (bread and milk). A combination of both reward and punishment was the most effective for bringing about accurate performance. Dodson (1918) obtained similar results, and in addition measured the effectiveness of various intensities of hunger and electric shock. The most effective condition of hunger was a period of from 41 to 48 hr. without food. Slight hunger and periods of starvation beyond 2 days were found to be much less effective. Similar results were obtained in the case of

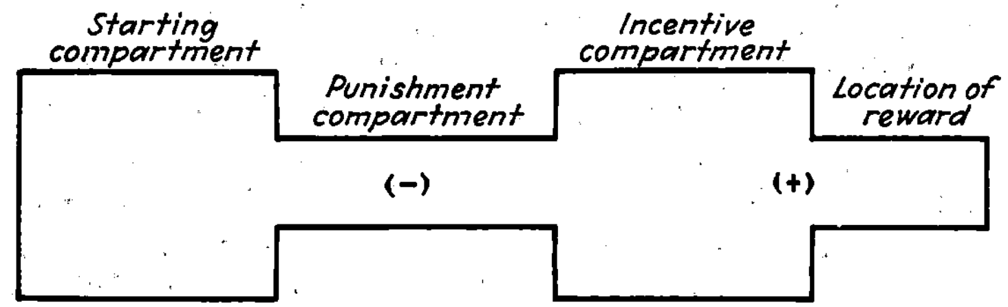

FIG. 92.—Plan of the Columbia obstruction apparatus. The animal must pass through the punishment compartment in order to reach the incentive. (*Modified from Warden*, 1931, *p.* 18.)

punishment. A relatively weak shock was greater in effect than a more intense shock. However, the relative effectiveness of reward and punishment as such can hardly be established because both depend on intensity, and it is difficult to know when they are equal in intensity. The maximum effectiveness of each was found to be about the same.

The Obstruction Method and the Study of Motivation.—Warden and his students (1931) have investigated the problem of motivation by the use of the obstruction method. A floor plan of the apparatus used in these studies is shown in Fig. 92. An incentive or positive attraction is located in a compartment, in front of which is an electric grill or negative attraction. In order to reach the desired object the animal is required to cross the grill and take an electric shock. If the goal object is more positive than the shock is negative, the animal should cross the grill. By leaving the "negative attraction" constant, the attractiveness of various incentives during various physiological needs can be measured by recording the number of times an animal will cross the grill in a given period of time. This period was arbitrarily set at 20 min. Each of the studies attempts to measure the incentive value of a particular goal object as the physiological need of the animals is varied.

Warner measured the incentive value of food under various degrees of starvation. In the case of male rats the number of times they crossed the grill in the 20-min. period increased with the period of starvation up to 4 days. Increasing the period of starvation up to 8 days resulted in a decrease in the number of crossings. In the case of female rats the maximum number of crossings was at the end of a 24-hr. period without food. This period is the usual feeding interval for rats.

In the case of maze performance Ligon (1929) found that after a 6-hr. period of starvation rats were actually superior to those which had been deprived of food during a 12-hr. period, though of the three starvation periods studied the 24-hr. period produced the best motivation. Ligon's study shows that the length of a starvation period, even when it does not result in suffering, does not determine the degree of motivation.

Hamilton (Warden, 1931) studied the effect of delayed feeding by the obstruction method and found that 15 sec. delay in feeding the rat after it arrived at the food compartment resulted in a 43 per cent decrease in the number or times that the grill was crossed. Delays of 30 sec. and of 1 min. showed no further decrease. These results are similar to the results she obtained by the use of the maze.

Warner (Warden, 1931) studied the incentive value of water under various degrees of thirst and found the number of crossings to be at a maximum after one day of water deprivation. Periods of water deprivation greater than one day caused a decrease in the number of crossings, until death intervened after about 6 days without water. The least number of crossings occurred after removal of a rat from a cage containing water. Unlike hunger, thirst showed no sex differences.

In the study of the sex drive in the male rat, Warner used a female in heat for an incentive. The least number of crossings occurred when the male had been removed from females for a period of 2 hr. or less. Periods of segregation up to 12 hr. showed a rapid increase in the number of crossings; still longer periods of segregation showed a further but less rapid increase in the number of crossings until the maximum interval of 24 hr. was reached.

The readiness with which a female crosses the grill to reach a male in the incentive compartment varies according to the stage of her oestrous cycle. There is a sharp rise in the number of crossings with the onset of the cycle, but the crossings show a gradual decrease as the cycle wanes. Warner points out that males and females are difficult to compare as to the relative intensities of their sex drives, because the response of the female is largely dependent upon her physiological condition, whereas the male is influenced by the conduct of the female in the incentive chamber as well as by his own condition.

Nissen (Warden, 1931) studied the maternal drive by placing a rat's litter in the incentive compartment and observing the number of times the mother crossed to it. The drive was varied by using litters of different ages. The results show that the incentive value of the litter decreases with the age of the litter and with the age of the mother. After she has been separated from her litter for 4 hr., there is a decrease in the number of crossings made by the mother.

Nissen also studied the incentive value of unexplored territory. In place of the incentive chamber he placed a large box fitted up like a maze and containing various objects. He found that the male rat will cross the grill to explore such an area.

The Adequacy of the Obstruction Method for Comparing the Intensity of Different Drives.—Warden (1931) compared the effectiveness of the various incentives by comparing the maximum number of crossings obtained with each incentive. Table 28 is a summary of these results. According to this table, the order of the intensity of the various needs is maternal, thirst, hunger, sex, and exploratory. If, instead of comparing the number of crossings in a 20-min. period, we compare the number of crossings occurring during the first 5 min. of each of the 20-min. periods, we find the order to be thirst, maternal, sex, hunger, and exploratory.

That the order of the intensity of the various needs under different periods of deprivation is dependent upon the length of the arbitrary period has been demonstrated by Leuba (1931). He points out, for example, that for two days of food deprivation the number of crossings in four consecutive 5-min. periods are 130, 102, 86, and 33, a total of 351, whereas for three days of food deprivation, the corresponding figures are 75, 91, 100, and 98, a total of 364. Thus for three days of food deprivation the total number of crossings is greater than for two days of food deprivation, but if the number of crossings in either a 5-, 10-, or a 15-min. period is compared, the number of crossings is greater for two days of food deprivation.

TABLE 28.—COMPARISON OF VARIOUS DRIVES (MODIFIED FROM WARDEN, 1931)

Drive	Condition of maximum need	Av. no. of crossings in 20-min. period	Av. no of crossings during first 5 min.
		Male	
Thirst	Second day	21.10	
Hunger	Fourth day	19.10	
Sex	First day	13.45	
Exploratory		6.00	
		Female	
Maternal	Young litter	22.40	
Thirst	First day	19.70	
Hunger	Third day	19.00	
Sex	Oestrum	14.14	
		Male and female results combined	
Maternal		22.40	5.60
Thirst		20.40	7.65
Hunger		18.20	3.75
Sex		13.80	3.92
Exploratory		6.00	2.40

A study by Holden (1926) shows that the intensity of the shock used also determines the nature of the results. She used three intensities of shock and counted the number of crossings in groups of rats deprived of food for 12, 24, 36, 48, 60, and 72 hr. For the least intense shock the maximum number of crossings occurred in the 48-hr. group; and for the most intense shock the maximum number of crossings appeared in the 24-hr. group. That the 24-hr. group crossed most often when the shock was intense indicates that starved animals are less able to take punishment. The length of the period of deprivation may therefore not only be a measure of the intensity of a need, but may indicate loss in ability to take punishment as well. Furthermore, 15 per cent of the animals in the 60-hr. group and 48 per cent of the animals in the 72-hr. group died before they were tested. If different periods of food deprivation produce different death rates it is obvious that the physical condition of the surviving animals which were tested must also have differed widely.

Ligon (1929) has pointed out that the obstruction method does not permit the measurement of all the characteristics of motivated behavior. In a learning situation incentives differ as to the trial on which the most accurate performance appears; the maximum speed at which an animal will perform; and the number of successive trips the animal will make with accuracy. Ligon found that the same incentive might be superior in one of these aspects and inferior in another.

There are a number of objections to the obstruction method which limit its value as a technique for the comparison of the intensity of various drives. These are as follows:

1. As has already been stated, the period of observation and the intensity of shock are arbitrary.

2. Some incentives satisfy certain forms of deprivation more effectively than others. Thus one kind of food may satisfy hunger more effectively than another kind. This would make a comparison of the number of crossings dependent upon the incentive used as well as the drive.

4. Certain needs may be satisfied more effectively with a few crossings than other kinds of deprivation. Thus water may alleviate thirst more quickly than food alleviates hunger. If this is true the number of crossings is not a true picture of the original intensity of the drive.

5. The degree of punishment may vary for various physiological states. (Water deprivation, for example, may change the effective intensity of the electric shock.)

6. The condition of the animal during various periods of deprivation may change and so make it less able to stand punishment and less able to cross the grill rapidly to obtain less punishment.

7. The method assumes that the drive increases quantitatively rather than qualitatively as deprivation increases.

8. Various drives involve emotional excitement in various ways so that the condition of an animal is not only a function of a particular form of deprivation but also of other physiological conditions which result from the deprivation.

The Importance of Motivation Studies to Animal Investigations.— The problem of motivation enters into almost every experimental situation because in order to control the animal's behavior sufficient motivation must be present. The foregoing studies have shown that the performance of an animal varies with the need of the animal, the nature of the incentive, and the promptness with which the animal receives the incentive after completing its performance. In order to make a comparative study of the ability of various animals, the degree of motivation must be the same for each of them. This perhaps seems to be an impossible task. However, it is reassuring to know that in most cases a considerable change in motivating conditions does not cause a great change in performance. Thus food and water each produce similar performance if the animals are hungry and thirsty, respectively. Further, the difference in the performance of animals with one or several days of food or water deprivation is not very great. If experiments are performed at regular intervals the routine also adds to the incentive value of the object.

From the experiments on maze performance it is difficult to learn whether the performance is at its maximum under any of the motivating conditions used. In some experiments two incentives were combined, with questionable improvement in performance. This would indicate that sufficient motivation is reached under the usual testing conditions.

In some experiments the question of motivation is of particular importance and distinct precautions must be taken. In comparing the abilities of animals of different ages; animals on normal and vitamin-deficient diets; animals with and without certain organs; and animals of different species, the differences in motivation may be very great.

For example, Lui (1928) in his study of the learning ability of rats of different ages found that young rats reached the criterion of learning more quickly than adult rats. As young rats require food more frequently than adult rats, the motivating conditions are quite different for rats of different ages. His conclusion that young rats learn better than adult rats must therefore be questioned. When Stone (1929*a* and *b*) partially controlled the difference in motivation by feeding the rats so as to maintain a normal weight for their age, the differences in maze performance of rats of various ages became negligible. In difficult test situations Maier (1932*c*) found young rats to be actually inferior to adult rats. Studies of the effects of drugs and vitamin-deficient diets on maze performance in animals persistently ignore the possible effect of these changes on motivation and attribute the difference in performance to a defect in learning ability.

The motivating conditions for different animal forms vary considerably. In higher animals, such as certain species of monkeys and children of certain ages, a satisfactory reward is often difficult to find. In the case of wild animals, fear is so great that controlled motivation is almost out of the question.

Attention and Distraction

Inattention Produced by Repetition.—In order that an animal respond to a particular characteristic of the experimental situation it is necessary that this characteristic be experienced. In the studies on form discrimination, for example (pp. 301*ff.*), it was found difficult to cause the rat to respond to form rather than to brightness. Different methods or techniques cause different factors in the

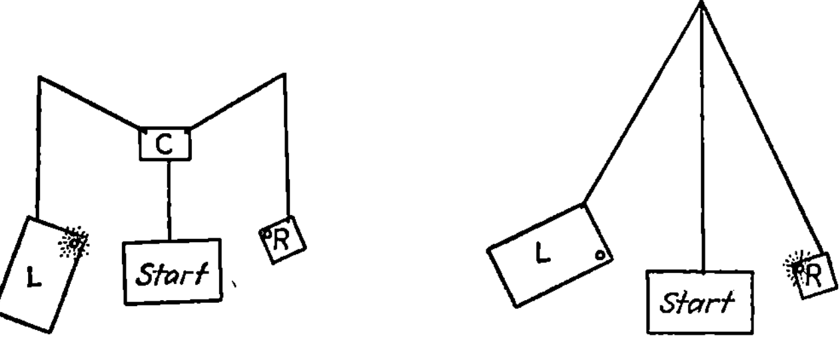

Fig. 93.—Apparatus used in attention studies. From the starting point the animal must either go to table *L* or *R*, depending on which of these has a light placed at *o*. (*Modified from Maier*, 1930, *p.* 291.)

experimental situation to predominate. For this reason the method used in an investigation becomes very important.

Even in situations where the problem of gaining the animal's attention has not been a difficulty, the problem of attention is present. Maier (1930) trained rats to go for food to an electric light wrapped in brown paper. After this preliminary training the animals were required to explore an arrangement of pathways and tables such as the two examples shown in Fig. 93. Following the period of exploration, food, as well as the light bulb, was placed on the inside corner of either the right or the left table. In turn, the rats were then placed on the center table from which the light was visible. In order to go to the food a rat had to respond to the position of the light before leaving the starting table. Ordinarily it ran to the corner of the starting table nearest the light and sniffed in that direction. For each trial, the food and light were on either the right or the left table. Nine trials were made each day and the experiment continued for varying periods of time on each of six different test situations similar to the one described.

The results may be summarized as follows:

1. In each of the test situations used, fewer errors were made on the first day of work than on succeeding days. The average number of errors for six test situations on different test days is shown in Fig. 94.

2. The number of errors made during the first four trials of a test period were only about one half as great as the number made during the last four trials of a day.

3. The errors made on runs after the food and light were changed from one side to the other were less than one half as great as the

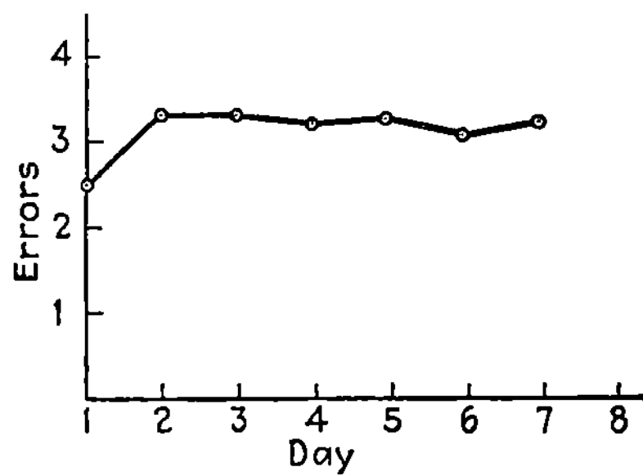

FIG. 94.—Curve showing the number of errors made on the same test given on successive days. The values are based upon the averages of six rats on six different experiments. *(From Maier, 1930, p. 298. By permission of the Journal of Genetic Psychology.)*

number of errors made on trials in which the food and light were left on the same table.

4. After a rest of several days, the performance rose to maximum efficiency.

These results are exactly the opposite of what one would expect on the basis of the usual understanding of learning. They become understandable when we consider the factor of attention. Attention is best in new situations and in situations involving a change, and it is in exactly these situations that performance was at its best. To respond correctly the rat had to attend to the light before starting.

Since repetition produces inattention, one might expect variability in behavior as an expression of a change in attention. Such behavior might be regarded as an escape from monotony.

Some observations of Lashley (1918) have a direct bearing on this point. He found that when first placed in a simple one-junction maze, 25 of his 52 rats took the correct turn. This is to be expected on the basis of chance. However, when the animals were again placed in the maze, only 8 of the 25 rats again took the correct turn. Most of the rats not only did not repeat their success, but did much more poorly than is to be expected on the basis of chance. The unexplored alley was apparently responded to on the second run.

Wingfield and Dennis (1934) found that in a maze which offered two routes to food, one of these routes was used predominantly when one trial was given per day, but when more trials were given per day this preference was lost and a tendency to alternate between the routes appeared. Repetition of trips apparently opposed the establishment of a preference for one route. Dennis and Sollenberger (1934) have shown that rats in a new maze tend to explore paths not recently occupied. As explorations continue, visits into all alleys decrease until activity ceases. Apparently a situation must offer variety or newness in order to produce responses from the animal.

These experiments support the contention that animals are subject to shifts in attention, and that consistent performance depends upon maintained attention.

In any experiment which requires attention, there is the possibility that performance will fall off if exposure to that situation continues. In maze learning, attention is required at the outset, but owing to the fact that the animal is learning the maze, its performance may show an improvement despite the fact that it is less attentive. When a maze is well learned, it tends to become automatized and attention becomes less and less important. This may explain why rats will make a large number of perfect runs in succession in simple mazes.

Inattention Produced by Distraction.—Distraction signifies that there has been a shift in attention. When a particular experience is dominant a distracting stimulus is one which will interfere with the previously dominant experience. Distraction is complete when the distracting stimulus becomes the dominant experience. A distracting stimulus may interfere with learning performance because an animal may cease to respond to that phase of the environment which results in adequate maze performance. In controlled experiments it is therefore necessary to exclude possible distractions of all sorts.

In a study of the effect of rickets on the maze performance of rats, Frank (1932) found a great variety of inconsistencies in her results. For example, when tests were made in one laboratory, rachitic rats were superior to underweight rats and these were superior to normal rats. In a different laboratory the rachitic rats were still superior, but the normal rats occupied the middle position. Further, all rats made decidedly better scores in the second laboratory than in the first, but all groups did not show the same amount of difference.

These and other inconsistencies became understandable when the possibility of distraction was considered. Since a distracting situation is one in which attention shifts from one aspect of the situation to another, the effects of distraction should be expressed by a variation in performance. By studying the degree of variation in the performance of these groups, it was found that all groups showed less variation in the second laboratory than in the first, but all did not show the same degree of variation. In fact, those that showed the least difference in the variability of their performance showed the least difference in score in the two laboratories. The apparent cause of the variation in performance was the laboratory environment. The first laboratory was near a shop which produced different kinds of noises, whereas the second laboratory was both quiet and isolated. Thus these two laboratories offered distinct differences in distractibility.

Further analysis of the data showed that normal rats were most subject to distraction and even showed variability in the second laboratory. Underweight rats were least distracted. If this distracting influence, as well as the difference in distractibility of the different groups, is taken into consideration, the above mentioned and other inconsistencies pointed out by Frank are easily understood. Because normal rats were most subject to distraction their performance in the two laboratories showed the greatest difference, and this resulted in a change in their relative position with respect to underweight rats whose scores in the two laboratories showed the least difference.

It is not surprising that underweight rats should be the least distracted. They are the most strongly motivated, and hence are least likely to be influenced by outside conditions. Rachitic rats are not underfed, but are on an inadequate diet. Their poor physical condition would tend to decrease the dominance of certain stimuli, but it would not make them as subject to distraction as normal animals. This analysis is further verified by the fact that after normal feeding, both the variability in performance and the difference between the groups decreased. Certainly it is obvious that the foregoing differences do not demonstrate differences in learning ability. Yet there are many studies which conclude that learning ability has been affected by diet on the basis of similar results.

Conclusion.—An error due to insufficient learning is one thing and an error due to inattention, distraction, or lack of motivation is another matter. Before it can be concluded that a difference in learning between groups has been found it must be shown that there is a difference in errors due to insufficient learning.

CHANCE

If one had two rats of exactly equal ability in every respect, they would not necessarily make the same score in the same maze test at a given time. Because equal scores are not necessarily obtained under such conditions, we recognize that to some extent the score is dependent upon unknown factors which we call chance. It is possible to conceive of one rat going through a maze without error, owing to a succession of lucky choices, and of another rat entering all blinds because of unlucky choices. Large mazes would, of course, reduce such probabilities. After the rat begins to learn the maze, the score becomes less a matter of chance, yet it must be granted that the score of any one rat is partly a picture of chance.

Because of this, a statistical treatment of data is essential to rule out the chance factor. Thus if we use two large groups of animals on a given problem rather than two animals, their scores tend to be equal. Not only does ability of large groups tend to be equal, as in the case of our two hypothetical rats, but also the average scores made by them. The extent to which chance determines a score may be detected by the lack of correlation between (1) the scores made on odd and even trials; (2) scores made during learning and relearning of the same maze after a certain interval of time; or (3) scores made by the same group of rats in two similar learning situations.

If the animal's behavior were entirely a matter of chance, the scores made on two successive trials would not be related. If both learning ability and chance played a role, the scores on successive trials would improve but not in a regular order. For this reason the improvement on the odd-numbered trials would not correspond to that on the even-numbered trials. The extent to which odd and even series of trials do not correspond thus becomes a measure of chance variation.

Learning and relearning records serve the same purpose in a similar manner. The testing instrument remains the same, but instead of comparing the trend of learning in odd and even trials, the trends of two learning occasions are compared. Chance factors would produce a lack of correspondence in these trends.

When two mazes are used to measure chance variation, two learning occasions are compared, but the material to be learned is different, and this difference may in part cause a lack of correspondence in the trends of learning. This method is therefore valid only when the learning of both mazes requires the same abilities.

Heron (1922) found chance to be a very important factor in determining the score in a problem box. It proved so important in the first two trials that he suggested that the scores made during these trials be disregarded. Hunter (1922) compared the learning scores made by rats in three different types of mazes, and found chance to be especially important during the early stages of learning.

On the other hand, Tryon (1930) found the correlation between the scores made by rats on odd and even trials to be as high as 0.9876 and 0.9682 on two different mazes. This means that the score made on each trial is a very reliable measure, and that chance variations in score are negligible. Tryon (1931*a*) also compared the scores made by rats on two similar mazes and found the correlation coefficient to be 0.8. In other words, the order of ability of the different rats as detected by one maze was very nearly equal to the order of ability as detected by the other maze. If chance contributed greatly to the score a marked correlation of this sort could not be obtained. When learning and relearning scores were studied (Tryon, 1931*b*), the correlation coefficient was found to be between 0.81 and 0.88. Tryon's studies indicate that when proper precautions are taken, the score made by any individual animal in a maze depends upon the animal's ability to learn that maze. Since Hunter used mazes which differed both in type and in complexity, the lack of correlation found by him may have been due to the fact that these different mazes did not measure exactly the same abilities. The evidence on the maze when considered as a whole seems to indicate that when it is carefully used, it is a consistent and reliable instrument, and that the performance of animals tested by it can be controlled. The problem box perhaps involves chance to a much greater degree, because the factor of selection is much greater. As we have already pointed out (p. 362), selection involves trial and error to a considerable extent, and for this reason chance plays a very important part.

Whether chance plays a major or a minor role in determining the score obtained by means of any testing instrument, it is always important to consider chance as a possible factor in an animal's performance when interpreting results. Whatever an animal does is of interest and importance to the animal psychologist and chance embodies the many variable factors which underlie an animal's behavior.

Suggested Readings

Maier, N. R. F. 1930. Attention and inattention in rats. *Jour. Genet. Psychol.*, 38, pp. 288–306.

Stone, C. P. 1934. Motivation: drives and incentives. *Comparative Psychology* (ed. Moss). Pp. 73–112. New York: Prentice-Hall.

Tolman, E. C. 1932. *Purposive behavior in animals and men.* Pp. 39–83. New York: Century.

Tryon, R. C. 1934. Individual differences. *Comparative Psychology* (ed. Moss). Pp. 409–448. New York: Appleton-Century.

CHAPTER XIX

FACTORS AFFECTING THE MASTERY OF A TASK AND THEIR RELATION TO LEARNING

We have found that certain conditions must be satisfied before learning takes place, and that after learning has taken place, certain other conditions determine an animal's performance in the learning situation. We must now consider the conditions which affect the extent or the degree of learning. Optimal conditions for learning may facilitate the mastery of a task by (1) increasing the rate of learning; (2) reducing the selective process through controlled attention (thereby making every exposure to a situation a dominant experience); or (3) eliminating the learning of certain elements which have no bearing on the task. Only the first of these possibilities is concerned with the actual learning process. Since learning tasks involve selection and intelligence, as well as the formation of associations, it is necessary to analyze the various conditions carefully in order to determine their effect upon the mastery of a task and their possible effect upon learning as such.

FREQUENCY

The Relation between Frequency and the Association Process.—Obviously many repetitions of an act to be learned are more effective than a few. The increase in learning resulting from the greater number of repetitions may be due either (*a*) to the learning of *more* items, or (*b*) to the better learning of *each* item. For example, if 10 trials in a maze make it possible for an animal to go through it with five errors, and 20 trials with two errors, this reduction in the number of errors may be due either to the learning of more parts of the maze, or to the fact that all parts have been better learned. If the first alternative holds, frequency is not a factor affecting the strength of an association but functions merely as a means for giving the animal more opportunities to form additional necessary associations. Undoubtedly this is often the case, especially in "trial-and-error" learning situations. For many trials there may be no progress, then after some particular trial a pertinent association is formed and a great deal of progress is evidenced.

However, there is also reason to believe that associations already in existence may be strengthened by further repetitions. In Pavlov's conditioned-reflex studies there is evidence of the formation of an association as soon as the unconditioned stimulus produces a flow of saliva, but further repetitions result in an increase in the quantity of saliva produced. This suggests that the association has increased in strength because of the further repetitions.

This suggestion is supported by the results of the studies on the "overlearning" of mazes. An animal that has learned the maze to the point of making no errors (thus having formed all necessary associations) will again make errors if tested some time later; but if the animal is given many extra trials after it has learned the maze to perfection, it will make fewer errors on the retention test. Apparently the extra repetitions ("overlearning") strengthen the learning of all items so that forgetting is not so evident.

Frequency Not a Basic Factor in Selection.—Frequency becomes a controversial factor in learning when it is used to account for *selection.* Watson (1914) used it to explain *how* an animal learned to open a puzzle box and to run a maze without errors. He argued that every time the animal opens a puzzle box, or reaches the food box in a maze, it must go through the correct acts, whether or not it has gone through the various incorrect ones. In the course of a sufficient number of trials the necessary acts must therefore be repeated more often than the unnecessary acts. This greater frequency of the necessary acts causes them to be more strongly associated with the situation; and because of their greater strength these acts dominate in behavior when the animal is placed in the test situation.

This theory makes learning purely a matter of dominance. All acts tend to be repeated, but as the unnecessary acts are opposed to the necessary ones and weakly associated with the situation, they do not come to expression. "Learning not to do a thing" plays no part in the scheme.

At first glance Watson's logic seems very plausible. Certainly the correct act *must* appear once in every trial. But an incorrect act may appear an indefinite number of times in a trial since it does not terminate the trial, as was pointed out by Thorndike (1915) in his review of Watson's book. Peterson (1917*b*) demonstrated that in maze experiments rats actually must eliminate incorrect responses in order that the correct ones can eventually become more frequent. Even at the very beginning of learning, frequency does not determine an animal's choice. As we have already stated in another connection, Lashley (1918) found that on the first trial in a maze with only one blind alley, 25 out of 52 rats made the correct turn, but only 8 of these 25 made the correct turn on the second trial. The unexplored turn which had no repetitions in its favor dominated the behavior on the second trial.

Maier (1929*a*) trained rats to go to whichever of two tables contained a light. These tables were accessible by way of elevated pathways which led from the starting table. The light was twice as often on one table as on the other and consequently the route to the one table was used twice as frequently as the route to the other table. Usually the rats made the correct choices on the basis of the light, but when errors were made it was found that more of them were due to the visiting of the table that was less frequently used. In this experiment both tables were familiar and both were associated with food. If frequency tended to produce a preference for a choice, it should have done so in this case. The opposite, however, was actually the case, showing that even here some other tendency dominated the preferences.

That the frequency with which an act is executed does not determine its preference in the animal's repertoire of acts is therefore obvious; but this does not eliminate it as an important factor in learning. When animals do not behave as one might expect on the basis of frequency, some other factors may be operating and determining the preference. If frequency by itself produces any preference, it is usually overshadowed by other factors which also determine the preference.

In all experiments in which the *consequences* of various choices differ, the preference of the animal is governed by the consequences of the acts rather than their frequency. For example, Kuo (1922) found that when there were various alternative routes to food, rats chose the short route in preference to the long route, the route involving an electric shock, or the route involving confinement for 20 sec., despite the fact that at the outset the other routes were more frequently taken.

Frequency May Be a Primitive Selective Factor.—Of interest is the fact that normal and partially decorticated rats differ in the way they are affected by frequency. Buytendijk (1932) found that normal rats trained to go to the right for food in a T maze easily learned to go to the left for food when it was changed to that side. Rats with cortical lesions, however, were given considerable trouble by the change in the position of the food. In their case the old habit persisted and was hard to break. Cameron (1928) found operated rats less adaptive to changes introduced in a previously learned maze. In the light of the above results this failure to adapt seems to be due to a persistence of old habits. Maier (1932*a*) has also shown that operated rats are much more stereotyped in their behavior than are normal rats. Stereotyped behavior indicates the dominance of frequency irrespective of the outcome of an act. A rat with an extensive lesion will repeat an error until it is exhausted without taking an alternative route which leads to food. Because rats with cortical lesions are more dominated by frequency than normal rats, it is probable that frequency is a primitive selective factor in learning.[1] When the animal has developed to the stage where it can effectively discriminate between various consequences, frequency ceases to be the important factor in selection.

In learning situations, such as conditioning, where consequences are not associated with the response it is probable that frequency plays a more important role as a selective factor.

RECENCY

The Relation between Recency and the Formation of Associations.—With the passing of time, learned acts (if not repeated) gradually become unlearned. Because *forgetting* takes place, recently acquired acts are more effectively reproduced than less recently acquired acts. A dog that has been conditioned to respond with 8 drops of saliva to the sound of a bell may only produce 4 drops if this response to the bell is left unused for a period of a month. Forgetting may thus be regarded as a gradual weakening of the changes produced through learning.[2]

[1] We have described the formation of an association in normal animals between the experience of a response to a certain stimulus and the experience of food. It is probable that in operated animals the dominant association is between the stimulus and the response to it.

[2] An alternative interpretation which undoubtedly holds in many instances is that forgetting involves the acquisition of opposed responses. An interval of time presents ample opportunities for such acquisition.

Recency is also important in determining the number of repetitions required for learning. As we have seen in Chap. XV, in order that a conditioned response be readily established the conditioned stimulus must not precede the unconditioned stimulus by more than a few seconds. When the interval between the presentation of the two stimuli is very short, fewer trials are required to set up the conditioned response than when the interval is longer. It is possible to argue that when the conditioned stimulus is very recent, each trial produces more learning than when the conditioned stimulus is less recent. However, it may also be argued that only part of the trials are effective when the conditioned stimulus is not recent enough, but that each effective trial produces the same amount of learning no matter what the recency. The latter theory makes recency a *condition* for learning, in that it greatly determines whether the two stimuli to be associated are experienced together. Whether various degrees of proximity of stimuli produce various *degrees* of learning, or whether proximity determines the *number* of effective repetitions, cannot be determined from the experimental literature. For practical purposes the solution of this problem is unimportant, but for theoretical purposes a solution is very desirable.

Recency Not a Factor in Selection.—Like frequency, recency has been used as a factor to account for selection in learning. Watson (1914) argued that the act which results in a reward is the last or most recent one and because of its recency it becomes associated with the reward. On the basis of the same logic it is also reasonable to supplement the theory by adding that acts which are unnecessary in the obtaining of the reward become associated with "no reward" and consequently such acts are learned and avoided. This being the case, recency does not determine which associations will be formed since each act is recent with respect to some consequence. It is true that only some of these come to expression in a situation. Which associations are expressed is therefore not determined by recency but by the nature of the consequences with which they are associated. The animal probably learns what not to do as well as what to do in a preference situation.

To postulate that recency produces learning in some cases and not in others is open to criticism, unless it can be shown that the conditions for learning are not satisfied in all cases. Carr (1914) has suggested the *principle of intensity* to do this. If we regard intensity as a factor determining whether or not the stimuli will dominate, and hence satisfy one of the conditions necessary for learning, then we can explain why all recent stimuli are not associated with the situation. But with this addition, learning is still a matter of avoidance as well as attraction and is not the acquisition of necessary acts and the forgetting of unnecessary acts. In the next section intensity will be considered as a separate principle in selection.

Experimental results demonstrate that recency, even with the aid of frequency, is ineffective in determining what responses will dominate in a learning situation. In Kuo's study (1922) already referred to (p. 429) rats chose the short route to food even though food had been just previously obtained over a different route.

Conclusion.—To summarize, it may be said that the relation of recency to learning *depends upon the meaning with which the word is used.* When used to describe the temporal proximity between two experiences which are to be associated, it becomes a necessary prerequisite to learning. When it designates that an association is new, it is related to retention in that the degree of retention of a learned act is dependent upon the recency with which the act was learned. It probably plays no part in the process of selection.

EFFECT

Thorndike (1911) believes that the effect or the consequences of an act largely determine whether or not it will be repeated on later occasions. His statement of the *law of effect* is as follows:

> Of several responses made to the same situation, those which are accompanied or closely followed by satisfaction to the animal will, other things being equal, be more firmly connected with the situation, so that when it recurs, they will be more likely to recur; those which are accompanied or closely followed by discomfort to the animal will, other things being equal, have their connections with the situation weakened, so that, when it recurs, they will be less likely to occur. The greater the satisfaction or discomfort, the greater the strengthening or weakening of the bond (p. 244).

Thorndike, like Watson, regards the unnecessary acts as not coming to expression because they gradually lose their association with the situation. According to this theory, associations are formed only when some pleasant experiences are connected with the situation.

There is no evidence which shows that only pleasant consequences of acts produce learning, much less that unpleasant consequences weaken learning. Rather the animal learns in all cases to associate the consequences of an act with the act. This learning results in the avoidance of acts which produce unpleasantness and the repetition of acts which produce pleasantness. The consequences thus determine the preference and are not a condition for learning. This explains the better performance of animals in situations which are well motivated.

Muenzinger (1934) has recently presented evidence which shows that it is not the pleasantness or unpleasantness of an experience which determines an animal's efficiency in a learning situation, but rather the emphasis which either gives to the situation. In training rats to form a black-white discrimination habit, some of the rats were given a mild electric shock in connection with incorrect responses, some, in connection with correct responses; while the remaining rats were not punished at all. The two shocked groups of rats were equal to each other and superior to the unshocked group of rats. These results show that the emphasis produced more rapid mastery of the task either (1) because more learning resulted from each repetition; or (2) because better attention was obtained and more of the trials became a repetition in experience. The second of these alternatives is preferable in the light of the foregoing discussion.

That the first alternative is without support from experimental findings is indicated by the results of Ni (1934). He found that rats punished for errors in their learning of a first maze were not only superior to unpunished rats which learned the same maze, but were also superior to them in a second maze in which they were no longer punished for errors. Punishment could not have increased the learning rate of the second maze since it was not present. But punishment did undoubtedly emphasize the difference between correct and incorrect choices and thus caused the animals to be more cautious in future maze performance. For such rats each trial became an effective repetition, and so increased the rate at which they mastered the task without increasing their ability to more quickly form associations. Theoretically, the assumption that certain conditions produce more rapid learning and the assumption that certain conditions determine whether or not an exposure to a situation will produce a dominant experience, have quite different consequences. The first implies that the plasticity or modifiability of the nervous system varies under different conditions, whereas the second statement implies that the plasticity remains constant.

Motivation

In our discussion of motivation (Chap. XVIII) evidence was presented to show that it played an important role in determining performance. This does not exclude the possibility of an effect upon learning. Certainly a motivating situation produces attention and emphasis. Whether it actually affects the rate of formation of associations or whether its effect can be reduced to frequency must be determined by future investigation.

Past Experience

Experiments have shown that learning in one situation frequently results in a more rapid mastery of a succeeding learning situation. This influence of one situation on another is generally regarded as a *transfer of training*. But it is conceivable that the learning of one situation may be detrimental as well as beneficial to the acquisition of another and we may therefore speak of positive and negative transfer. The things learned in one situation which may effect the mastery of another situation may be of three kinds: (1) secondary, (2) specific, and (3) general.

Secondary Effects.—In Chap. XVI we pointed out *what* an animal learns in various learning situations. In every learning situation certain secondary features of the situation are learned and these may be common to all learning situations. Wiltbank (1919) regards the recession of timidity and the formation of the food association as important factors determining some of his transfer effects in mazes. Dashiell (1920*b*) experimentally tested the influence of mere adaptation to the general learning situation (by permitting rats to experience pushing open a door into a food box) and adaptation to the alley situation (by letting rats run about in alleys), and found that such experiences reduced the learning time of mazes.

The foregoing examples of changes produced in the animal, which favorably affect the mastery of a task are not true cases of transfer. Adaptation to a maze situation and mastering a route to food are different things, and if part of the adaptation has previously taken place, this cannot be construed as an effect on the acquisition of the route to food. All laboratory situations have certain features in common. As far as these features are concerned, the situation remains the same as we go from one learning situation to another.

Before one can speak of transfer from one situation to another, the learning which is transferred must affect the actual task to be learned.

Specific Effects.—There are specific elements in the learning of any task which may influence the acquisition of a difference task. Wiltbank (1919) points out that after learning one maze, his rats had tendencies to enter some pathways and to avoid others. Any new learning situation which is similar in certain respects to an old one might be expected to be favorably influenced by it, because certain parts of the new situation are not new. Webb (1917) found that learning one maze greatly facilitated the mastery of similar mazes. There was also some saving when the mazes were just opposite in the reactions they required. Because a saving was present in both types of mazes, it is obvious that much of the transfer was due to secondary effects. Since Webb found a correlation between the degree of transfer and the extent of similarity in pattern between the first and second mazes, the indication is that some specific reactions acquired in one situation are applicable to other situations.

As the maze habit is an integration of a whole series of reactions, this application of specific details may not be in proportion to the number of common details. In fact, if two mazes are alike in all respects except one, great confusion arises because the whole maze habit tends to be transferred.

A number of studies intending to show a negative transfer of specific elements have been performed. In such studies the animal is trained to go toward one member of a pair of stimuli (such as visual or auditory stimuli) in order to find food. After the response is learned the conditions are reversed and the animal is rewarded only when it goes to the stimulus which it has been trained to avoid. These studies cannot be regarded as studies on transfer because the whole stimulus-situation remains unchanged. Rather the animal is required *to learn to change* its reaction. That animals actually do learn to reverse their reactions is indicated by the results of Buytendijk (1930) who found that when he continued to reverse the conditions, the rats made the adjustment more and more readily. Schneirla (1929) reported the same findings for ants.

General Effects.—In Dashiell's (1920*b*) list of factors accounting for transfer, two are of a general nature. The first of these is a tendency not to repeat the same error in a single run. When rats were caused to run a different maze every day but learned none of the mazes completely, they learned to reduce their errors by not making the same mistake twice in one trial. Experience in mazes thus taught the animals a technique which was useful. The second of these factors is a tendency on the part of animals to orient themselves with reference to the food box. As the starting point and the food box occupied the same relative position in all of the mazes that the rats ran through on different days, a general impression of the position of the food box was gained. This was indicated by the fact that the rats tended to turn into alleys which opened in the direction of the food box. In this case the animals gained knowledge of a general form which was applied to new situations. To what extent animals are capable of forming generalizations will be discussed under abstraction in Chap. XX. At present it is necessary only to point out that generalizations formed in previous situations will affect the learning behavior in other situations.

Neither of the above types of factors can be construed as affecting learning ability. Rather they cause an animal to attend to certain important features of a learning situation. (In problem-box learning similar techniques are developed.) In this manner the process of selection is reduced, but the actual formation of associations need not be affected. With a reduction in selection the mastery of a task is more readily achieved.

Tendencies developed in animals such as taking all turns in a certain direction, alternately taking turns to the right and left, avoiding all turns and continuing forward at junctions, making all turns, and so forth, are cases of the application of generalized experiences which may reduce the selective process or increase it in case they are wrong.

Transfer of Training and Its Relation to the Equivalence of Stimuli.—Testing the influence of past experience on the mastery of new situations by measuring the saving in learning in the new situation does not permit a very satisfactory analysis of what is actually taking place. It seems that before we determine whether the learning acquired in one situation affects other situations we should first determine *what* has been learned in the first situation. Klüver (1933) has made extensive investigations on monkeys in order to determine what characteristics make various stimulus situations equivalent to one another. He finds that if monkeys have learned to react to certain stimuli in a particular way, this learning determines their reaction to a large number of other stimuli. All new stimulus situations which call out a reaction previously learned are regarded as equivalent with the original stimulus situation. When one approaches the problem in this manner, the question of transfer does not arise. The animal has learned to give a certain reaction to a particular aspect of a sensory pattern and whenever this aspect is reproduced by a new situation, the learned reaction is reproduced. Past experience thus affects an animal's reactions in future situations whenever these future situations contain the sensory aspects to which the animal has learned to react.

In Fig. 76, all of the stimuli labeled "equivalent" produced the same reaction, those labeled "nonequivalent" produced a different reaction. To produce the same reaction all equivalent stimuli must contain the common sensory aspect which is the basis for the response.

Meaning and Organization

In human psychology a distinction is made between sense and nonsense learning. We are told that words are easier to memorize than nonsense syllables because they have sense or meaning. This, of course, is not an explanation but merely raises a problem. Köhler has raised the same problem in the case of apes. In an experiment in which a banana was placed outside the cage and out of reach, the chimpanzees failed to solve the problem because they failed to fit together two hollow sticks of wood (which were lying inside the cage) in order to make a long stick which could have been used to pull in the banana. After failing in the problem, one of the animals was seen to play with the sticks. In its play it happened to fit one stick inside the other. The long stick produced accidentally in this manner was immediately used to obtain the banana. On future occasions this solution was readily repeated. After but one repetition the solution was learned. The accident, one might say, formed a meaningful organization. As these animals solve new problems without any previous repetitions it is difficult to know whether certain combinations are quickly learned because they form meaningful organizations or whether the formation of such organizations is a product of intelligence and therefore not a matter of learning.

A chimpanzee will solve in much less time, a problem in which the releasing of a loop causes a banana to fall to the ground when the manner in which the banana is suspended is visible, than it will open a problem box which requires that a loop be unfastened. A situation in which certain relations are visible to the animal, Köhler maintains, has "sense" for the animal.

Because situations which have "sense" for the animal are more readily mastered than situations which do not, it is probable that intelligence is often playing a part in learning. When a series of associations are formed in a learning problem, the elements in the pattern of associations acquire a function or a meaning in the same way that jumping at a triangle means food to the rat in a discrimination experiment. The function or meaning of elements in a pattern may therefore be regarded as dependent upon learning. If the elements to be associated have sense or meaning before learning has begun, then this meaning must either (*a*) have been previously acquired in other association patterns, or (*b*) it must have arisen from a pattern spontaneously formed at that moment. If the former possibility is true, we are dealing with the transfer of past experience; if the latter is true, we are dealing with a process other than learning. Transfer has already been discussed and as the formation of spontaneous patterns of behavior will be treated in the next chapter, it is not necessary to consider it at this time except to indicate that meaning, no matter how it arises, aids in the mastery of certain types of tasks.

The Distribution of Effort

The learning of a task may be confined to a short period of time by concentrating or accumulating the repetitions; or it may be spread over a period of time by inserting rest periods between some or all of the repetitions, thus dividing or distributing the repetitions. In the case of human beings, it is quite generally agreed that distributing the repetitions in learning is more economical (in that fewer trials are required to learn) than accumulating or concentrating the repetitions within a short period of time. As more physical work can be done before fatigue is reached, when effort is gradually rather than rapidly expended, it might be supposed that the gradual expenditure of learning effort leads to more learning per unit of effort than does the rapid expenditure of learning effort. Experimental results alone can determine the correctness of this supposition.

Massed versus Distributed Repetitions.—For problem-box learning Ulrich (1915) found that the order of efficiency of his different distribution of trials was as follows: 1 trial every third day;

1 trial every other day; 1 trial every day; 3 trials in succession every day; and 5 trials in succession every day. The most favorable distribution of trials required from 11 to 17 trials, the least favorable from 35 to 50 trials before the rat was able to open a latch box in 1 sec. on several successive trials. Similar differences in efficacy of the above distributions of trials were obtained for the learning of a maze.

TABLE 29.—AVERAGE NUMBER OF TRIALS REQUIRED TO LEARN A MAZE WHEN EFFORT IS DISTRIBUTED IN DIFFERENT WAYS
(Data from Mayer and Stone, 1931)

Number of cases	Distribution of trials	Average number of trials required to reach criterion of learning	
		Lenient criterion	Severe criterion
Young rats			
47	1 per day	13.2	21.6
40	3 per day	13.5	22.5
29	5 per day	15.0	20.8
31	10 per day	13.0	21.3
Adult rats			
17	1 per day	17.0	23.3
34	3 per day	22.5	28.0
23	5 per day	20.9	30.0

Warden (1923*b*) likewise found that in maze learning an interval between each trial was more efficient than a like interval placed between groups of 3 or 5 trials. He tested various intervals and found that an interval of 12 hr. was more efficient in all cases than longer and shorter intervals.

On the other hand, Pechstein (1921) found that massed trials resulted in the mastery of a small maze more readily than did distributed trials, and attributed this effect to his use of very simple mazes. Cook (1928) found no reliable difference between the procedures of massed and distributed effort. He believes that the inconsistencies in the various studies can be accounted for if we regard massed effort the better procedure in simple problems; distributed effort the better for complex problems; and neither superior for problems of intermediate difficulties. He considers his maze intermediate in difficulty between those of Warden and Pechstein.

Other Conditions Vary When Effort Is Differently Distributed.—Mayer and Stone (1931), using a maze which was perhaps as complex as used in any of the above studies, found that the relative efficiency of massed and distributed repetitions was different for adult than for young (about one month old) rats. Trials were given at the rate of 1, 3, 5, or 10 per day. On these various distributions of repetitions the young rats did equally well. This was true whether one used a lenient criterion (3 or less errors in 3 consecutive trials) or a severe criterion (1 or no errors in 4 consecutive trials). In the case of adult rats the lenient criterion likewise showed no difference between groups of rats learning by different distributions of trials, but when the severe criterion was used, the efficiency became less as greater numbers of trials were given per day. Table 29 gives the number of trials required by young and adult rats to reach the two criteria of learning. It was further observed that both young and adult rats tended to retrace parts of the maze when successive trials were given.

The results of Mayer and Stone suggest that the difference between the effect of massed and distributed repetitions may be purely one of performance. Several trials in succession may lead to inattention if the animal is not sufficiently motivated. Young rats are likely to be more interested in food and so maintain a high efficiency of performance. (From Table 29 it will be seen that under identical learning conditions young require fewer trials than adult rats.) It seems that adult rats, being less motivated, become inattentive to the extent of making an error now and then. When a high criterion of learning is used, such slips in attention become important. We have seen in Chap. XVIII that several trials in succession favor inattention, and as a consequence, the accumulated repetitions result in less efficient maze performance. The fact that massed repetitions make for retracing also indicates inattention. In order for an animal not to enter an alley it has only previously left, maintained attention is required. The difference found by Warden in the effectiveness of various intervals is likely to be due to the differences in motivation which necessarily must accompany the use of different intervals.

Whether the complexity of the maze is actually a factor cannot be determined. No complex learning problem has been used in any of the studies. It is to be expected that complex learning situations require a greater amount of sustained attention, and consequently the method of massed trials in such situations should result in somewhat inefficient performance; and perhaps in inefficient learning in that attention is a required condition for the formation of associations as well as for selection.

When we consider that massed and distributed practice are associated with many indirect influences such as those having to do with differences in attention, motivation, age, the complexity of the learning task, and the severity of the criterion used, there seems to be little reason to believe that the distribution of effort as such is a factor concerned with the mastery of a learning problem. Ulrich (1915) believed that the learning rate actually was better when effort was distributed, and suggested that massed effort perhaps involved rapid metabolic changes which hindered learning. Although studies on distributions of effort may give us no insight into the mechanisms involved in learning, they have a practical value in that they may indicate which procedures introduce an unknown group of factors which increase the efficiency of mastering a task.

Part-whole Learning

Like the problem of the distribution of effort, the order of presentation of material is a practical problem which concerns primarily the educational psychologist. Is it better from the standpoint of efficiency to repeat the whole of the material to be learned before repeating any part twice; or is it better to repeat a certain part until it is learned, then go on to the next part, and so on until all the parts have been separately learned? It is obvious that the second method requires some extra repetitions in which the parts separately learned are fitted together. These two types of approaches to a task to be learned have in themselves no effect upon learning, but each method is necessarily associated with different factors which may have a bearing on learning efficiency. Since these factors have not been isolated and measured in any of the investigations, a comparison of the two methods to determine which leads to better mastery of a task gives us the resultant effect of these several unknown factors.

Accompanying conditions which probably favor the *whole* method are: (1) the total required integration is built up from the outset and there is no need for learning to combine separately learned parts; (2) sustained attention because a part is not repeated over and over; and (3) the terminal and the starting points remain the same throughout the learning and there is less likelihood of forming unnecessary associations with various beginnings and starting points.

The *part* method may involve such favorable conditions as: (1) sustained attention on a limited portion of the material; (2) the possibility for more frequent repetitions of the more difficult parts; and (3) less new material to learn at a time and therefore less possibility for confusion.

In studies using human beings as subjects the *whole* method is generally regarded as superior to the *part* method. Pechstein (1917) raised the issue with the rat in maze learning by comparing not only *part* and *whole* methods, but also combinations of these (such as learning part 1, then part 2, then parts 1 and 2 together, then part 3, then parts 1, 2, and 3 together, etc., until four parts had been learned and integrated). He found that a combination of the part-whole procedure was the most efficient; that learning by the whole procedure was the least efficient; and that learning by part was between these in efficiency. When retracing in the maze was not permitted, however, the *whole* method was superior to the *part* method.

Hanawalt (1931, 1933) found the order of merit of these three procedures to be exactly the reverse of Pechstein's order. She found the *whole* method the most efficient, the *part* method the least, and a combination of the methods intermediate. Hanawalt's experiments were more carefully controlled than those of Pechstein in that she (1) used maze-experienced animals; (2) excluded individual differences by having all animals learn one maze by each method; and (3) used mazes in which the parts were decidedly different and hence less likely to cause confusion (or negative transfer) when the animal ran from one part of the maze to another. Her results, therefore, carry more weight and indicate that the resultant of the factors introduced by each of the respective methods favors the *whole* method.

It is obvious that although the mastery of a task is affected by the different procedures just discussed, the actual factors which cause the differences are unknown. Even such differences in procedure as employed by Pechstein and Hanawalt lead to reversed results. Pechstein found in his own results that the presence or absence of retracing altered his results. What effect these various procedures have on (1) tasks having various degrees of complexity (2) tasks involving greater or lesser degrees of selection, and (3) tests made on animals under different degrees of motivation, is still unknown.

Suggested Readings

HUNTER, W. S. 1934. Experimental studies of learning. *Handbook of general experimental psychology* (ed. by Murchison). Pp. 497-570. Worcester: Clark Univ. Press.

KLÜVER, H. 1933. *Behavior mechanisms in monkeys*. Pp. 19–226. Chicago: Univ. Chicago Press.

MUNN, N. L. 1933. *An introduction to animal psychology*. Pp. 233–290. Boston: Houghton Mifflin.

WATSON, J. B. 1914. *Behavior: an introduction to comparative psychology*. Pp. 251–276. New York: Holt.

CHAPTER XX

HIGHER MENTAL PROCESSES

INTRODUCTION

Previous to the statement of the theory of evolution lower animals were regarded as living machines having no soul or reason. Man was considered qualitatively different from them in that he possessed a soul and was a rational being. With the advent of the theory of evolution the gap between man and other animals became less marked; and it was held that all of the characteristics of man had their beginnings in lower forms. This was taken to mean that "mind" and "reason" must also have sprung from less gifted beings and must therefore be present to some degree in some animal forms at least. Many psychologists as a consequence became interested in discovering the evidence of mind in animals, as is shown for example, by the title of Washburn's textbook of animal psychology, *The Animal Mind*, first published in 1908. The term *mind*, which to philosophers meant something nonmaterial and was related to the soul, was largely accepted by psychologists to designate consciousness. Although the study of conscious states occupied psychologists in their study of human beings, it was rather empty as far as the study of animals was concerned. Mind could be inferred in their case, but it could not be investigated. Interest in animal psychology was therefore confined to behavior, and through the influence of Watson (1914) the emphasis on behavior even influenced psychological investigations of man.

During the last 35 years a number of controlled experimental techniques have been developed which represent attempts to throw light upon the higher mental capacities of animals. Since learning consists in the formation of associations through repetition, higher processes must involve more than this; and for this reason, experiments demonstrating the existence of higher processes must be designed to exclude learning as the determining process in the animal's behavior.

The attempt at solving a problem is usually characterized by a certain amount of stereotyped behavior. An animal in a problem box repeatedly makes the same errors. This behavior is generally traceable in the animal's past. Struggling is a primitive reaction to confinement and is often successful. It is called out in the problem box and it appears in all novel situations in which the animal's behavior is obstructed. Responses acquired in past situations also are applied to new situations. All of this activity may be termed *trial-and-error* behavior in the sense that it is not a specific response to the situation.

We have already referred to Krechevsky's studies (Chap. XVI) in which he found that rats respond rather systematically, first to one characteristic of the situation, then to another, and often return to an earlier response. Variation in behavior appears in this manner and increases the possibility of success in a novel situation. Of interest is the fact that operated rats (Krechevsky, 1934) show such variation to a much lesser degree. Generally speaking, cortical destruction increases the amount of stereotypy in behavior. Cameron (1928) and Buytendijk (1931) found operated rats less able to break a habit and form a new one than normal rats. Maier (1932*a*) found that operated rats which made an error in one trial tended to repeat that error on following trials, whereas if they ran correctly on the first trial, following runs were also likely to be correct. This tendency increases with the amount of cortical destruction. Some rats have been observed to repeat an error 30 times. This persistence in errors not only characterizes the behavior of mammals with cortical lesions but also that of the lower forms treated in Part I.

Stereotypy in behavior shows the persistence of past experience. In novel problems the organism must often inhibit the influence of past experience. Inability to inhibit previously acquired tendencies often prevents the animal from obtaining a goal which might have been reached had there been more variety in the animal's repertoire of acts.

Although variation in behavior increases the animal's chances of solving a problem, the solution is still a matter of "trial and error" and does not satisfy the usual criteria for an "intelligent" solution. Since man was considered a rational being and since his most pronounced conscious states were characterized by thoughts, it is natural that the presence of "ideas" should become identified with the higher processes. The thought processes have always been regarded as higher and different from the learning process. It was generally conceded that man solved problems by "ideas." If the presence of "ideas" could be demonstrated in animals below man, then one would have evidence of higher processes in them. As we shall see, most of the tests of higher processes assume the validity of this criterion, but they differ in the method of demonstrating "ideas."

The foregoing discussion assumes that conscious states are determining factors in behavior, but this assumption is without foundation. It is therefore necessary to be critical of all experimental approaches which make this assumption. A system of animal psychology must be constructed from behavior data, and complex processes are no exception to this rule. It is therefore our task to develop an objective definition of, as well as objective criteria for, higher processes. Before doing this, however, we shall examine the various experimental approaches to the study of higher processes.

The Delayed Reaction

Introduction.—A method for determining the presence of ideas in animals was developed by Carr at Chicago and has become known as the *delayed-reaction method.* The first study reported which utilized this method is that of Hunter (1913). Since then the method and many variations of it have been used to measure the relative mentality of mammals of the same and of different species. It is widely accepted as a valid measure of symbolic behavior.

An animal can be trained to react in a certain way to the presentation of a stimulus. If, however, this stimulus is removed before the animal has been permitted to react, is the animal capable of reacting as if the stimulus were still present? If so, the reaction cannot be in response to a stimulus which is not present but must be in response to something which has taken the place of the stimulus, *i.e.*, an "idea." A reaction which is delayed for a period after the stimulus has been removed is a delayed reaction. The length of time that an animal can delay its reaction then becomes a quantitative measure of the animal's capacity to substitute something for the stimulus.

Hunter (1913) devised the delayed-reaction situation by first training animals to go to the one of three boxes which contained a lighted electric lamp. His apparatus is shown in Fig. 95. The lighted lamp was always in the box containing food, but the food-box was varied from trial to trial so that each box contained food at one time or another. This preliminary training established the light as the stimulus to which the animal reacted. After the reaction to the light was mastered, the stimulus (indicating which of the three boxes contained food) was presented only momentarily and the animal was restrained in a glass starting box. At the end of the restraining period the animal was released in order to determine whether or not it could go directly to the correct box. By varying the length of confinement in the starting box, the extent of an animal's ability to delay its reaction was measured.

According to Hunter (1917, p. 76) the only necessary features of an apparatus for measuring the delayed reaction are as follows:

1. It must be adapted to the size of the subject and to its mode of response—walking, reaching, swimming, or flying.

2. It must provide a means for presenting a stimulus in one of several places.

3. These stimulus positions must be equally accessible to the response.

4. The stimulus and the method employed should be such as to present no differential cues to the subject during the interval of delay.

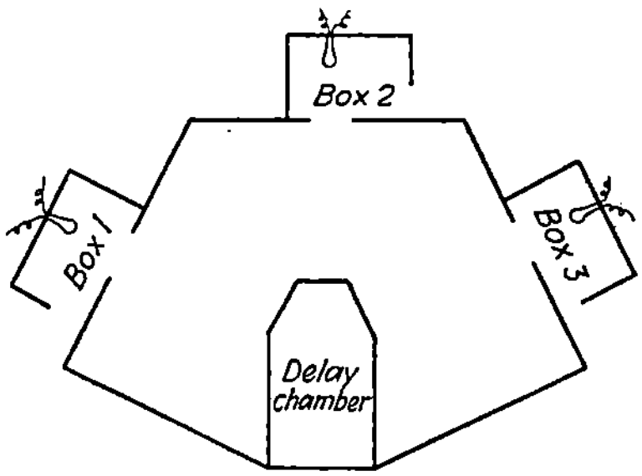

FIG. 95.—Plan of Hunter's "delayed-reaction" apparatus. While the animal is confined in the glass delay chamber an electric light is momentarily turned on in one of the boxes. Upon release, the animal is required to go to the box previously lighted. (*Modified from Hunter*, 1913, *p.* 24.)

Hunter (1913) found that rats could delay their responses for only 10 sec., and dogs for as long as 5 min. Both animal forms, however, maintained a posture in which they kept their heads pointed in the direction of the box in which the light had appeared. This postural set was called an *overt orienting attitude*. The sensations arising from this postural set were regarded as being a substitute for the visual sensations produced by the light in a certain box. Because the delayed response could be explained as a reaction to the maintained posture, Hunter believed it was not necessary to suppose that an "idea" had been substituted for the visual sensation.

Raccoons were able to delay their reactions for 25 sec., and two-and-one-half-year-old children, for 25 min. without resorting to overt orienting attitudes. Because of the absence of orientation in these cases, Hunter believed that a central process (an "idea") must have been substituted for the stimulus. Only when orienting postures were absent, was the length of a delayed reaction regarded as a measure of central or mental processes.

In a later experiment with a child, Hunter (1917) placed a toy in one of two boxes and after a delay period the child was permitted to reach for it. Because the desired object was itself the stimulus, no preliminary training was required. By a similar method Köhler (1926) found that the chimpanzee could delay its reaction 16½ hr. He permitted the animal to watch him bury a pear in the sand. After the delay period the animal was brought back to the scene and it successfully dug up the fruit. Köhler considered his method as more satisfactory than Hunter's first method because, *seeing food buried at some particular point*, is more meaningful to the chimpanzee than *seeing a light in a certain box.* However, it would seem that the preliminary training should make the latter meaningful also. In any case, the two methods have been distinguished from each other and on the whole yield different results. When preliminary training is required, the method is known as *indirect;* when the stimulus itself is the desired object, the method is known as *direct.*

The Results Obtained in Different Delayed-reaction Experiments.—The results obtained by different experimenters working on the mammals most widely used are given in Table 30. The presence or absence of overt orientation, and the type of method used, are indicated. Careful inspection of the table shows the strikingly great variation in the results of animals of the same species, despite the fact that in all cases the necessary conditions laid down by Hunter have been complied with. Most striking is the fact that Adams (1929) and Maier (1929*b*) have obtained delays in cats and rats (without overt orientation) which are as long as those found by Köhler in the chimpanzee.

Because of the great differences in results obtained by different methods and by different experimenters, and especially because lower mammals do as well as higher mammals, several writers have explained these differences in results as dependent upon the use of the *direct* and *indirect* methods. Yet our table does not bear out the validity of this distinction.[1] Furthermore, in both methods the

[1] Maier's method is usually classified as direct. However, it is plausible to argue that his method was indirect since (as we shall see later) training was

TABLE 30.—RESULTS OF DELAYED-REACTION TESTS

Subject	Maximum delay	Overt orientation	Method	Experimenter
Rats	10 sec.	With	Indirect	Hunter, 1913
	11½ sec.	Without	Indirect	McAllister, 1932
	40 sec.	With	Indirect	Ulrich, 1921
	45 sec.	Without	Indirect	Honzik, 1931
	7–24 hr.	Without	Direct	Maier, 1929
Cats	18 sec.	With	Indirect	Yarbrough, 1917
	30 sec.	Without	Direct	Cowan, 1923
	3–16 hr.	Without	Direct	Adams, 1929
Dogs	5 min.	With	Indirect	Hunter, 1913
	5 min.	Without	Indirect	Walton, 1915
Raccoons	25 sec.	Without	Indirect	Hunter, 1913
Lemur	30 sec.	Without	Direct	Harlow, Uehling, and Maslow, 1932
Monkeys	60 sec.	Without	Indirect	Harlow, Uehling, and Maslow, 1932
	2 min.	Without	Direct	Harlow, Uehling, and Maslow, 1932
	7 min.	Without	Direct	Buytendijk, 1921
	15–20 hr.	Without	Direct	Tinklepaugh, 1928
Chimpanzees	2 min.	Without	Direct	Maslow and Harlow, 1932
	4 hr.	Without	Direct	Yerkes and Yerkes, 1928
	16½ hr.	Without	Direct	Köhler, 1926
	48 hr.	Without	Direct	Yerkes and Yerkes, 1928
Orang-outan	5 min.	Without	Direct	Harlow, Uehling, and Maslow, 1932
Gorilla	2 min.	Without	Direct	Maslow and Harlow, 1932
	3 hr.	Without	Direct	Yerkes, 1927
	48 hr.	Without	Direct	Yerkes, 1927

For bibliography see: Tinklepaugh, (1928); McAllister, (1932); Foley and Warden (1934).

required in order to produce a response to the stimulus. If Maier's method is conceded to be indirect, then the distinction between the methods ceases to have any value for explaining the variation in results.

stimulus is experienced indirectly by the animal. In the direct method past experience causes the visual appearance of food to be associated with its taste. By the indirect method, the animal is given the past experience by means of which a light or some other stimulus becomes associated with food. The difference in the methods, therefore, lies only in the time at which the food association was formed and has nothing to do with delaying a reaction. The results obtained are undoubtedly a function of the method, but even the direct method, when used by Harlow, Uehling and Maslow, (1932) yields delays as short as 2 min. in the chimpanzee. It seems unlikely that such a distinction aids in accounting for the lack of agreement in results.

Analysis of Method Producing Long Delays.—In Maier's (1929*b*) technique the presentation of the stimulus was more elaborate than in any other, and the delayed reactions obtained from the rat were unusually long. The results obtained indicate either that (1) the rat has "ideation" comparable to that of the chimpanzee; that (2) the delayed reaction is invalid as a test of higher processes; or that (3) the test is not a true measure of delayed reactions. Because it is the only experiment which attempts a critique of the delayed-reaction method, it is important to analyze the experiment in detail.

The apparatus was placed in a room previously explored and thoroughly familiar to the rats. In Fig. 96, points *R*1, *R*2, and *R*3 represent ringstands with ladders attached. The animals were trained to climb these in order to arrive at an elevated pathway which led to a corner of a table indicated by *F*. This corner contained food and was separated from the remainder of the table by a wire screen which the animals were unable to pass. Before the experiment began, the rats were permitted the freedom of the room and often climbed up and down the ladders. The pathways leading from the ladders to the food table were, however, not present. Consequently no food was obtained in this preliminary exploration.

In the actual experiments the *presentation* of the stimulus consisted of placing a rat at the base of one of the ringstands and urging it to climb the ladder. At the top of the ladder it found a pathway which led to food. The experience of a particular route to food, rather than the experience of the light in a particular place, is a rather elaborate stimulus experience. However, the elaborateness of a stimulus is not a feature upon which the delayed reaction is supposed to depend. Further, the stimulus was repeated three times in order to insure the experience of the stimulus, a feature which is assumed in Hunter's technique.

After the third trip the animal was removed from the situation for a certain length of time and this constituted the *delay period.*

At the end of the delay period the animal was *tested* by being returned to the situation and allowed to react. Point *A* indicates the starting place. It is near the food and consequently serves to motivate the animal. In order to make a correct response the rat

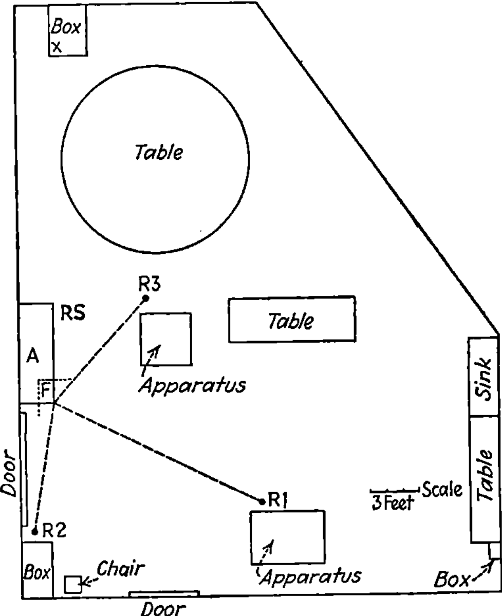

FIG. 96.—Maier's "delayed-reaction" apparatus. Three routes (*R*1, *R*2, and *R*3) lead to food *F*. When placed at *A*, the animal must choose (after a delay interval) the route which last led to food. *(From Maier,* 1929, *p.* 540. *By permission of the Journal of Genetic Psychology.)*

had to descend ringstand *RS* and go to the ringstand which contained the path to food. On different tests, different ringstands were used. The animal was therefore required, after a period of delay, to respond to one of three ringstands, all of which were equally accessible.

Control experiments demonstrated that no sensory aspect of the situation indicated to the animal which ringstand was to be taken. Rather the choice depended upon previously experiencing the stimulus (*i.e.*, the ringstand) which was associated with food. Overt orientation during the delay period was impossible because the animal was removed from the situation during this period. The third of the three alternatives mentioned above seems therefore to be excluded. There remains only a choice between granting the rat an ability for "ideas" which challenges the chimpanzee's or questioning the delayed-reaction technique.

An Evaluation of the Delayed Reaction.—The value of the delayed reaction as a technique is very questionable, considering the results obtained. Its apparent importance is largely dependent upon Hunter's logical deductions from the assumption that if a response appears in an animal, the stimulus for it must be active immediately before. If these assumptions are questioned, it becomes unnecessary to suppose that some overt posture or an internal set (such as an "idea"), which has been maintained during the delay period, has replaced the stimulus and so made possible the delayed response. There is not even a reason for supposing that a response has been held back or delayed.

The results may be satisfactorily explained by regarding the total situation as one favorable to recall. After the delay period the animal must merely reproduce the missing part of the stimulus situation and react to it. If a lighted box was used as the stimulus, then the appearance of the box must merely rearouse the light which was previously associated with it. In our discussion of conditioning we found that the animal salivated to the sound of a bell, because on previous occasions the bell has been associated with the taste of meat. In the maze the animal reacts to a reward it has received on previous occasions. Reacting to a box because of its association with light therefore seems merely to demonstrate an ability to form associations.

Tinklepaugh (1928) describes a monkey as sulking for a time when brought back into the experimental room, then suddenly observing the cups under one of which food had been placed during the stimulus period. After glancing from one cup to the other, the correct one was chosen. This behavior seems to indicate that the animal was recalling the events which occurred before the delay. The delayed response thus becomes a reaction to the associations which have been rearoused.

The situations in which long delays have been obtained have had stimulus places which were very different for the animal, both in appearance and in associative value. Thus Yerkes and Yerkes (1928) used the four sides of a room; Tinklepaugh (1928), the two ends of a stick (each end having a different relationship with different parts of the room); Adams (1929), four very different boxes placed in different parts of a room; and Maier (1929*b*), different parts of a room having decided characteristics because of its furnishings. In all of these cases the room and all of its contents were familiar to the animal through previous exploration. During this exploration, different parts of the rooms and their furnishings must have become rich in associations. These associations undoubtedly favor the rearousal of past experiences which may have been "enjoyed" in the room (*e.g.*, having experienced food, visually or otherwise, in a box next to some familiar part of the room).

Yerkes and Yerkes found that shifting the boxes during the delay period greatly reduced performance. In such cases the rearousal was entirely dependent upon the characteristics of the box. Tinklepaugh, as well as Carpenter and Nissen (1934), found that increasing the distance between the possible locations of the food increased performance. This condition brings the position of food in closer relation with parts of the room and so increases the rearousal value of the situation by increasing the distinctiveness of its several parts.

Considered in this light, differences in results obtained in the various experiments on delayed reaction are artifacts and not measures of a special ability to delay a reaction. As a result, the delayed reaction cannot be regarded as a measure of some higher process, or a demonstration of ideational behavior. This does not mean that intelligence and learning ability may not function in the delayed-reaction experiments. Certainly the ability to reproduce is important in adaptive behavior. But the ability to reproduce is not the same as the ability to hold a response in abeyance or to solve problems by means of "ideas."

A Sudden Drop in the Learning Curve as Evidence of Intelligence

Thorndike (1898) presented strong evidence to show that animals often meet new situations by displaying activity which is relatively random and unadaptive, and that the eventual solution only appears intelligent because unnecessary activity has been eliminated through learning. He did not show, as is often supposed, that the animal is *incapable of intelligent solutions*. However, his demonstration that cats solve puzzle boxes by means of learning has forced the psychologist to be critical of any apparently intelligent behavior on the part of the animal until he is convinced that past experience may not have furnished the solution. The exclusion of past experience as a possibility is not always easy. When a chimpanzee uses a stick of wood to knock down a piece of food from a high place, it is difficult to know whether this behavior was new in that situation or whether it was learned in the past through trial and error.

Because it is difficult to exclude the possible influence of past experience in meeting problem situations, the sudden drop in the

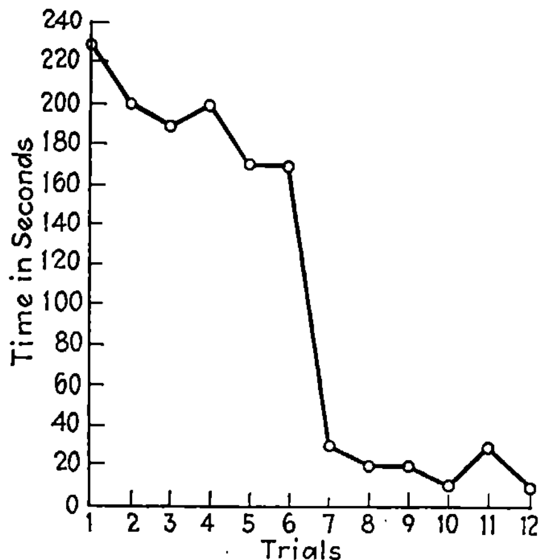

FIG. 97.—Curve showing a *sudden drop* in the time required to open a problem box. The first six trials show little improvement, but on the seventh trial the improvement is very marked.

learning curve has been treated as holding significance. For example, if an animal required a relatively long time to get out of a puzzle box on its first six trials, then on its seventh trial and all trials thereafter it gains its freedom in a minimum amount of time (see Fig. 97), it would be supposed that the animal has experienced a new relationship, formed an "idea," or had "insight." In other words, it is assumed that at a particular time, a process other than learning has intervened and has caused the sudden drop in time. However, as we have pointed out (pp. 359*f.*) the mastery of such a task is largely dependent upon selection. The essential act must be isolated and associated with the situation. This isolation may depend on the precision of the animal's motor adjustments as well as on the acuity of its sensory equipment, but once the elements to be learned have been isolated, the learning takes place rapidly. It is therefore unnecessary to assume the intervention of a higher function. Because explanations in terms of simpler processes are possible, such a criterion of intelligent behavior is unsatisfactory.

Abstraction as a Measure of Higher Processes

Introduction.—Concept formation has been regarded as dependent upon higher mental processes. It is argued that if a number of percepts (such as different geometrical figures) have something in common (such as triangularity), then this common factor becomes a characteristic of a group of percepts and may be called a "concept." In order for an animal to react to a concept it must be trained to react to some particular trait (*e.g.*, figures having three sides). To do this, it is supposed that a higher process is involved, because some common element must be found. Since triangles may vary greatly, it is believed that the "idea" of triangularity must be formed. But another view of the whole matter deserves consideration. When we remember that animals always react to certain characteristics, and not to the total appearance of a stimulus, it would seem that their reaction is an abstraction to begin with. If they react to the size of a pattern, this is just as much a concept as triangularity. The difficulty that animals have in reacting to triangularity seems not to be their inability to respond to it as an abstraction, but rather to find, by trial and error, the characteristic that is effective.

Krechevsky (1932 *a* and *b*) has shown that in discrimination situations, rats respond successively to certain characteristic aspects such as brightness, leftness, etc., at the outset (see p. 350). It seems, therefore, that in concept-formation experiments the animal is merely required to learn which concept will work; it is not required to do anything other than what it always does in the most simple learning situations, *e.g.*, react to a certain hue or brightness in a discrimination-box window, irrespective of its position. Of course, the concept to which an animal is to respond may be made so difficult that the animal fails to find it; but it may also be caused to fail to open a problem box or to learn a discrimination habit.

Analysis of Studies on Concept Formation.—Fields (1932) made an extensive study of the ability of rats to react to "triangularity." He found that after he had trained rats to jump to a triangle and to avoid other figures such as circles, rectangles, crosses, dots, etc., they still responded to the triangle when this figure was changed in size. The new triangles were equivalent to the first in that the aspects of the figures determining the response had not been affected by the change. However, when the triangle was tilted, the response broke down. This demonstrates that the tilted triangles were not equivalent to the original. An important aspect of the stimulus had been changed. With training, the animals learned to respond to tilted triangles, so that the last eight of the 24 positions were reacted to without training. By this additional training, new aspects were reacted to, and position dropped out as a factor influencing the response. Altering the proportions of the triangle within the limits tested had no effect. In other words, the reactions of the rats were not dependent upon the original triangles having been equilateral.

When using outlines of triangles instead of solid figures, and when using three dots to indicate a triangle, the resulting reactions were similar to those obtained by the original triangles, without further training. Through training, the equivalent stimuli became more and more specifically dependent upon fewer characteristics. However, even this training was not sufficient to cause a black and a white triangle to be equivalent to each other. When white triangles on black backgrounds were exchanged for black triangles on white backgrounds, the preference for triangles broke down. It is probable that further training would have made the response independent of features of the background.

The experiments indicate that, through training, a response to some particular characteristic of a stimulus can be developed. A rat's response to a limited aspect of the situation does not require the intervention of processes other than those involved in association formation.

Gengerelli (1930*b*) found that rats were able to learn a maze in which all correct choices were to the right (or left) and in which the distance between the choices was altered from time to time. This shows that it was possible for rats to learn the maze without depending on the distance between choices, but entirely in terms of turns to the right (or left). After the animals had learned this response, they were tested in a new maze pattern, and it was found that the tendency to turn right (or left) at all junctions was carried over. He concludes that the rats had generalized their response, but this seems to be nothing other than saying that the rats were able to respond in a characteristic way to a junction in a maze and were uninfluenced by the pathways themselves. In other words they responded to one characteristic aspect of a junction and this aspect was equivalent in all the junction situations used.

Results similar to those of Fields have been reported even in the case of chickens. Kroh and his students (1927) were able to train chickens to peck corn from triangular papers and not to peck from circular papers when the latter were arranged in rows. At first, changes in the order of the figures in the row caused errors, but additional training eliminated them. The introduction of new figures, such as squares and pentagons, resulted in responses to these as well as to triangles. With further training in which the new figures were used, it was found that triangles of any shape and papers of different colors could be introduced into the series, with the result that grain was taken only from three-sided figures. Although direct comparisons between the studies of Fields and Kroh cannot be made, the indication is that the chickens responded to triangularity more readily than the rat. Since birds are much less plastic than rats in their behavior, as far as adjustments to new situations are concerned, this comparison further strengthens the contention that abstraction is not a measure of higher processes.

Robinson (1933) succeeded in training a monkey always to pull one of three boxes toward itself. The boxes were either gray or contained a black paper disc. Two were always alike, and the box containing food was always different. The progress of the monkey in making correct choices was gradual, much like the progress in any other learning situation. As in the case of experiments on rats, this type of response does not prove the presence of higher processes.

Of interest is the fact that after a chimpanzee has learned to choose a triangle, it continues to choose triangles even when the form and color are radically changed. According to Gellermann (1933), the chimpanzee reacts to triangularity without additional training. Merely because the ape is more intelligent than the rat, it cannot be concluded that such a response is due to its higher intelligence. The response to the particular aspect of a situation which happens to be characteristic of a class of objects may be due to better vision as well as better learning ability, or to a natural tendency to respond to the same details as those which were outstanding to the human being who classified the figures.

Of importance, however, is the demonstration that various animals have tendencies to respond to different specific aspects of a situation.

What constitutes equivalence among stimuli is therefore a variable factor among different animal species. This being the case it is probable that perceptual organization is a variable factor in animal behavior and must be taken into account when comparisons in concept formation are made.

Conclusion.—In the light of the foregoing discussion, abstraction may be regarded as the ability to learn to respond to certain limited aspects of a situation. All situations which contain this limited aspect become equivalent to each other and call out the same response. This does not mean that such learning is not of a higher order than learning which depends upon all characteristics of a situation. We have already pointed out (pp. 163*f.*) that the ant's modification in behavior through experience is limited to the situation in which the learning has taken place. Testing situations which differ even in the motivating conditions are nonequivalent with the original learning situation, and fail to call out the ant's learned response. Learning which becomes more and more dependent upon specific features of a situation allows for much wider range of adaptation.

It is our contention that the abstraction experiments do not demonstrate the presence of functions other than the processes of association and selection. It does not follow from this that higher processes may not facilitate the mastery of an abstraction problem. Tests of abstraction merely fail to be indices and measures of a specific higher process.

Multiple Choice as a Test of Abstraction

The Ability of Animals to Learn Which Choice Not to Make.—Hamilton (1911) developed a method for analyzing animal behavior which has become known as the *multiple-choice* method. His apparatus consisted of a box with four exits placed side by side in an arc and equally distant from the animals' starting point, as shown in Fig. 98. Food was obtained upon getting out of the box. Three doors were always locked and one unlocked. The unlocked door was never one which had been unlocked in the previous trial. If this is learned by the animal, its efforts to escape from the box would be confined to the three other doors. Hamilton believes that to avoid the last open door involves "inference," in that discovering such a principle depends on several previous trials. In other words the animal must "abstract." According to our preceding analysis this means that the animal is required to learn to avoid a certain door, the door which has the characteristic of having been unlocked on the previous trial.

Hamilton tested humans, monkeys, dogs, cats, and a horse, and counted the number of incorrect doors opened in a hundred trials. The order of superiority is as we have listed the animals. Young animals were also found to be inferior to adults. Of interest is the procedure used by various animals. In humans, and to some

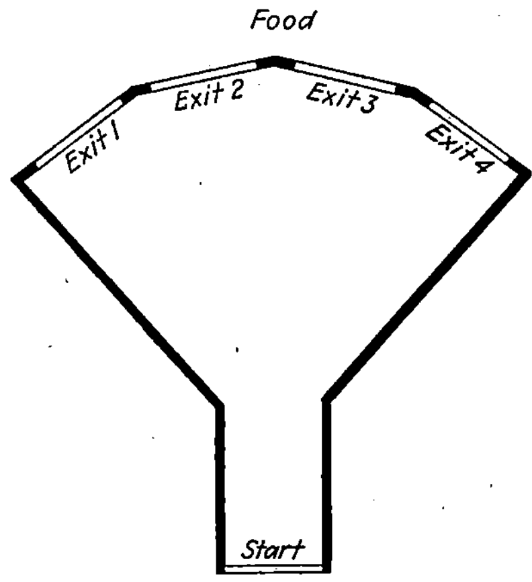

FIG. 98.—The multiple-choice apparatus. One door of the box is unlocked. The animal must discover the unlocked door on each trial in order to obtain food outside the box. If the experimenter follows a system which determines which door is to be unlocked on a given trial, the animal has an opportunity to learn to avoid locked doors. (*Modified from Hamilton*, 1911, *p*. 34.)

extent in monkeys, the characteristic reaction was to avoid the door which had been open in the previous trial. The characteristic reaction of some of the other animals was to try each door once before trying any one twice, the door to be avoided being reacted to as the others. A still less intelligent response was to try any door twice before all had been tried once. The horse did this a great deal; cats and dogs to some extent. In the 100 trials allotted, humans learned what door to avoid, and monkeys learned this to a lesser extent. Only the horse demonstrated what may be called stereotyped behavior, in that it repeatedly tried the same door.

Learning Which Choice to Make.—The method has been further developed by Yerkes so as to make possible the learning of a correct response rather than the learning of what is not correct. In the study of two pigs, Yerkes and Coburn (1915) used an apparatus in which the animal was confronted with nine compartments. On any trial, from three to nine of the compartments would have the doors open, but only one of these contained food. The animal was required to learn which of the open series of compartments contained the food. This arrangement makes possible any number of complicated learning tasks. Thus food may be found in (1) the open compartment at the left (or right); (2) the middle compartment; (3) the open compartment second from the left (or right); (4) alternately, the first compartment on the right and the first on the left; etc. By changing the number and position of the open series of compartments a response to a particular compartment was excluded.

The pigs learned to react on the basis of the open compartment at the left or right end of the series as well as to alternate between them. They failed to learn to react to the middle compartment when seven were used in the series, although they were quite successful when a smaller group of open compartments was used.

Two crows (Coburn and Yerkes, 1915) learned to react to either of the end compartments, but failed to learn to react to the second from the end. A bird studied by Miss Sadovinkova (1923) learned to react to the middle door under the same conditions in which the pigs had failed. The rats studied by Burtt (1916) succeeded in reacting to the end compartments, but failed to react to the second from the end. Yerkes (1916) found the monkey able to react to the second from the end, although an orang-outan did more poorly than the pigs. Of the four chimpanzees tested by Yerkes (1934) all failed to learn to react to the middle door. Three failed to respond alternately to the right- and left-end compartments and three also failed to learn to choose the second from the right end. All succeeded in reacting to the end compartments.

In general, higher vertebrates are more capable in multiple-choice problems than lower forms, but the order of ability is inconsistent. The method is inadequate for demonstrating the presence or nature of higher processes, but it is valuable for the study of equivalent stimuli, and for the study of abstraction when this is regarded as the ability to respond to a certain aspect of the stimulus pattern which is common to a number of stimulus patterns.

Combining Isolated Experiences (Reasoning)

A Problem Requiring the Combination of Isolated Experiences.— Since learning is the ability to combine contiguous experiences (see p. 343) it would seem that if higher functions exist, they should exhibit themselves in situations where learning has been excluded. In his experiments with rats Maier (1929*a*) endeavored to investigate this point.

A typical experimental situation is shown in Fig. 99. Table *A* occupies the center of a room. *RS* represents a ringstand by means of which the rat can go from the floor to the table and back to the floor again. One corner of table *A*, marked *F*, is fenced off from the remainder of the table by means of a wire obstruction, so that *F*

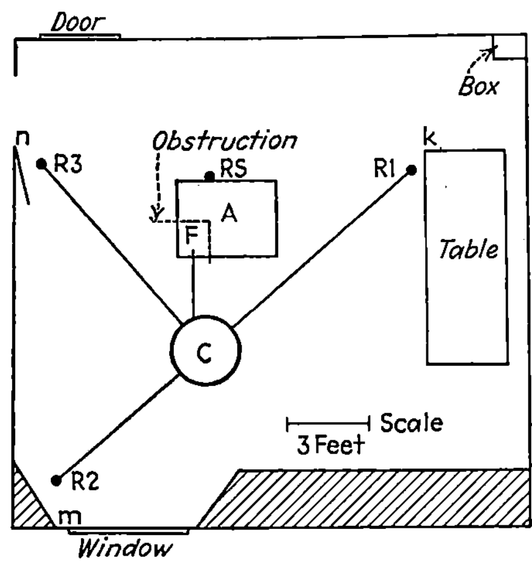

FIG. 99.—Setting of the "reasoning" test. When placed at *A* the animal is required to combine different past experiences in order to achieve an indirect route to the food *F* which is placed behind a wire obstruction. (*Modified from Maier*, 1929, *p.* 23.)

is not accessible to the rat. This much of the situation is present during the preliminary part of the experiment. A group of rats having the freedom of the room for several days have thoroughly explored it and have become well habituated.

The situation is now complicated by the addition of a second table *C* which is connected to point *F* by means of a pathway. Three other pathways also lead from *C* in various directions. These pathways terminate with ringstands *R*1, *R*2, and *R*3, each of which forms a ladder to the floor. The addition of these new structures makes point *F* of table *A* accessible.

The experimental procedure is as follows: (1) the rats have already experienced table *A* and various parts of the room. The learning acquired in this connection may be called *experience* I. (2) After the addition of the structures, a rat is placed at the base of *R*1, *R*2, or *R*3, and is encouraged to ascend. When it does so it is guided until it reaches *F*, where it finds food. A second and a third time the animal is placed at the base of the same ringstand, but on these trips it needs neither a push to encourage it nor guidance. It has thus learned to go from the base of one of the ringstands to *F*. This learning may be called *experience* II. (3) The animal is *tested* by being placed at *A*, which is in front of the obstructing fence that separates this place from the food.

In the test situation the rat is confronted with the problem of how to attain the food behind the obstruction. No single past experience will be sufficient to guide it there. From past experience it has learned to go from *A* to other parts of the room. It has also learned to go from the base of a ringstand to *F*. But it is now at *A*. If it uses the pertinent part of *experience* I and combines it with *experience* II, it can obtain the food. This requires the combining of parts of separate past experiences. No single previous experience, if reproduced, is adequate to take the animal to the food.

The typical behavior of the rat in this situation involves (1) a period of struggling at the cage; (2) running back and forth between the cage and the edges of the table; (3) descending to the floor by means of *RS*; (4) going directly to the ringstand it used in *experience* II; and (5) proceeding up the ringstand and to the food. Other rats that have experienced the route from one of the other ringstands to the food behave in a similar manner but reach the food by way of the other ringstands. For example, three rats trained to reach food from the three different ringstands all arrive at the food by different routes when placed at *A*.

By means of various controls it has been demonstrated that rats solve the problem by combining the two separate experiences and do not find the route by chance. With *experience* II omitted, or with *experience* I eliminated by placing the same situation in a strange room, the animals fail. Both *experience* I and *experience* II must be given before an animal can solve the problem.

Tests for Isolating Abilities R and L.—If we assume that rats have the ability to combine parts of separate or isolated experiences as well as contiguous experiences, and if these abilities are different psychological functions, it should be possible to place them in opposition with each other and determine their relative importance.

Maier (1932*a*) therefore devised three tests in order to determine the validity of the above assumption. Figure 100 shows three tables with three connecting pathways. Exploration of these tables and pathways corresponds to *experience* I in the above experiment. The rat is then placed on one of the tables (*e.g.*, table *B*) and permitted to feed. This corresponds to *experience* II. The animal is tested by being placed on one of the other tables (*e.g.*, table *A*). By repeating this test, using different combinations of tables, a score is obtained. According to chance, half of the test runs would be cor-

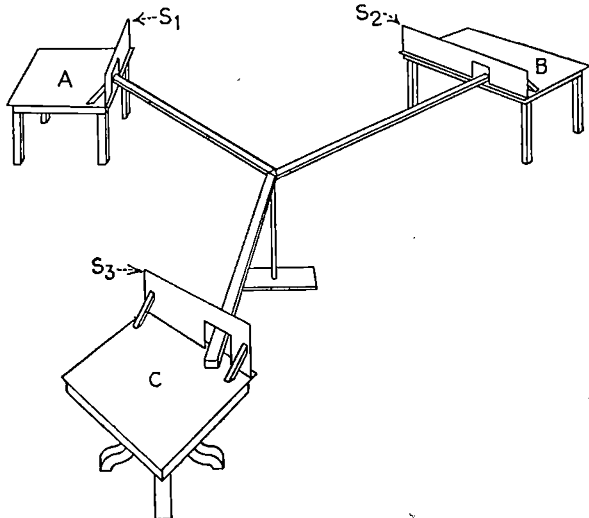

FIG. 100.—A simple test of "reasoning." After the animal has explored the apparatus it is fed on one of the tables. It is then carried to a different table and must find its way back to he food table. Because of screens S_1, S_2, and S_3 the rat cannot see the food. (*Modified from Maier*, 1932*b*, *p*. 181.)

rect. To the extent that the animal goes correctly more frequently than chance would warrant, the ability to combine two isolated experiences is involved. This test was called a measure of *ability R*. In order statistically to eliminate the chance factors, the difference between the number of correct and incorrect runs was divided by the number of tests. The score is then entirely a measure of *ability R*.

A second test measured *ability R* acting in conjunction with learning ability (*ability L*). When *ability R* was tested on one day by feeding the rat at table *B* and starting from table *A*, it was tested the following day in a like manner. The previous day the rat had gone from A to B. This was repeated several times so that learning the route to food was possible. By repeating the procedure of feeding the animal at B and starting it at A the animal may now reproduce what it learned (*ability L*) or it may again combine the two separate experiences (*ability R*). Thus either *ability L* or *ability R* will produce the same responses. This therefore becomes a test for $R + L$, and when corrected for chance becomes a measure of the two abilities functioning together.

TABLE 31.—PERCENTAGE OF CORRECT RESPONSES IN TESTS INVOLVING R AND L. (MAIER, 1932*a*)
(Chance Score = 0 Per Cent)

Extent of lesion	0%	1–10%	11–17%	18–24%	25–41%
Score in test R	80.7	63.9	43.2	8.3	1.1
Score in test $R + L$	89.1	78.1	70.8	35.9	34.1
Score in test $R - L$	70.1	53.1	41.7	18.8	3.8

TABLE 32.—RELATIVE VALUES FOR R AND L (MAIER, 1932*a*)

Amount of cortical destruction, per cent	0	1–10	11–17	18–24	25–41
Calculated value of R	79.6	65.6	56.3 (72.5)	27.4	19.0
Calculated value of L	9.5	12.5	14.5 (12.5)	8.5	15.1
Experimental value of R	80.7	63.9	43.2 (69.0)	8.3	1.1

The two abilities may be placed in opposition to each other by feeding the above animal at C and testing it by starting it at A. *Ability L* in this case will cause the animal to go to B and make an error, whereas *ability R* will cause it to go to C. This is the test for $R - L$, and when corrected for chance becomes a measure of these two abilities when acting in opposition to each other. Table 31 shows the results obtained from normal and partially decorticated rats in these three tests.

From the scores made in tests for $R - L$ and $R + L$, the values for R and L may be calculated for each group of rats. As the value for R was also obtained experimentally, it serves as a check on the validity of this method for isolating the function of abilities *R* and *L*. The results of these calculations are shown in Table 32.

It will be seen that for rats with no cortical destruction or with less than 11 per cent of the cortex destroyed, the experimental and calculated values for *R* are approximately equal. Rats with more than 18 per cent destruction all made chance scores in tests involving *R*. For this reason the values obtained are a matter of chance and a correspondence is not to be expected. The group with 11 to 17 per cent destruction contained a number of cases which made chance scores. When these are excluded and only records of rats able to make better than a chance score on the test for *R* are used, a correspondence between the calculated and experimental values for *R* is again obtained. These corrected values are shown in parentheses in Table 32. The table shows that for rats in which *ability R* is still functional, there is a striking correspondence between the experimental and calculated values for *R*. The table also shows that when cortex is destroyed, *ability R* suffers greatly, being completely excluded when the lesions extend beyond 18 per cent. *Ability L*, on the other hand, remains fairly constant, despite the cortical destruction. Because cortical lesions have a different effect on abilities *R* and *L*, and because these abilities can be treated as separate unknowns, it is probable that *R* and *L* are separate psychological functions. The assumption of two different abilities which are involved in adaptive behavior is thus verified by the experiment.

Campbell (1935) tested the validity of this distinction between *R* and *L* by a different method. He obtained the score made by rats on two different mazes and the test for *R*. The correlation between the maze scores was 0.55 and the correlations between the score on test *R* and each of the mazes were −.37 and −.22, respectively. The marked positive correlation between the two maze scores shows that mazes primarily measure the same ability, but the lack of such a correlation between either maze score and the score on test *R* shows that different abilities are measured by these tests.

The Combination of Isolated Experiences as a Characteristic Process in Reasoning.—*Ability R* may therefore be regarded as a function distinct from that of association by contiguity. As it concerns the ability to form a spontaneous combination of parts of separate experiences, it becomes a fundamental process in the formation of new patterns of behavior. In this sense it is an ability which makes it possible for an animal to meet new situations, not by trial and error, but by a reorganization of parts of its past experiences. It is a capacity which is akin to reasoning or creativeness in man, and may therefore be regarded as the fundamental characteristic of reasoning. Therefore we shall regard *ability R* as the psychological function which characterizes the reasoning process.

Because of its existence, it is necessary that we consider the possibility of its functioning in ordinary learning situations. Its presence will undoubtedly greatly facilitate the mastery of a task which allows for the reorganization of experience, but to what extent it plays a role is often difficult to determine.

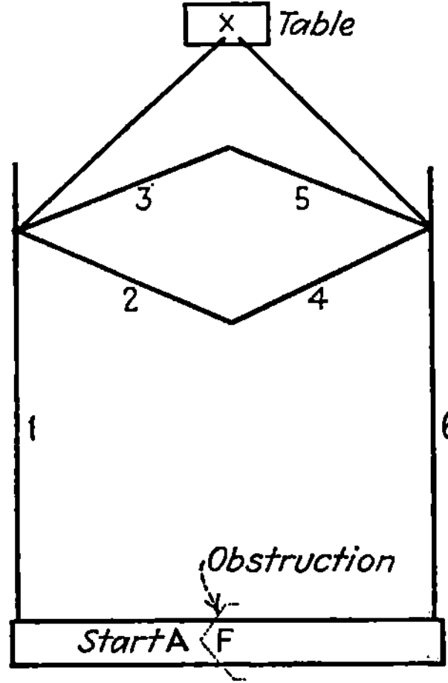

FIG. 101.—Maze in which shortest route to food is selected without previous trials. After the rat has learned the route from the starting point to *F* (food place), it is capable of taking the shortest route from the starting point to *X* when food is placed there. (*Modified from Maier*, 1929, *p.* 65.)

REASONING IN MAZE SITUATIONS

That the ordinary maze may involve more than the formation of associations is suggested by Lashley's study (1929) of maze learning by rats with brain lesions. Although the learning of discrimination habits was little affected by brain injuries, maze learning was greatly retarded, but in no case abolished. Since ability *L* does not appear to be affected by lesions, and since maze learning is affected, it is possible to regard the reduction in maze learning as due to handicapped reasoning ability. In other words, the effect of the brain injury was to force rats to depend more on learning in order to master the maze. The mastery of discrimination habits depends largely on selection, and the presence of reasoning ability is of little advantage. Loss of reasoning ability, therefore, does not handicap the animal in the acquisition of simple discrimination habits.

Maze situations which more definitely involve higher mental processes have been designed. We have already discussed experiments (pp. 408*ff.*) in which the animals ran the maze without food at the end. The learning which takes place in such trials is therefore not dependent on an association with food. Because the introduction of food suddenly makes for accurate performance, it is likely that the past experiences in the maze have become organized in terms of food.

The ability of the rat to choose the shorter route to food may in some cases involve reasoning. Certainly it is the case when the shorter route is chosen without first trying the alternatives. Fig. 101 shows a maze used by Maier (1929*a*) in which this was demonstrated. Rats freely explored all parts of the maze, and also were taught to go from *A* to *F* by means of pathways 2-4 and 3-5. They

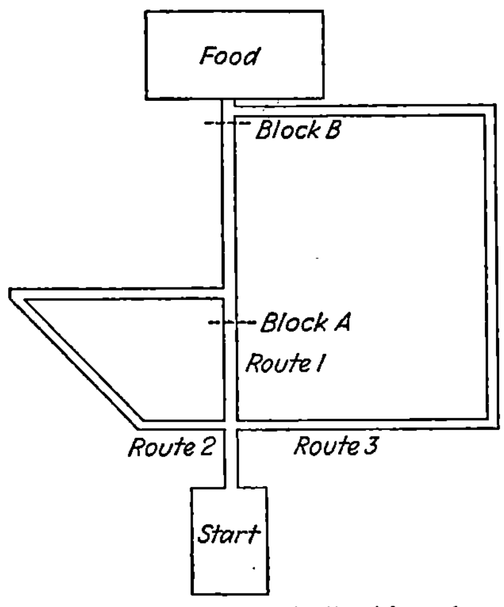

FIG. 102.—Apparatus for testing "insight." After the animal has learned the three routes to food, it is required to choose route 3 when a block is placed at *B*, but to choose route 2 when the block is placed at *A*. (*Modified from Tolman and Honzik*, 1930*b*, *p*. 223.)

were not permitted to go to the food by way of *X*, however. The pathways leading to *X* were experienced only during an exploration period. After the rats had learned to go from *A* to *F* for food and showed no inclination to visit *X*, the food was moved to *X*. The rats experienced the food in the new location only by being placed there to eat. They were then removed and placed at *A*, one at a time. Of 7 rats, 4 went directly from *A* to *X*; 3 went to *F* and then from *F* directly to *X*. On the second run all rats went from *A* directly to *X*. Thus despite the habit of going from *A* to *F*, some of the rats were able not only to refrain from going to *F*, but also were able to take the shortest route to *X* without ever having gone to *X* for food previously. Since previous visits to table *X* were part of the general exploratory behavior with no particular route followed, and since the learned pattern would cause the rat to go from *A* to *F*, the above performance can only be explained by assuming a reorganization of experience.

Tolman and Honzik (1930*b*) used a maze, the principle of which is shown in Fig. 102. It will be seen that there are three routes to food, all of different length. A preference for the shorter routes soon developed, but by blocking these, the long route was also taken. All routes thus became learned. In the figure two blocks are

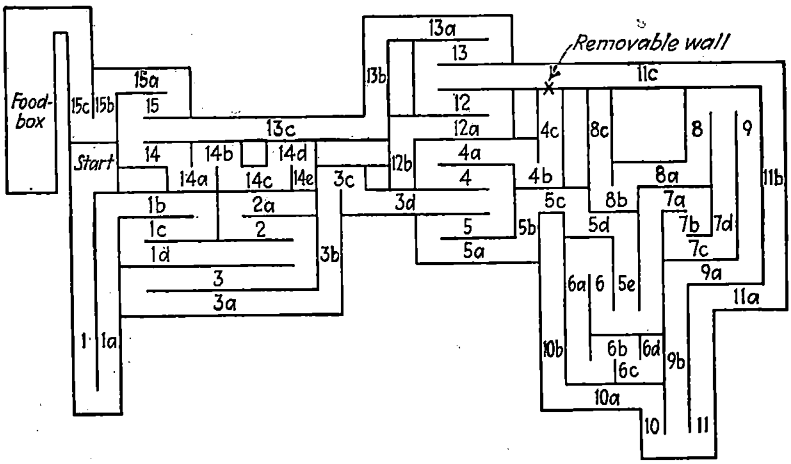

FIG. 103.—Shepard's test of "reasoning." The odd numbers indicate true path; the even, blind alleys. After rats have learned the maze, the section indicated by *X* is removed, thereby causing a previous blind alley to become a short cut through the maze. On discovering the change, rats immediately use the short cut. (*Courtesy of J. F. Shepard.*)

indicated. By means of block *A*, route 1 can be barred, and by means of block *B* both routes 1 and 2 can be barred. As the rats prefer route 1, what will be their next choice if this route is blocked at *B*? Of the 25 rats tested, 21 took route 3 on the first test after finding route 1 blocked at *B*. When the block is placed at *A*, however, the majority of the rats take route 2, which is then usable and shorter.

Shepard (1933) used mazes such as the one shown in Fig. 103. The starting point and the food box are to the left. The odd numbers indicate the true pathways; the even numbers, the blind alleys; and the letters of the alphabet indicate different sections of the alleys. Such a maze is readily learned by the rat. After mastery of the route from starting point to food box, a change is made in the maze by removal of the wall indicated by *X* in the blind alley 4*c*. This change, being at the end of a blind alley, remains unnoticed by the rat until it reaches alley 11*c* which is part of the true path. The characteristic response of the rat is to explore this new opening for a short distance, then to return to alley 11*c* and complete the run. When next placed at the starting point, the rat shows a marked change in its performance. It chooses the former blind alley 4 and uses it as a short cut to 11*c*. This experiment gives a remarkable illustration of the rat's ability to reorganize its experience. Shepard points out that an animal's ability to make such a short cut is not highly correlated with its ability to learn the maze. This lack of relationship between learning and ability to reorganize experience substantiates the belief that these are qualitatively different functions. Different situations involve them in different proportions.

Problems Solved by Other Animal Forms

With a considerable body of evidence demonstrating processes more complex than association formation in the rat, there is little doubt that cats, dogs, raccoons, and primates have similar and more highly developed capacities along the same line. We must now analyze the behavior of some of these forms in various situations and determine how adequately our concept of reasoning applies to the solutions they achieve.

Köhler's Experiments with Chimpanzees.—The classical experiments of Köhler (1926) first reported in 1917 demonstrated the remarkable capacity of the chimpanzee to meet problem situations intelligently. Taken as a whole, the behavior of the chimpanzee in solving these problems excludes interpretation in terms of accident or previous learning. In many cases these possible interpretations are not excluded to the satisfaction of the skeptic, but his alternative explanations are convincing only when isolated cases are examined. When all of the evidence is considered, one becomes convinced that the ape behaved intelligently or had what Köhler called *insight*[1] into the situation. Since rats have abilities other than learning,

[1] The term *insight* is used to describe the animal's behavior rather than its experience. It is true that the term ordinarily refers to a characteristic experience. Such experiences are, however, accompanied by characteristic behavior. If an animal behaves as if it had *insight*, this behavior can be described as insightful, whether or not its experience corresponds to that of a human being.

FIG. 104.—Photograph of chimpanzee obtaining food by the use of a long pole. (*From Köhler, 1926, opposite p. 72. By permission of Harcourt, Brace & Company.*)

it is futile to argue in favor of solutions by accident or trial and error in the case of the chimpanzee.

Köhler found the chimpanzee very adept at using a stick to haul in food lying out of reach outside the cage. Some of them also succeeded in using a short stick to pull in a larger one which was adequate for reaching food a long distance from the cage. Others failed to make the intermediate step and tried in vain to obtain the food with the short stick. When food was suspended in mid-air, the animals used long poles, but instead of using them to knock down the food, they used them to climb up to the food. They were able to climb the pole and grab the food before the pole fell over. Figure 104 is a photograph of such a solution. This solution surprised Köhler, as he had expected them to use the poles to knock down the food, but poles had different associative connections for Köhler than they had for the apes. The solution, being an integration of different past experiences, was therefore different for the experimenter and his subjects.

The animals also succeeded in obtaining suspended food by stacking boxes. Some of them had difficulty in making their structures stand, but falls did not discourage them. Figure 105 is a photograph of three boxes successfully stacked. One ape achieved a four-box structure as shown in Fig. 106. Solutions requiring a stack of boxes and a pole to knock down food were also achieved. The use of the human as something to climb upon occurred in the absence of other material. The keeper was instructed not to resist when Sultan (an ape) led him under the food, but was told to kneel down as soon as the animal climbed to his shoulder. Köhler describes what happened as follows:

> Sultan climbs on his shoulders, after he has dragged him underneath the objective, and the keeper quickly bends down. The animal gets off, complaining, takes hold of the keeper by his seat with both hands, and tries with all his might to push him up. A surprising way to try to improve the human implement! [P. 146.]

Solutions of the above types appear even when the equipment and the position of the food are greatly varied. The animal has therefore not learned how to obtain food from a certain position, but its behavior tends to be adapted to meet the difficulties in each situation. In other words, we may say that the experiences with boxes, poles, and the positions of the food are obtained in various connec-

Fig. 105.—Chimpanzee obtaining food by means of a stack of boxes. (*From Köhler*, 1926, *opposite p.* 142. *By permission of Harcourt, Brace & Company.*)

FIG. 106.—A chimpanzee which successfully stacked four boxes. (*From Köhler, 1926, opposite p. 144. By permission of Harcourt, Brace & Company.*)

tions. To solve a problem, these experiences must be reorganized so as to overcome the difficulty. Without reorganization, certain mistakes would necessarily arise. The experience of a box in an altogether different connection will cause the animal to recall the box when the need arises. Thus a box seen in a building on some other occasion will be dragged out into the open when some feature of the problem situation recalls the "box experience." Mistakes that are made by the animals from time to time clearly show the need of the proper reorganization of experience to meet the various situations. For example, one of the duller chimpanzees often stacked boxes below points where the food had previously hung, and not under the new position of the food. He also stacked boxes inside the cage, instead of using a stick, when food was lying on the ground outside the cage. In many such instances, the recurrence of habitual or learned acts prevented the animal from making clear-cut solutions.

The Element of Learning in Box Stacking.—Bingham (1929*a*) attempted to make a standardized analysis of box stacking. He found that chimpanzees without experience in stacking, but with previous experience with boxes in other situations, readily move them fairly accurately under a suspended incentive. They also stack boxes when first confronted with a situation which requires it, but the adequacy of the structure improves with experience in such situations. Apparently certain things must be learned before an organization of experience, which involves several boxes, becomes efficient. That two boxes give more height than one seems to require no separate experience, but that two boxes must be placed in certain ways in order to remain stacked must be learned. When once learned, this experience can be organized with other experiences without previous repetition. The stacking of four boxes, for example, readily develops from the stacking of two.

The selection of a box large enough to meet the demands of a situation adequately also seems to improve with practice. The up-ending of boxes to obtain greater height, however, appeared rather readily, which shows considerable ability to select adequate dimensions. Up-ending seemed to be easier than the stacking of two boxes, although the latter developed readily from the former. Bingham points out that " . . . insight involves little if any behavior that is strictly new. More likely, it is composed of response factors which have been combined, perhaps practiced, in other fusions" (pp. 90–91). In other words, we may regard the intelligent behavior of chimpanzees as new organizations of elements of past experience. In another study, Bingham (1929*b*) points out that solutions depend upon the animal's having had adequate past experience. This means that parts of several past experiences cannot be integrated until all parts have been previously encountered. He also finds cases in which past habits blocked solutions. A tendency to go directly to food, for example, tends to block the formation of a solution which is a roundabout way to food. Thus apes as well as rats make habitual attempts to obtain food before a new combination is formed.

Instrumentation in Monkeys.—The use of tools in problem solving is usually regarded as a mark of intelligence. This form of activity is primarily confined to primates, and the manner in which the tools are used suggests that the presence of a high degree of intelligence is necessary. Nevertheless, it must not be overlooked that primates have a highly developed sensory and motor equipment which is very essential for the manipulation of tools. Since monkeys are similar in structure to chimpanzees, any possible difference in tool manipulation between these forms would suggest a basis for a comparison in intelligence. Nellmann and Trendelenburg (1926) tested monkeys in a variety of problem situations and found them markedly inferior to apes in the use of tools. For example, a box was used to reach food only under most favorable conditions. A box out of sight was never used, and boxes were never stacked. Poles were used successfully for pulling in food only when the pole was arranged with a cross piece (forming a sort of rake) so that simply pulling the stick in was required. Their use of poles did not indicate any understanding of how to arrange the tool to bring the food toward them.

Yerkes and Yerkes (1929) in their monumental work concluded that only apes and man were able to use implements in any complex fashion in problem solving. Monkeys were capable only of simple forms of instrumentation. They regarded the difference in instrumentation between monkeys and apes as indicative of a "great gulf" between these forms. However, Klüver (1933) in his extensive study of the behavior of monkeys found a *Cebus* monkey capable of a high degree of instrumentation. His monkey used sticks and other objects to pull in food lying out of reach. It even used a short stick or other objects to obtain a long stick which would reach a goal object a long distance away. When the stick was fastened to a post by means of a rope and thus could not be carried to the scene of the food, the monkey at first failed to unhook the rope from the post. On a later occasion the animal did succeed in freeing the stick by unhooking the rope. It is not clear, however, to what extent good fortune may have entered into this solution.

The monkey failed when the solution depended upon stacking two boxes. It solved problems, however, by combining the use of one box and a stick. The monkey had to find the box and place it fairly accurately under the food. Only after several trials was the box placed so that by standing on it the food could be knocked down with the stick. Practice greatly improved instrumentation.

Bierens de Haan (1931) succeeded in getting a monkey to stack boxes. The first problem required that the animal move a box under the suspended fruit. In the next test, with the fruit suspended still higher and one box placed directly under it, the monkey stacked a second box on top of the first. After this test, the animal succeeded in stacking the two boxes when both were placed in different parts of the room. With two boxes stacked under the food by the experimenter, a third was placed on top of these by the monkey. Then with only one box properly placed, the monkey succeeded in stacking the other two boxes on top of it. Finally, the animal stacked all three boxes from the ground up. It also achieved the solution when some of the boxes were not in view at the time of the test.

In many cases the placement of the boxes under the fruit was inaccurate, and the boxes often were so poorly stacked that they fell over before the fruit was obtained. Practice improved these details. The production of these solutions was greatly assisted by learning. The simple problems at the beginning taught the animal how to place and stack boxes. It is therefore difficult to know to what extent the solutions were spontaneous. In any case the monkey cannot be regarded as markedly inferior to the chimpanzee on the basis of instrumentation.

Problems Solved by Raccoons.—The early experiments of Cole (1907) suggest that raccoons are rather apt at problem solving. They mastered complex problem boxes with a combination of locks almost as well as did monkeys. They were also trained to climb to a platform for food when the experimenter pushed a lever which raised a certain card, and not to respond when some other card was exposed. Of interest is the fact that the raccoons soon began clawing at the cards. If a "no-food" card appeared, it was ignored or clawed down and another card attacked until a "food" card was clawed up. Cole argues that the raccoon has an "image" in mind and seeks the stimulus which corresponds to it. The experiments suggest that the raccoon is intelligent, but they do not give us any idea as to the nature of the ability required. No ability other than learning need be postulated to account for the behavior.

McDougall and McDougall (1931) used the situation illustrated in Fig. 107. When the raccoon was at *D*, food was placed at *C*. As it attempted to go from *D* to *C*, the animal was restrained because of

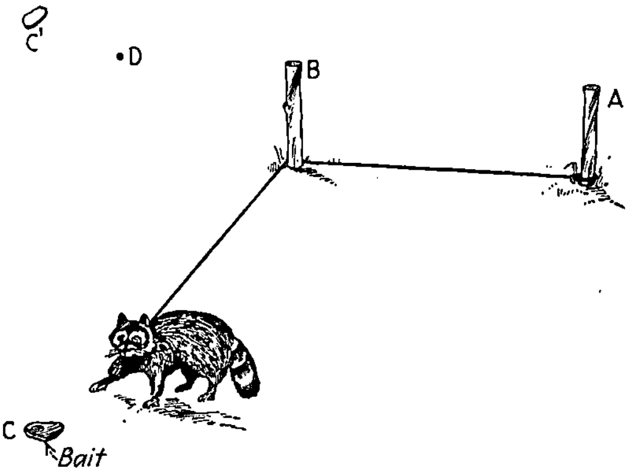

Fig. 107.—The chain and post problem. The raccoon must circle post *B* in a clockwise direction in order to obtain the food. (*Redrawn from McDougall and McDougall*, 1931, *p.* 257. *By permission of the Journal of Comparative Psychology.*)

post *B*. After struggling for the food for some time, the animal wandered off apparently disinterested. In doing so, it walked around post *B*. A little later it went in the direction of the food and succeeded in getting it. On the next four trials it obtained food by apparently wandering about *B*. Only on the sixth trial did it make a direct trip around *B* to *C*. From the description given one cannot determine whether or not these solutions were accidental. That the animal finally attained a clear-cut solution proves nothing, because it had a suitable opportunity to learn. A monkey in a similar situation lifted its chain over *B*, but again it is difficult to determine whether or not this was accidental. The monkey was restrained by the chain and tugged at it. It did not go directly to the point of difficulty and lift the chain over the post.

It is clear that experiments of this sort are not convincing. An animal may apparently be demonstrating unusual capacities, but accident and trial and error learning may account for their success. On the other hand, animals capable of complex processes may appear stupid in a problem because the situation encourages a number of stupid acts. One cannot tell the animal to remain seated until it has inhibited its first impulses. Only clear-cut solutions carry weight, and even then one knows only that the animal has solved the problem, but little of the nature of the processes involved in solving it.

Experiments on Birds Fail to Meet Criterion for Reasoning. Evidence for reasoning thus far presented has been obtained from animals having a well-defined cortex. The question which now arises is, to what extent do animals below the rodents in the series show an ability to reorganize experience?

We have already seen that birds do very well in problem-box and discrimination situations. In tests involving intelligence, however, they seem to behave rather stupidly. Köhler (1926) describes hens as struggling at a wire fence when food is on the other side. Although familiar with the environment, they are unable to use their past experience to achieve an indirect route around the fence. By running back and forth in an excited fashion they may finally get around the fence if the detour is not too great.

Miss Hertz (1928), however, attempted to make a case for intelligent behavior in birds. She presented various problems to a crow. The best of these were her obstacle experiments. After the bird had learned to open a door for food, three stones were placed on a ledge on the front side of the cage. One of the stones prevented the door from opening. This test was to determine whether the bird would attack the offending stone. In the first experiment the bird walked back and forth before the door, stumbling over the stones. It then attacked the offending stone (the middle stone, nearest the door) which resisted, and so the other two were attacked and finally removed. Only later did the bird move the critical stone. The next day all stones were again attacked and removed by the crow. On the fifth day the critical stone was moved first, *but on this day the other stones were 70 cm. away and so out of the food situation.* The evidence here does not indicate an ability to solve the problem by reasoning. The stones were new and may have been attacked for reasons other than because they blocked the door. Further, the offending stone was only removed, without a general attack on all the stones, on the fifth day, and then only when the accompanying stones were some distance away. None of Hertz's experiments showed clear-cut responses, and in all cases the situation was repeated several times before there was an indication of problem solving. Before birds can be credited with ability to reorganize experience, much more convincing evidence must be obtained.

In the ant we have found evidence of learning rather complicated mazes. Schneirla's (1934) evidence, however, indicates that such learning is stereotyped and not subject to spontaneous reorganization. The bird, though very highly developed in many ways, has a rudimentary cortex, and it is probable that it is largely incapable of reorganizing its experiences in order to adapt to a new situation. Certainly no convincing evidence to the contrary has yet been obtained.

The Role of Reasoning in Adaptive Behavior

We have seen how learning releases an animal from the limited forms of behavior furnished by maturation. Learning likewise has its limitations. At best the animal (through learning) can respond to a range of equivalent stimuli. This range is increased to its maximum through abstraction, in which case the response is made to a very limited part of the situation. By this process the number of equivalent stimuli is increased. But mere learning cannot fit an animal for meeting entirely new situations. (In abstraction experiments, new situations are not met because they are equivalent to the old.) New situations may release a number of unspecific responses which may be useful, but a specific integration must eventually be attained, and the process of association cannot furnish this without previous repetitions.

In the ability to reorganize experience, the animal is released from stereotyped forms of behavior which are laid down by learning. With the development and dominance of this ability, learning takes on a new role. Past experience ceases to furnish the patterns of response, and instead furnishes the data from which new patterns may be formed. This ability frees the animal from a particular bit of learning and makes possible almost unlimited patterns of response.

BIBLIOGRAPHY

ABBOTT, C. 1882. The intelligence of batrachians. *Science*, **3**, 66–67.

ADAMS, D. K. 1929. Experimental studies of adaptive behavior in cats. *Comp. Psychol. Mono.*, **6**, pp. 168.

AGAR, W. E. 1927. Regulation of behavior in water-mites and some other arthropods. *Jour. Comp. Psychol.*, **7**, 39–74.

ALLABACH, L. F. 1905. Some points regarding the behavior of Medridium. *Biol. Bull.*, **10**, 35–43.

ALLEE, W. C. 1931. *Animal aggregations.* Chicago: Univ. Chicago Press. Pp. 431.

ALVERDES, F. 1922. *Neue Bahnen in der Lehre vom Verhalten der niederen Organismen.* Berlin: J. Springer. Pp. 62.

———. 1925. *Tiersoziologie.* Leipzig: Hirschfeld. (tr. 1927, *Social life in the animal world.* New York: Harcourt, Brace. Pp. 216.)

AMELN, P. 1930. Der Lichtsinn von *Nereis diversicolor* O. F. Müller. *Zool. Jahrb., Zool. Physiol.*, **47**, 685–722.

ANGULO Y GONZÀLEZ, A. W. 1932. The prenatal development of behavior in the albino rat. *Jour. Comp. Neurol.*, **55**, 395–442.

———. 1933. Development of somatic activity in the albino rat. *Proc. Soc. Exper. Biol. and Med.*, **31**, 111–112.

ANREP, G. V. 1920. Pitch discrimination in the dog. *Jour. Physiol.*, **53**, 367–385.

AUDUBON, J. J. 1835. *Ornithological Biography.* II, pp. 33–49. Edinburgh: Black.

BAIER, L. 1930. Contribution to the physiology of the stridulation and hearing of insects. *Zool. Jahrb., Zool. Physiol.*, **47**, 151–248.

BAJANDUROW, B. 1932. Zur Physiologie des Sehanalysators bei Vögeln. *Zsch. vergl. Physiol.*, **18**, 288–306.

BALLACHEY, E. L. 1934. Variations in maze length as a factor influencing rate of learning in the white rat. *Jour. Comp. Psychol.*, **17**, 23–45.

———, and BUEL, J. 1934. Centrifugal swing as a determinant of choice-point behavior in the maze running of the white rat. *Jour. Comp. Psychol.*, **17**, 201–223.

———, and KRECHEVSKY, I. 1932. "Specific" versus "general" orientation factors in maze running. *Univ. Calif. Publ. Psychol.*, **6**, 83–97.

BANTA, A. M. 1910. A comparison of the reactions of a species of surface isopod with those of a subterranean species. *Jour. Exper. Zool.*, **8**, 243–310; 439–488.

BARBOUR, T. 1934 (rev. ed.). *Reptiles and Amphibia, their habits and adaptations.* Boston: Houghton Mifflin. Pp. 129.

BARROWS, W. 1915. The reactions of an orb-weaving spider, *Epeira sclopeteria* Clerck to rhythmic vibrations of its web. *Biol. Bull.*, **29**, 316–332.

BARTELS, M. 1929. Sinnesphysiologische und psychologische Untersuchungen an der Trichterspinne *Agelena labyrinthica*. *Zsch. vergl. Physiol.*, **10**, 527–593.

BATESON, W. 1889. Notes on the senses and habits of some Crustacea. *Jour. Mar. Biol. Assn.*, n.s. **1**, 211–214.

BAUMANN, F. 1929. Experimente über den Geruchssinn und den Beuteerwerb der Viper (*Vipera aspis* L.) *Zsch. vergl. Physiol.*, **10**, 36–119.

BAYLISS, W. M. 1924. *Principles of general physiology.* London: Longmans, Green. Pp. 882.

BERTHOLF, L. M. 1932. The extent of the spectrum for Drosophila and the distribution of stimulative efficiency in it. *Zsch. vergl. Physiol.*, **18**, 32–64.

BETHE, A. 1897. Vergleichende Untersuchungen über die Funktionen des Centralnervensystems der Arthropoden. *Pflüg. Arch. ges. Physiol.*, **68**, 449–545.

———. 1898. Dürfen wir den Ameisen und Bienen psychische Qualitäten zuschreiben? *Pflüg. Arch. ges. Physiol.*, **70**, 15–100.

———. 1931. Plastizität und Zentrenlehre. *Handb. d. Norm. u. Path. Physiol.*, **15** (zweite Hälfte), 1175–1220.

BIERENS DE HAAN, J. A. 1925. Experiments on vision in monkeys. I. The colour-sense of the pig-tailed macaque (*Nemestrimus nemestrimus L*). *Jour. Comp. Psychol.*, **5**, 417–453.

———. 1926. Versuche über den Farbensinn und das psychische Leben von *Octopus vulgaris*. *Zsch. vergl. Physiol.*, **4**, 766–796.

———. 1927. Versuche über das Sehen der Affen. IV. Das Erkennen gleichförmiger und ungleichförmiger Gegenstände bei niederen Affen. V. Erkennen Affen in zweidimensionalen Abbildungen ihnen bekannte Gegenstände wieder? *Zsch. vergl. Physiol.*, **5**, 699–729.

———. 1931. Werkzeuggebrauch und Werkzeugherstellung bei einem niederen Affen (*Cebus hypoleucus Hump*). *Zsch. vergl. Physiol.*, **13**, 639–695.

BINGHAM, H. C. 1913. Size and form perception in *Gallus domesticus*. *Jour. Anim. Behav.*, **3**, 65–113.

———. 1922. Visual perception of the chick. *Behav. Mono.*, **4**, pp. 104.

———. 1929*a*. Chimpanzee translocation by means of boxes. *Comp. Psychol. Mono.*, **5**, pp. 91.

———. 1929*b*. Selective transportation by chimpanzees. *Comp. Psychol. Mono.*, **5**, pp. 45.

BIRD, C. 1925. The relative importance of maturation and habit in the development of an instinct. *Ped. Sem.*, **32**, 68–91.

———. 1933. Maturation and practise; their effects upon the feeding reaction of chicks. *Jour. Comp. Psychol.*, **16**, 343–366.

BLANCHARD, F. 1930. The stimulation to the breeding migration of the spotted salamander. *Amer. Nat.*, **64**, 154–167.

BLODGETT, H. C. 1929. The effect of the introduction of reward upon the maze performance of rats. *Univ. Calif. Publ. Psychol.*, **4**, 113–134.

BOGARDUS, E. S., and HENKE, F. G. 1911. Experiments on tactual sensations in the white rat. *Jour. Anim. Behav.*, **1**, 125–137.

BOHN, G. 1907. Le rhythme nychthéméral chez les actinies. *Compt. rend. Soc. Biol.*, Paris, **1**, 473–476.

———. 1907. Introduction à la psychologie des animaux à symétrie rayonée. *Bull. Inst. gén. Psych.*, 7, 81–129.

——— and DRZEWINA, A. 1928. Les Convoluta. *Ann. Sci. Nat. Zool.*, 10 sér., 11, 299–398.

BONNET, C. 1779–1783. *Oeuvres d'histoire naturelle et de philosophie.* II. Neuchâtel: Fauche.

BOROVSKI, W. M. 1927. Experimentelle Untersuchungen über den Lernprozess. *Zsch. vergl. Physiol.*, 6, 489–529.

BOSE, J. 1926. *The nervous mechanism of plants.* London: Longmans, Green. Pp. 224.

BOULENGER, E. G. 1914. *Reptiles and batrachians.* New York: Dutton. Pp. 278.

BOUVIER, E. L. 1922. *The psychic life of insects.* (L. O. Howard, tr.) New York: Appleton-Century. Pp. 377.

BOVARD, J. F. 1918. The transmission of nervous impulses in relation to locomotion in the earthworm. *Univ. Calif. Publ. Zool.*, 18, 103–134.

BOZLER, E. 1926. Sinnes- und Nervenphysiologische Untersuchungen an Scyphomedusen. *Zsch. vergl. Physiol.*, 4, 37–80.

———. 1926. Weitere Untersuchungen . . . Erregungsleitung, Funktion der Randkörper, Nahrungsaufnahme. *Zsch. vergl. Physiol.*, 4, 797–817.

BREDER, C., BREDER, R., and REDMOND, A. 1927. Frog tagging: A method of studying anuran life habits. *Zoologica*, 9, 201–229.

BREED, F. 1911. The development of certain instincts and habits in chicks. *Behav. Mono.*, 1, pp. 78.

BROWN, C. R., and HATCH, M. 1929. Orientation and "fright" reactions of whirligig beetles (Gyrinidae). *Jour. Comp. Psychol.*, 9, 159–189.

BRUN, R. 1914. *Die Raumorientierung der Ameisen.* Jena: Fischer. Pp. 234.

———. 1916. Weitere Untersuchungen über die Fernorientierung der Ameisen. *Biol. Centrlbl.*, 35, 261–303.

BRUYN, E., and VAN NIFTERIK, C. 1920. Influence du son sur le réaction d'une excitation tactile chez les grenouilles et les crapauds. *Arch. néerl. Physiol.*, 5, 363–379.

BÜCHNER, L. 1880. *Mind in animals* (tr. A. Besant). London: Freethought. Pp. 359.

BUDDENBROCK, W. v. 1915. Die Statocyste von Pecten, ihre Histologie und Physiologie. *Zool. Jahrb., Zool. Physiol.*, 35, 301–356.

BUEL, J. 1934. The linear maze. I. "Choice-point expectancy," "correctness," and the goal-gradient. *Jour. Comp. Psychol.*, 17, 185–199.

BULL, H. 1928. Studies on conditioned responses in fishes. *Jour. Mar. Biol. Assn.*, n.s., 15, 485–533.

BURNETT, T. C. 1912. Some observations on decerebrate frogs with especial reference to the formation of associations. *Amer. Jour. Psychol.*, 30, 80–87.

BURTT, H. E. 1916. A study of the behavior of the white rat by the multiple choice method. *Jour. Anim. Behav.*, 6, 222–246.

BUYTENDIJK, F. J. 1919. Acquisition d'habitudes par des êtres unicellulaires. *Arch. néerl. Physiol.*, 3, 455–468.

———. 1921. Une formation d'habitude simple chez le limaçon d'eau douce (Limnaeus). *Arch. néerl. Physiol.*, 5, 458–466.

———. 1924. Über die Formwahrnehmung beim Hunde. *Pflüg. Arch. Physiol.*, **205**, 4–14.

———. 1930. Über Hemmungen gewöhnter Bewegungen bei Tieren. *Arch. néerl. de Physiol.*, **15**, 381–401.

———. 1931. Le cerveau et l'intelligence. *Jour. de Psychol.*, **28**, 345–371.

———. 1932. An experimental investigation into the influence of cortical lesions on the behavior of rats. *Arch. néerl. de Physiol.*, **17**, 370–434.

——— and REMMERS, J. 1923. Nouvelles recherches sur la formation d'habitudes chez les poissons. *Arch. néerl. de Physiol.*, **8**, 165–184.

CAMERON, N. 1928. Cerebral destruction in its relation to maze learning. *Psychol. Rev.*, **39**, pp. 68.

CAMPBELL, A. A. 1935. Community of function in the performance of rats in alley mazes and the Maier reasoning apparatus. *Jour. Comp. Psychol.*, **19**, 69–76.

CARMICHAEL, L. 1926. The development of behavior in vertebrates experimentally removed from the influence of external stimulation. *Psychol. Rev.*, **33**, 51–58.

———. 1927. A further study of the development of behavior in vertebrates experimentally removed from the influence of external stimulation. *Psychol. Rev.*, **34**, 34–47.

CARPENTER, C. R. 1933. Psychobiological studies of social behavior in Aves. *Jour. Comp. Psychol.*, **16**, 25–97.

———. 1934. A field study of the behavior and social relations of howling monkeys (*Alouatta palliata*). *Comp. Psychol. Mono.*, **10**, pp. 168.

——— and NISSEN, H. W. 1934. An experimental analysis of some spatial variables in delayed reactions of chimpanzees. *Psychol. Bull.*, **31**, 689.

CARR, H. A. 1914. Principles of selection in animal learning. *Psychol. Rev.*, **21**, 157–165.

———. 1917*a*. The distribution and elimination of errors in the maze. *Jour. Anim. Behav.*, **7**, 145–159.

———. 1917*b*. Maze studies with the white rat. I. Normal animals. *Jour. Anim. Behav.*, **7**, 259–275.

———. 1917*c*. Maze studies with the white rat. II. Blind animals. *Jour. Anim. Behav.*, **7**, 277–294.

———. 1917*d*. Maze studies with the white rat. III. Anosmic animals. *Jour. Anim. Behav.*, **7**, 295–306.

——— and WATSON, J. B. 1908. Orientation in the white rat. *Jour. Comp. Neurol. and Psychol.*, **18**, 27–44.

CASTEEL, D. 1911. The discriminative ability of the painted turtle. *Jour. Anim. Behav.*, **1**, 1–28.

CHAPMAN, F. M. 1929. *My tropical air castle.* New York: Appleton. Pp. 416.

CHILD, C. M. 1921. *The origin and development of the nervous system.* Chicago: Univ. Chicago Press. Pp. 296.

———. 1924. *Physiological foundations of behavior.* New York: Holt. Pp. 330.

CHURCHILL, E. 1916. The learning of a maze by goldfish. *Jour. Anim. Behav.*, **6**, 247–255.

CLARK, L. B. 1928. Adaptation versus experience as an explanation of modification in certain types of behavior (circus movements in Notonecta). *Jour. Exper. Zool.*, **51**, 37–50.

CLARKE, G. L. 1930. Change of phototropic and geotropic signs in Daphnia induced by changes of light intensity. *Jour. Exper. Biol.*, **7**, 109–131.

COBURN, C. 1914. The behavior of the crow, *Corvus americanus* Aud. *Jour. Anim. Behav.*, **4**, 185–201.

——— and YERKES, R. M. 1915. A study of the behavior of the crow *Corvus americanus* Aud. by the multiple choice method. *Jour. Anim. Behav.*, **5**, 75–114.

COGHILL, G. E. 1929. *Anatomy and the problem of behaviour.* London: Cambridge Univ. Press. Pp. 113.

———. 1930*a*. The genetic interrelation of instinctive behavior and reflexes. *Psychol. Rev.*, **37**, 264–266.

———. 1930*b*. The structural basis of the integration of behavior. *Proc. Nat. Acad. Sci.*, **16**, 637–643.

COLE, L. J. 1913. Direction of locomotion in the starfish (*Asterias forbesii*). *Jour. Exper. Zool.*, **14**, 1–32.

COLE, L. W. 1907. Concerning the intelligence of raccoons. *Jour. Comp. Neurol. and Psychol.*, **17**, 211–261.

COLEMAN, T. B., and HAMILTON, W. F. 1933. Color blindness in the rat. *Jour. Comp. Psychol.*, **15**, 177–181.

COLVIN, S. S., and BURFORD, C. C. 1909. The color perception of three dogs, a cat, and a squirrel. *Psychol. Rev. Mono.*, **11**, 1–48.

COMMINS, W. D., MCNEMAR, Q., and STONE, C. P. 1932. Intercorrelations of measures of ability in the rat. *Jour. Comp. Psychol.*, **14**, 225–235.

COMSTOCK, J. H. 1912. *The spider book.* Garden City: Doubleday. Pp. 707.

COOK, S. A. 1928. The effect of various temporal arrangements of practice on the mastery of an animal maze of moderate complexity. *Arch. Psychol.*, **15**, pp. 33.

COPELAND, M. 1913. The olfactory reactions of the spotted newt, *Diemyctylus viridescens* (Rafinesque). *Jour. Anim. Behav.*, **3**, 260–272.

———. 1918. The olfactory reactions of the marine snails *Alectrion obsoleta* (Say) and *Busycon canaciculatum* (Linn.). *Jour. Exper. Zool.*, **25**, 177–227.

———. 1930. An apparent conditioned response in *Nereis virens*. *Jour. Comp. Psych.*, **10**, 339–354.

——— and WIEMAN, H. 1924. The chemical sense and feeding behavior of *Nereis virens* Sars. *Biol. Bull.*, **42**, 231–238.

CORNETZ, V. 1910. Trajets de fourmis et retours au nid. *Inst. gén. Psychol.*, Mém. no. 2, pp. 167.

CORONIOS, J. D. 1933. Development of behavior in the fetal cat. *Genet. Psychol. Mono.*, **14**, 283–286.

COWDRY, E. V. (ed.) and others. 1924. *General cytology.* Chicago: Univ. of Chicago Press. Pp. 754.

COWLES, R. F. 1910. Stimuli produced by light and by contact with solid walls as factors in the behavior of ophiuroids. *Jour. Exper. Zool.*, **9**, 387–416.

CRAIG, W. 1908. The voices of pigeons regarded as a means of social control. *Am. Jour. Sociol.*, **14**, 86–100.

———. 1914. Male doves reared in isolation. *Jour. Anim. Behav.*, **4**, 121–133.

CRAWFORD, S. 1934. The habits and characteristics of nocturnal animals. *Quart. Rev. Biol.*, **9**, 201–214.

CROSBY, ELIZABETH C. 1917. The forebrain of *Alligator mississippiensis*. *Jour. Comp. Neurol.*, **27**, 325–402.

CROZIER, W. J. 1920. Notes on some problems of adaptation. 2. . . . *Biol. Bull.*, **39**, 116–129.

———. 1926. Note on the critical temperatures for biological processes. *Jour. Gen. Physiol.*, **9**, 525–528.

——— and COLE, W. 1929. The phototropic excitation of Limax. *Jour. Gen. Physiol.*, **12**, 669–674.

——— and FEDERIGHI, H. 1925. Phototropic circus movements of Limax as influenced by temperature. *Jour. Gen. Physiol.*, **7**, 151–169.

CULLER, E., and METTLER, F. A. 1934. Conditioned behavior in a decorticate dog. *Jour. Comp. Psychol.*, **18**, 291–303.

CUMMINS, H. 1920. The rôle of voice and coloration in spring migrations and sex recognition in frogs. *Jour. Exper. Zool.*, **30**, 325–343.

CURTIS, Q. F. 1931. *Control of floor cues in the maze learning of white rats.* Thesis, Ohio Wesleyan Univ.

DAKIN, W. 1910. The visceral ganglion of Pecten, with some notes. *Mitt. Zool. Sta. Neapel*, **20**, 1–40.

DANISCH, F. 1921. Ueber Reizbiologie und Reizempfindlichkeit von Vorticella nebulifera. *Zsch. allg. Physiol.*, **19**, 133–190.

DASHIELL, J. F. 1920*a*. The need for analytical study of the maze problem. *Psychobiol.*, **2**, 181–186.

———. 1920*b*. Some transfer factors in maze learning by the white rat. *Psychobiol.*, **2**, 329–350.

———. 1930. Direction orientation in maze running by the white rat. *Comp. Psychol. Mono.*, **7**, pp. 72.

——— and BAYROFF, A. G. 1931. A forward-going tendency in maze running. *Jour. Comp. Psychol.*, **12**, 77–94.

DAVIS, H. B. 1907. The raccoon: a study of animal intelligence. *Am. Jour. Psychol.*, **18**, 447–489.

DAWSON, JEAN. 1911. The biology of Physa. *Behav. Mono.*, **1**, pp. 120.

DAY, LUCY, and BENTLEY, M. 1911. A note on learning in Paramecium. *Jour. Anim. Behav.*, **1**, 67–73.

DECAMP, J. E. 1920. Relative distance as a factor in the white rat's selection of a path. *Psychobiol.*, **2**, 245–253.

DENNIS, W. 1929. The sensory control of the white rat in the maze habit. *Jour. Genet. Psychol.*, **36**, 59–89.

——— and SOLLENBERGER, R. T. 1934. Negative adaptation in the maze exploration of albino rats. *Jour. Comp. Psychol.*, **18**, 197–206.

DESCY, A. 1925. La vie social chez les insectes. *Ann. Sci. Nat. Zool.*, 10 ser., **8**, 87–104.

DEVOSS, J. C., and GANSON, ROSE. 1915. Color blindness of cats. *Jour. Anim. Behav.*, **5**, 115–139.

Dodson, J. D. 1918. Relative values of reward and punishment in habit formation. *Psychobiol.*, **1**, 231–276.

Donaldson, J. C. 1928. Adrenal gland in wild gray and albino rat: cortico medullary relations. *Proc. Soc. Exper. Biol. and Med.*, **25**, 300–301.

Dunlap, K. 1931. Standardizing electric shocks for rats. *Jour. Comp. Psychol.*, **12**, 133–135.

———, Gentry, Evelyn, and Zeigler, T. W. 1931. The behavior of white rats under food and electric shock stimulation. *Jour. Comp. Psychol.*, **12**, 371–378.

Dusser de Barenne, J. G. 1933. Welche Elemente der Grosshirnrinde bringen bei ihrer elektrischen Reizung die motorischen Reaktionen hervor? *Pflüg. Arch. ges. Physiol.*, **233**, 529–536.

Dykgraaf, S. 1933. Untersuchungen über die Funktion der Seitenorgane an Fischen. *Zsch. vergl. Physiol.*, **20**, 162–214.

Edinger, L. 1908. The relations of comparative anatomy to comparative psychology. (H. Rand, tr.) *Jour. Comp. Neur. Psychol.*, **18**, 437–457.

Edwards, J. 1923. The effect of chemicals on locomotion in Amoeba. *Jour. Exper. Zool.*, **38**, 1–43.

Eidmann, H. 1925. Das Mitteilungsvermögen der Ameisen. *Die Naturwiss.*, Berl., **13**, 126–128.

———. 1927. Die Sprache der Ameisen. *Rev. Zool. Russe*, **7**, 39–47.

Elder, J. H. 1934. Auditory acuity of the chimpanzee. *Jour. Comp. Psychol.*, **17**, 157–183.

Elliott, M. H. 1929*a*. The effect of appropriateness of reward and of complex incentives on maze performance. *Univ. Calif. Publ. Psychol.*, **4**, 91–98.

———. 1929*b*. The effect of change of "drive" on maze performance. *Univ. Calif. Publ. Psychol.*, **4**, 185–188.

Emerson, A. E. 1926. Development of soldier termites. *Zoologica*, **7**, 69–100.

———. 1928. Communication among termites. IV. Int. Cong. Entom., II.

Engelmann, T. 1882. Ueber Licht und Farbenperception niederster Organismen. *Pflüg. Arch. ges Physiol.*, **29**, 387–400.

Engelmann, W. 1928. Untersuchungen über die Schallokalisation bei Tieren. *Zsch. Psychol.*, **105**, 317–370.

Espinas, A. 1878. *Les sociétés animales, étude de psychologie comparée.* Paris: Baillière. Pp. 588.

Evans, H. M. 1931. A comparative study of the brains in British cyprinoids in relation to their habits of feeding, with special reference to the anatomy of the Medulla oblongata. *Proc. Roy. Soc. Lond.*, **108B**, 233–257.

Ewald, W. 1910. Über Orientierung, Lokomotion und Lichtreaktion einiger Cladoceren und deren Bedeutung für die Theorie der Tropismen. *Biol. Centbl.*, **30**, 1–16; 49–63; 379–384; 385–399.

Fabre, J. H. 1918. *The wonders of instinct.* (Tr.) New York: Appleton-Century. Pp. 322.

Fielde, Adele. 1901. Further study of an ant. *Proc. Acad. Nat. Scs. Phila.*, pp. 521–544.

———. 1904. The power of recognition among ants. *Biol. Bull*, **7**, 227–250.

Fields, P. E. 1928. Form discrimination in the white rat *Jour. Comp. Psychol.*, **8**, 143–158.

———. 1929. The white rat's use of visual stimuli in the discrimination of geometrical figures. *Jour. Comp. Psychol.*, **9**, 107–122.

———. 1932. Studies in concept formation. *Comp. Psychol. Mono.*, **9**, pp. 70.

FISCHEL, W. 1930. Weitere Untersuchung der Ziele der tierischen Handlung. *Zsch. vergl. Physiol.*, **11**, 523–548.

———. 1931. Dressurversuche mit Schnecken. *Zsch. vergl. Physiol.*, **15**, 50–70.

———. 1932. Methoden zur psychologischen Untersuchung der Wirbeltiere. *Handb. biol. Arbeitsmethoden.* Abt. IV. Teil D. Heft 3, 233–338.

FLEURE, H., and WALTON, C. 1907. Notes on the habits of some sea anemones. *Zool. Anz.*, **31**, 212–220.

FOCKE, F. 1930. Experimente und Beobachtungen über die Biologie des Regenwurms, unter besonderer Berücksichtigung der Frage nach der Raumorientierung. *Zsch. wiss. Zool.*, **136**, 376–421.

FÖH, H. 1932. Der Schattenreflex bei *Helix pomatia* nebst Bemerkungen über den Schattenreflex bei *Mytilus edulis*, *Limnaea stagnalis* und *Testudo iberz.* *Zool. Jahrb.*, *Zool. Physiol.*, **52**, 1–78.

FOLEY, J. 1923. Reflexes of spinal frogs as induced by chemical stimulation. *Jour. Comp. Neurol.*, **35**, 15–20.

FOLEY, J. P. 1934. First year development of a rhesus monkey (*Macaca mulatta*) reared in isolation. *Jour. Genet. Psychol.*, **45**, 39–105.

——— and WARDEN, C. J. 1934. The effect of practice on the delayed reaction in the rhesus monkey. *Jour. Genet. Psychol.*, **44**, 390–413.

FOLGER, H. T. 1927. The relation between the responses by Amoeba to shock and to sudden illumination. *Biol. Bull.*, **53**, 405–412.

FORBES, A., MILLER, R. H., and O'CONNOR, J. 1927. Electric responses to acoustic stimuli in the decerebrate animal. *Am. Jour. Physiol.*, **80**, 363–380.

FOREL, A. 1908. *The senses of insects.* (tr. Yearsley). London: Methuen. Pp. 324.

———. 1928. *The social world of the ants compared with that of man.* (tr.) New York: Putnam's. 2 vols.

FRAENKEL, G. 1928. Über den Auslösungsreiz des Umdrehreflexes bei Seesternen und Schlangensternen. *Zsch. vergl. Physiol.*, **7**, 365–378.

———. 1932. Untersuchungen über die Koordination von Reflexen und automatisch-nervösen Rhythmen bei Insekten. I. Die Flugreflexe der Insekten und ihre Koordination. *Zsch. vergl. Physiol.*, **16**, 371–393.

FRANK, MARGARET. 1932. The effect of a rickets-producing diet on the learning ability of white rats. *Jour. Comp. Psychol.*, **13**, 87–105.

FRANZ, S. I. 1907. On the functions of the cerebrum. The frontal lobes. *Arch. Psychol.*, **1**, 1–64.

———. 1915. On the functions of the cerebrum. II. Variation in distribution of motor centers. *Psychol. Mono.*, **19**, 80–162.

FRANZ, V. 1927. Zur tierpsychologischen Stellung von *Rana temporaria* und *Bufo calamita*. *Biol. Zentrlbl.*, **47**, 1–12.

FRIEDLÄNDER, B. 1894. Beiträge zur Physiologie des Zentralnervensystems und Bewegungsmechanismus der Regenwürmer. *Pflüg. Arch. ges. Physiol.*, **58**, 168–207.

FRIEDRICH, H. 1932. Studien über die Gleichgewichtserhaltung und Bewegungs-physiologie bei Pterotrachea. *Zsch. vergl. Physiol.*, **16**, 345–361.

FRISCH, K. v. 1914. Der Farbensinn und Formensinn der Biene. *Zool. Jahrb., Zool. Physiol.*, **35**, 1–182.

———. 1919. Über den Geruchsinn der Bienen und seine blutenbiologische Bedeutung. *Zool. Jahrb., Zool. Physiol.*, **37**, 1–238.

———. 1923. Über die Sprache der Bienen. *Zool. Jahrb., Zool. Physiol.*, **20**, 1–186.

——— und RÖSCH, G. 1926. Neue Versuche über die Bedeutung von Duftorgan und Pollenduft für die Verständigung im Bienenvolk. *Zsch. vergl. Physiol.*, **4**, 1–21.

——— und STETTER, H. 1932. Untersuchungen über den Sitz des Gehörsinnes bei der Elritze. *Zsch. vergl. Physiol.*, **17**, 686–801.

FROLOFF, J. 1925. Bedingte Reflexe bei Fischen. *Pflüg. Arch. ges. Physiol.*, **220**, 339–349.

FUERTES, L. W. 1916. Impressions of the voices of tropical birds. *Smiths. Inst. Ann. Rep.* 1915, 299–323.

GARREY, W. E., and MOORE, A. R. 1915. Peristalsis and coordination in the earthworm. *Amer. Jour. Physiol.*, **39**, 139–148.

GARTH, T., and MITCHELL, MARY. 1925. The learning curve of a land snail. *Jour. Comp. Psychol.*, **6**, 103–113.

GEE, W. 1913. Modifiability in the behavior of the California shore-anemone *Cribrina xanthogrammica* Brandt. *Jour. Anim. Behav.*, **3**, 305–328.

GELLERMAN, L. W. 1933. Form discrimination in chimpanzees and two-year old children. I. Form (triangularity) *per se*. II. Form versus back ground. *Jour. Genet. Psychol.*, **42**, 3–50.

GENGERELLI, J. A. 1930*a*. The principle of maxima and minima in animal learning. *Jour. Comp. Psychol.*, **11**, 193–236.

———. 1930*b*. Studies in abstraction in the white rat. *Jour. Genet. Psychol.*, **38**, 171–202.

GIERSBERG, H. 1926. Über den chemischen Sinn von *Octopus vulgaris*. *Zsch. vergl. Physiol.*, **3**, 827–838.

GLASER, O. C. 1907. Movement and problem solving in Ophiura. *Jour. Exper. Zool.*, **4**, 203–219.

GOETSCH, W. 1928. Beiträge zur Biologie körnersammelnder Ameisen. *Zsch. Morph. Oekol. Tiere*, **10**, 353–419.

———. 1930. Beiträge zur Biologie körnersammelnder Ameisen. *Zsch. Morph. Oekol. Tiere*, **16**, 371–452.

GOLTZ, F. 1869. *Beiträge zur Lehre von den Funktionen der Nervencentren des Frosches*. Berlin: Hirschwald. Pp. 130.

GREENE, C. 1926. The physiology of the spawning migration. *Physiol. Rev.*, **6**, 201–241.

GRINDLEY, G. C. 1929. Experiments on the influence of the amount of reward on learning in young chickens. *Brit. Jour. Psychol.*, **20**, 173–180.

GROSS, A. 1921. The feeding habits and chemical sense of *Nereis virens* Sars. *Jour. Exper. Zool.*, **32**, 427–442.

GULLIKSEN, H. 1932. Studies of transfer of response. I. Relative versus absolute factors in the discrimination of size by the white rat. *Jour. Genet. Psychol.*, **40**, 37–51.

GUNDLACH, R. 1932. A field study of homing in pigeons. *Jour. Comp. Psychol.*, **13**, 397–402.

———. 1933. The visual acuity of homing pigeons. *Jour. Comp. Psychol.*, **16**, 323–342.

HAECKER, V. 1912. Über Lernversuche bei Axolotln. *Arch. ges. Psychol.*, **25**, 1–35.

HAMILTON, EDNA L. 1929. The effect of delayed incentive on the hunger drive in the white rat. *Genet. Psychol. Mono.*, **5**, 137–207.

HAMILTON, G. V. 1911. A study of trial and error reactions in mammals. *Jour. Anim. Behav.*, **1**, 33–66.

HAMILTON, W. F. 1921. Coordination in the starfish. I. Behavior of the individual tube feet. *Jour. Comp. Psychol.*, **1**, 473–488.

———. 1922*a*. Coordination in the starfish. II. Locomotion. *Jour. Comp. Psychol.*, **2**, 61–76.

———. 1922*b*. Coordination in the starfish. III. The righting reaction as a phase of locomotion (righting and locomotion). *Jour. Comp. Psychol.*, **2**, 81–94.

——— and COLEMAN, T. B. 1933. Trichromatic vision in the pigeon as illustrated by the spectral hue discrimination curve. *Jour. Comp. Psychol.*, **15**, 183–191.

——— and GOLDSTEIN, J. L. 1933. Visual acuity and accommodation in the pigeon. *Jour. Comp. Psychol.*, **15**, 193–197.

HANAWALT, ELLA M. 1931. Whole and part methods in trial and error learning. *Comp. Psychol. Mono.*, **7**, pp. 65.

———. 1933. Whole and part methods in trial and error learning. A supplementary study. *Jour. Comp. Psychol.*, **15**, 395–406.

HARGITT, C. 1907. Notes on the behavior of sea anemones. *Biol. Bull.*, **12**, 274–284.

HARLOW, H. F., UEHLING, H., and MASLOW, A. H. 1932. Comparative behavior of primates. I. Delayed reaction tests on primates. *Jour. Comp. Psychol.*, **13**, 313–343.

HARPER, E. H. 1905. Reactions to light and mechanical stimuli in the earthworm, *Perichaeta bermudensis* (Beddard). *Biol. Bull.*, **10**, 17–34.

HAUG, G. 1933. Die Lichtreaktionen der Hydren (*Chlorohydra viridissima* und *Pelmatohydra oligactis* (P.) *typica*). *Zsch. vergl. Physiol.*, **19**, 246–303.

HECHT, S. 1919*a*. Sensory equilibrium and dark adaptation in *Mya arenaria*. *Jour. Gen. Physiol.*, **1**, 545–558.

———. 1919*b*. The nature of the latent period in the photic response of *Mya arenaria*. *Jour. Gen. Physiol.*, **1**, 657–666.

———. 1920. The photochemical nature of the photosensory process. *Jour. Gen. Physiol.*, **2**, 229–246.

———. 1929. *The nature of the photoreceptor process.* Found. Exper. Psychol., Chap. 5 (ed. by Murchison). Worcester: Clark Univ. Press.

——— and WOLF, E. 1929. The visual acuity of the honey bee. *Jour. Gen. Physiol.*, **12**, 727–760.

HECK, L. 1920. Über die Bildung einer Assoziation beim Regenwurm auf Grund von Dressurversuchen. *Lotos Naturwiss. Zsch.*, **68**, 168–189.

HELSON, H. 1927. Insight in the white rat. *Jour. Exper. Psychol.*, **10**, 378–396.

HEMPELMANN, F. 1926. *Tierpsychologie vom Standpunkte des Biologen.* Leipzig: Akad. Verlagsges. Pp. 676.

HERING, E. 1897. (2d ed.) *On memory and the specific energies of the nervous system* (tr.). Chicago: Open Court. Pp. 50.

HERINGTON, G. B., and GUNDLACH, R. H. 1933. How well can guinea pigs and cats hear tones? *Jour. Comp. Psychol.*, **16**, 287–303.

HERON, W. T. 1922. The reliability of the inclined plane problem box as a method of measuring the learning ability of the rat. *Comp. Psychol. Mono.*, **1**, 1–36.

HERRICK, C. J. 1903. The organ and sense of taste in fishes. *Bull. U. S. Fish. Comm.*, **22**, 237–272.

———. 1924. *Neurological foundations of animal behavior.* New York: Holt. Pp. 334.

———. 1926. *Brains of rats and men.* Chicago: Univ. Chicago Press. Pp. 382.

———. 1933. The amphibian forebrain. VI. *Jour. Comp. Neurol.*, **58**, 1–288.

———. 1934. The amphibian forebrain. X. *Jour. Comp. Neurol.*, **59**, 239–266.

HERRICK, F. 1901. *The home life of wild birds.* New York: Putnam's. Pp. 148.

———. 1911. Nests and nest-building in birds. *Jour. Anim. Behav.*, **1**, 151–192; 244–276; 336–373.

HERTER, K. 1926. Versuche über die Phototaxis von *Nereis diversicolor* O. F. Müller. *Zsch. vergl. Physiol.*, **4**, 103–141.

———. 1932. Beiträge zur Zentrenfunktion zehnfüssiger Krebse. *Zsch. vergl. Physiol.*, **17**, 209–266.

HERTZ, MATHILDE. 1928. Weitere Versuche an der Rabenkrähe. *Psychol. Forsch.*, **10**, 111–141.

———. 1931. Die Organisation des optischen Feldes bei der Biene. *Zsch. vergl. Physiol.*, **14**, 629–674.

———. 1933. Über das Verhalten des Einsiedlerkrebses *Clibanarius misanthropus* gegenüber verschiedenen Gehäuseformen. *Zsch. vergl. Physiol.*, **18**, 597–621.

HESS, W. 1924. Reactions to light in the earthworm, *Lumbricus terrestris* L. *Jour. Morph. Physiol.*, **39**, 515–542.

HEYDE, KAETHE. 1924. Die Entwicklung der psychischen Fähigkeiten bei Ameisen und ihr Verhalten bei abgeänderten biologischen Bedingungen. *Biol. Zentrlbl.*, **44**, 624–654.

HINSCHE, G. 1926. Über Brunst- und Kopulationsreaktion von *Bufo vulgaris*. *Zsch. vergl. Physiol.*, **4**, 564–606.

HOAGLAND, H. 1928. On the mechanism of tonic immobility in vertebrates. *Jour. Gen. Physiol.*, **11**, 715–742.

——— and CROZIER, W. 1931. Geotropic excitation in Helix. *Jour. Gen. Physiol.*, **15**, 15–28.

HOGE, MILDRED A., and STOCKING, RUTH J. 1912. A note on the relative value of punishment and reward as motives. *Jour. Anim. Behav.*, **2**, 43–50.

HOLDEN, FRANCES. 1926. A study of the effect of starvation upon behavior by means of the obstruction method. *Comp. Psychol. Mono.*, **3**, pp. 45.

HOLMES, S. J. 1905. The selection of random movements as a factor in phototaxis. *Jour. Comp. Neurol. Psychol.*, **15**, 98–112.

———. 1905. The reactions of Ranatra to light. *Jour. Comp. Neurol. Psychol.*, **15**, 305–349.

———. 1906. *The biology of the frog.* (1927, 4th ed.) New York: Macmillan. Pp. 386.

———. 1916. *Studies in animal behavior.* Boston: Badger. Pp. 266.

——— and MCGRAW, K. 1913. Some experiments on the method of orientation to light. *Jour. Anim. Behav.*, **3**, 367–373.

HOLST, E. V. 1934. Zwei Versuche zum Hirnproblem der Arthropoden. *Pflüg. Arch. ges. Physiol.*, **234**, 114–123.

HOLT, E. B. 1931. *Animal drive and the learning process.* New York: Holt. Pp. 307.

HONIGMANN, H. 1921. Zur Biologie der Schildkröten. *Biol. Zentrlbl.*, **41**, 241–250.

HOOKER, D. 1911. Certain reactions to color in the young loggerhead turtle. (Papers Tortugas Lab.) *Carn. Inst. Wash.*, **3**, No. 132, 71–76.

HOPKINS, A. E. 1927. On the physiology of the central nervous system in the starfish, *Asterias tennispinosa. Jour. Exper. Zool.*, **46**, 263–275.

HORTON, G. P. 1933. A quantitative study of hearing in the guinea pig (*Cavia coboya*). *Jour. Comp. Psychol.*, **15**, 59–73.

HOVEY, H. B. 1929. Associative hysteresis in flatworms. *Physiol. Zool.*, **2**, 322–333.

HOWARD, H. E. 1920. *Territory in bird life.* London: Murray. Pp. 308.

———. 1929. *An introduction to bird behavior.* New York: Macmillan. Pp. 146.

HOWLAND, RUTH B. 1924. Dissection of the pellicle of Amoeba verrucosa. *Jour. Exper. Zool.*, **40**, 263–270.

HUBBERT, HELEN B., and LASHLEY, K. S. 1917. Retroactive association and the elimination of errors in the maze. *Jour. Anim. Behav.*, **7**, 130–138.

HUBER, P. 1810. *Recherches sur les moeurs des fourmis indigenes.* Paris: Paschaud. Pp. 328.

HULL, C. L. 1932. The goal gradient hypothesis and maze learning. *Psychol. Rev.*, **39**, 25–43.

———. 1934. The factor of the conditioned reflex. *Handb. Gen. Exper. Psychol.* (ed. Murchison). Worcester: Clark Univ. Press, 382–455.

HUNTER, W. S. 1911. Some labyrinth habits of the domestic pigeon. *Jour. Anim. Behav.*, **1**, 278–304.

———. 1913. The delayed reaction in animals and children. *Behav. Mono.*, **2**, pp. 86.

———. 1914. The auditory sensitivity of the white rat. *Jour. Anim. Behav.*, **4**, 215–222.

———. 1917. The delayed reaction in a child. *Psychol. Rev.*, **24**, 74–87.

———. 1922. Correlation studies with the maze in rats and humans. *Comp. Psychol. Mono.*, **1**, 37–56.

———. 1927. Further data on the auditory sensitivity of the white rat. *Jour. Genet. Psychol.*, **34**, 177–187.

———. 1929. The sensory control of the maze habit in the white rat. *Jour. Genet. Psychol.*, **36**, 505–537.

HYMAN, LIBBIE. 1917. Metabolic gradients in Amoeba and their relation to the mechanism of amoeboid movement. *Jour. Exper. Zool.*, **24**, 381–407.

———. 1919. Physiological studies in Planaria. III. *Biol. Bull.*, **37**, 388–403.

IMMS, A. 1931. *Social behavior in insects.* New York: Dial Press. Pp. 117.

INGEBRITSEN, O. C. 1932. Maze learning after lesions in the cervical cord. *Jour. Comp. Psychol.*, **14**, 279–294.

JACOBSEN, C. F. 1931. A study of cerebral function in learning. The frontal lobes. *Jour. Comp. Neurol.*, **52**, 271–340.

JACOBSEN, C. J., and WOLFE, J. B. 1934. An experimental analysis of frontal lobe function in monkeys and chimpanzees. *Psychol. Bull.*, **31**, 692.

JANZEN, W. 1933. Untersuchungen über Grosshirnfunktionen des Goldfisches (*Carassius auratus*). *Zool. Jahrb.*, *Zool. Physiol.*, **52**, 591–628.

JARMER, K. 1928. *Das Seelenleben der Fische.* München, Berlin: Oldenbourg. Pp. 131.

JENKINS, T. N. 1927. A standard problem box of multiple complexity for use in comparative studies. *Jour. Comp. Psychol.*, **7**, 129–144.

JENNINGS, H. S. 1905. Modifiability in behavior. I. Behavior of sea anemones. *Jour. Exper. Zool.*, **2**, 447–472.

———. 1906*a*. *The behavior of lower organisms.* (1923 ed.) New York: Columbia Univ. Press. Pp. 366.

———. 1906*b*. Modifiability in behavior. II. Factors determining direction and character of movement in the earthworm. *Jour. Exper. Zool.*, **3**, 435–455.

———. 1907. Behavior of the starfish, *Asterias forreri* de Loriol. *Univ. Calif. Publ. Zool.*, **4**, 53–185.

———. 1909. The work of J. v. Uexkuell on the physiology of movements and behavior. *Jour. Comp. Neurol. Psychol.*, **19**, 331–336.

———. 1930. *The biological basis of human nature.* New York: Norton. Pp. 384.

JOHNSON, H. M. 1913. Audition and habit formation in the dog. *Behav. Mono.*, **2**, pp. 78.

———. 1914. Visual pattern-discrimination in the vertebrates. II. Comparative visual acuity in the dog, the monkey, and the chick. *Jour. Anim. Behav.*, **4**, 340–361.

JORDAN, D. S. 1907. *Fishes.* New York: Appleton. Pp. 773.

JORDAN, H. 1917. Das Wahrnehmen der Nahrung bei *Aplysia limacina* und *Aplysia depilans*. *Biol. Zentrlbl.*, **37**, 2–9.

———. 1929. *Allgemeine vergleichende Physiologie der Tiere.* Berlin, Leipzig: de Gruyter. Pp. 761.

KAFKA, G. 1922. *Handbuch der vergleichenden Psychologie.* T. I. München: Reinhardt. Pp. 144.

KAHMAN, H. 1932. Sinnesphysiologische Studien an Reptilien. I. Experimentelle Untersuchungen über das Jakobson'sche Organ der Eidechsen und Schlangen. *Zool. Jahrb.*, *Zool. Physiol.*, **51**, 173–238.

KARN, H. W., and MUNN, N. L. 1932. Visual pattern discrimination in the dog. *Jour. Genet. Psychol.*, **40**, 363–374.

KATZ, D., und TOLL, A. 1923. Die Messung von Charakter- und Begabungsunterschieden bei Tieren (Versuch mit Hühnern). *Zsch. Psychol.*, **93**, 287–311.

KELLOGG, R. 1929. Habits and economic importance of alligators. U. S. Dept. Agric., *Tech. Bull.* **147**.

KELLOGG, V. L. 1906. Some silkworm moth reflexes. *Biol. Bull.*, **12**, 152–154.

KENNEDY, C. H. 1927. Some non-nervous factors that condition the sensitivity of insects to moisture, temperature, light and odors. *Ann. Ent. Soc. Amer.*, **20**, 87–106.

KEPNER, W. A., and RICH, A. 1918. Reactions of the proboscis of *Planaria albissima* Vejdovsky. *Jour. Exper. Zool.*, **26**, 83–100.

KIKUCHI, K. 1930. Diurnal migration of plankton Crustacea. *Quart. Rev. Biol.*, **5**, 189–206.

KINNAMAN, A. J. 1902. Mental life of two Macacus rhesus monkeys in captivity. *Am. Jour. Psychol.*, **13**, 98–148.

KIRK, S. A. 1935. *The effect of unilateral lesions on handedness, pattern vision, and reasoning in the albino rat.* Thesis, Univ. Mich. Library.

KLÜVER, H. 1933. *Behavior mechanisms in monkeys.* Chicago: Univ. Chicago Press. Pp. 387.

KOEHLER, O. 1932. Beiträge zur Sinnesphysiologie der Süsswasserplanarien. *Zsch. vergl. Physiol.*, **16**, 606–756.

KÖHLER, W. 1918. *Nachweis einfacher Strukturfunktionen beim Schimpansen und beim Haushuhn.* Abh. König. preus. Akad. Wiss., phys.-math. Klasse. No. 2, 3–101.

———. 1926. *The mentality of apes.* New York: Harcourt, Brace. Pp. 342.

KRASNOGORSKI, N. 1926. Die letzten Fortschritte in der Erforschung der Methodik der bedingten Reflexe an Kindern. *Jahrb. f. Kinderhk.*, **114**, 255–267.

KRECHEVSKY, I. 1932*a*. "Hypotheses" versus "chance" in the pre-solution period in sensory discrimination-learning. *Univ. Calif. Publ. Psychol.*, **6**, 27–44.

———. 1932*b*. The genesis of "hypotheses" in rats. *Univ. Calif. Publ. Psychol.*, **6**, 45–64.

———. 1934. Brain mechanisms and "hypotheses" in the rat. *Psychol. Bull.*, **31**, 193.

KRIBS, H. G. 1910. The reactions of Aelosoma (Ehrenberg) to chemical stimuli. *Jour. Exper. Zool.*, **8**, 43–74.

KROH, O. 1927. Weitere Beiträge zur Psychologie des Haushuhns. *Zsch. Psychol.*, **103**, 203–227.

KRÖNING, F. 1925. Über die Dressur der Biene auf Töne. *Biol. Zentrlbl.*, **45**, 496–507.

KROPP, B., and ENZMANN, E. 1933. Photic stimulation and leg movements in the crayfish. *Jour. Gen. Physiol.*, **16**, 905–910.

KÜHN, A. 1919. *Die Orientierung der Tiere im Raum.* Jena: Fischer. Pp. 71.

KUO, Z. Y. 1922. The nature of unsuccessful acts and their order of elimination in animal learning. *Jour. Comp. Psychol.*, **2**, 1–27.

———. 1924. A psychology without heredity. *Psychol. Rev.*, **31**, 427–448.

———. 1930. The genesis of the cat's responses to the rat. *Jour. Comp. Psychol.*, **11**, 1–35.

———. 1932*a*. Ontogeny of embryonic behavior in Aves. I. The chronology and general nature of the behavior of the chick embryo. *Jour. Exper. Zool.*, **61**, 395–430.

———. 1932*b*. Ontogeny of embryonic behavior in Aves. II. The mechanical factors in the various stages leading to hatching. *Jour. Exper. Zool.*, **62**, 453–489.

———. 1932*c*. Ontogeny of embryonic behavior in Aves. III. The structure and environmental factors in embryonic behavior. *Jour. Comp. Psychol.*, **13**, 245–272.

———. 1932*d*. Ontogeny of embryonic behavior in Aves. IV. The influence of embryonic movements upon the behavior after hatching. *Jour. Comp. Psychol.*, **14**, 109–122.

———. 1932*e*. Ontogeny of embryonic behavior in Aves. V. The reflex concept in the light of embryonic behavior in birds. *Psychol. Rev.*, **39**, 499–515.

LANGENDORFF, O. 1877. Die Beziehungen des Sehorgans zu den reflexhemmenden Mechanismen des Froschgehirns. *Arch. Anat. Physiol.*, Physiol. Abt., Jg. 1877, 435–442.

LASHLEY, K. S. 1912. Visual discrimination of size and form in the albino rat. *Jour. Anim. Behav.*, **2**, 310–331.

———. 1916. The color vision of birds. I. The spectrum of the domestic fowl. *Jour. Anim. Behav.*, **6**, 1–26.

———. 1918. A simple maze: with data on the relation of the distribution of practice to the rate of learning. *Psychobiol.*, **1**, 353–367.

———. 1920. Studies of cerebral function in learning. *Psychobiol.*, **2**, 55–135.

———. 1923. Temporal variation in the function of the gyrus precentralis in primates. *Am. Jour. Physiol.*, **65**, 585–602.

———. 1923. The behavioristic interpretation of consciousness. *Psychol. Rev.*, **30**, 237–277; 329–353.

———. 1924. Studies of cerebral function in learning. V. The retention of motor habits after destruction of the so-called motor areas in primates. *Arch. Neurol. & Psychiat.*, **12**, 249–276.

———. 1926. Studies of cerebral function in learning. VII. The relation between cerebral mass, learning, and retention. *Jour. Comp. Neurol.*, **41**, 1–58.

———. 1929. *Brain mechanisms and intelligence.* Chicago: Univ. Chicago Press. Pp. 186.

———. 1930*a*. The mechanism of vision. I. A method for rapid analysis of pattern-vision in the rat. *Jour. Genet. Psychol.*, **37**, 353–460.

———. 1930*b*. The mechanism of vision. II. The influence of cerebral lesions upon the threshold of discrimination for brightness in the rat. *Jour. Genet. Psychol.*, **37**, 461–480.

———. 1933. Integrative functions of the cerebral cortex. *Physiol. Rev.*, **13**, 1–42.

———. 1934*a*. The mechanism of vision. VII. The projection of the retina upon the primary optic centers in the rat. *Jour. Comp. Neurol.*, **59**, 341–373.

———. 1934*b*. The mechanism of vision. VIII. The projection of the retina upon the cerebral cortex of the rat. *Jour. Comp. Neurol.*, **60**, 57–79.

———. 1935*a*. Studies of cerebral function in learning. XI. The behavior of the rat in latch box situations. *Comp. Psychol. Mono.*, **11**, 5–42.

———. 1935*b*. The mechanism of vision. XII. Nervous structures concerned in the acquisition and retention of habits based on reactions to light. *Comp. Psychol. Mono.*, **11**, 43–79.

———, and FRANZ, S. I. 1917. The effects of cerebral destruction upon habit-formation and retention in the albino rat. *Psychobiol.*, **1**, 71–140.

———, and MCCARTHY, DOROTHEA A. 1926. The survival of the maze habit after cerebellar injuries. *Jour. Comp. Psychol.*, **6**, 423–433.

———, and BALL, JOSEPHINE. 1929. Spinal conduction and kinaesthetic sensitivity in the maze habit. *Jour. Comp. Psychol.*, **9**, 71–105.

———, and FRANK, MARGARET. 1932. The mechanism of vision. VI. The lateral portion of the area striata in the rat: a correction. *Jour. Comp. Neurol.*, **55**, 525–529.

——— ———. 1934. The mechanism of vision. X. Postoperative disturbances of habits based on detail vision in the rat after lesions in the cerebral visual areas. *Jour. Comp. Psychol.*, **17**, 355–391.

———, and WILEY, L. E. 1933. Studies of cerebral function in learning. IX. Mass action in relation to the number of elements in the problem to be learned. *Jour. Comp. Neurol.*, **57**, 3–55.

———, and RUSSELL, J. T. 1934. The mechanism of vision. XI. A preliminary test of innate organization. *Jour. Genet. Psychol.*, **45**, 136–144.

LEBOUR, MARIE. 1923. The food of plankton organisms. (II). *Jour. Mar. Biol. Assn.*, **13**, 70–92.

LEUBA, C. J. 1931. Some comments on the first reports of the Columbia study of animal drives. *Jour. Comp. Psychol.*, **11**, 275–279.

LIDDELL, H. S., and ANDERSON, O. D. 1931. A comparative study of the conditioned motor reflex in the rabbit, sheep, goat, and pig. *Am. Jour. Physiol.*, **97**, 539–540.

———, and BAYNE, T. L. 1927. The development of "experimental neurasthenia" in the sheep during the formation of difficult conditioned reflexes. *Am. Jour. Physiol.*, **81**, p. 49.

———, JAMES, W. T., and ANDERSON, O. D. 1934. The comparative physiology of the conditioned motor reflex. *Comp. Psychol. Mono.*, **11**, pp. 89.

LIEBERMANN, A. 1925. Correlation zwischen den antennalen Geruchsorganen und der Biologie der Musciden. *Zsch. Morph. Oekol. Tiere*, **5**, 1–97.

LIGGETT, J. R. 1928. An experimental study of the olfactory sensitivity of the white rat. *Genet. Psychol. Mono.*, **3**, 1–64.

LIGON, E. M. 1929. A comparative study of certain incentives in the learning of the white rat. *Comp. Psychol. Mono.*, **6**, pp. 95.

LILLIE, R. S. 1918. Heredity from the physico-chemical point of view. *Biol. Bull.*, **34**, 65–90.

———. 1923. *Protoplasmic action and nervous action.* Chicago: Univ. of Chicago Press. Pp. 417.

LISSMANN, H. W. 1932. Die Umwelt des Kampffisches (*Betta splendens* Regan). *Zsch. vergl. Physiol.*, **18**, 65–111.

LOEB, J. 1893. Über künstliche Umwandung positiver heliotropischer Thiere in negative heliotropische und umgekehrt. *Pflüg. Arch. ges. Physiol.*, **54**, 81–107.

———. 1900. *Comparative physiology of the brain and comparative psychology.* New York: G. P. Putnam's. Pp. 309.

———. 1918. *Forced movements, tropisms, and animal conduct.* Philadelphia: Lippincott. Pp. 209.

LOSINA-LOSINSKY, L. K. 1931. Zur Ernährungsphysiologie der Infusorien. Untersuchungen über die Nahrungsauswahl und Vermehrung bei *Paramaecium caudatum. Arch. Prot.*, **74**, 18–120.

LOTMAR, RUTH. 1933. Neue Untersuchungen über den Farbensinn der Bienen, mit besonderer Berücksichtigung des Ultravioletts. *Zsch. vergl. Physiol.*, **19**, 673–723.

LOUCKS, R. B. 1931. Efficacy of the rat's motor cortex in delayed alternation. *Jour. Comp. Neurol.*, **53**, 511–567.

LUBBOCK, J. (Avebury). 1887. *Ants, bees, and wasps.* (1932, rev. ed.) New York: Appleton. Pp. 448.

LUCANUS, F. 1922. *Die Rätsel des Vogelzuges.* Langensalza: Beyer. Pp. 226.

LUI, S. Y. 1928. The relation of age to the learning ability of the white rat. *Jour. Comp. Psychol.*, **8**, 75–85.

LUTZ, F. 1924. Apparently non-selective characters and combinations of characters, including a study of ultra-violet in relation to the flower-visiting habits of insects. *Ann. N. Y. Acad. Sci.*, **29**, 181–283.

LYON, E. P. 1904. On rheotropism. I. Rheotropism in fishes. *Amer. Jour. Physiol.*, **12**, 149–161.

MCALLISTER, W. G. 1932. A further study of the delayed reaction in the albino rat. *Comp. Psychol. Mono.*, **8**, pp. 103.

MCCRACKEN, I. 1907. The egg-laying apparatus in the silkworm (*Bombyx mori*) as a reflex apparatus. *Jour. Comp. Neurol. Psychol.*, **17**, 262–285.

MCDOUGALL, K. D., and MCDOUGALL, W. 1931. Insight and foresight in various animals—monkey, raccoon, rat, and wasp. *Jour. Comp. Psychol.*, **11**, 237–273.

MACGILLIVRAY, M. E., and STONE, C. P. 1931. The incentive value of food and escape from water for albino rats forming the light discrimination habit. *Jour. Comp. Psychol.*, **11**, 319–324.

MCNAIR, G. T. 1923. Motor reactions of the fresh-water sponge, *Ephydatia fluviatilis. Biol. Bull.*, **44**, 153–166.

MAIER, N. R. F. 1929*a*. Reasoning in white rats. *Comp. Psychol. Mono.*, **6**, pp. 93.

———. 1929*b*. Delayed reaction and memory in rats. *Jour. Genet. Psychol.*, **36**, 538–550.

———. 1930. Attention and inattention in rats. *Jour. Genet. Psychol.*, **38**, 288–306.

———. 1932*a*. The effect of cerebral destruction on reasoning and learning in rats. *Jour. Comp. Neurol.*, **54**, 45–75.

———. 1932*b*. Cortical destruction of the posterior part of the brain and its effect on reasoning in rats. *Jour. Comp. Neurol.*, **56**, 179–214.

———. 1932c. Age and intelligence in rats. *Jour. Comp. Psychol.*, **13**, 1–6.

———. 1932d. A study of orientation in the rat. *Jour. Comp. Psychol.*, **14**, 387–399.

———. 1934. The pattern of cortical injury in the rat and its relation to mass action. *Jour. Comp. Neurol.*, **60**, 409–436.

———. 1935. The cortical area concerned with coordinated walking in the rat. *Jour. Comp. Neurol.*, **61**, 395–405.

MANGOLD, E. 1908. Studien zur Physiologie des Nervensystems der Echinodermen. I. Die Füsschen der Seesterne und die Koordination ihre Bewegungen. *Pflüg. Arch. ges. Physiol.*, **122**, 315–360.

———. 1908. II. Über das Nervensystem der Seesterne und über den Tonus. *Pflüg. Arch. ges. Physiol.*, **123**, 1–39.

MASLOW, A. H., and GROSHONG, ELIZABETH. 1934. Influence of differential motivation on delayed reactions in monkeys. *Jour. Comp. Psychol.*, **18**, 75–83.

MAST, S. O. 1911. *Light and the behavior of organisms.* New York: Wiley. Pp. 410.

———. 1923. Photic orientation in insects, with special reference to the drone-fly, *Eristalis tenax*, and the robberfly *Erax rufibarbis. Jour. Exper. Zool.*, **38**, 109–205.

———. 1931. The nature of response to light in *Amoeba proteus* (Leidy). *Zsch. vergl. Physiol.*, 15, 139–147.

———, and PUSCH, L. 1924. Modification of response in Amoeba. *Biol. Bull.*, **46**, 55–59.

MATTHES, E. 1924. Die Rolle des Gesichts-, Geruchs- und Erschütterungssinnes für den Nahrungserwerb von Triton. *Biol. Zentrlbl.*, **44**, 72–87.

MAXWELL, S. 1897. Beiträge zur Gehirnphysiologie der Anneliden. *Pflüg. Arch. ges. Physiol.*, **67**, 263–297.

———. 1923. *Labyrinth and equilibrium.* Philadelphia: Lippincott. Pp. 163.

MAYER, A. 1906. Rhythmical pulsation in Scyphomedusae. *Carn. Inst., Wash.*, Publ. No. 47; 1908, *ibid.*, Publ. No. 102, 1, 113–131.

MAYER, BARBARA A., and STONE, C. P. 1931. The relative efficiency of distributed and massed practice in maze learning by young and adult albino rats. *Jour. Genet. Psychol.*, **39**, 28–49.

METALNIKOW, S. 1912. Contributions à l'étude de la digestion intracellulaire chez les protozoaires. *Arch. de Zool.*, **49**, 373–498.

MINNICH, D. E. 1926. The chemical sensitivity of the tarsi of certain muscid flies. *Biol. Bull.*, **51**, 166–178.

———. 1931. The chemical senses of insects. *Quart. Rev. Biol.*, **4**, 100–112.

MOODY, P. 1929. Brightness vision in the deer-mouse, *Peromyscus maniculatus gracilis. Jour. Exper. Zool.*, **52**, 367–405.

MOORE, A. R. 1910. On the righting movements of the starfish. *Biol. Bull.*, **19**, 235–239.

———. 1910. On the nervous mechanism of the righting movements of the starfish. *Amer. Jour. Physiol.*, **27**, 207–211.

———. 1923. Muscle tension and reflexes in the earthworm. *Jour. Gen. Physiol.*, **5**, 327–333.

———. 1924. The nervous mechanism of coordination in the crinoid *Antedon rosaceus*. *Jour. Gen. Physiol.*, **6**, 281–288.

Moseley, Dorothy. 1925. The accuracy of the pecking response in chicks. *Jour. Comp. Psychol.*, **5**, 75–97.

Muenzinger, K. F. 1934. Motivation in learning. I. Electric shock for correct response in the visual discrimination habit. *Jour. Comp. Psychol.*, **17**, 267–277.

———, and Gentry, Evelyn. 1931. Tone discrimination in white rats. *Jour. Comp. Psychol.*, **12**, 195–205.

———, and Mize, R. H. 1933. The sensitivity of the white rat to electric shock: threshold and skin resistance. *Jour. Comp. Psychol.*, **15**, 139–148.

———, and Walz, F. C. 1932. An analysis of the electrical stimulus producing a shock. *Jour. Comp. Psychol.*, **13**, 157–171.

Munn, N. L. 1929. Concerning visual form discrimination in the white rat. *Jour. Genet. Psychol.*, **36**, 291–300.

———. 1931. The relative efficacy of form and background in the chick's discrimination of visual patterns. *Jour. Comp. Psychol.*, **12**, 41–75.

———. 1932. An investigation of color vision in the hooded rat. *Jour. Genet. Psychol.*, **40**, 351–362.

Murchison, Carl. 1935. The experimental measurement of a social hierarchy in *Gallus domesticus:* I. The direct identification and direct measurement of Social Reflex No. 1 and Social Reflex No. 2. *Jour. General Psychol.*, **12**, 3–39.

Nagel, W. 1894. Experimentelle Sinnesphysiologische Untersuchungen an Coelenteraten. *Pflüg. Arch. ges. Physiol.*, **57**, 494–552.

———. 1894. Beobachtungen über den Lichtsinn augenloser Muscheln. *Biol. Cent.*, **14**, 385–390.

Neet, C. C. 1933. Visual pattern discrimination in the macacus rhesus monkey. *Jour. Genet. Psychol.*, **43**, 163–196.

Nellmann, H., and Trendelenburg, W. 1926. Ein Beitrag zur Intelligenzprüfung niederer Affen. *Zsch. vergl. Physiol.*, **4**, 142–200.

Ni, C. F. 1934. The influence of punishment for errors during the learning of the first maze upon the mastery of the second maze. *Jour. Comp. Psychol.*, **18**, 23–28.

Nicholas, J. 1922. The reactions of *Amblystoma tigrinum* to olfactory stimuli. *Jour. Exper. Zool.*, **35**, 257–281.

Nielsen, E. 1932–1934. Sur les habitudes des hymenoptères aculeates solitaires. *Ent. Meddel.* (Copenhagen), **18**, 1–57; 84–174; 259–336.

Noble, G. K. 1931. *The biology of the Amphibia.* New York: McGraw-Hill. Pp. 577.

———, and Farris, E. J. 1929. The method of sex recognition in the woodfrog, *Rana sylvatica* Le Conte. *Amer. Mus. Nov.*, No. 363, 1–17.

Nolte, W. 1932. Experimentelle Untersuchungen zum Problem der Lokalization des Assoziationsvermögens im Fischgehirn. *Zsch. vergl. Physiol.*, **18**, 255–273.

Ogden, R., and Franz, S. I. 1917. On cerebral motor control: the recovery from experimentally produced hemiplegia. *Psychobiol.*, **1**, 33–50.

OLMSTED, J. 1922. The role of the nervous system in the locomotion of certain marine polyclads. *Jour. Exper. Zool.*, **36**, 57–66.

OPFINGER, E. 1931. Über die Orientierung der Biene an der Futterquelle (Die Bedeutung von Anflug und Orientierungsflug für den Lernvorgang bei Farb-, Form-, und Ortsdressuren). *Zsch. vergl. Physiol.*, **15**, 431–487.

ORTON, J. H. 1922. The mode of feeding of the jellyfish Aurelia. *Nature*, **110**, 178–179.

PADILLA, S. G. 1930. *Further studies on the delayed pecking of chicks.* Thesis: Univ. of Mich. Library.

PANTIN, C. F. 1924. On the physiology of amoeboid movement. I. *Jour. Mar. Biol. Assoc.*, **13**, 24–69.

———. 1935. The nerve net of the Actinozoa. I. Facilitation. II. Plan of the nerve net. III. Polarity and after discharge. *Jour. Exper. Biol.*, **12**, 119–138, 139–155, 156–164.

PARKER, G. H. 1904. The function of the lateral-line organs in fishes. *Bull. Bur. Fish.*, **24**, 185–207.

———. 1911. The olfactory reactions of the common killifish. *Jour. Exper. Zool.*, **10**, 1–5.

———. 1917. Actinian behavior. *Jour. Exper. Zool.*, **22**, 193–230.

———. 1919. *The elementary nervous system.* Philadelphia: Lippincott. Pp. 229.

———. 1922. The crawling of young loggerhead turtles toward the sea. *Jour. Exper. Zool.*, **36**, 323–331.

———. 1927. Locomotion and righting movements in echinoderms, especially in Echinarachnius. *Amer. Jour. Psychol.*, **39**, 167–180.

———, and BURNETT, F. 1901. The reactions of planarians with and without eyes to light. *Amer. Jour. Physiol.*, **4**, 373–385.

———, and PARSHLEY, H. 1911. The reactions of earthworms to dry and to moist surfaces. *Jour. Exper. Zool.*, **11**, 361–364.

———, and SHELDON, R. E. 1913. The sense of smell in fishes. *Bull. Bur. Fish.*, **32**, 33–46.

PARSCHIN, A. 1929. Bedingte Reflexe bei Schildkröten. *Pflüg. Arch. ges. Physiol.*, **222**, 328–333.

PATRICK, J. R., and LAUGHLIN, R. M. 1934. Is the wall-seeking tendency in the white rat an instinct? *Jour. Genet. Psychol.*, **44**, 378–389.

PATTERSON, T. 1933. Comparative physiology of the gastric hunger mechanism. *Ann. N. Y. Acad. Sci.*, **34**, 55–272.

PAVLOV, I. P. 1927. *Conditioned reflexes.* (Tr. G. V. Anrep.) London: Oxford Univ. Press. Pp. 430.

PEARL, R. 1903. The movements and reactions of fresh-water planarians. *Quart. Jour. Micr. Sci.*, **46**, 509–714.

PECHSTEIN, L. A. 1917. Whole vs. part methods in motor learning. *Psychol. Mono.*, **23**, pp. 80.

———. 1921. Massed vs. distributed effort in learning. *Jour. Educ. Psychol.*, **12**, 92–97.

PECKHAM, G. W., and ELIZABETH G. 1905. *Wasps, social and solitary.* Boston and New York: Houghton, Mifflin. Pp. 311.

———. 1887. Some observations on the mental powers of spiders. *Jour. Morph.*, **1**, 383–419.

PETERSON, G. M. 1934. Mechanisms of handedness in the rat. *Comp. Psychol. Mono.*, **9**, pp. 67.

PETERSON, J. 1917*a*. The effect of length of blind alleys on maze learning; an experiment on twenty-four rats. *Behav. Mono.*, **3**, pp. 53.

———. 1917*b*. Frequency and recency factors in maze learning by white rats. *Jour. Anim. Behav.*, **7**, 338–364.

PETRUNKEVITCH, A. 1911. Sense of sight, courtship and mating in *Dugesiella heutzi* (Girard), a theraphosid spider from Texas. *Zool. Jahrb., Syst.*, **31**, 355–376.

———. 1926. The value of instinct as a taxonomic character in spiders. *Biol. Bull.*, **50**, 427–432.

PHIPPS, C. 1915. An experimental study of the behavior of amphipods with respect to light intensity, direction of rays, and metabolism. *Biol. Bull.*, **28**, 210–223.

PIÉRON, H. 1909. La loi de l'oublie chez la limnée. *Arch. Psychol.*, **9**, 39–50

———. 1927. Le rôle des statocystes chez les mollusques et les données fournies par l'étude du géotropism des limaces. *C. rend. Soc. Biol.*, **97**, 1390–1392.

———. 1928. Gravitational sensitivity and geotropic reactions in slugs; an analysis of the laws of excitation and of the internal factors in reaction. *Jour. Genet. Psychol.*, **35**, 3–17.

POLTYREW, S., and ZELONY, G. 1930. Grosshirnrinde und Assoziationsfunktion. *Zsch. Biol.*, **90**, 157–161.

PORTER, J. 1904. A preliminary study of the psychology of the English sparrow. *Amer. Jour. Psychol.*, **15**, 313–346.

———. 1906. Further study of the English sparrow and other birds. *Amer. Jour. Psychol.*, **17**, 248–271.

———. 1910. Intelligence and imitation in birds, a criterion of instinct. *Amer. Jour. Psychol.*, **21**, 1–71.

PRATT, K. C., NELSON, A. K., and SUN, K. H. 1930. The behavior of the newborn infant. *Ohio State Univ. Stud. Psychol.*, No. 10, pp. 237.

PREYER, W. 1886. Ueber die Bewegungen der Seesterne. *Mitt. Zool. Sta. Neapel*, **7**, 27–127; 8, 191–233.

PROSSER, C. 1934. The nervous system of the earthworm. *Quart. Rev. Biol.*, **9**, 181–200.

RÁDL, E. 1903. *Untersuchungen über den Phototropismus der Tiere.* Leipzig: Engelmann. Pp. 188.

RANEY, E. T., and CARMICHAEL, L. 1934. Localizing responses to tactual stimuli in the fetal rat in relation to the psychological problem of space perception. *Jour. Genet. Psychol.*, **45**, 3–21.

RAU, P., and RAU, NELLIE. 1918. *Wasp studies afield.* Princeton Univ. Press. Pp. 372.

———. 1929. Experimental studies in the homing of carpenter bees. *Jour. Comp. Psychol.*, **9**, 35–70.

———. 1931. Additional experiments on the homing of carpenter- and mining-bees. *Jour. Comp. Psychol.*, **12**, 257–261.

REEVES, CORA D. 1919. Discrimination of light of different wave-lengths by fish. *Behav. Mono.*, **4**, pp. 106.

REGEN, J. 1914. Untersuchungen über die Stridulation und das Gehör vom *Thamnotrizon apterus* Fab. ♂. *Sitz. Akad. Wiss., Wien, math.-nat. Kl.*, **123**, 853–892.

REIGHARD, J. 1920. The breeding behavior of suckers and minnows. I. The suckers. *Biol. Bull.*, **38**, 1–32.

RÉVÉSZ, G. 1924. Experiments on animal space perception. *Brit. Jour. Psychol.*, **14**, 387–414.

RICH, W., and HOLMES, H. 1928. Experiments on marking young Chinook salmon on the Columbia River, 1916 to 1927. *Bull. Bur. Fish.*, **44**, 215–264.

RILEY, C. F. 1913. Responses of young toads to light and contact. *Jour. Anim. Behav.*, **3**, 179–214.

RILEY, C. V. 1895. The senses of insects. *Nature*, **52**, 209–212.

ROBINSON, E. W. 1933. A preliminary experiment in abstraction in a monkey. *Jour. Comp. Psychol.*, **16**, 231–236.

ROGERS, C. G. 1927. *Textbook of comparative physiology.* New York: McGraw-Hill. Pp. 635.

ROMANES, G. J. 1885. *Jelly-fish, star-fish, and sea-urchins—being a research on primitive nervous systems.* New York: Appleton. Pp. 323.

RÖSCH, G. A. 1925. Untersuchungen über die Arbeitsteilung im Bienenstaat. I Teil. Die Tätigkeiten im normalen Bienenstaate und ihre Beziehungen zum Alter der Arbeitsbienen. *Zsch. vergl. Physiol.*, **2**, 571–631.

———. 1927. Über die Bautätigkeit im Bienenvolk und das Alter der Baubienen. *Zsch. vergl. Physiol.*, **6**, 265–298.

———. 1930. Untersuchungen über die Arbeitsteilung im Bienenstaat. II. Die Tätigkeiten der Arbeitsbienen unter experimentelle veränderten Bedingungen. *Zsch. vergl. Physiol.*, **12**, 1–71.

ROSE, M. 1930. *La question des tropismes.* Paris: Presses Universitaires. Pp. 469.

ROUBAUD, E. 1910. Recherches sur la biologie des Synagris. Evolution de l'instinct chez les guêpes solitaires. *Ann. Soc. Ent. Fr.*, **79**, 1–21.

ROULE, L. 1928. *Fishes, their journeys and migrations.* New York: Norton. Pp. 270.

ROUSE, J. E. 1905. Respiration and emotion in pigeons. *Jour. Comp. Neurol. Psychol.*, **15**, 494–513.

ROWAN, W. 1929. Experiments in bird migration. I. Manipulation of the reproductive cycle: seasonal histological changes in the gonads. *Proc. Bost. Soc. Nat. Hist.*, **39**, 151–208.

———. 1931. *The riddle of migration.* Baltimore: Williams and Wilkins. Pp. 151.

RUCH, F. L. 1934. Kinesthesis, motivation and transfer. I. Preliminary experiments and statement of a problem. *Jour. Comp. Psychol.*, **18**, 259–269.

SACKETT, L. W. 1913. The Canada porcupine: a study of the learning process. *Behav. Mono.*, **2**, pp. 84.

SADOVINKOVA, MARY P. 1923. A study of the behavior of birds by the multiple choice method. *Jour. Comp. Psychol.*, **3**, 249–282.

SANTSCHI, F. 1911. Observations et remarques critiques sur le mécanisme de l'orientation chez les fourmis. *Rev. Suisse Zool.*, **19**, 303–338.

———. 1923. Les différentes orientations chez les fourmis. *Rev. Zool. Afric.*, **11**, 10–144.

———. 1930. Nouvelles expériences sur l'orientation des Tapinoma par sécrétions dromographiques. *Arch. de Psychol.*, **22**, 348–351.

SCHAEFFER, A. 1916. On the behavior of Amoeba toward fragments of glass and carbon and other indigestible substances and toward some very soluble substances. *Biol. Bull.*, **31**, 303–328.

———. 1920. *Ameboid movement.* Princeton Univ. Press. Pp. 156.

SCHARRER, E. 1932. Experiments on the function of the lateral-line organs in the larvae of *Amblystoma punctatum*. *Jour. Exper. Zool.*, **61**, 109–114.

SCHEURING, L. 1921. Beobachtungen und Betrachtungen über die Beziehungen der Augen zum Nahrungserwerb bei Fischen. *Zool. Jahrb., Zool. Physiol.*, **38**, 113–136.

SCHJELDERUP-EBBE, T. 1922. Beiträge zur Sozialpsychologie des Haushuhns. *Zsch. Psychol.*, **88**, 225–252.

SCHMIDT, J. 1925. The breeding places of the eel. *Smiths. Inst. Ann. Rep.*, 1924, Wash., 279–316.

SCHNEIDER, G. 1905. Die Orientierung der Brieftauben. *Zsch. Psychol. Physiol. Sinnesorg.*, **40**, 252–279.

SCHNEIRLA, T. C. 1929. Learning and orientation in ants. *Comp. Psychol., Mono.*, **6**, pp. 143.

———. 1933*a*. Motivation and efficiency in ant learning. *Jour. Comp. Psychol.*, **15**, 243–266.

———. 1933*b*. Some comparative psychology. *Jour. Comp. Psychol.*, **16**, 307–315.

———. 1933*c*. Some important features of ant learning. *Zsch. vergl. Physiol.*, **19**, 439–452.

———. 1934*a*. The process and mechanism of ant learning. *Jour. Comp. Psychol.*, **17**, 303–328.

———. 1934*b*. Raiding and other outstanding phenomena in the behavior of army ants. *Proc. Nat. Acad. Sci.*, **20**, 316–321.

SCHWARTZ, B., and SAFIR, S. R. 1915. Habit formation in the fiddler crab. *Jour. Anim. Behav.*, **5**, 226–239.

SCHWARZ, H. F. 1931. The nest habits of the diplopterous wasp *Polybia occidentalis* variety *scutellaris* (White) as observed at Barro Colorado, Canal Zone. *Amer. Mus. Nov.*, No. 471, 1–27.

SCHWITALLA, A. 1924. The influence of temperature on the rate of locomotion in Amoeba. I. Locomotion at diverse constant temperatures. *Jour. Morph.*, **39**, 465–514.

SEARS, R. 1934. Effect of optic lobe ablation on the visuo-motor behavior of goldfish. *Jour. Comp. Psychol.*, **17**, 233–265.

SHELFORD, V., and POWERS, E. 1915. An experimental study of the movement of herring and other marine fishes. *Biol. Bull.*, **28**, 315–445.

SHEPARD, J. F. 1911. Some results in comparative psychology. *Psych. Bull.*, **8**, 41–42.

———. 1929. An unexpected cue in maze learning. *Psychol. Bull.*, **26**, 164–165.

———. 1931. More learning. *Psychol. Bull.*, **28**, 240–241.

———. 1933. Higher processes in the behavior of rats. *Proc. Nat. Acad. Sci.*, **19**, 149–152.

———, and Breed, F. S. 1913. Maturation and use in the development of an instinct. *Jour. Anim. Behav.*, **3**, 274–285.

Sherrington, C. S. 1906. *The integrative action of the nervous system.* (1923 ed.) New Haven: Yale Univ. Press. Pp. 411.

———. 1925. Remarks on some aspects of reflex inhibition. *Proc. Roy. Soc.*, London, **97*B***, 519–545.

Shull, A. F. 1907. The stridulation of the snowy tree-cricket (*Oecanthus niveus*). *Canad. Entom.*, **39**, 213–225.

Simmons, Rietta. 1924. The relative effectiveness of certain incentives in animal learning. *Comp. Psychol. Mono.*, **2**, pp. 79.

Small, W. S. 1899. An experimental study of the mental processes of the rat. *Am. Jour. Psychol.*, **11**, 133–165.

———. 1901. Experimental study of the mental processes of the rat. II. *Am. Jour. Psychol.*, **12**, 206–239.

Smith, A. C. 1902. The influence of temperature, odors, lights and contact on the movements of the earthworm. *Amer. Jour. Physiol.*, **6**, 459–486.

Smith, E. M. 1912. Some observations concerning colour vision in dogs. *Brit. Jour. Psychol.*, **5**, 119–202.

Smith, S. 1908. The limits of educability in Paramecium. *Jour. Comp. Neurol. Psychol.*, **18**, 499–510.

Spence, K. W. 1932. The order of eliminating blinds in maze learning by the rat. *Jour. Comp. Psychol.*, **14**, 9–27.

———, and Shipley, W. C. 1934. The factors determining the difficulty of blind alleys in maze learning by the white rat. *Jour. Comp. Psychol.*, **17**, 423–436.

Spragg, S. D. S. 1933. Anticipation as a factor in maze errors. *Jour. Comp. Psychol.*, **15**, 313–329.

———. 1934. Anticipatory responses in the maze. *Jour. Comp. Psychol.*, **18**, 51–73.

Steche, C. 1911. *Hydra und die Hydroiden.* Leipzig: Klinkhardt. Pp. 162.

Steiner, A. 1932. Die Arbeitsteilung der Feldwespe *Polistes dubia*. *Zsch. vergl. Physiol.*, **17**, 101–152.

Stone, C. P. 1922. The congenital sexual behavior of the young male albino rat. *Jour. Comp. Psychol.*, **2**, 95–153.

———. 1926. The initial copulatory response of female rats reared in isolation from the age of twenty days to the age of puberty. *Jour. Comp. Psychol.*, **6**, 73–83.

———. 1927. The retention of copulatory ability in male rats following castration. *Jour. Comp. Psychol.*, **7**, 369–387.

———. 1929*a*. The age factor in animal learning. I. Rats in the problem box and the maze. *Genet. Psychol. Mono.*, **5**, 1–130.

———. 1929*b*. The age factor in learning. II. Rats on a multiple light discrimination box and a difficult maze. *Genet. Psychol. Mono.*, **6**, 125–202.

———, and Sturman-Huble, Mary. 1927. Food vs. sex as incentives for male rats on the maze learning problem. *Am. Jour. Psychol.*, **38**, 403–408.

———, Darrow, C. W., Landis, C., and Heath, L. L. 1932. *Studies in the dynamics of behavior* (ed. Lashley). Chicago: Univ. Chicago Press, pp. 332.

Stough, H. 1926. Giant nerve fibers of the earthworm. *Jour. Comp. Neurol.*, **40**, 409–463.

Strasburger's Text-book of Botany. 6th Eng. Ed., 1930. London: Macmillan. Pp. 818.

Swann, H. G. 1933. The function of the brain in olfaction. I. Olfactory discrimination and an apparatus for its test. *Jour. Comp. Psychol.*, **15**, 229–241.

———. 1934. The function of the brain in olfaction. II. The results of destruction of olfactory and other nervous structures upon the discrimination of odors. *Jour. Comp. Neurol.*, **59**, 175–201.

Swartz, Ruth. 1929. Modification of behavior in earthworms. *Jour. Comp. Psychol.*, **9**, 17–34.

Szymanski, J. S. 1912. Modification of the innate behavior of cockroaches. *Jour. Anim. Behav.*, **2**, 81–90.

———. 1918. Versuche über die Wirkung der Faktoren, die als Antrieb zum Erlernen einer Handlung dienen können. *Pflüg. Arch. ges. Physiol.*, **171**, 374–385.

Talbot, Mary. 1934. Distribution of ant species in the Chicago region with reference to ecological factors and physiological toleration. *Ecology*, **15**, 416–439.

Taliaferro, W. H. 1920. Reactions to light in *Planaria maculata*, with special reference to the function and structure of the eyes. *Jour. Exper. Zool.*, **31**, 59–116.

Taylor, W. 1932. The gregariousness of pigeons. *Jour. Comp. Psychol.*, **13**, 127–132.

Thompson, Elizabeth. 1917. An analysis of the learning process in the snail, Physa gyrina, Say. *Behav. Mono.*, **3**, pp. 89.

Thorndike, E. L. 1898. Animal intelligence; an experimental study of the associative processes in animals. *Psychol. Rev. Mono.*, **2**, pp. 109.

———. 1911. *Animal intelligence.* New York: Macmillan. Pp. 297.

———. 1915. Watson's "Behavior." *Jour. Anim. Behav.*, **5**, 462–467.

Thuma, B. D. 1932. The response of the white rat to tonal stimuli. *Jour. Comp. Psychol.*, **13**, 57–86.

Tinbergen, N. 1932. Über die Orientierung des Bienenwolfes (*Philanthus triangulum* Fabr.). *Zsch. vergl. Physiol.*, **16**, 315–334.

Tinklepaugh, O. L. 1928. An experimental study of representative factors in monkeys. *Jour. Comp. Psychol.*, **8**, 197–236.

Tolman, E. C. 1924. The inheritance of maze learning ability in rats. *Jour. Comp. Psychol.*, **4**, 1–18.

———. 1932. *Purposive behavior in animals and men.* New York: Appleton-Century, pp. 463.

——— and Honzik, C. H. 1930*b*. "Insight" in rats. *Univ. Calif. Publ. Psychol.*, **4**, 215–232.

———, and Honzik, C. H. 1930*a*. Introduction and removal of reward, and maze performance in rats. *Univ. Calif. Publ. Psychol.*, **4**, 257–275.

Tonner, F. 1933. Das Problem der Krebsschere. *Zsch. vergl. Physiol.*, **19**, 762–784.

TORREY, H. B. 1904. On the habits and reactions of *Sagartia davisi*. *Biol. Bull.*, **6**, 203–216.

TRYON, R. C. 1930. Studies in individual differences in maze ability. I. The measurement of the reliability of individual differences. *Jour. Comp. Psychol.*, **11**, 145–170.

———. 1931*a*. Studies in individual differences in maze ability. III. The community of function between two maze abilities. *Jour. Comp. Psychol.*, **12**, 95–115.

———. 1931*b*. Studies in individual differences in maze ability. IV. The constancy of individual differences: correlation between learning and relearning. *Jour. Comp. Psychol.*, **12**, 303–345.

TSAI, L. S. 1932. The laws of minimum effort and maximum satisfaction in animal behavior. *Mono. Nat. Resear. Inst. Psychol.* (Peiping, China) No. 1, pp. 49.

TURNER, C. 1913. Behavior of the common roach (Periplaneta orientalis) in an open maze. *Biol. Bull.*, **25**, 348–365.

UEXKÜLL, J. V. 1891. Physiologische Untersuchungen an *Eledone moschata*. *Zsch. Biol.*, **28**, 550–566; 1894, *ibid.*, **30**, 179–183; 1895, *ibid.*, **31**, 584–609.

———. 1899. Die Physiologie der Pedicellarien. *Zsch. Biol.*, **37**, 334–403.

———. 1900. Die Physiologie des Seeigelstachels. *Zsch. Biol.*, **39**, 73–112.

———. 1905. Studien über den Tonus. II. Die Bewegungen der Schlangensterne. *Zsch. Biol.*, **46**, 1–37.

———. 1912. Studien über den Tonus. VI. Die Pilgermuschel. *Zsch. Biol.*, **58**, 305–332.

UFLAND, I. 1929. Die Reflexerregbarkeit des Frosches während des Umklammerungsreflexes. *Pflüg. Arch. ges. Physiol.*, **221**, 605–622.

ULRICH, J. L. 1915. Distribution of effort in learning in the white rat. *Behav. Mono.*, **2**, pp. 51.

UPTON, M. 1929. The auditory sensitivity of the guinea pig. *Am. Jour. Physiol.*, **41**, 412–422.

VEN, C. D. 1921. Sur la formation d'habitudes chez les astéries. *Arch. néerl. Physiol.*, **6**, 163–178.

VERRIER, M. L. 1928. Recherches sur les yeux et la vision des poissons. Suppl. X, *Bull. Biol. fr. Belg.*, Paris, pp. 222.

VERWORN, M. 1889. *Psycho-physiologische Protistenstudien.* Jena: Fischer. Pp. 219.

VINCENT, STELLA B. 1912. The function of the vibrissae in the behavior of the white rat. *Behav. Mono.*, **1**, pp. 86.

———. 1915*a*. The white rat and the maze problem. I. The introduction of a visual control. *Jour. Anim. Behav.*, **5**, 1–24.

———. 1915*b*. The white rat and the maze problem. II. The introduction of an olfactory control. *Jour. Anim. Behav.*, **5**, 140–157.

———. 1915*c*. The white rat and the maze problem. III. The introduction of a tactual control. *Jour. Anim. Behav.*, **5**, 175–184.

WAGNER, G. 1904. On some movements and reactions of Hydra. *Quart. Jour. Micr. Sci.*, **48**, 585–622.

WAGNER, H. 1932. Über den Farbensinn der Eidechsen. *Zsch. vergl. Physiol.*, **18**, 378–392.

WALLS, G. L. 1934. The visual cells of the white rat. *Jour. Comp. Psychol.*, **18**, 363–366.

WALTON, A. 1930. Visual cues in maze running by the albino rat. *Jour. Genet. Psychol.*, **38**, 50–77.

WALTON, W. E. 1933. Color vision and color preference in the albino rat. II. The experiments and results. *Jour. Comp. Psychol.*, **15**, 373–394.

WARD, H. B. 1921. Some of the factors controlling the migration and spawning of the Alaska red salmon. *Ecology*, **2**, 235–254.

WARDEN, C. J. 1923*a*. Some factors determining the order of elimination of culs-de-sac in the maze. *Jour. Exper. Psychol.*, **6**, 192–210.

———. 1923*b*. The distribution of practice in animal learning. *Comp. Psychol. Mono.*, **1**, pp. 64.

———. 1929. A standard unit animal maze for general laboratory use. *Jour. Genet. Psychol.*, **36**, 174–176.

———. 1931. *Animal motivation: experimental studies on the albino rat.* New York: Columbia Univ. Press. Pp. 502.

———, and CUMMINGS, S. B. 1929. Primacy and recency factors in animal motor learning. *Jour. Genet. Psychol.*, **36**, 240–256.

———, and HAMILTON, EDNA L. 1929. The effect of variations in length of maze pattern upon the rate of fixation in the white rat. *Jour. Genet. Psychol.*, **36**, 229–237.

———, JENKINS, T. N., and WARNER, L. H. 1934. *Introduction to comparative psychology.* New York: Ronald. Pp. 581.

WARDEN, F., and ROWLEY, J. 1929. The discrimination of absolute versus relative brightness in the ring dove, *Turtor risorius. Jour. Comp. Psychol.*, **9**, 317–337.

WARNER, L. H. 1931. The present status of the problems of orientation and homing in birds. *Quart. Rev. Biol.*, **6**, 208–214.

WASHBURN, MARGARET F. 1926. *The animal mind.* New York: Macmillan. Pp. 431.

———, and ABBOTT, E. 1912. Experiments on the brightness value of red for the light-adapted eye of the rabbit. *Jour. Anim. Behav.*, **2**, 145–180.

WASMANN, E. (S. J.) 1899. Die psychischen Fähigkeiten der Ameisen. *Zoologica*, **11**, pp. 132.

WATSON, J. B. 1907. Kinaesthetic and organic sensations: their role in the reactions of the white rat to the maze. *Psychol. Mono.* **8**, pp. 100.

———. 1909. Some experiments bearing upon color vision in monkeys. *Jour. Comp. Neurol. & Psychol.*, **19**, 1–28.

———. 1914. *Behavior: an introduction to comparative psychology.* New York: Holt. Pp. 439.

———, and LASHLEY, K. S. 1915. Homing and related activities of birds. (Papers Dept. Mar. Biol., 7) *Carn. Inst. Wash.*, Publ. No. 211, 9–60.

———, and WATSON, MARY I. 1913. A study of the responses of rodents to monochromatic light. *Jour. Anim. Behav.*, **3**, 1–14.

WEBB, L. W. 1917. Transfer of training and retroaction. *Psychol. Mono.*, **24**, pp. 90.

WELSH, J. H. 1932. The nature and movement of the reflecting pigment in the eyes of crustaceans. *Jour. Exper. Zool.*, **62**, 173–182.

WENDT, G. R. 1934. Auditory acuity of monkeys. *Comp. Psychol. Mono.*, **10**, pp. 51.

WENRICH, D. H. 1916. Notes on the reactions of bivalve mollusks to changes in light intensity: image formation in Pecten. *Jour. Anim. Behav.*, **6**, 297–318.

WENT, F. 1930. Les conceptions nouvelles sur les tropismes des plantes. *Rev. gén. Sci.*, **41**, 631–643.

WESTERFIELD, F. 1922. The ability of mud-minnows to form associations with sound. *Jour. Comp. Psychol.*, **2**, 187–190.

WETMORE, A. 1926. *The migration of birds.* Cambridge: Harv. Univ. Press. Pp. 217.

WEVER, E. G., and BRAY, C. W. 1930. The nature of acoustic responses; the relation between sound frequency and frequency of impulses in the auditory nerve. *Jour. Exper. Psychol.*, **13**, 373–387.

WHEELER, W. M. 1910. *Ants, their structure, development and behavior.* New York: Columbia Univ. Press. Pp. 663.

———. 1923. *Social life among the insects.* New York: Harcourt, Brace. Pp. 375.

———. 1928. *The social insects.* New York: Harcourt, Brace. Pp. 378.

———. 1933. *Colony-founding among ants.* Cambridge: Harv. Univ. Press. Pp. 179.

WHITE, A. E., and TOLMAN, E. C. 1923. A note on the elimination of short and long blind alleys. *Jour. Comp. Psychol.*, **3**, 327–331.

WHITMAN, C. O. 1919. *Behavior of pigeons.* v. 3, posthumous works (H. A. Carr, ed.) *Carn. Inst. Publ.* No. 257.

WIEDEMANN, E. 1932. Zur Biologie der Nahrungsaufnahme der Kreuzotter *Vipera berus* L. *Zool. Anz.*, **97**, 278–286.

WILEY, L. E. 1932. The function of the brain in audition. *Jour. Comp. Neurol.*, **54**, 143–172.

WILLIAMS, J. A. 1926. Experiments with form perception and learning in dogs. *Comp. Psychol. Mono.*, **4**, pp. 70.

WILSON, E. B. 1891. The heliotropism of *Hydra*. *Am. Nat.*, **25**, 413–433.

WILSON, H. 1910. A study of some epithelioid membranes in monaxonid sponges. *Jour. Exper. Zool.*, **9**, 537–577.

———, and MILUM, V. 1927. Weather protection for the honey-bee colony. *Wis. Univ., Agr. Exp. Sta. Res. Bull.* No. 75, pp. 47.

WILTBANK, R. T. 1919. Transfer of training in white rats upon various series of mazes. *Behav. Mono.*, **4**, pp. 65.

WINGFIELD, R. C., and DENNIS, W. 1934. The dependence of the rat's choice of pathway upon the length of the daily trial series. *Jour. Comp. Psychol.*, **18**, 135–147.

WOLF, E. 1925. Physiologische Untersuchungen über das Umdrehen der Seesterne und Schlangensterne. *Zsch. vergl. Physiol.*, **3**, 209–224.

———. 1933. Das Verhalten der Bienen gegenüber flimmernden Feldern und bewegten Objekten. *Zsch. vergl. Physiol.*, **20**, 151–161.

WOLFLE, HELEN M. 1930. Time factors in conditioning finger-withdrawal. *Jour. Gen. Psychol.*, **4**, 372–378.

WOOD, A. B. 1933. A comparison of delayed reward and delayed punishment in the formation of a brightness discrimination habit in the chick. *Arch. Psychol.*, **24**, pp. 40.

YERKES, R. M. 1901. The formation of habits in the turtle. *Pop. Sci. Mo.*, **58**, 519–525.

———. 1902*a*. A contribution to the physiology of the nervous system of the medusa *Gonionemus murbachii*. I. The sensory reactions of Gonionemus. *Amer. Jour. Physiol.*, **6**, 434–449.

———. 1902*b*. A contribution to the physiology of the nervous system of the medusa *Gonionemus murbachii*. II. The physiology of the central nervous system. *Amer. Jour. Physiol.*, **6**, 181–198.

———. 1903. The instincts, habits, and reactions of the frog. I. Associative processes of the green frog. *Psych. Rev. Mon.*, Suppl. IV, *Harv. Psychol. Stud.*, **1**, 579–597.

———. 1904*a*. The reaction time of *Gonionemus murbachii* to electric and photic stimuli. *Biol. Bull.*, **6**, 84–95.

———. 1904*b*. Inhibition and reinforcement of reactions in the frog *Rana clamitans*. *Jour. Comp. Neurol. Psychol.*, **14**, 124–137.

———. 1905. The sense of hearing in frogs. *Jour. Comp. Neurol. Psychol.*, **15**, 279–304.

———. 1907. *The dancing mouse*. New York: Macmillan. Pp. 290.

———. 1912. The intelligence of earthworms. *Jour. Anim. Behav.*, **2**, 332–352.

———. 1916. The mental life of monkeys and apes; a study of ideational behavior. *Behav. Mono.*, **3**, pp. 145.

———. 1934. Modes of behavioral adaptation in chimpanzees to multiple choice problems. *Comp. Psychol. Mono.*, **10**, pp. 108.

———, and COBURN, C. A. 1915. A study of the behavior of the pig *Sus scrofa* by the multiple choice method. *Jour. Anim. Behav.*, **5**, 185–225.

———, and HUGGINS, G. 1903. Habit formation in the crawfish, *Cambarus affinis*. *Psychol. Rev. Mono.*, Suppl. IV, *Harv. Psychol. Stud.*, **1**, 565–577.

———, and MORGULIS, S. 1909. The method of Pavlov in animal psychology. *Psychol. Bull.*, **6**, 257–273.

———, and WATSON, J. B. 1911. Methods of studying vision in animals. *Behav. Mono.*, **1**, pp. 90.

———, and YERKES, D. N. 1928. Concerning memory in the chimpanzee. *Jour. Comp. Psychol.*, **8**, 237–271.

——— and YERKES, A. W. 1929. *The great apes*. New Haven: Yale Univ. Press. Pp. 652.

YOAKUM, C. S. 1909. Some experiments upon the behavior of squirrels. *Jour. Comp. Neurol. & Psychol.*, **19**, 541–568.

YOCOM, H. B. 1918. The neuromotor apparatus of *Euplotes patella*. *Univ. Calif. Publ. Zool.*, **18**, 337–396.

YONGE, C. M. 1928. Feeding mechanisms in the invertebrates. *Biol. Rev. (Camb. Phil. Soc.)*, **3**, 21–76.

YOSHIOKA, J. G. 1928. A note on a right or left going position habit with rats. *Jour. Comp. Psychol.*, **8**, 429–433.

———. 1929. Weber's law in the discrimination of maze distance by the white rat. *Univ. Calif. Publ., Psychol.*, 4, 155–184.

———. 1933. A study of orientation in a maze. *Jour. Genet. Psychol.*, 42, 167–183.

Ziegler, H. E. 1910. *Der Begriff des Instinktes einst und jetzt.* Jena: Fischer. Pp. 110.

SUPPLEMENT TO DOVER EDITION

ARTICLE I

AN EVOLUTIONARY AND DEVELOPMENTAL THEORY OF BIPHASIC PROCESSES UNDERLYING APPROACH AND WITHDRAWAL[1]

T. C. SCHNEIRLA

APPROACH-WITHDRAWAL MECHANISMS AND PSYCHOLOGICAL LEVELS

The aspect of *towardness* or *awayness* is common in animal behavior. Our problem is to consider, from the phylogenetic and the ontogenetic approaches, the question of how animals generally manage to reach beneficial conditions and stay away from the harmful, that is, how *survivors* do this. Although this valuable series of conferences has by now covered nearly all phases of the motivation problem, the evolutionary and developmental aspects are perhaps the ones touched on lightly in evidence and theory. My purpose is to discuss some promising theoretical ideas and evidence bearing on these questions.[2]

Motivation, broadly considered, concerns the causation and impulsion of behavior. The question here is what impels the approach and withdrawal reactions of very different animals from protozoans

[1] Chapter 1, pp. 1–42, in *Nebraska Symposium on Motivation*, 1959, M. R. Jones (ed.). Lincoln: University of Nebraska Press. Reprinted by permission of the Editor and the publishers. A discussion of the concept of approach-fixation, as applied to approach and following behavior in neonatal birds, will be found in "Comments on Dr. Hess's paper," pp. 78–81, *Nebraska Symposium on Motivation.*

[2] The theory presented in this article was first developed in a book on comparative psychology (75); its application to behavioral phylogeny and ontogeny was sketched a few years later (99) and discussed in subsequent publications (101, 102, 103). The concept of biphasic adaptive mechanisms has been developed on the basis of evidence or theoretical contributions from the work of Charles Darwin, H. S. Jennings, W. Cannon, C. S. Sherrington, J. B. Watson, E. B. Holt, and F. H. Allport, in particular. The concept of antagonistic functional mechanisms has been discussed by Kurt Goldstein (44) with reference to human behavior, and recently E. J. Kempf (68) has developed a theory of conflict and neurosis based on a wholistic study of biphasic mechanisms in man. All of these sources may be consulted for detailed references on this ramifying subject.

to man and how each level develops its characteristic pattern. Have these levels anything in common, or does each have a basis very different from the others?

In studying this broad problem of behavior, an objective methodology is indispensable. To be specific, I submit that *approach* and *withdrawal* are the *only* empirical, objective terms applicable to *all* motivated behavior in *all* animals. Psychologically superior types of adjustment are found, but only on higher psychological levels and after appropriate individual development.

As elementary definitions, an animal may be said to *approach* a stimulus source when it responds by coming nearer to that source, to *withdraw* when it increases its distance from the source. This point is not sufficiently elementary, however, to escape confusion. Confusion is indicated when the term "approach" is combined with "avoid," as if these were opposite concepts for motivation. This practice, common in the literature, is indulged in even by psychological dictionaries. But however conventional, it is psychologically wrong, for *withdrawal* is the conceptual opposite of *approach*, and the opposite of avoidance is seeking, which means "to look or search for" something. Seeking and avoidance are of a higher evolutionary and developmental order than approach and withdrawal, and these terms should not be mismated.

To put it differently, whereas all behavior in all animals tends to be adaptive, only some behavior in some animals is purposive. Behavior is *adaptive* when it contributes to individual or to species survival, especially the latter. To take a contrast, when an amoeba is stimulated by directed weak light, typically there occurs a local flow of protoplasm toward the source, perhaps followed by a general movement in that direction: an approach. This is a forced protoplasmic reaction, very different at basis from behavior observed in a rat (75, frontispiece). Uneasy about coming into the open, this animal does so, however, when food is placed there. But instead of eating in the open, it manipulates the dish across to and over the barrier, and ends by feeding in its retreat. On inductive grounds, both amoeba and the rat act adaptively, but only the rat accomplishes goal-directed responses to an incentive, and therefore behaves *purposively*. The response of the amoeba is energized directly by protoplasmic processes set off by the stimulus—that of the rat involves specialized, higher-level processes not indicated in the protozoan.

The real issue concerns what assumptions we made about teleology or the problem of final causes. Vitalists such as Driesch and McDougall settle the problem of behavior causation by assuming *a priori* that all adaptive behavior is "aimed at" resolving a difficulty, and that all animals, at whatever level, can "see" the beneficial and "avoid" the injurious. Warburton (118), however, believes that telic usage of the term "goal," a frequent indulgence in biology, implies only that the phenomenon is held selectively advantageous to the reacting organism. If this is really so, confusion can be avoided by using the term "adaptive," unless psychological "purposing" is adequately demonstrated.[3] It is becoming more and more evident that adaptive behavior, which has attained many specialized forms through natural selection, is basically natural and organic in all animals, from those of lowest psychological status to those capable of reaching solutions by specialized anticipative processes.

How, then, does behavior in surviving animals, even the lowliest, come to have the character of adaptiveness, centering on efficient approach and withdrawal responses? The main reason seems to be that behavior, from its beginning in the primitive scintilla many ages ago, has been a decisive factor in natural selection. For the haunts and the typical niche of any organism must depend on what conditions it approaches and what it moves away from—these types of reaction thereby determine what future stimuli can affect the individual, its life span, and the fate of its species.

The principle may be stated roughly as follows: *Intensity of stimulation, basically, determines the direction of reaction with respect to the source, and thereby exerts a selective effect on what conditions generally affect the organism.* This statement derived from the generalization that, for all organisms in early ontogenetic stages, *low intensities of stimulation tend to evoke approach reactions, high intensities withdrawal reactions with reference to the source* (99). Doubtless the highroad of evolution has been littered with the remains of species that diverged too far from these rules of effective adaptive relationship between environmental conditions and response.

The chief difficulty of the vitalists seems to arise from an unwillingness to accept the possibility that mechanisms correlated with physical reality could start very simply as forced-reaction types and then change through long evolution to provide a basis for psychologically

[3] A positivistic generalization of "purpose" was used by Bertocci (3) in arguing against Gordon Allport the problem of causal factors in motivation. The present writer finds himself in agreement with Allport that if these problems are to be investigated scientifically, a really objective set of concepts is needed, not contaminated by unexamined *a priori* assumptions. The latter form of obfuscation frequently is accompanied by disarming protests such as that of the biologist McKinley (79), who, while holding that purposive striving underlies the adaptive behavior of all organisms through evolution, denies that his assumption is teleological (i.e., that it is purposivistic!).

advanced types such as seeking or avoiding. Although terms such as "appetite-aversion" imply that all animals know what to look for and what to avoid, this is not so in all cases. Jennings' (67) discovery that protozoans such as Paramecium approach the source of potentially lethal chemicals and withdraw from potentially beneficial or innocuous chemicals obviously does not prove that the animal has made a mistake, but only that the experimenter has been ingenious enough to find stimulative situations reversing the normal formula that low-energy stimulation leads to food or other benefits, including no harm, and high-energy stimulation leads to harm or death. Although at first glance all animals seem to act as though they could seek or avoid things in view of benefits or dangers, it is the methods of objective science that must be directed at this question, and not the forensic arts.

Much evidence shows that in *all* animals the species-typical pattern of behavior is based upon biphasic, functionally opposed mechanisms insuring approach or withdrawal reactions according to whether stimuli of low or of high intensity, respectively, are in effect. This is an over-simplified statement; however, in general, what we shall term the *A-type* of mechanism, underlying approach, favors adjustments such as food-getting, shelter-getting, and mating; the *W-type*, underlying withdrawal, favors adjustments such as defense, huddling, flight, and other protective reactions. Also, through evolution, higher psychological levels have arisen in which through ontogeny such mechanisms can produce new and qualitatively advanced types of adjustment to environmental conditions. Insects are superior to protozoans, and mammals to insects, in that ontogeny progressively frees processes of individual motivation from the basic formula of prepotent stimulative-intensity relationships.

Mistakes in the psychological level of adaptive processes are common risks (101). What this means is illustrated by a controversy over the status of sea-anemone motivation (75, p. 45). Jennings found that after he had dripped meat juice for a time over one side (A) of an individual's crown of tentacles, neither the tentacles of side A nor those of the unstimulated side B would respond by extending and waving. The results suggested to Jennings a general motivational change in the animal, which he characterized as a "loss of hunger." Parker repeated this test, but with an additional control, that of injecting a current of meat juice into the body cavity through the body wall. Here also the entire disc presently became quiescent, leading Parker to conclude that the meat juice had adapted all tentacles alike by entering the hollow interiors and diffusing through the walls to the receptor cells. On this low level, peripheral (i.e., molecular) types of change seem dominant rather than central (i.e., molar) types produced through organized neural conditions as in mammals. In contrast, a satiated rat can become unresponsive to food through an avoidance process involving perceptual rejection, and not just local or restricted physiological changes in some property such as sensitivity. The difficulty with terms such as *sensitization* and *satiation* for animal drive is that they are vague and may run a wide range of qualitatively different processes in different animals.

Thus, although acceptance and rejection, beneception and nociception, love and hate, and other motivational dichotomies are based on perhaps similar organic properties arising through evolution, they involve widened potentialities for behavior organization and superior capacities such as anticipation demonstrable only in higher mammals.

I recall a psychologist expressing surprise that a worker bee participates in comb-building although, as he supposed, it is "not hungry at the time." This type of case must remain enigmatic for those who assume without question that the insect is motivated by an anticipation of the over-all result to which its act contributes. But I believe the relationships of food and actions leading to it are rather different for insects than for mammals (**75**, **100**, **104**). The insect builds its maze habit slowly and by stages: first a general habituation to the situation develops, then a process of very gradually mastering the local choice-points, and in a final stage an over-all integration of the habit comes about only through stereotyped interactions between segmentally anchored habits. The rat masters the "same" maze pattern very differently, starting its local learning and over-all integration together even on the first run. Unlike the ant, he can soon anticipate distant parts of the maze, and this may get him into trouble at times. The ant's habit is built up in sections, without any indications that the subject is capable at any time of specifically anticipating the end-point from a distance. Her drive processes are evidently rather different from those of the rat, which can anticipate either by stages as a habit builds up or by leaps to the end-point, depending on the circumstances.

Our theory of animal motivation is likely to be influenced strongly by the views we hold as to whether learning can differ qualitatively at different phyletic levels. Even the sparse existing evidence indicates that differences of a fundamental nature exist, not to be understood in terms of "complexity." Basic differences in the nature of learning seem involved in the fact that ants do not reverse a learned maze readily, although rats do, as also in the fact that the two animals initially acquire the habit very differently. Accordingly, it is significant that repeatedly reversing alternative responses differing greatly in difficulty can result in neurotic behavior in rats (124) but not in ants (100, 104). The results indicate that *drive* in the ant's habit is much more strictly bound to specific effects from extrinsic stimulation (e.g., from food carried) and from internal condition than in the rat's. The finding of Bitterman *et al.* (5) that fish progress very differently from rats in a reversal problem, with slight indications of day-to-day progress in only a few of the fish and strong evidence for such progress in all of the rats, indicates basic differences in how lower vertebrates and mammals learn the "same" problem.

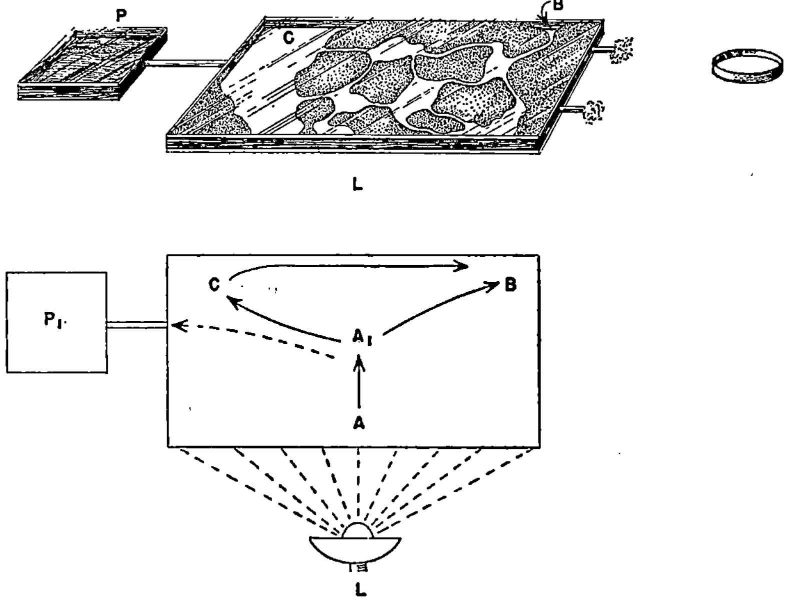

FIG. 1. Nest setup for comparison of reactions of Formica ants and inquiline beetles to intense stimulative conditions. *Top:* Ant nest, water-dish removed; L, lamp; P, food-box; *bottom:* A,A1, B, C—see text.

Concerning the extent to which reasoning may differ qualitatively from learning, much evidence points to the existence of a fundamental distinction, which discrimination-generalization methods for example may not be valid to test (75, Chap. 20).

I believe that a strong case can be made for differentiating the following levels in capacity to modify behavior (101):

(1) fluctuant changes, through peripheral effects (e.g., protozoans);
(2) sensory integration, habituation, based on central trace effects (e.g., flatworms);

(3) contiguity-type conditioned responses (worms);
(4) selective learning (different in insects; mammals);
(5) insight, reasoning (higher mammals).

On different phyletic levels, therefore, the relation of the fundamental biphasic mechanisms to drive may differ significantly according to the scope and nature of capacities for modifying behavior through ontogeny.

Such differences may be crucial for survival in emergency situations. As an example, one of my nests of Formica ants had been overfed and developed a large population of tiny parasitic beetles. To get rid of the beetles, I put a lamp on side A (Figure 1). At first both ants and beetles gave strong taxis reactions to the light and heat, all withdrawing towards side B-C. But soon certain of the ants began running to the chamber P, taking a previously learned foraging route, then returned to the nest and carried their more sluggish species-mates into the box and safety. Meanwhile, the beetles collected increasingly in corner B, the last part of the nest to dry out in the heat. At length all of the beetles were at B and were dead or dying; all of the ants were in the box P and alive. The predominant initial drive process of both animals was of the *taxis* type, governed by intense extrinsic stimulation, with varying A- and W-type adjustments as the nest situation changed. This pattern led to the extinction of the beetles, whereas a qualitatively different adjustment process admitted by ontogenetic specialization saved the ants.

Biphasic Mechanisms Basic to Behavioral Ontogeny

Through biphasic adjustive mechanisms, stimulative energy fundamentally dominates the approach and withdrawal responses of all animals. A simple, inclusive pattern is that involved in the gel-sol cycle of the protozoan amoeba. In this animal, stimulus energy, whether in the form of light, chemical, or tactual effects, if weak, sets up solation in the locality, perhaps with pseudopod formation and extension toward the source following; if strong, it sets up gelation or contraction on the stimulated side or perhaps widely, then solation may occur on the opposite side with movement away from the source. Mast (77) demonstrated that with different intensities of light the protoplasmic effects graduate from solation and prompt forward locomotion to delays in movement based on different degrees of reversal in the protoplasmic flow and, finally, a movement from the source. Such differences may be seen clearly under the microscope. The intervening variables are here exposed as different degrees of change ranging to a complete reversal of the gel-sol cycle; the drive is generalized in protoplasmic activity.

Biphasic mechanisms facilitating approach-withdrawal adjustments are present in all functional systems, from sensory to motor, in multicellular animals. One important type of biphasic afferent process enters with what Hecht (55) called the "duality of the sensory processes." In the clam *Mya* he demonstrated two opposed photochemical processes, one (A-type) aroused in weak light, the other (W-type) in strong light, the A-process eliciting siphon extension and feeding, the W-type siphon retraction and valve closure. Hecht generalized this dual-process formula for the photoreceptors of widely different animals, including man.

The biphasic properties of receptor mechanisms are sometimes not apparent until their relations to general function are analyzed. Earthworms, for example, lack specialized eyes but have light-sensitive cells distributed in the skin. In W. Hess' (59) experiment, dark-adapted earthworms turned from a light source of 0.3 mc. in 99 per cent of 144 trials but turned from a source of 0.001 mc. in just 43 per cent of the trials (Figure 2). As earthworms lack specialized eyes, this difference is due to the differential arousal of light-sensitive cells in the skin. The weak light evidently stimulated relatively few of these generalized receptors, those of lowest threshold, producing a low-energy neural discharge with approach the dominant response, whereas the stronger light aroused also many of the cells with higher thresholds, resulting in a stronger neural discharge forcing W-type reactions.

Specialized eyes of many types exist in the animal series, their properties bearing on action in diversified ways. In many invertebrates, the sectors of the compound eye are known to differ significantly in the sensitivity and relation to action of their afferent elements (75, Chap. 6). In Clark's studies on water beetles, as an example, specimens with left eyes covered turned to the left when light reached anterior sections of the open compound eye, but to the right when sections farther back were stimulated by light of the same intensity. The reason, effective for many animals from flatworms to arthropods, is that sensillae of lower threshold in certain zones of the eye discharge neural patterns of higher energy, inducing the animal to react as to a more intense stimulus than when sensillae of higher thresholds in other zones are aroused. That a forced difference in action results is indicated for example by the finding that varnishing over the lower part of both eyes in certain flies causes extension of the anterior legs and raising of the body, whereas covering the upper parts causes flexion of anterior legs and lowering of the body (Fraenkel and Gunn [38]). Research such as that of Mittelstaedt (83) with insects shows that such relationships are often complexly involved in the control of action.

According to the type of eye possessed by an animal, basic differences may arise in the relative potency of stimuli to excite neural

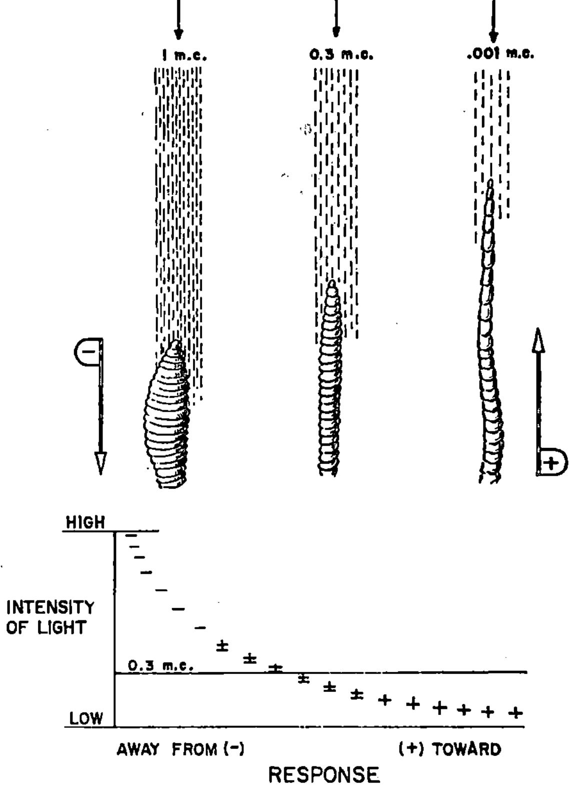

FIG. 2. Reactions of dark-adapted earthworms to light (W. Hess, 59). *Top:* at the high intensity, prompt anterior shortening, movement from source predominate; at the middle intensity, less prompt shortening, increasing movement towards source; at the low intensity, thinning and movement towards the source predominate. Intensity and response relationships are graphed *below.*

discharges affecting action mechanisms. As we shall see, certain lower vertebrates, with lens eyes, are directly opposite to the bee, with compound eyes, in their tendency to approach figures of simpler rather than more complex outlines. Hertz reported that the honeybee's approach to visual figures varied in proportion to the heterogeneity of outlines and internal articulation, e.g., a star was approached more frequently than a square, and both of these more frequently than a circle (57a). Wolf (125), analyzing such behavior, found that bees approached visual areas in frequencies directly related to the rate of flicker. From such results he concluded that responses to flickering fields and to articulated figures had a common basis, namely, the number of pulses per unit time produced in the sensillae of the eye as the bee flies across the stimulus fields. Lower vertebrates, which instead, as we shall find, approach circles more readily than figures of more complex outlines, evidently have afferent properties opposite to those of the compound eye as concerns action.

Thus, in vision and other modalities, types of mechanism have evolved which, by regulating the potency of neural discharge in specialized ways, serve to canalize the stimulative conditions under which animals approach or withdraw. The vertebrate ear, considered a specialization of the cutaneous system, significantly extends reactivity to vibratory stimuli. This has occurred through the evolution of afferent specializations, as of a cochlea from the lagena of lower vertebrates, and of changes in the anatomy of the auditory ossicles affecting both the range and the acuity of hearing (Tumarkin, 115). In chemoception, also, many end-organ specializations have arisen which admit species-typical thresholds and other differences in sensitivity. Striking cases may be found in the responses of insects to chemicals (Dethier, 25), from approaches to odor of food plants or of animal hosts to withdrawals from other plant or insect odors. In many cases, as with affinities for colony odors in social insects, species secretory functions as well as earlier feeding experience also influence reactivity; in others, as when a former food odor becomes repellent, changes in susceptibility clearly occur in dependence upon organic condition or age. Many types of secretory and other organic specializations have evolved, variously significant for such reactions. One interesting type of adaptive effect was discovered by von Frisch (41), who extracted an essence from the skin of the minnow which proved strongly repellent to species mates. This phenomenon is evidently widespread, as Heintz (57) found it in worms and insects also.

A great diversity of mechanisms affecting attraction or repulsion along species-typical lines is evidenced in the natural behavior of animals.

Correlated with biphasic afferent functions are neural mechanisms with distinctive arousal thresholds or other functional properties dependent on stimulus intensity. One interesting type of specialization is known for the earthworm and certain other invertebrates, in which intense stimulation of the head elicits a sudden jerk of the entire animal backward. In this response, the function of regular short-relay conductors underlying smooth locomotion is interrupted by the arousal of W-type effectors through the "giant fibers" (Stough, **112**; Bullock, **11**), a specialized high-threshold conductor especially responsive to intense stimulation.

Whatever their equipment of specialized neurones may be, the nervous systems of multicellular animals are all capable of discharging differentially according to afferent intensity. The general formula supported by much neurophysiological research on both invertebrates (e.g., **51**) and vertebrates is that, with increasing stimulus intensity in any modality, successive increases are evoked not only in the volley rate of discharge into the central nervous system but also in the number of neurones conducting (Brink, **10**). Consequently, instead of type-A actions, predominant at lower stimulus intensities, type-W actions of higher threshold can be aroused increasingly.

Primary to all such phenomena is a dependence of the neural discharge on the quantitative aspects of stimulation. For particular types of discharge, Sherrington (**109**) remarked that "The accuracy of grading within a certain range of intensities is so remarkable that the ratio between stimulus intensities and response-intensity has by some observers been assigned mathematical precision." From Sherrington's pioneer work on the differential arousal of flexion and extension reflexes through the spinal cord, with his associates (Creed *et al.*, **17**) he derived the principle of "half centers," which may be characterized as *reciprocally functioning ganglionic subcenters.* This principle outlines the functional differentiation of partially overlapping neurone groups adjacent to each other in the same neural center, differing in their arousal thresholds and capable of working reciprocally in the excitation of antagonistic effector systems. The view that such reciprocally related neurone groups normally operate to oppose the strong arousal of A-type and W-type effectors simultaneously, but favor the predominance of one or the other type in a smoothly working pattern, is well supported (Eccles, **31**). This principle, first demonstrated for antagonistic neurone groups in the cord, may have a considerably wider application to neural correlation in the control of biphasic effector systems.

A finding of significance for mammalian ontogeny is that of Adrian (**1**), that impulses from deep or protopathic and impulses from epicritic or superficial cutaneous receptors, selectively aroused according to stimulus intensity, may take different paths and have typical discharge-pattern differences, as in the lesser opportunities for summation in epicritically aroused than in protopathically aroused patterns. The quantitative aspects of stimulation and of conduction therefore would tend to favor a routing of impulses in the first case predominantly through that neurone group in the spinal level discharging to A-type effectors, and in the second case through its antagonist discharging to W-type effectors. This may be considered the paradigm for neural discharge to common centers in early ontogeny, dependent on stimulus energy delivered to any modality.

Although most of the research on the functions of the autonomic nervous system has been done with *adult* mammals, and the evidence is understandably complex, the generalization seems justified that, according to stimulus magnitudes acting upon the central nervous system, the two principal sections of the autonomic system are selectively aroused to characteristically different functions. Through the cranial and sacral regions of the central system, the *parasympathetic* outflow tends to arouse visceral and skeletal functions of an A-type, of vegetative character; through the thoracic and lumbar regions the *sympathetic* outflow arouses functions of a W-type, of interruptive nature. The antagonism of these systems may be illustrated by Kuntz's (70) description of the selective arousal by weak stimuli of impulses over a dorsal cord depressor path of long fibers and low resistance, associated with the parasympathetic system and arousing vasodilitation, and by strong stimuli of impulses over a ventral cord pressor relay of short fibers and high resistance, associated with the *sympathetic* system and eliciting vasoconstriction.

Significantly, although the autonomic nervous system is old in vertebrates, in its full biphasic function in higher mammals it is highly specialized. Recently, Pick (88) has concluded that whereas primitively in vertebrates the dorsal-outflow (*parasympathetic*) system is the predominant system of visceral arousal from the cord, the ventral outflow, although at first negligible, from amphibians to man increases steadily in potency and specialization as the *sympathetic* system, and becomes progressively more effective as the antagonist of the parasympathetic.

Evidence concerning metabolic processes effected largely through smooth muscles and the autonomic nervous system widely supports the Cannon-Langley theory of antagonistic functioning dependent on opposition of the *parasympathetic* and *sympathetic* autonomic divisions. As Darling and Darrow (20) demonstrated, the biphasic character of human visceral function shows through even in the *adult* in blood pressure and other physiological aspects of emotional reactivity, notwithstanding certain exceptions (e.g., galvanic skin response) in which the divisions of the autonomic tend to reinforce each other. Starting with Cannon's research, the patterning of opposed function has been revealed in all types of visceral actions, including smooth muscle and the neurohumoral. Neurophysiological research gives a familiar picture of *vegetative* (A-type) changes (homeostatic, smooth-running processes occurring within a range characteristic of species and developmental stage), aroused characteristically by low-intensity stimulation through the parasympathetic system, and of *interruptive* (W-type) changes supported by secretion of adrenin and aroused characteristically by high-intensity stimulation through the *sympathetic* system. With experience, the functional relationships of these systems become more complex and varied, perhaps often to the point of cancelling or reversing aspects of their basic antagonistic, reciprocal relationship.

The effectors, also, seem to function biphasically in all animals. In the earthworm, under weak stimulation the circular muscles are dominantly aroused, thinning and extending the body and facilitating movement toward the source; under strong stimulation the longitudinal muscles are dominant, thickening the body and facilitating withdrawal from the source. A striking interruption of normal smooth reciprocal function of the two systems occurs when high-intensity stimulation causes the body to contract suddenly backward through a rapid maximal innervation of the W-system *via* giant fiber conduction (91). In other invertebrates, through corresponding mechanisms in tentacles and other local structures, the mobile part typically swings toward the side of weak stimulation, away from that of strong stimulation, and through comparable synergic action the entire organism may react equivalently.

For vertebrates, the principle of differential and antagonistic systems in skeletal muscle function was established by Sherrington (109), who demonstrated it in the action of sets of muscles working against each other at the same joint or on different joints of the same limb. Between extensor dominance with weak stimulation and flexor dominance with strong stimulation, a gradation of functional combinations was demonstrated both by myographic recordings and by direct palpation of muscles to intact limbs. In the developing organism, relationships of threshold may be stated conveniently in terms of the Lapique concept of chronaxie, as a measure of arousal threshold. Bourguignon (6) found that adult mammalian extensors tend to have a low chronaxie and flexors a high chronaxie, particularly in the limbs. These two muscular systems commonly act concurrently in the general behavior of animals (Levine and Kabat, 74); however, summaries of the evidence (e.g., Fulton, 42; Hinsey, 60) leave no question that their antagonistic function in synergic patterns is a primary fact in mobility and action, and that their reciprocal arousal is essential to the development of coordinated patterns of behavior (Hudgins, 65). For research on vertebrates generally, flexion of a limb to strong stimulation is a response taken for granted as available for conditioning from early stages. Although Sherrington described extension as postural and flexion as a "reflex to noxious stimulation," their biphasic function, as we shall see, constitutes a mechanism of expanding relationships basic to ontogeny.

In sum, diverse biphasic mechanisms of the receptors, central and auxiliary nervous systems, and effectors are fundamental to ontogeny in all animals. They have in common, generally, the property of A-type arousal by weak stimuli, facilitating local or general approach to the source, and W-type arousal by strong stimuli, facilitating local or general withdrawal. Recently Kempf (68) has reviewed the evidence for such antagonistic functional systems in man, in relation to problems of conflict and neurosis. The thesis here is that relationships between stimulus magnitude and the degree and direction of response, although different on the various psychological levels as to their form and the extent to which their function may be modified, are always critical for the determination in ontogeny of what conditions may attract and what may repel members of a given species. These matters must not be oversimplified, of course, as many conditions besides experience, including adaptation, fatigue, and health, may affect the potency and directness of the biphasic processes.

The property of differential thresholds in biphasic systems may take a more specific origin in early ontogeny when the probing termini of developing neuroblasts react selectively to the biochemical properties of tissue fields nearby, and may influence their environs in turn. Threshold differences evidently progress throughout the organism along biphasic lines. Many types are as yet little understood, as for example that involved in Eccles' (31) distinction between "fast" and "slow" muscles in the cat. Such differences, paralleling biphasic differences in afferent, visceral, and neural systems, underlie the rise of specialized patterns of approach and withdrawal, the detailed character of which is a matter of phyletic level and ontogenetic attainments.

Species-Typical Processes and Specialization of Approach-Withdrawal Reactions

It should be emphasized that whether the described biphasic processes influence behavior in rudimentary or in specialized ways depends particularly upon species capacities for their modification through ontogeny into correlations and organized adaptive systems.

In early ontogeny in all animals, the quantitative aspects of stimulation evidently dominate both the direction and vigor of action. To illustrate the effectiveness of this relationship on the invertebrate level, let us consider a phenomenon known as the "reversal of taxis." Here the animal is reported as reversing its normal responses to stimulus intensity, or in conventional terms it becomes "positive" to bright light rather than "negative" as usual. In W. Hess' (59) experiments, light-adapted earthworms approached a source of 0.3 mc. in 55 per cent of the trials, in contrast to only 1 per cent for dark-adapted worms to the same intensity (cf. Figure 2). It seems likely, however, that no reversal of response polarity has occurred, as the light-adapted worm does not withdraw from weak light. Instead, the intensity threshold has been raised so that physically strong light exerts an effect equivalent to that normal for weak light. Significantly, removal of the earthworm's brain comparably brought a greater readiness to approach bright light. As any means taken to raise the animal's threshold to anterior stimulation has this effect, the so-called reversal of taxis to extrinsic stimulation involves instead a temporary decrease in sensitivity.

It is well known that the "sign" of the taxis responses of many insects and other animals varies according to age or excitability, developmental stage, and external conditions (e.g., Dolley and Golden [27], Dolley and White [28], Fraenkel and Gunn [38]). These phenomena appear to have in common an effect on metabolism, decreasing or increasing functional thresholds according to their direction. There are many physiological means of changing an animal's readiness to respond, which may acquire more or less involved relationships in the motivational pattern according to species and ontogeny. Consider two rather different examples, both more specialized with respect to motivation than taxis-change cases of the type mentioned. Evans and Dethier (35) analyzed the rise through feeding of the blowfly's tarsal thresholds to four sugars, all of which had been found tasteless for the fly and two of them completely non-nutritious. By eliminating factors such as blood chemistry, the experimenters narrowed the locus of the inhibitory effect to some part of the gut other than the crop. In contrast, Harris *et al.* (50) trained vitamin-B-deficient rats to take a stock food containing this vitamin, as against the same food without the vitamin, by adding a trace of cocoa to the former. With this subtle cue, the B-deprived rats discriminated between foods critically different for their organic deficiency, although normal rats ate these foods equally. In these instances, worm, blowfly, and rat are clearly motivated at very different psychological levels, at which the term "sensitization" would have very different meanings. There are many ways of increasing or decreasing the responsiveness and drive of any organism, which, particularly at higher levels, may have very different relationships to the directionality and plasticity of response.

Lower vertebrates present many striking species-typical approach and withdrawal tendencies. Many reptiles respond differently to long and short rays of the spectrum, a tendency attributed to aspects of their retinal chemistry. The nocturnal approach of newly hatched loggerhead turtles to the sea, which Daniel and Smith (18) attribute to the reflection of moonlight from the surf, may thus be analogous to the proneness of *Lacerta* lizards to approach green as against other spectral stimuli. These are strong tendencies, since Wagner (117) found lizards persisting through several hundred trials in snapping at a green disc bearing a quinine-soaked mealworm as against one of another hue bearing an untreated mealworm. Ehrenhardt (32) investigated with similar results another of Lacerta's persistent responses, its approach to figures such as discs with smooth outlines as against figures such as squares with broken outlines. The possibility that such tendencies relate to characteristic properties of the receptor and afferent system, governing neural discharge to viscera and effectors, is suggested by the fact that the honeybee approaches figures of *broken* outlines as against those with smooth outlines, due, as Wolf (126) demonstrated, to compound-eye properties making visual-flicker effects attractive. Without doubt they bear a significant relationship to phenomena of approach described for birds (36, 58) as "imprinting" and "following."

The general disposition of lower vertebrates to approach the source of low-energy stimulation is clear. The fish *Coris*, studied by von Holst (61), reacts to weak contact or light by bending towards the source and by spreading dorsal and anal fins, to strong stimuli by bending away and *folding* these fins. In many fishes, changes in skin coloration also occur, differentiated as to stimulus intensity, indicating disparate visceral components in the excitation processes. Many fishes, amphibians, and reptiles respond rather consistently to moving stimuli of small area by lunging forward. The snapping responses of toads depend upon quantitative properties of the stimulus such as rate of movement rather than its qualitative nature, as Honigmann (63) and also Freisling (40) demonstrated in tests controlling size, intensity, and rate of movement. These stimulus characteristics seem to be effective as quantitative values, hence essentially equivalent to one another in their control over neural dis-

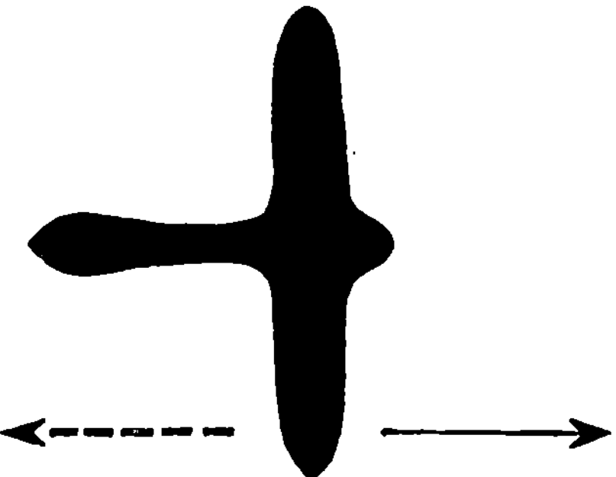

Fig. 3. From Eibl-Eibesfeld and Kramer (34): "Model of bird of prey releases escape reactions (flight) among young gallinaceous birds, ducks, and geese when moved to the right. Moved to the left (broken arrow) it has no releasing function. In this case shape in relation to the direction of movement provides the active sign stimulus. (After Tinbergen, 1948.)"

This caption presents the ethologists' hypothesis of "innate schemata." For a different interpretation, see text.

charge in the visually naive animal. With any one of these as independent variable, the lunge characteristically occurs to stimulus intensities below a given magnitude, above which turning and withdrawal increase in frequency. Although toads flick the tongue alone at very small moving objects, they lunge at somewhat larger ones and use the jaws, but still larger stimuli bring retreat (33).

Such response tendencies, dominated by stimulus magnitude, may be basic to the differentiation of patterns such as feeding, mating, and flight in the lower vertebrates. In these animals the naive responses to quantitative sensory effects have a compulsive character and, as we have seen, tend to resist change through experience. At best they can be changed only within strict limits, as Schneider (98) found in the extensive training required by the toad to inhibit snapping at edible objects presented behind glass, a training much less effective in young animals than in adults. Deficiencies in central nervous correlation evidently restrict the extent to which such processes can be changed through ontogeny. Results of this general type, signifying resistance to the modification of stimulus-intensity-dominated reactions, are obtained characteristically with cold-blooded vertebrates having a low supply of internuncial neurones.

From these considerations, it is more likely that, for lower vertebrates predominantly and for mammals at earlier stages, approach or withdrawal depends on generalized effects governed by stimulus energy, rather than on what ethologists have called an "inborn schema." Lorenz and Tinbergen's interpretation (Figure 3) of the effect of their hawk-goose figure (cf. 34), which reportedly caused goslings to crouch in disturbance when it was pulled blunt end first above them, but to show no disturbance or extend when it was pulled across narrow end first, has been attributed to innate figural effects. I have suggested (101, 102) that instead the results may have been due to a sudden retinal change in the first case *vs.* a gradual one in the second, and not to innate figural effects as such. The corresponding opposite behavioral effects should then be produced by triangles drawn across the visual field base or apex first, sounds breaking suddenly or gradually, and the like. Thus an "enemy" may exert an initial flight-provoking effect because of its size and brusque movements rather than its specific, qualitative appearance. In vertebrates such as birds, however, the qualitative appearance of objects may soon become effective according to adjustive circumstances in which they appear. Lehrman (72) suggests that the gaping responses of young perching birds are given to artificial stimuli resembling the general shape of the parent sitting at the nest edge, not because these approximate a native releaser pattern, as some ethologists maintain, but because they have been glimpsed briefly through opening of the eyes during earlier gaping responses to weak mechanical stimuli. Significantly, the "flight distance," or minimal space within which animals in the wild tolerate a strange object without retreating (56), generally corresponds roughly to body size, to which through natural selection the threshold properties of the species may have become adjusted.

Light is generally the most potent extrinsic stimulus influencing movements of animals in space, and has therefore been featured in this discussion. Consequently it is important to distinguish from typical orienting responses, in which directionality of response tends to correspond to the intensity of the neural effect, a type which Verheijen (116) has called the "trapping effect." This arises through the animal's encountering an isolated powerful light in a dark field, when the stimulus disorients the animal and literally forces a tonic fixation to the source. In this way, intense lights on dark nights cause injury and death to countless insects and birds. Had such effects been common as natural hazards during evolution, the course of natural selection in flying animals might have been somewhat different. No species could survive long with processes facilitating a compulsive approach to the source of potentially destructive stimuli, were such conditions common in its environment.

An elementary type of orientation is that in which an animal approaches a low-energy stimulus source, as a neonate kitten nears the abdomen of its mother through crudely directionalized responses to thermal and tactual stimuli from her. In more specialized cases, if the source moves off, *following reactions* occur. In the simple type of following reaction illustrated in Figures 4 and 5, an army ant trails one of another species in the column, evidently adjusting her locomotory movements to a stimulative field in which weak olfactory and tactual effects predominate. Comparably, the neonates of many birds and precocial young of many mammals approach an object

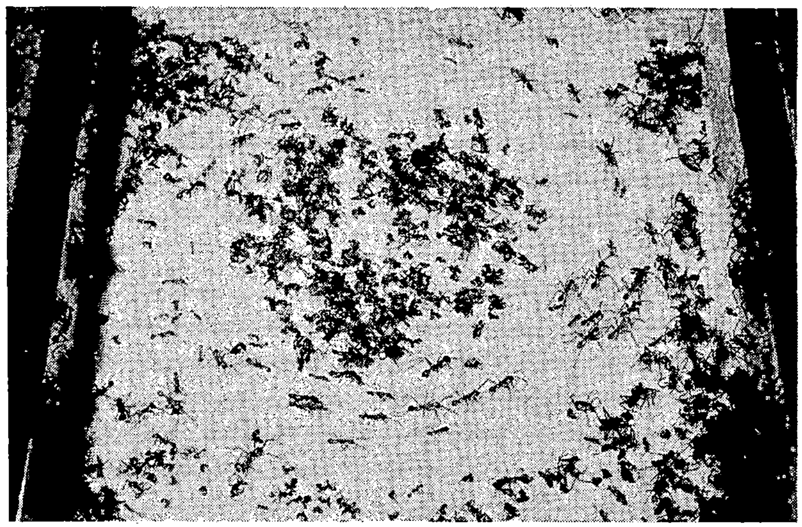

FIG. 4. Circular column in army ants of the species *Eciton burchelli* and *E. hamatum*, 35 minutes after these ants (previously separated by gauze for mutual olfactory habituation) were released together into a common area. While the excited ants circle, *following* reactions predominate, based on common tactual and olfactory stimuli. When circling stops, disturbed reactions to species odor differences appear.

centered in their visual field, evidently through responses to low-intensity stimulation from it, and subsequently may follow if the object moves away. The initially generalized character of the stimulus is suggested by the finding of Fabricius (36) and E. Hess (58) investigating "imprinting," that auditory and other stimuli also influence the reaction.

An important instance from natural behavior is the activity of a blind altricial neonate mammal readjusting to a changed stimulus gradient created when its mother disengages from the litter but remains nearby. The weakly directionalized stimulus pattern still promotes A-responses but is suboptimal for the A-type organic set, and thereby energizes effectively (i.e., facilitates) reactions maintaining or increasing this set. The function of the stimulus at this stage is suggested by the fact that, if it drops suddenly to zero, the rapid change acts as a disturbance which is likely to inervate interruptive visceral processes, mass action, and loud vocalization.

It would appear that following may be considered an extension of approaching in which locomotion is adjusted in such a way that a given range of stimulus magnitude prevails. The basis may be postulated as the arousal of low-energy stimulation of both A-type extension or approach reactions and A-type (vegetative) organic processes. Through this concurrence of events, there becomes established (according to species capacities) an organic set in terms of

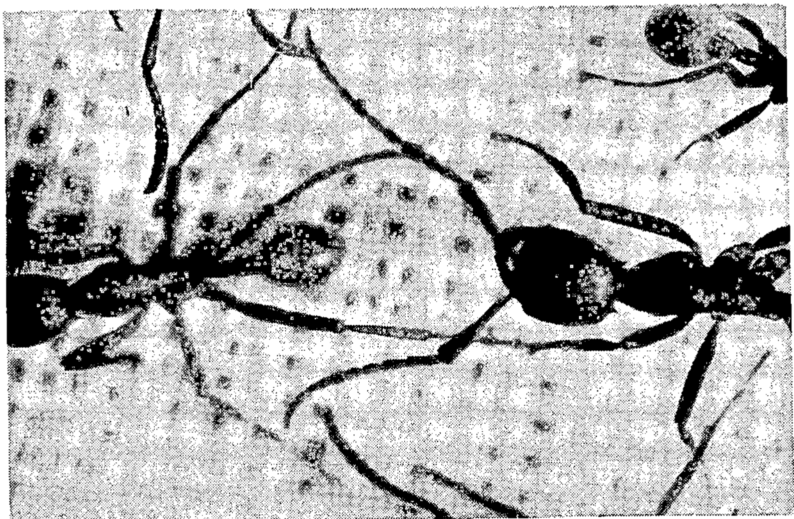

Fig. 5. Close-up of a worker of *E. burchelli* following one of *hamatum* in the circular column. (See text.)

which the animal can readjust its position so that the external stimulus is maintained within an intensity range optimal for this set. The attractive stimulus then not only arouses approach movements, but also the A-type set capable of facilitating recovery movements when the stimulus changes away from the optimum. The status of the extrinsic stimulus then governs whether approach continues into following, and whether following accelerates, changes direction, or stops. Disparities between the organic set and the stimulus magnitude can thereby be readjusted reciprocally, and following reactions are balanced off against this pattern according to the direction of its variations. Ontogenetic changes in these relations vary in nature and extent according to species capacities. Leuba's

(73) "optimal-stimulation" theory for mammalian learning represents one potentially useful way of studying such phenomena.

In the altricial young of certain mammals, the species-typical pattern of relationships between stimulus magnitude, organic set, and action is established much more slowly than in the precocial young of other mammals, and on a higher level than in lower vertebrates. Investigations in our laboratory (95) show that neonate kittens improve from the first hour in reaching the mother after separations and soon begin to establish individually differentiated suckling positions: fore, aft, or between. In the same period orientation in the nest vicinity, first established on the basis of proximal stimulation, decreases somewhat in efficiency after the eyes open at about one week, then rises steadily to a new pattern of organization based on visual perception. Regular progress of the kitten after about the tenth day in close approaches leading into suckling as well as in following the mother when she disengages indicates a growing capacity to hold or augment cues from a meaningful object. Then, after about the eighteenth day when the mother no longer "presents" regularly to the young and initiates feeding less and less, the kittens become steadily more proficient in a process of mutual approach (Figure 6). At length, after about the thirtieth day, the kittens themselves not only initiate feeding increasingly, but can follow the mother more efficiently and persistently as she decreases in responsiveness to them. A complex and changing set of seeking-avoidance relationships is indicated, doubtless basic and important to weaning. For the kittens, these studies reveal a behavioral development in which perceptual learning undergoes a succession of qualitative adjustments to objects which are first approached as naively attractive, then sought out as meaningful patterns in relation to organic condition.

In early stages, then, the approaching and following reactions are generalized; when and how far they specialize to environmental and species cues depends on properties of the species, the individual, and the changing developmental situation. It is often claimed that the mating patterns of lower vertebrates are natively given and little influenced by experience. But Shaw (106), working in our laboratory, has found that platyfishes raised from birth apart from species mates and surrounded by plates of ground glass scored minimally in mating tests at maturity. In contrast, individuals raised similarly in isolation during the first month but thereafter with species mates made much higher scores, though still inferior to those made by individuals normally raised with species mates. In this species, therefore, the elicitation of approach and other reactions in the complex repertoire of mating depends to a major extent on processes of perceptual development requiring earlier social experience as one factor.

Group adjustments such as those of schooling in fishes and flock-flying in birds are specialized following reactions of obvious adaptive

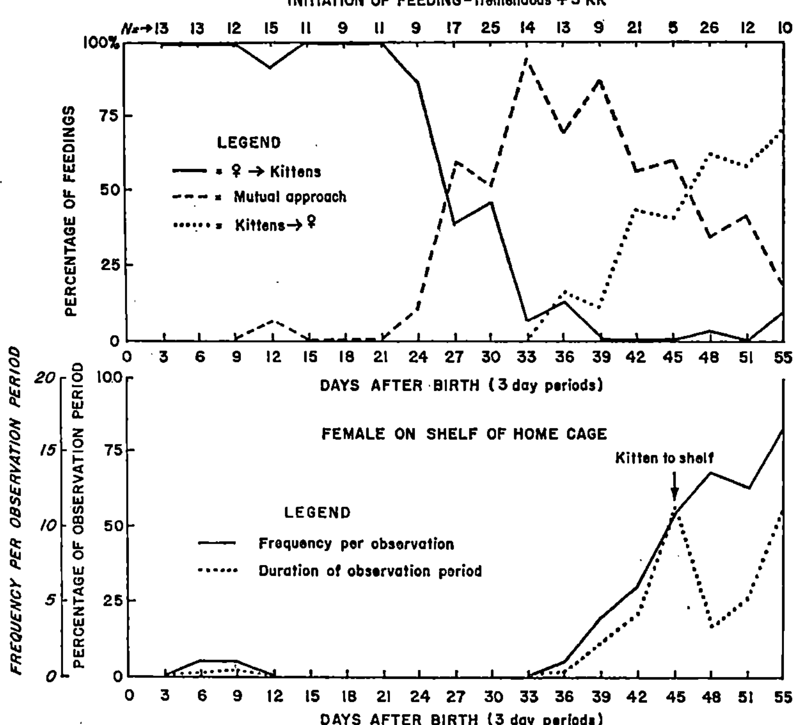

FIG. 6. *Top:* Graphic summary of the frequency of different patterns of initiating feeding in a representative litter of domestic cats consisting of the mother ("Tremendous") and her three kittens (94). *Legend:* Initiations by the female alone, *solid line;* by mutual approach, *broken line;* and by kittens alone, *dotted line.*

Below: Graph of results for female's reactions to a wall shelf in the living cage. Note the sharp drop in the female's "time on shelf" when the kittens begin to reach the shelf at about 45 days of age.

significance. To their development, early behavioral relationships with species mates may contribute, judging by preliminary findings in our research. In the schools of fishes such as mackerel, and in the circling mills these species may form, Parr (86) finds evidence that each participant fixes others unilaterally, and in the circular movement accelerates, slows, or turns in such a manner as to hold the size of a retinal image relatively constant. Such adjustments, in which nonvisual factors also play a part, doubtless have a specialized ontogeny appropriate to species mechanisms, capacities, and typical experiences.

As it is likely that the ontogeny of such adjustments is characteristic of the phyletic level, the claim that species behavior patterns represent an "original nature," later changed more or less through learning, is misleading. In what particular developmental stage is "original nature" to be identified? For vertebrates, evidence from the work of Carmichael (13) and others suggests the possibility that in early embryonic stages integrations may form between postural and visceral stimulation, characteristically diffuse, and cutaneous stimulation, localized in its effects when at low intensities. On such a basis in the embryo, a local response such as raising and pushing forward the head might gradually come under the control of local tactual stimulation when linked with vegetative visceral states. The careful research of Kuo (see 75) on the chick embryo indicates that such integrations in the embryo, although still unstable, may set up trace effects furnishing a basis for later reactions such as the head lunge, a response critical for feeding, and numerous other postnatal reactions in vertebrates.

There may be a significant clue here to the understanding of initial visual reactions in vertebrates. Carmichael and Smith (14) established for the early guinea pig embryo that lightly touching a limb typically arouses local responses, often with limb extension. Comparably, with appreciable frequency in the embryos of this and other mammals, weak stimulation of one cheek elicits a turning of the head toward the touched side. Strong local stimulation, in contrast, arouses generalized and variable activity. Not only may the integration of postural impulses with cutaneous stimuli provide a basis for progressive specialization of such actions, but an important part of the complex may be the role of organic processes of low threshold. Maier and Schneirla (75) suggested that at hatching or birth, such local action patterns, then tactually controlled, might be directly accessible to control by low-energy visual stimulation. The idea is that, when the animal is first visually stimulated, a diffusion of neural impulses from optic centers could arouse adjacent centers in mid-brain through which might be activated one or another tactuomotor pattern already partially integrated. The first visual stimuli might then, through their partial equivalence to weak anterior contact, elicit local or even general approach responses. This phenomenon may be a crude early form of the effect well known for mammals in later stages and termed "sensory pre-conditioning"

(76); also, it bears a distinct resemblance to the spontaneous integrative effects demonstrated for fishes and other vertebrates and termed "inter-sensory transposition" by von Schiller (97). On such a basis, local or even general approach responses to visual stimulation could occur in visually naive vertebrates, or local extension movements in neonates incapable of general progression at birth.

Stimulus Magnitude, Experience, and Motivational Development

These considerations suggest that ontogenetic attainments in motivational and perceptual adjustment arise through biphasic processes dominated in early stages by stimulus magnitude, progressing later as species capacities may permit. Lower invertebrates seem to maintain a species *status quo* close to minimal deviation from this pattern; insects vary more from it, although their motivation is dominated by specific extrinsic and organic conditions; it is higher mammals that deviate most from the biphasic pattern through ontogeny. Psychologists who emphasize disproportions, reversals, and exceptions between stimulus magnitude and response properties are therefore talking of adult stages at higher psychological levels.

Our review has stressed the existence of interrelationships between biphasic afferent, organic, and response processes at all phyletic levels. Because such organizations underlie ontogeny in all animals, the concept of stimuli as having a "drive" status seems inadequate for any animal at any stage. Neither is it adequate to stress endogenous processes without recognizing their changing role in ontogeny, as do the ethologists, who with their "releaser" concept minify changing stimulus relevances in development.

From the phyletic and ontogenetic perspectives, the changing relationships of stimulation, organic processes, and reaction in animal motivation, although grounded in biphasic conditions in all phyla, are seen as different kinds of processes at each level. For the blowfly, Dethier and Bodenstein (26) discovered that although sucking is aroused and controlled in rate by afferent input from tarsal chemoreceptors, continuation and termination of the act comes about through impulses relayed to the brain from the foregut, hence "hunger can be equated [in this insect] with the absence of stimulating fluids in the foregut. . . ." To illustrate the complex relationships of afferent to organic factors in a higher vertebrate, consider pigeon behavior. Light at dawn arouses mature pigeons to strut and coo, both actions known to stimulate endogenous functions such as crop manufacture of "pigeon's milk." Sight of the strutting male also facilitates ovulation in the female, with related organic changes promoting activities such as nest-making (72). At various stages in this behavior, extrinsic stimuli play the role of arousers, general organic excitants, specific excitants, or perceptual cues. Extrinsic stimuli, endogenous processes, and action can play such versatile roles in these patterns, with action differences so dependent on what conditions may affect each and what stage the sequence of events has reached, as to make the classical distinction between stimulus, drive, and mechanism an oversimplified and misleading thought pattern.

The role of early experience in motivational development is still inadequately understood; however, our evidence indicates that the part played by stimulative gains in development enlarges and specializes qualitatively from insects to mammals. Although in early stages, reactivity is dominated in all by stimulus intensity, mammals progress most in changing the relationships between biphasic processes and extrinsic conditions. The concept of "original nature" is vitiated by the fact that, on all levels, ontogeny constitutes a series of adjustive processes elaborating and changing under the conditions of each stage and not a process beginning suddenly at some one stage; moreover, attempts to distinguish sharply between the unlearned and the learned in development are unrealistic (102, 103).

Through early experience, particularly in mammals, A-type processes become cued to special stimuli such as odors accompanying food, W-type processes to the special stimuli of predators, and the like. But the relevance of stimulation to motivation is not episodic; rather, as sensory-deprivation research shows clearly (54), one prerequisite is a continuous afferent input maintaining a metabolic pattern essential to all activities from nesting to problem-solving. From our research with cats (95), it is evident that from birth the regular presence of a complex of chemical and tactual stimuli from the nest environment and maternal and litter associations is needed for normal perceptual development. Through certain types of focal experience, in which mild stimuli are contiguous with optimal metabolic or trophic processes, A-type patterns become environmentally conditioned, and certain intense, disturbing stimuli contiguous with interruptive processes become environmental controllers of W-type patterns. Focal stimuli in the litter situation thereby facilitate mating reactions at maturity, focal predator odors help put the animal on guard wherever encountered. The pattern of contiguity or S-R learning evidently is elementary for the early stimulus-cueing of organic sets of A-types and W-types.

For later developments, however, selective-learning processes must be postulated. Specifically, Maier and Schneirla (75, 76) suggested that whereas the contiguity formula could account for conditioning withdrawal or escape reactions to intense stimuli, a selective-learning formula is needed when changing relationships of action to stimuli and to endogenous conditions affect further adjustments. Accordingly, Roberts (94) discovered that although cats learned to avoid "an apparently strong noxious motivational set elicited by hypothalamic stimulation," with the strongest or "flight" intensity there were no signs of *avoidance* training within as many as 270 trials. The strong shock, by forcing an immediate withdrawal, may have admitted only contiguity learning, whereas the two lesser intensities, by producing milder disturbances, admitted selective learning. Furthermore, concepts of "tension reduction" are much too general to deal adequately with ontogenetic progress. From the results of Thompson (114) and others, it appears that from early infancy mammals establish approach and withdrawal habits differently. Accordingly, the theory of "optimal stimulation" recently advanced by Leuba (73) admits processes of selective cue adjustments made in relation to organic optima of different types.

The practice of combining the terms "approach" and "avoidance," although rather prevalent, is misleading. Whereas in its earliest stages a mammal evidently approaches or withdraws according to sheer stimulus intensity, operations of seeking or of avoiding under particular conditions can be accomplished only later, depending on capacity and opportunity to learn perceptual adjustments to qualitatively specific conditions. For reasons stated, the term "tension reduction" is deemed unsuitable for endogenous processes in learning, as processes of A-types and of W-types differ significantly in their relevance both to contiguity and to selective-learning. In fact, ontogenetic considerations suggest that specialized seeking patterns typically are mastered by young mammals *before* withdrawal reactions have well begun to differentiate into avoidances. The literature accordingly suggests that the human infant specializes perceptually in reaching and smiling before he avoids and sulks discriminatively.

Processes of behavioral organization and motivation cannot be dated from any one stage, including birth, as each stage of ontogeny constitutes the animal's "nature" at that juncture and is essential for the changing and expanding accomplishments of succeeding stages. Such improvements arise through an intimate interrelationship of "maturation" and "experience," representing all conditions introduced in any stage through tissue growth and through stimulative effects, respectively (103). Thus we have suggested that early integrations between postural, tactual, and organic conditions may be basic to later adaptive specializations as in feeding. Also, the concept "organic set" has been used to represent the gains of early stages underlying an endogenously facilitated readiness to adjust to something (e.g., to approach) without demonstrable perception of object qualities as such. I believe that ants do not advance appreciably beyond this level in their ontogeny, whereas mammals clearly do. Mammalian ontogeny leads into further stages in which "attitudes" become effective, involving readiness to organize adjustments relevantly to discriminated situational meanings. In mammalian ontogeny, it is likely that motivated learning of the anticipative type enters soon after distance (i.e., more comprehensive) cues, effective through progressive visual perception, really begin to modify and dominate the more restricted proximal processes of orientation and adaptation carried over from earlier stages.

A-System and W-System Processes in the Ontogeny of Emotion and Motivation

The view has been followed here that processes of perception, motivation, and emotion progress in an intimate relationship in animal ontogeny. Modification of early biphasic excitatory states inevitably overlaps the processes of perceptual development and anticipative motivated adjustments. Consider the frequency with which terms such as "pain" or "pleasure" are used to designate excitatory effects of stimuli through brainstem-implanted electrodes in mammals. It appears from such studies that disturbed responses termed "pain" or "fear" can be elicited through stimulating specific areas, "pleasure" or "reward" effects from other areas (82, 23). In the Olds and Milner (85) experiment, rats stimulated through the septal area acted as though rewarded, repeatedly operating a lever turning on the shock, as though in "strong pursuit of a positive [i.e., A-type] stimulus," whereas through areas nearby the stimuli seemed to have a "neutral or punishing effect" instead (cf. 43). These tests may have "cut in" on A- or W-type patterns, tapping critical way-stations in circuits of distinctively different arousal thresholds. But also, Bower and Miller (8) made implantations in the middle to anterior forebrain bundle of the rat through which a reward effect was obtained at the onset of stimulation and aversive effects with continued stimulation. Evidently A-type patterns were first aroused here, but were displaced by W-type patterns when the effective intensity reached their threshold through summation. It is important to emphasize the fact that *adult* animals were used in this research.

Although in recent years the term "anxiety" has been much used in psychology to denote a so-called "secondary drive" reinforcing learning, the ontogenetic background of this type of process is really no better understood than that of "hope" or "faith" drives. Actually, the nature of such processes may vary from organic set to anticipation, with corresponding differences in relevance to learning, according to ontogenetic stage and situation. "Fear" also raises many problems. Hebb's (52) conclusion from tests of captivity-raised chimpanzees that many primate fears are "spontaneous" may involve both (1) stimulus equivalences, in seemingly very different object characteristics and types of movement, referable to stimulus intensity-effects, and (2) effects of the "strange" on partially socialized animals, with unidentified relationships to earlier experience. Other problems which may be investigated to advantage through considering interrelated perceptual, motivational, and emotional processes in development concern the nature of "uncertainty" and "danger," and the ontogeny of "psychological distance under threat" (81, 122).

The interrelationships of these processes of perceptual, motivational, and emotional ontogeny are suggested if we define "emotion" broadly as: (1) episodes or sequences of overt and incipient somatic adjustment, (2) often loosely patterned and variable, (3) usually with concurrent exciting sensory effects, perhaps also perceptual attitudes characterizable as desirable or undesirable, pleasant or unpleasant, (4) related to the intensity effects or perceptual meaning of a stimulus, (5) synergic with organic changes of A- or W-types.

The cautious but significant acknowledgment by W. A. Hunt (66), after Woodworth, of the relevance of attitudes of acceptance or rejection to "feelings of pleasantness or unpleasantness with respectively different organic backgrounds," suggests a synthesis of the concepts of motivation, emotion, and perception in human ontogeny significant for comparisons with lower animals. One crucial property of these processes is that, from fishes to man, as Pick has emphasized, the *sympathetic* nervous system increases in prepotency as an energy-expending system, antagonistic to the *parasympathetic* as an energy-conserving system.

This view agrees with Leeper's (71) point that a motivational theory of emotion is needed, and that Hebb's concept of emotion as disorganized response is misleading. In A-type adjustments as ontogeny progresses, it is not so much disorganization as a *re*-organization of organic processes that occurs, depending on advances in motivated perception. Studies such as Rowland's (96) revealing a low correlation between "intensity" and type of organic changes in emotion are from adult subjects, for whom object-meaning tends to overweigh stimulus magnitude. With ontogenetic progress in higher vertebrates, perceptual meanings often modify, occlude, or reverse the earlier effects of stimulus intensity.

I therefore conclude that although the James-Lange type of theory provides a useful basis for studying the early ontogeny of mammals, in which A-type or W-type patterns dominate behavior according to stimulus magnitude, a Cannon-type theory of higher-center control is *indispensable* for later stages of perceptual and motivational development. If ontogeny progresses well, specialized patterns of A- and W-types, or their combinations, perceptually controlled, often short-circuit or modify the early viscerally dominated versions. Proposals like that of Harlow and Stagner (49) that thalamic centers are functionally differentiated as to processes of pleasure *vs.* pain, and the like, seem applicable as ontogeny progresses, suggesting for later stages *brainstem* patterns such as implanted-electrode results indicate. Comparisons with results from needed studies on earlier stages will be most interesting. But, from this approach, recent proposals, e.g., Golightly (46), that the James-Lange theory be abandoned as outmoded must be held premature until the ontogeny of emotion is better understood.

Doubtless, as Bousfield and Orbison (7) state, the infant mammal is "essentially precorticate at birth." Also, as a perceptually naive animal, his emotional-motivational processes seem diffuse and dominated by stimulus magnitude swaying autonomic-visceral susceptibilities. But, contrary to impressions that seem general in the child-psychology literature, there is significant evidence that at birth the infant mammal already has a crudely dichotomized organic basis for his perceptual-motivational-emotional ontogeny. In this context, however, the question, "Are there any innate emotions?" should be dismissed as posing the false alternatives of *finding* or *not finding* adultlike patterns in psychologically barren early stages.

This interpretation differs sharply from that of Bridges (9), that an "undifferentiated excitement" prevails for the human neonate (Figure 7), differentiated only later into conditions of "delight" and "distress." Textbook writers who adopt this conclusion seem to have overlooked the fact that Bridges' research on infant emotionality did not involve intensive research on neonates. Many psychologists seem also to have been influenced in their views by a conventional rejection of Watson's (119) theory of infant emotions. Watson, as many will recall, concluded from studies with neonate infants that their emotions may be differentiated as "love," "fear," and "rage," arousable by stimuli such as stroking and patting, loud noises, and restraint, respectively. Sherman (107), however, found competent judges unable to distinguish the fear and rage patterns without knowing the stimuli, and Taylor (113) reported that Watson's conditions did not initiate "constant pattern responses" in infants he

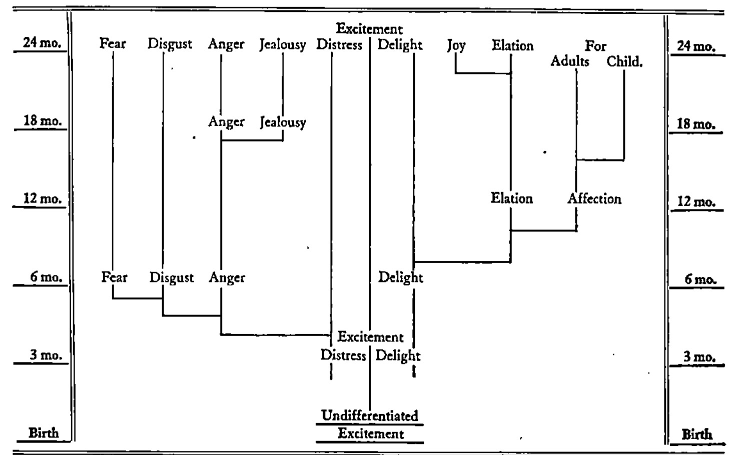

FIG. 7. Bridges' (8) schema of infant behavioral ontogeny based on the postulation of an initial condition of "undifferentiated emotional excitement" (cf. Figure 8 and text).

tested. Watson would have been well advised to call his emotional reactions X, Y, and Z (as he himself once suggested), differentiating them by effective stimulus intensities rather than arousal patterns, for then the differentiation can be made of *X* as relatively unexcited responses of A-type from *XY* as excited responses of W-type, to weak and strong stimuli, respectively (Figure 8). With gentle tactual stimulation of neonates, Pratt *et al.* (90) obtained pacifying reactions and relaxation of limbs; with stroking, Watson obtained extension of limbs and quiescence as a rule; with intense light Canestrini (12) observed typically an intake of breath and sudden contraction of the abdominal wall, rising fenestral and breathing curves, and external signs of "fear"; with strong contact Peiper (87) noted a withdrawal of the part affected, general action, and crying; with sudden intense auditory stimuli the Shermans (108) noted arm flexion, crying, and signs of visceral disturbance or "colic"; with intense stimuli a "startle pattern" is obtained involving strong general limb flexion and signs of visceral disturbance (79). In the general evidence, the biphasic aspects of neonate reactions, and

INITIAL STAGES	Adequate s'imuli	High intensities	Low intensities
Incidental "approach-withdrawal"	Reaction: Somatic component	Vigorous mass action Flexion, adduction Crying, tension	Pacified, local action Extension, abduction Turning-to, "smile"
	Visceral component	"Interruptive" effects Sympathetic-autonomic predominant	"Vegetative" effects Parasympathetic-autonomic predominant
SOCIAL ADAPTATION	Adequate stimuli	Focal to disturbing situations	Focal to "gratifying" situations
Purposive approach-withdrawal	Reaction: Perceptual adjustment	Specialized withdrawal: pulling . . . walking off Indirect modes of escape, negation, aggression "Sulk," gestures, sounds symbolizing NO	Specialized approach: reaching . . . walking to Indirect modes of approval, acquisition Social smile, gestures, sounds, sympolizing YES
	Visceral background	Condition interruptive patterns Sympathetic-automatic facilitation	Conditioned vegetative patterns Parasympathetic-autonomic facilitation

Schema: Ontogeny of approach-withdrawal and emotional differentiation in man

FIG. 8. Schema of theory outlining biphasic functional conditions basic to the ontogeny of early approach or withdrawal and of related subsequent perceptual and emotional differentiation in man (cf. Figure 7).

their general correlation with stimulus intensity, seem clearly indicated.

The conclusion seems warranted that in neonate mammals generally these early biphasic processes of a physiological order, aroused according to stimulus magnitude, furnish a basis for individual perceptual, motivational, and emotional development. The theory is schematized in Figures 8 and 9. The foregoing discussion, in this sense, supports Leeper's (71) position that the processes of emotion and motivation are fundamentally related. The socialization of early physiologically given biphasic excitatory states and the specialization of motivation and emotion seem to advance hand in hand in the education of the infant mammal in perceptual processes and action. To suggest the relevance for human psychological development of the theory of biphasic processes sketched in this article, I shall outline evidence bearing on the ontogenic course of two adjustive processes unquestionably crucial in man's perceptual and motivational adjustment to his world: *smiling* and *reaching*.

BIPHASIC BASES OF HUMAN APPROACH AND WITHDRAWAL RESPONSES

After Darwin, the general recognition advanced that primate evolution must have involved concurrent organic changes admitting an upright posture, an increase in visual range, a freeing of the limbs for specialized prehension and the face for social expression, and a specialization of the brain admitting conceptual plasticity. These superior human assets for perceptual-motivational development may be considered marks of a stage in natural selection far advanced beyond that at which lower vertebrates with far less advanced properties for cephalic dominance are limited to the simple lunge as an approach. Significant also in this impressive evolutionary progression are changes such as Pick (88) has described, underlying a steady

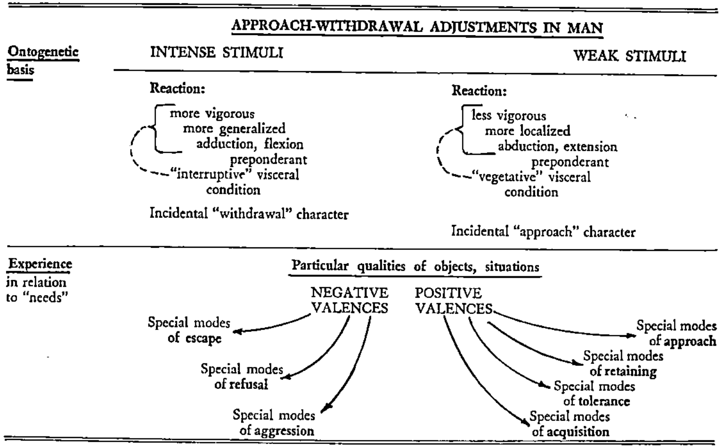

FIG. 9. Schema of theory basing approach and withdrawal reactions on biphasic processes dependent in early stages (*top*) on effective stimulative intensity, subsequently (*bottom*) on developmental processes related particularly to experience (cf. Figures 7 and 8).

specialization of the *parasympathetic* autonomic division, an energy-conserving system, correlated with an even more rapid advance of its antagonist, the sympathetic system, an energy-expending system in adjustive behavior. In the organismic setting available on the human level, emphasizing adaptive plasticity, the potential ontogenetic relevance of the autonomic divisions, as with that of other biphasic processes, reaches far beyond an elementary status of simple antagonism.

Consistent with Darwin's (21) concept of *antithesis* in mammalian emotional expression, anatomists find evidence for the rise of two antagonistic systems of facial muscles in mammals, one elevating the lip corners, the other pulling them down, in close correlation with progressive elaboration of the trigeminal or facial-cutaneous sensory system (64). Significantly, physiological and behavioral evidence indicates that the *levator* system has a lower arousal threshold than the *depressor* system. With low-intensity shocks to cheek or mastoid regions of adult human subjects, Dumas (29, 30) noted a mechanical "smile" resembling that of hemiplegics; with higher intensities, a different grimace involving facial musculature more widely.

Now O. Koehler (69) represents the view of many who consider smiling innate as a social pattern in man, pointing out that a semblance of this reaction can occur in premature and neonate infants (Figure 10). Strictly considered, however, this early response is a

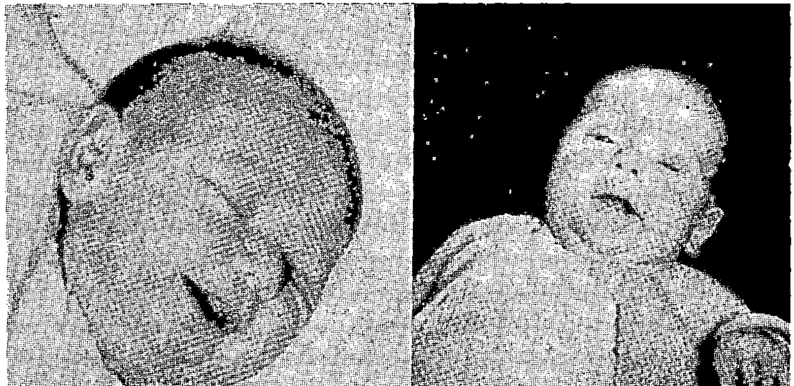

FIG. 10. Photographs of facial reactions in a neonate infant, from O. Koehler (69), who describes this reaction as an "innate smile." Unilateral "smile" on the fourth postnatal day (left) and on the seventh postnatal day (right).

grimace, a physiologically forced response to low-intensity stimulation, and the "smile" is a specialized social adjustment appearing only after much perceptual-motivational development. Although light facial contact, particularly near the lip corners, is generally most effective in producing the neonate grimace, Dennis (24) obtained it with other weak stimuli such as light patting of the chest and asafetida released at the nostrils. In the early neonate this action is infrequent and has a fleeting character, occurring "incidentally with other facial activities in an overflow reaction," according to Spitz and Wolf (111). A concurrence with vegetative visceral conditions is indicated by its appearance when the infant is calming down, relieved from disturbance, or recently satiated.

Neither Dennis nor Spitz and Wolf, however, reported any "unconditioned" stimulus for this reaction. The difficulty seems to have been that they looked for a *specific* unconditioned stimulus. For there is evidence, including their own, that although a variety of stimuli can arouse this reaction, *all* have in common the property of *low intensity*. Even interoceptive changes after feeding may be effective, as suggested by the traditional "gas smile," which seems more often a response to a gentle stomach pressure (as during "burping") than to "belly-ache" as Spitz and Wolf maintain. The conclusion seems in order that this action first specializes through a gradual conditioning process, in which the unconditioned stimulus is a facial-nerve arousal by low-intensity stimulation.[4] Such a process, elicited most frequently by tactual stimuli when the neonate is in a vegetative organic condition, would later be available for control by visual stimuli of low intensity, when visual centering of stimuli has improved sufficiently.

A diffuse, generalized conditioning process is indicated initially. Although within the first two months Spitz and Wolf obtained few grimace reactions to visual stimuli, these came increasingly after the third month, when Dennis noted also a growing reactivity by grimacing to a face presented over the crib. The face *as such* cannot be credited for this effect, as Spitz and Wolf obtained the response with a variety of stimuli, including sounds and horror masks. The visual effect evidently is initially very general and diffuse, but in time becomes more specific. A crude contiguity-conditioning process is indicated, through the concurrence of certain visual effects with satiation or relief from disturbance. By six months, as a rule, Spitz and Wolf observed definite advances in the specificity of stimulus control, "in the direction of increased discrimination," with the response not only elicited selectively by faces rather than masks, but also changed in pattern from the "fleeting grimace" of the first three months, "not well differentiated from general facial action," to a "more definite expression."

Next indicated is a discriminative "social smile," to familiar as against unfamiliar faces. That the "smile" comes about through a progressive perceptual-motivational learning process dependent on the developmental situation to an important extent is indicated by Spitz and Wolf's finding that this response developed precociously in infants who had the "best relations with attentive mothers," but was delayed, infrequent, and unspecialized in infants who had experienced poor relations with indifferent mothers. From its initial status as a crude reaction forced through low-intensity arousal, under optimal

[4] Spitz and Wolf conclude from the fact that they found the nursing bottle ineffective as a stimulus for this response during the first six months, that "this disposes of the hypothesis that smiling is a conditioned reflex established in response to gratification of being satiated with food." This does not follow, however, as the nursing bottle usually is seen only briefly before feeding, as a rule, and thereafter is touched rather than seen. An experiment on this point seems in order.

conditions this response appears to specialize simultaneously in its afferent, organic, and motor aspects, first with a general set to associated stimuli, then with a rudimentary perceptual anticipation of more specifically "desirable" stimuli. It is difficult to fit these results within the framework of contiguity conditioning alone; rather, from the circumstances of development, it is probable that a process of selective learning underlies the metamorphosis of this and related aspects of the infant's expanding repertoire of "seeking" adjustments.

The *reach*, another significant approach response, seems to develop concurrently in the infant with the "smile," as a related adjustment. These early stages, as Frank (39) has pointed out, involve a progressive self-contact process of individual adjustment, which, as Nissen, Chow, and Semmes (84) demonstrated for chimpanzees, progresses badly without adequate opportunities for tactual experience through limb mobility. The matrix of this process in earlier stages evidently consists in the characteristically variable and diffuse reactions of relaxation and local limb extension to gentle limb contact, when a vegetative organic condition prevails. These types of reaction occur characteristically with low-intensity stimulation of many kinds, usually tactual but also at times auditory or olfactory. The local reactions typically elicited by such stimuli stand in definite contrast to the mass-action type obtained with strong stimulation. To stimulation of the foot with a needle, termed "pain," the Shermans (108) usually obtained a "pulling away of the leg or the face, associated with movements of the arms and crying." Although neonate behavior seems chaotic and often unpredictable, limb flexion is typically obtained with strong stimulation, and this was the unconditioned response used both by Wenger (120) and by the Wickens (123), who conditioned it to previously ineffective stimuli within the first ten postnatal days. Watson, who noted limb flexion as a typical response to loud sounds or dropping, conversely identified arm extension as a frequent response to gentle stroking of the neonate's arm or cheek.

The infant's approach or A-type reactions normally seem to change from soon after birth in a more regular manner than do the more episodic, excited W-type reactions. At first the face takes the lead, with close functional relationships between sucking, lip-contact, and hand movements. As Davis (22) concluded, the initial afferent control seems to be tactuo-postural, with vision entering gradually into a steadily more dominant role. A progressive self-communication through tactuo-motor channels, emphasized by E. B. Holt (62) as basic to motivation and learning, underlies gradually developing integrations between proximal sensory control and visual control in early space perception.

Initially the infant gets his arm out incidentally, as part of a forced overt A-type response to *low*-energy stimulation. One significant aspect of the tactuo-visual integration that gradually assumes control of this reaction is the concurrence of head-turning to the side of gentle stimulation with progressive efficiency in centering the eyes on the focal area. McGinnis (78) reported for the first six weeks an increased frequency of head turning towards small objects seen in motion. Centering objects in the visual field, as it improves, elaborates into an eye-following of both the infant's hand and objects.

Advancing the anterior body or its parts extends the infant's "space," as Werner (121) puts it, and in its early posturo-tactual basic may relate to reactive trends begun *in utero*. Progress through tactuo-visual integration represents complex psychological developments in which maturation *and* experience factors are doubtless interrelated inextricably (103). A dependence of progress in learning and motivation on the trophic aspects of reciprocal stimulative exchanges with self and environs is emphasized by Spitz and Wolf's finding that smiling is likely to be retarded when infants have an inferior maternal adjustment, and without much doubt this is also true of the closely related response of reaching.

Halverson's (48) studies on grasping and reaching began with the infant at about 12 weeks, supported in a chair with a small cube of white sugar on the table before him. At this stage the infants tended to "regard" this stimulus or center it variably in the visual field, with frequent head-bending forward. Responses were quantified in terms of the nearest approach of the hand to the cube in each trial. Early deviations were large, as the infant's arms moved in various directions from his vertical axis but not particularly towards the exciting stimulus. From the data, these responses may be interpreted as crude arm extensions forced by low-intensity stimulation, a getting-out-the-arm reaction rather than a reaching for something. Significantly, in infants from 12 to 24 weeks there was an actual increase in the average digression from the stimulus, evidently due mainly to an accelerated maturation of shoulder muscles, so that the arm struck out laterally rather than forward in the trials.

It is significant that during this period the infant continued to respond to the stimulus by extending and abducting his arms, although he seldom touched or grasped the object and hardly ever got it into his mouth. But, as Castner (15) noted, in these months the duration of visual "regard" increases steadily, along with the frequency of arm extensions to visually centered stimuli. Daily life of course offers many experiences to the infant, in which he handles just-seen objects in A-type situations, as when (after feeding) the waving hand is seen, then gets into his mouth and stays there. It is a question of equivalent processes facilitating one another in the learning of related adjustments, and Halverson's results show steady progress after about 24 weeks in arm extension through elbow straightening, paralleled by increasing efficiency in getting the hand to the pellet. In the test situation the infant progressed steadily in directing his arm *to* a seen object, rather than just extending the limb as at first. Around six months, therefore, as the infant starts his reach, we may infer the presence of a simple anticipation of having the object in hand, *then* a different anticipation of mouthing it.

The significant point is that forelimb extension, the local adjustment central to the infant's perceptual-motivational development in the first year, begins as an incidental response to low-energy stimulation but, through progressive integrations of vision with pre-existing tactuo-proximal adjustments, becomes first a specific approach response, then a perceptually directed seeking response (Figure 9).

Motivation and learning theory meet difficulties here, of course, partly through insufficient evidence and partly through logical shortcomings. It is plain that research on infant perception and motivation has taken us beyond the stage of Watson's kinaesthetic-motor theory and Holt's reflex-circle theory. But for reasons advanced before, S-R or connectionistic theories seem limited to early stages, when events are dominated by what contiguities of stimuli and A-type processes are sufficiently experienced. Such occurrences might explain attainment of a simple visually directed arm approach based on earlier tactuo-proximal adjustments. But, to understand progress into the stage of perceptually directed reaching, we must postulate qualitatively new processes of selective learning developing through the situation-altering effect of action (76). Related aspects concern relevance of stimulus properties to the optimal organic condition (73), changing relationships between action and its organic consequences (76, 105), and qualitative differences between nonperceptual and perceptual learning (92).

Reaching may well be the most significant indicator of early approach-responsiveness in man. Fundamental to processes changing incidental limb extension to perceptual seeking is an intimate relationship to A-type organic processes, first as concomitant events, then in versatile roles influencing the plastically utilized relationships of selective learning (Figures 8 and 9).

In these ways, the human infant normally acquires in his first half year an anticipative way of dealing with objects at a distance. Anticipative reaching, in *its* turn, is basic first to approach and *then* to seeking by crawling and walking respectively. These developments themselves provide a perceptual-motivational basis for approaches and seeking of the *conceptual* type. What earlier stages may contribute to the more vicarious forms of reaching is suggested by the difficulties experienced in completing approach reactions in the period when "the infant must touch everything." Richardson's (93) study of infant learning showed that infants must pass through many months of progressive toil before they can perceive a string and lure as related, with the former a means of getting the latter. From Fishel's (37) tests it appears that dogs never deal with such relationships as well as do young children.

Early progress in motivated reaching, providing a basis for perceiving and using the arm as an *extension* of the body towards incentives, thereby leads in some primates to the use of objects as means to increase still more the scope of approaching and seeking responses. Such attainments are limited in monkeys, but accelerate from chimpanzee to man. This view of motivated problem-solving is supported by Birch's (4) evidence that chimpanzees achieve few insight solutions in the Köhler stick-combination test unless sticks have been handled previously in play. After the animal has perceived a stick in an arm-extending pattern, a reasoned combining of sticks evidently can arise through anticipating further extensions of a single held stick. Dependence of this indirect *seeking* response on the animal's motivated condition was demonstrated when Birch made the subject's hunger excessive before tests, whereupon chimpanzees capable of insight solutions limited themselves to reaching *directly* for the distant food with the arm alone.

By means of comprehensive rules of conduct, inhibitive procedures and the like, man is capable of extending his motivated approach adjustments under conditions of stress beyond the limitations of direct response. But even man, schooled as he can be in ways of seeking what may lie in hyperspace, at times follows the wrong rules or for other reasons reaches short through petty motivation.

References

1. Adrian, E. D. *The mechanism of nervous action.* Philadelphia: Univ. Pennsylvania Press, 1935.
2. Allport, F. H. *Social psychology.* New York: Houghton Mifflin, 1924.
3. Bertocci, P. A. A critique of G. W. Allport's theory of motivation. *Psychol. Rev.*, 1940, 47, 501–532. (Cf. Allport's reply, *ibid.*, pp. 533–553.)

4. Birch, H. The relation of previous experience to insightful problem-solving. *Jour. Comp. Psychol.*, 1945, **38**, 367–383.

5. Bitterman, M. E., Wodinsky, J., and Candland, D. K. Some comparative psychology. *Amer. Jour. Psychol.*, 1958, **71**, 94–110.

6. Bourguignon, G. Classification fonctionnelle des muscles par la chronaxie. *Traité Physiol. Norm. Path.*, 1929, **8**, 157–238.

7. Bousfield, W. A., and Orbison, W. D. Ontogenesis of emotional behavior. *Psychol. Rev.*, 1952, **59**, 1–7.

8. Bower, G., and Miller, N. E. Rewarding and punishing effects from stimulating the same place in the rat's brain. *Jour. Comp. Physiol. Psychol.*, 1958, **51**, 669–674.

9. Bridges, Katharine M. B. *The social and emotional development of the pre-school child.* London: Kegan Paul, 1931.

10. Brink, F., Jr. Excitation and conduction in the neuron. Chap. 2, pp. 50–93.—Synaptic mechanisms. Chap. 3, pp. 94–120. In S. S. Stevens (ed.), *Handbook of experimental psychology*, New York: Wiley, 1951.

11. Bullock, T. H. Physiological mapping of giant fiber systems in polychaete annelids. *Physiol. Comp. Oecol.*, 1948, **1**, 1–14.

12. Canestrini, S. Über das sinnesleben des Neugeborenen. *Monogr. Neurol. Psychiatr.*, No. 5, Berlin: Springer, 1913.

13. Carmichael, L. A re-evaluation of the concepts of maturation and learning as applied to the early development of behavior. *Psychol. Rev.*, 1936, **43**, 440–470.

14. ———, and Smith, M. F. Quantified pressure stimulation and the specificity and generality of response in fetal life. *Jour. Genet. Psychol.*, 1939, **54**, 425–434.

15. Castner, B. M. The development of fine prehension in infancy. *Genet. Psychol. Monogr.*, 1932, **12**, 105–193.

16. Craig, W. Appetites and aversions as constituents of instincts. *Biol. Bull.*, 1918, **34**, 91–108.

17. Creed, R. S., Denny-Brown, D., Eccles, J. C., Liddell, E. G. T., and Sherrington, C. S. *Reflex activity of the spinal cord.* London: Oxford Univ. Press, 1932.

18. Daniel, R. S., and Smith, K. U. The migration of newly hatched loggerhead turtles toward the sea. *Science*, 1947, **106**, 398–399.

19. Darrow, C. W. Neural mechanisms controlling the palmar galvanic skin reflex and palmar sweating. *Arch. Neurol. Psychiatr.*, 1937, **37**, 641–663.

20. Darling, R., and Darrow, C. W. Determining activity of the autonomic nervous system from measurements of autonomic change. *Jour. Psychol.*, 1938, **5**, 85–89.

21. Darwin, Charles. *Expression of the emotions in man and animals.* New York: Appleton, 1873.

22. Davis, R. C. The genetic development of patterns of voluntary activity. *Jour. Exper. Psychol.*, 1943, **33**, 471–486.

23. Delgado, J. M. R., Roberts, W. W., and Miller, N. E. Learning motivated by electrical stimulation of the brain. *Amer. Jour. Physiol.*, 1954, **179**, 587–593.

24. Dennis, W. Experimental test of two theories of social smiling in infants. *Jour. Soc. Psychol.*, 1935, **6**, 214–223.

25. Dethier, V. G. *Chemical insect attractants and repellents.* Philadelphia: Blakiston, 1947.

26. ———, and Bodenstein, D. Hunger in the blowfly. *Zeitschr. Tierpsychol.*, 1958, **15**, 129–140.

27. Dolley, W. L., Jr., and Golden, L. H. The effect of sex and age on the temperature at which reversal in reaction to light in *Eristalis tenax* occurs. *Biol. Bull.*, 1947, **92**, 178–186.

28. ———, and White, J. D. The effect of illuminance on the reversal temperature in the drone fly *Eristalis tenax. Biol. Bull.*, 1951, **100**, 84–89.

29. Dumas, G. L'expression des émotions. *Revue philosophique*, 1922, **47**, 32–72, 235–258.

30. ———. *Le sourire, psychologie et physiologie.* (3me ed.) Paris: Presses Univer., 1948.

31. Eccles, J. C. *The neurophysiological basis of mind.* Oxford: Clarendon, 1953.

32. Ehrenhardt, H. Formensehen und Sehscharfebestimmungen bei Eidechsen. *Zeitschr. vergl. Physiol.*, 1937, **24**, 258–304.

33. Eibl-Eibesfeldt, I. Nahrungserwerb und Beuteschema der Erdkröte (*Bufo bufo* L.). *Behaviour*, 1951, **4**, 1–35.

34. ———, and Kramer, S. Ethology, the comparative study of animal behavior. *Quart. Rev. Biol.*, 1958, **33**, 181–211.

35. Evans, D. R., and Dethier, V. G. The regulation of taste thresholds for sugars in the blowfly. *Jour. Ins. Physiol.*, 1956, **1**, 3–17.

36. Fabricius, E. Some experiments on imprinting phenomena in ducks. *Proc. Xth Internatl. Ornith. Cong., Uppsala*, 1951, 375–379.

37. Fischel, G. *Die Seele des Hundes.* Berlin: Parey, 1950.

38. Fraenkel, G. S., and Gunn, D. L. *The orientation of animals.—Kineses, taxes, and compass reactions.* New York: Oxford Univ. Press, 1940 (Dover Publications, Inc., 1961).

39. Frank, L. Tactile communication. *Genet. Psychol. Monogr.*, 1957, **56**, 209–255.

40. Freisling, J. Studien zur Biologie und Psychologie der Wechselkröte (*Bufo viridis* Laur.). *Öst. Zool. Zeitschr.*, 1948, **1**, 383–440.

41. Frisch, Karl von. Über einen Schreckstoff der Fischhaut und seine biologische Bedeutung. *Zeitschr. vergl. Physiol.*, 1941, **29**, 46–145.

42. Fulton, J. F. *Muscular contraction and the reflex control of movement.* Baltimore: Williams and Wilkins, 1926.

43. Glickman, S. E. Deficits in avoidance learning produced by stimulation of the ascending reticular formation. *Canad. Jour. Psychol.*, 1958, **12**, 97–102.

44. GOLDSTEIN, K. *The organism.* New York: American Book Co., 1939.

45. ———. On emotions: considerations from the organismic point of view. *Jour. Psychol.*, 1950, **31**, 37–49.

46. GOLIGHTLY, C. L. The James-Lange theory: a logical post-mortem. *Phil. Sci.*, 1953, **20**, 286–299.

47. GREENFIELD, A. Alteration of a conditioned avoidance tendency by a procedure of adaptation to the unconditioned stimulus. *Ann. N. Y. Acad. Sci.*, 1955, **62**, 277–294.

48. HALVERSON, H. M. Studies of the grasping response of early infancy. *Jour. Genet. Psychol.*, 1937, **51**, I, 371–392; II, 393–424; III, 425–449.

49. HARLOW, H., and STAGNER, R. Psychology of feelings and emotions: I. Theory of feelings. *Psychol. Rev.*, 1932, **39**, 570–589.

50. HARRIS, L. J., CLAY, J., HARGREAVES, J., and WARD, A. Appetite and choice of diet. The ability of the vitamin-B deficient rat to discriminate between diets containing and lacking the vitamin. *Proc. Roy. Soc., B*, 1933, **113**, 161–190.

51. HARTLINE, H. K. The discharge of nerve impulses from the single visual sense cell. *C. Spr. Harb. Symp. Quant. Biol.*, 1935, **3**, 245–250.

52. HEBB, D. O. On the nature of fear. *Psychol. Rev.*, 1946, **53**, 259–276.

53. ———. *The organization of behavior.* New York: Wiley, 1949.

54. ———. The motivating effects of exteroceptive stimulation. *Amer. Psychol.*, 1958, **13**, 109–113.

55. HECHT, S. Vision: II. The nature of the photoreceptor process. Chap. 4, pp. 169–272, in C. Murchison (ed.), *Foundations of experimental psychology*, Clark Univ. Press, 1929.

56. HEDIGER, H. *Wild animals in captivity.* London: Butterworths, 1950.

57. HEINTZ, E. Actions attractives et répulsives spécifiques de broyats et de substances extraites de broyats chez *Apis mellifica* et *Tubifex tubifex*. *C. R. Soc. Biol.*, 1955, **149**, 2224–2227.

57*a*. HERTZ, MATHILDE. Die Organisation des optischen Feldes bei der Biene. *Zeitschr. vergl. Physiol.*, 1933, **14**, 629–674.

58. HESS, E. "Imprinting" in animals. *Sci. Amer.*, 1958, **198**, 81–90.

59. HESS, W. Reactions to light in the earthworm, *Lumbricus terrestris* L. *Jour. Morph. Physiol.*, 1924, **39**, 515–542.

60. HINSEY, J. C. The innervation of skeletal muscle. *Physiol. Rev.*, 1934, **14**, 514–585.

61. HOLST, E. VON. Bausteine zu einer vergleichenden Physiologie der lokomotorischen Reflexe bei Fischen. II. Mitteilung. *Zeitschr. vergl. Physiol.*, 1937, **24**, 532–562.

62. HOLT, E. B. *Animal drive and the learning process.* New York: Holt, 1931.

63. HONIGMANN, H. The visual perception of movement by toads. *Proc. Roy. Zool. Soc., B*, 1945, **132**, 291–307.

64. HUBER, E. Evolution of facial musculature and cutaneous field of trigeminus. *Quart. Rev. Biol.*, 1930, **5**, 133–188, 389–437.

65. HUDGINS, C. V. The incidence of muscular contraction in reciprocal movements under conditions of changing loads. *Jour. Gen. Psychol.*, 1939, **20**, 327–338.

66. HUNT, W. A. A critical review of current approaches to affectivity. *Psychol. Bull.*, 1939, **36**, 807–828.

67. JENNINGS, H. S. *The behavior of lower organisms.* New York: Columbia Univ. Press, 1906.

68. KEMPF, E. J. Neurosis as conditioned, conflicting, holistic, attitudinal, acquisitive-avoidant reactions. *Ann. N. Y. Acad. Sci.*, 1953, **56**, 307–329.

69. KOEHLER, O. Das Lächeln als angeborene Ausdrucksbewegung. *Zeitschr. menschl. Vererb.-u. Konstitutionslehre*, 1954, **32**, 390–398.

70. KUNTZ, A. *The autonomic nervous system.* Philadelphia: Lea and Febiger, 1929.

71. LEEPER, R. W. A motivational theory of emotion to replace "emotion as disorganized response." *Psychol. Rev.*, 1948, **55**, 5–21.

72. LEHRMAN, D. A critique of Lorenz's "objectivistic" theory of animal behavior. *Quart. Rev. Biol.*, 1953, **28**, 337–363.

73. LEUBA, C. Toward some integration of learning theories: the concept of optimal stimulation. *Psychol. Reports*, 1955, **1**, 27–33.

74. LEVINE, M. G., and KABAT, H. Cocontraction and reciprocal innervation in voluntary movement in man. *Science*, 1952, **116**, 115–118.

75. MAIER, N. R. F., and SCHNEIRLA, T. C. *Principles of animal psychology.* New York: McGraw-Hill, 1935.

76. ———, ———. Mechanisms in conditioning. *Psychol. Rev.*, 1942, **49**, 117–134.

77. MAST, S. O. The nature of response to light in *Amoeba proteus* (Leidy). *Zeitschr. vergl. Physiol.*, 1931, **15**, 139–147.

78. MCGINNIS, J. M. Eye-movements and optic nystagmus in early infancy. *Genet. Psychol. Monogr.*, 1930, **8**, 321–430.

79. MCGRAW, M. The Moro reflex. *Amer. Jour. Dis. Children*, 1937, **54**, 240–251.

80. MCKINLEY, G. M. *Evolution: the ages and tomorrow.* New York: Ronald, 1956.

81. MEYER-HOLZAPFEL, M. Unsicherheit und Gefahr im Leben höherer Tiere. *Schweiz. Zeitschr. Psychol. Anwend.*, 1955, **14**, 171–194.

82. MILLER, N. E. Central stimulation and other new approaches to motivation and reward. *Amer. Psychol.*, 1958, **13**, 100–108.

83. MITTELSTAEDT, H. Prey capture in mantids. *Rec. Adv. Invert. Physiol.*, Univ. Oregon Publ., 1957, 51–71.

84. NISSEN, H., CHOW, K. L., and SEMMES, JOSEPHINE. Effects of restricted opportunity for tactual, kinesthetic, and manipulative experience on the behavior of a chimpanzee. *Amer. Jour. Psychol.*, 1951, **64**, 485–507.

85. OLDS, J., and MILNER, P. Positive reinforcement produced by electrical stimulation of septal area and other regions of rat brain. *Jour. Comp. Physiol. Psychol.*, 1954, **47**, 419–427.

86. PARR, A. E. A contribution to the theoretical analysis of the schooling behavior of fishes. *Occas. Papers Bingham Oceanogr. Coll.*, 1927, No. 1, 1–32.

87. PEIPER, A. Untersuchungen über die Reaktionzeit im Säuglingsalter: II. Reaktionzeit auf Schmerzreiz. *Monatschr. Kinderheilk.*, 1926, **32**, 136–143.

88. PICK, J. The evolution of homeostasis. *Proc. Amer. Phil. Soc.*, 1954, **98**, 298–303.

89. PRATT, K. C. The neonate.—Chap. 4, pp. 190–254 in L. Carmichael (ed.), *Manual of child psychology*. New York: Wiley, 1946.

90. ———, NELSON, A. K., and SUN, K. H. The behavior of the newborn infant. *Ohio State Univ. Stud., Contr. Psychol.*, 1930, No. 10.

91. PROSSER, C. L. Effect of the central nervous system on responses to light in *Eisenia foetida* Sav. *Jour. Comp. Neurol.*, 1934, **59**, 61–91.

92. RAZRAN, G. Conditioning and perception. *Psychol. Rev.*, 1955, **62**, 83–95.

93. RICHARDSON, H. M. The growth of adaptive behavior in infants: An experimental study of seven age levels. *Genet. Psychol. Monogr.*, 1932, **12**, 195–359.

94. ROBERTS, W. W. Rapid escape learning without avoidance learning motivated by hypothalamic stimulation in cats. *Jour. Comp. Physiol. Psychol.*, 1958, **51**, 391–399.

95. ROSENBLATT, J. S., WODINSKY, J., TURKEWITZ, G., and SCHNEIRLA, T. C. Analytical studies on maternal behavior in relation to litter adjustment and socialization in the domestic cat. II. Maternal-young relations from birth to weaning. III. Development of orientation. (In ms.)

96. ROWLAND, L. W. The somatic effects of stimuli graded in respect to their exciting character. *Jour. Exper. Psychol.*, 1936, **19**, 547–560.

97. SCHILLER, PAUL VON. Intersensorielle Transposition bei Fischen. *Zeitschr. vergl. Physiol.*, 1933, **19**, 304–309.

98. SCHNEIDER, D. Beitrag zu einer Analyse des Beute und Fluchverhaltens einheimischer Anuren. *Biol. Zentralbl.*, 1954, **73**, 225–282.

99. SCHNEIRLA, T. C. A theoretical consideration of the basis for approach-withdrawal adjustments in behavior. *Psychol. Bull.*, 1939, **37**, 501–502.

100. ———. Ant learning as a problem in comparative psychology. Pp. 276–305 in P. Harriman (ed.), *Twentieth century psychology*. New York: Philosophical Library, 1946.

101. ———. Levels in the psychological capacities of animals. Pp. 243–286 in R. W. Sellers *et al.* (eds.), *Philosophy for the future*. New York: Macmillan, 1949.

102. ———. Interrelationships of the "innate" and the "acquired" in instinctive behavior. Pp. 387–452 in *L'instinct dans le comportement des animaux et de l'homme*. Paris: Masson et Cie, 1956.

103. ———. The concept of development in comparative psychology. Pp. 78–108 in D. B. Harris (ed.), *The concept of development*. Minneapolis: Univ. Minnesota Press, 1957.

104. ———. L'apprentissage et la question du conflit chez la fourmi.—Comparaison avec le rat. *Jour. de Psychol.*, 1959, **57**, 11–44.

105. SEWARD, J. P. Introduction to a theory of motivation in learning. *Psychol. Rev.*, 1952, **59**, 405–413.

106. SHAW, EVELYN. Environmental conditions and the appearance of sexual behavior in the platyfish. Chapter in *Roots of behavior*, E. L. Bliss (ed.). New York: Harper (Hoeber), 1962.

107. SHERMAN, M. The differentiation of emotional responses in infants. I. Judgments of emotional responses from motion picture views and from actual observation. *Jour. Comp. Psychol.*, 1927, **7**, 265–284.

108. ———, and SHERMAN, IRENE C. Sensory-motor responses in infants. *Jour. Comp. Psychol.*, 1925, **5**, 53–68.

109. SHERRINGTON, C. S. *The integrative action of the nervous system.* (1923 edn.) New Haven: Yale Univ. Press, 1906.

110. SPITZ, R. A. *No and yes—on the genesis of human communication.* New York: International Univ. Press, 1957.

111. ———, and WOLF, K. M. The smiling response: a contribution to the ontogenesis of social relations. *Genet. Psychol. Monogr.*, 1946, **34**, 57–125.

112. STOUGH, H. B. Giant nerve fibers of the earthworm. *Jour. Comp. Neurol.*, 1936, **40**, 409–464.

113. TAYLOR, J. H. Innate emotional responses in infants. *Ohio State Univ. Stud., Contrib. Psychol.*, 1934, No. 12, 69–81.

114. THOMPSON, R. Approach versus avoidance in an ambiguous-cue discrimination problem in chimpanzees. *Jour. Comp. Physiol. Psychol.*, 1954, **47**, 133–135.

115. TUMARKIN, A. On the evolution of the auditory conducting apparatus: A new theory based on functional considerations. *Evol.*, 1955, **9**, 221–243.

116. VERHEIJEN, F. J. The mechanisms of the trapping effect of artificial light sources upon animals. *Arch. Néerl. Zool.*, 1958, **13**, 1–107.

117. WAGNER, H. Über den Farbensinn der Eidechsen. *Zeitschr. vergl. Physiol.*, 1932, **18**, 378–392.

118. WARBURTON, F. E. Feedback in development and its evolutionary significance. *Amer. Nat.*, 1955, **89**, 129–140.

119. WATSON, J. B. *Psychology from the standpoint of a behaviorist.* Philadelphia: Lippincott, 1919.

120. WENGER, M. A. An investigation of conditioned responses in human infants. *Univ. Iowa Stud. Child Welfare*, 1936, **12**, 8–90.

121. WERNER, H. *Comparative psychology of mental development.* New York: Harper, 1940.

122. ———, and WAPNER, S. Changes in psychological distance under conditions of danger. *Jour. Pers.*, 1955, **24**, 153–167.

123. WICKENS, D. D., and WICKENS, C. A study of conditioning in the neonate. *Psychol. Bull.*, 1939, **36**, 599.

124. WITKIN, H. A. Restriction as a factor in adjustment to conflict situations. *Jour. Comp. Psychol.*, 1942, **33**, 41–74.

125. WOLF, E. Das Verhalten der Bienen gegenüber flimmernden Feldern und bewegten Objekten. *Zeitschr. vergl. Physiol.*, 1933, **3**, 209–224.

ARTICLE II

INSTINCTIVE BEHAVIOR, MATURATION—EXPERIENCE AND DEVELOPMENT[1]

T. C. SCHNEIRLA

"Instinctive Behavior" a Problem of Development

All animals display patterns of adaptive action characteristic of their species, normally appearing within a typical range of developmental conditions. Moth caterpillars at maturity spin cocoons, female carnivores commonly wean and leave their young when these near the end of infancy, and human babies commonly begin to smile at faces around three months after birth. Advocates of "instinct" theory think of such actions as due to inborn, innate impulses blindly impelling actions appropriate to reaching the respective ends. Others (e.g., Schneirla, 1948, 1956), skeptical of the instinct dogma, prefer the less presumptive term "instinctive behavior" for a set of problems concerning the evolutionary and ontogenetic origins of species-typical behavior. These are the real problems raised by the instinct tradition, and the fact is that no common theoretical formula is yet available to meet them on all functional levels.

It is a strong probability that heredity is basically involved in the development of all behavior. The issue is not whether heredity or environment dominates in particular cases, nor is it a matter of their relative strengths. It is very likely that the real contributions of American anti-instinctivists have been obscured by the impression that their goal was to lock out genetics from the discussion. Although they may have gone "too far," as Hunter (1947) concluded, in the case of Watson (1914), for example, I believe an objective historical appraisal raises his pioneer studies on terns and on neonate human infants to a significance well above that of his post-experimental lectures questioning the role of heredity in human development. The principal contribution of this group rather was its emphasis on objective, experimental investigations of behavior

[1] Reprinted from *Perspectives in Psychological Theory—Essays in Honor of Heinz Werner*, B. Kaplan and S. Wapner (eds.), 1961, New York: International Universities Press, by permission of the Editors and the publishers.

development to replace the phenomenistic, descriptive type characteristic of the classical instinctivists.

To an alarming extent scientists are still bound up in the main confusion of the traditional instinct problem, through a tendency to substitute for the "outworn" (Beach, 1947) nature-nurture dichotomy equivalent dogmas such as the distinction of the innate and the acquired in behavior. The problem is thus often stated as one of innate behavior, and a common answer among ethologists is that the genes pass on a code which through encapsulated intraneural processes directly determines development of instinctive patterns. Another variation, not uncommon among child psychologists, is that genic determiners install an organic growth through which innate behavior arises, to be modified through learning as infancy advances. The objective of this essay is to mark out, from the standpoint of comparative psychology, some concepts that promise to facilitate the avoidance of instinctivistic obscurantism and that may improve our perspective in the investigation of development (Schneirla, 1956).

How are these problems to be solved? Of course, the criterion of *early appearance* has been weakened by evidence for prenatal conditioning in certain birds and mammals. On the other hand, the changes so produced lack clarity, and thus far their postnatal effects have not been traced. Of course, *universality* is doubtful as a criterion, in view of the fact that common habits tend to prevail in species living under similar natural conditions; it is doubtful, however, that beavers learn to build dams, and patterns such as hoarding in rats and nesting in birds seem very difficult to explain except as innate. To many, the conventional criteria of the innate seem thus defensible; also, perhaps most impressive, characteristic patterns often appear in animals *isolated* from their kind and from their species environment from birth. And although it is admitted, with Hebb (1953), that the nature of *learning* is not yet sufficiently well understood to be distinguished readily from the hypothetical innate in behavior, the problem of demonstrating how learning influences early stages of ontogeny still seems very hazy, and the possibility of assigning a role for learning in many species-typical patterns such as the predatory patterns of solitary wasps seems negative. Even the hardest-headed epigeneticists may be impressed by the genic-behavior correspondences described by the psychogeneticists (Hall, 1951), although such evidence may not do much more than emphasize the problems of ontogeny without solving them. Even if all of these difficulties are considered, there are nevertheless grounds for holding that no infallible rules exist for differentiating the innate from the acquired, as both are hypothetical, and that questions of instinctive behavior must be investigated along other lines.

The fact cannot be emphasized too often that the central problem of instinctive behavior is one of development (Maier and Schneirla, 1935, Chaps. VI, XI), working comparatively not just with higher mammals and a few other convenient types but through the widest possible range. Furthermore, experience shows that the gains will come not only through comparisons of ontogeny in closely related animals, as is usual in psychogenetics, for example, but also in very different phyletic types. Jellyfish take food with tentacles and manubrium, carnivores with forepaws and jaws—how different are these processes in their development? It is doubtful, for example, that a perception of "food" arises in both cases. How are we to account for the development of functional differences often more striking in closely related species than are those between more distantly related types? Clark, Aronson, and Gordon (1954) have shown that the "isolating mechanisms" opposing the crossing of two closely related species of fish involve not only structural and physiological but also behavioral processes. As yet there have been very few experiments on this problem, so that, for example, we cannot say whether the behavioral components of isolating mechanisms in lower vertebrate and in mammalian species are similar or are very different.

Investigators are still hampered by various types of reductionistic tendencies, a common one being the generalization of "learning" and "perception" through the animal series, with all behavior modifications classed as the former and all sensitivity classed as the latter. For these matters, comparisons should be developmental and go far beyond the scope of adult properties. If scientists are to compare lower and higher animals insightfully, the entire causal nexus intervening between the fertilized egg and later stages in ontogeny must be studied analytically, with the functional properties of each stage always examined in relation to those of the preceding and the following stages. The concept of *intervening variables*, denoting all types of processes mediating between initial and later stages in development, has great potential utility for the study of instinctive behavior.

Appropriate methods are needed for studying development and the involved causal nexus both analytically and synthetically in each important species, to discern what common processes and what specialized differences prevail in the ontogeny of each. It is now recognized that what the individual really inherits in each species is a characteristic set of genes, the *genotype*, with an accompanying cytoplasm, as distinguished from possible *phenotypes*, the developing organism and its properties at successive stages in ontogeny. It is of course the phenotype that can be studied more or less directly, whereas investigations of the genotype are necessarily indirect. This unavoidable fact unfortunately encourages nativistic conceptions of species templates in the genes, directly uncoded or translated through growth into the functional patterns of later stages.

But in reality the idea of a direct uncoding of ontogenetic patterns from the genes remains hypothetical. Its application even to early stages seems unclear, as may be gathered from statements such as the following by Weiss, an embryologist:

> It is evident that every new reaction must be viewed in terms of the cellular system in its actual condition at that particular stage, molded by the whole antecedent history of transformations and modifications, rather than solely in terms of the unaltered genes at the core . . . no cell develops independently, but . . . all of them have gone through a long chain of similar environmental interactions with neighboring cells and the products of distant ones. (1954, p. 194.)

The genes are involved in interactions with their medium from the start, and the fact is that science has barely touched the problem of how genic factors influence organic development (Stern, 1954). The most comprehensive and reasonable attitude for investigators of functional development therefore seems to lie along the lines of Dobzhansky's (1950) statement to the effect that all development is both genotypic and environmental, rather than preformistic. Our premise consequently is that the hypothetical genic effects are mediated at each ontogenetic stage by systems of intervening variables characteristic of that stage in that species under prevalent developmental conditions, and that from initial stages these variables include both factors indirectly dependent on the genotype and others primarily dependent on the situation and environs of development.

Developmental Factors Differently Related to Behavior According to Phyletic Level

Very probably most if not all functional and behavioral properties as well as structural properties of organisms have figured in minor or major ways, separately or in patterns, as factors in natural selection, related *pro* or *con* to survival. In the list of factors having relatively greater weighting in species evolution must be included the intervening variables of all developmental stages, for, as Orton (1955) has pointed out, the evolutionary role of developmental properties at earlier stages may often have exceeded that of adaptive properties at mature stages. A complex array of very different types of organisms has resulted, with contrasting adaptive properties from those of the viruses to those of the primates. It is advantageous as a conceptual procedure to characterize and differentiate these major adaptive types as *levels* (Needham, 1929; Werner, 1940; Schneirla, 1947), in that they may be classed as higher or lower with respect to the status of abilities and types of organization underlying their adaptive adjustments.

The concept of levels has great potential significance for comparative psychology, as a comprehensive basis for the analysis and synthesis of evidence concerning similarities and differences among the varied adaptive patterns of major phyletic types. What, for example, have similar or different ontogenetic processes to do with the manner in which insects, fishes, and mammals modify their adaptive adjustments under somewhat equivalent conditions? But considerations of the "molecular" as in specific movements and the "molar" as in mating patterns (e.g., Roe and Simpson, 1958) are a different type of theoretical exercise, and do not answer such questions. They are directed at the analysis of functional properties at different stages in individual ontogeny, and should not be confused with "levels" theory, which is directed at phyletic comparisons. For excellent reasons, on both biological (Allee *et al.*, 1950) and psychological (Werner, 1940) grounds, recapitulation doctrine has only a limited significance for the study of ontogeny at any phyletic level in relation to its evolutionary background.

In psychological literature, complexity is often used as a criterion to distinguish the adaptive patterns of higher and lower animals. But although animals may be differentiated to some advantage in terms of the complexity of cell aggregates, levels theory also concerns psychological properties, and its hypothetical units of ability and organization can differ greatly among the levels. The definition of "simple" or "complex" in the psychological sense at different levels, therefore, must depend upon the nature of adaptive processes and the status of organization attained in the respective animals. This cannot be a matter of what totals of hypothetical molecular components (or complexities) are attained, but of how the subsystems of each pattern are interrelated, what degrees of plasticity are possible among them, the nature of perceptual re-arousal and of self-excitation processes and the like. As an example, I have found that *Formica* ants and pigmented rats learn the "same" maze problem very differently, the former making a really *complex* problem of it and advancing gradually by distinct stages, the latter behaving as though the problem were much *simpler* and mastering it in less than half the trials (Schneirla, 1959*a*). What is much more important, the rats exhibit a superior organization in the learning, whereby processes (distantly) similar to those appearing in the ant's second and third stages appear even in the first runs. How are we to determine what the "units" are, so that the learning of these two animals may be compared as to "complexity"? The dilemma can be solved only by viewing the two adjustments as qualitatively different, if we are to derive a theory appropriate to the wholistic nature and the psychological properties of the different animal types.

Notwithstanding a contemporary tendency among behavior investigators to disavow teleology, the practice persists of naming reactions in terms of their adaptive results (e.g., "courtship," "threat reactions"). Subjective impressions rather than objective criteria tend to dominate when common descriptive terms are used —the adjustments of insects and apes, for example, seem to become more alike psychologically—and the possibility wanes of distinguishing functional levels. By hypothesis, however, all existing animals are successful in adapting to their respectively different environments, and the efficiency of adaptive actions to which common names (such as "feeding") are applied therefore cannot be a valid criterion of psychological levels. For example, insects in general in all probability are able to carry out the reproductive adjustments ("courtship," "mating") at least as efficiently as man and often more smoothly, but without proving thereby any appreciable equivalence in the underlying psychological processes.

The task of the developmental psychologist, therefore, is to find formulae appropriate to the ontogenetic patterns of the respective phyletic levels. One important disadvantage of instinct theory, in scientific as in popular writing, is that the same terms are used for adjustments on very different levels, with the result that important differences are obscured. To use a broad contrast, feeding in a jellyfish and in a cat might both be called instinctive, since both acts are products of ontogeny in the normal species habitat, both are species-typical and both adaptive. Yet the developmental processes underlying the acts are relatively brief in the former and lengthy in the latter, and the two are organized very differently from a psychological standpoint.

In the lower vertebrates, patterns such as food taking are stereotyped in that they are obtainable through adequate arousal of species-characteristic tissues in routine ways. The component processes of feeding in the jellyfish depend upon the functional properties of tentacles, manubrium, and bell as affected by adequate stimulation, and with suitable techniques comparable reactions may be obtained even from the operatively separated parts. The integration of these part processes into a serial action pattern depends on the conductile functions of the nerve net; their order of participation in action thus depends chiefly on their spatial locations as determined through growth. By virtue of arousal-threshold properties, on adequate stimulation (glutathione at a low concentration), the manubrium (mouth tube) bends adaptively to the given side; on adequate (i.e., high intensity) stimulation both manubrium and bell contract strongly. Although behavior in this animal is variable, stereotypy predominates in the essential components, and modifications in the pattern seem to be limited to relatively short-lived changes in the nature of sensory adaptation or muscular tonus.

Feeding in the domestic cat also is a species-typical pattern developing in the standard habitat. But the chief properties and integration of this mammalian act differ vastly from those in coelenterates. The ontogeny of feeding in kittens passes through long and involved stages, that in the jellyfish is relatively short and essentially a product of tissue growth. In his well-known experiments with feline rodent killing, Kuo (1930) found evidence that development normally brings several part processes into a functional pattern. Organic components were indicated such as those adapting the cat to being excited by small moving objects, others to pouncing on, to seizing, and to devouring small animals. Other ways of feeding developed instead of this pattern under appropriate extrinsic conditions, and by suitably regulating the kitten's experiences with rodents Kuo produced subjects that preyed upon, feared and avoided, or lived placidly with rodents.

It is apparent that in lower invertebrates and in mammals, patterns such as feeding or mating develop through the mediation of very different intervening variables in radically different ontogenetic progressions. In coelenterates, growth processes impose relatively narrow limits upon action at all stages; in mammals, the involved factors greatly exceed the limits of growth and exploit them very differently according to the developmental conditions. The mammalian pattern is *plastic* in the sense that, under appropriate conditions, ontogenetic systems rather divergent from the species norm may arise; but in coelenterates, change is a matter of variability on a relatively fixed basis, and is not to be confused with the characteristic of plasticity.

Research on instinctive behavior generally has not advanced much beyond an introductory or descriptive stage of study, in which the investigator deals with his problem in terms of vague, categorical terms (e.g., "displacement reaction"). To illustrate the difficulty, we might say that the adaptive patterns of both coelenterates and insects are stereotyped, without recognizing the significant fact that the term "stereotypy" may have significantly different meanings in the two cases.

Interrelationships in Complex, Stereotyped Behavior Patterns

Because insects are often cited for their instinctive, stereotyped behavior, investigations going beyond the descriptive into the analytical stage are needed for insight into the developmental background of typical patterns. It is convenient to point to the fact that the arrangement of tracts and centers, together with properties of rapid conduction over short arcs, promotes a domination of activities by organic mechanisms and by sensory conditions. Among insect species, systems are found in a bewildering variety which illustrate this generalization broadly, but typically these have involved organizations resisting analytical investigation. As an example, Spieth (1952) found the mating patterns of closely related *Drosophila* species so well differentiated as to be taxonomically reliable. In such cases, it is conventional to emphasize the role of innate determination as effective through endogenous factors; however, the ontogenetic background of such factors remains obscure. Investigating such behavioral phenomena in relation both to their organic setting and to their developmental situation promises to be fruitful, as the following studies have shown.

One of the outstanding investigations of behavior in a solitary insect was the study of spinning activities in the *Cecropia* silkworm by Van der Kloot and Williams (1953). Under natural conditions, the mature caterpillar of this lepidopteran stops feeding and wanders for a time, then posts itself in a twig crotch and spins a cocoon with a thin apex through which the adult escapes at the end of pupation. The experimenters found the termination of feeding in this caterpillar due to glandular changes marking the period, as was also shown for the wandering, which was absent in caterpillars deprived of their silk glands. Furthermore, the removal of these glands eliminated two movements, the stretch-bend and the swing-swing, basic to normal spinning which was thereby shown to depend at least partly on sensory input from the spinning apparatus itself.

Although the genesis and temporal order of activities essential to the typical *Cecropia* spinning pattern thereby are shown to depend to an important extent upon processes in the individual's changing internal condition, their occurrence and function in the pattern involves relationships of these processes with tactual and gravitational cues dependent on relationships with the external situation. Thus it was found that the initiation of spinning requires contact of mouth parts with a surface, also that the termination of either of the two critical spinning movements depends upon contact of spinnerets with a surface when the caterpillar is at full flexion or full extension. Accordingly, the pattern of the species-typical cocoon may be considered a compound resulting from the merging of processes based on metamorphosis with sensory-input processes depending on the spatial and other properties of the extrinsic situation. Consequently, in a uniform environment such as the interior of a balloon, the caterpillar spun a flat layer of silk as a lining; tethered to a peg, the larva spun a cone-shaped tent—both structures arising by virtue of the very different physical properties of the extrinsic situation. The typical behavior pattern thus arises through a complex progressive relationship between changing organic conditions and the sensory input from these states, fused with that from sets of external circumstances which may change in their effects as the act progresses.

For the social insects, much evidence indicates that morphological and functional properties of the individual contribute critically to relationships arising in the group situation. Thus, in ants such as *Formica*, it is possible that early mouth-part reactions may be modified from initial passive feeding, through group interactions, into active feeding of others and then into foraging outside the nest (Schneirla, 1941). In these events a simple conditioning process, perhaps beginning with larval feeding in the nest situation, may implement and enlarge the function of organic factors involved in mouth responses and in regurgitation. In the social situation, other individuals function as key agents facilitating such integrations, thus in a sense equivalent to (but far more specialized in their developmental role than) certain properties of the general environment (e.g., food-plant odors) in solitary insects.

In the social insects, different specialized processes have evolved concurrently in the various types of individuals within a community, indispensable for the rise and maintenance of functional patterns in the colony as a unit. The army ants, for example, exhibit regular functional cycles with phases characteristic of the species, which owe both their timing and their integrity to the degree and nature of stimulative effects exerted on the adult population by the great developing broods (Schneirla, 1957). The massive broods appear in regular succession, and each, as it develops, is the source of excitatory tactual and chemical stimulation which dominates adult behavior but varies radically in intensity according to the developmental stage. A brood when in the larval stage arouses the workers maximally, so that large daily raids and nightly emigrations then are carried out regularly by the colony; but when this brood enters the quiescent pupal stage, and social stimulation falls abruptly to a low level, daily raids are small and emigrations cease.

At first sight, from the fact that the single colony queen lays the eggs producing successive broods in regular periods somewhat more than one month apart, one might think that a rhythmic process endogenous to the queen must serve as "clock" for the cycle. But the pacemaker is not in the queen, as investigation reveals the presence of recurrent changes in the extrinsic colony situation capable of influencing the queen's reproductive condition in critical ways. The relationship is complex, as these extrinsic changes concern the amount of tactual and chemical stimulation and of food received by the queen from the workers, and periodic changes in these factors depend in their turn upon developmental brood changes. When a brood is passing through the larval stage and its stimulative effect on the workers is high, responses of the workers to the queen fall off and her reproductive processes advance only slowly; when, however, this larval brood is mature, ceases to feed, and begins spinning cocoons, workers continue to be excited by brood activities but divert their responses appreciably to the queen, whose reproductive processes at once accelerate toward maturation and laying of the eggs. The large-scale reproductive processes of the army-ant queen are thus controlled extrinsically, running their course according to the level of stimulative and nutritive conditions in the colony.

The necessary cause evidently lies in complex stimulative relationships depending on the brood and changing with its development, but depending indirectly on the queen as the broods initiate with the queen. Each re-arousal of the queen is timed rather precisely through changes in this set of factors. Clearly, the army-ant cyclical pattern does not pre-exist in the genes of any one type of individual—workers, brood, or queen—nor is it additive from these sources. Rather, organic factors basic to the species pattern have evolved in close relationship with extrinsic conditions which in the evolved pattern supply key factors essential for integrating processes from all sources into a functional system (Schneirla, 1957).

"Maturation" and "Experience" as Factors in Basic Vertebrate Ontogeny

Whatever general principles may hold for behavior development, it is likely that they express themselves very differently according to the phyletic level. Traditional sharp distinctions such as that between the native and the acquired in ontogeny now seem untenable for any level. Instead, conceptual tools are needed to mark off the principal sources of ontogenetic factors, without the implication that either is ever independent of or altogether distinct from the other in normal development in any animal (Schneirla, 1956).

The conceptual terms to be used on this basis are "maturation" and "experience." The differentiation of these terms is considered here a conceptual convenience rather than one assuming a dichotomy of factors clearly separable in reality. It is improbable that any clear distinction can be made between what heredity contributes and environment contributes to the ontogeny of any animal, and no such distinctions are implied here. As a guide to investigation, the view that development is a natively determined unfolding of characters and integrations is no more valid than that development is directed only by extrinsic forces or by learning. Instead, concepts are needed which do not carry traditional biases into research on the real problem, namely, how development occurs in individuals at each level under conditions both typical and atypical of the species "niche."

The contention here is that, in all animals, individual genic equipment influences ontogeny throughout its course, but indirectly, and in relation with the developmental situation according to a formula characteristic of functional level and of species. The distinction between the sets of processes denoted by these terms is therefore a relative one, in view of the impossibility of separating the intrinsic strictly from the extrinsic (McGraw, 1946). *Maturation*, therefore, designates factors contributed to ontogeny essentially through processes of growth and differentiation at all stages; *experience* designates factors contributed through the effects of stimulation at all stages.

"Maturation" as used in this sense includes the effects of growth and differentiation without implying that these effects are exerted in a pattern directly derived from genic templates, as does Gesell (1950) in speaking of "the innate processes of growth called maturation." The dogma of a direct hereditary determination of development is thus excluded, and instead we emphasize the principles of genic factors indirectly influencing growth in different ways and of growth and differentiation as differently related to extrinsic conditions according to phyletic level, developmental situation, and ontogenetic stage. The doctrine of a genic "code" that is directly "uncoded" in development is opposed by evidence concerning genotypes, not only with respect to organic growth and differentiation but also with respect to behavioral ontogeny. As an example, in experiments with pure strains of fruitflies, Harnly (1941) found that the same genes may influence differently the development of wings and other structures affecting locomotion, according to what temperature prevails during ontogeny. With different temperatures stimulating growth differently, phenotypes capable of normal flight, erratic flight, or no flight were produced from the same genotypic strain. Under the different conditions, factors such as wing size, wing articulation, and neuromuscular control were indicated as variables affecting behavior differently in the phenotypes. Accordingly, the term *maturation* denotes here an expanding system of effects dependent on growth but operating at all stages intimately in connection with effects from the developmental situation; not a closed system of forces operative exclusively through organic growth.

Nativists frequently consider the nervous system as encapsulating a "code," derived from the genes, which through passive growth processes directly controls the development of specific behavior patterns. Coghill's (1929) results with larval salamanders are conventionally regarded as strongly supporting this theory. Yet these studies involved no analysis of behavior, but rather demonstrated histologically and descriptively a correspondence between successive new types of neural connections and changes in motility. Carmichael's (1928) experiment with larval salamanders is accepted by many as further proof. In this investigation, embryos immobilized with chloretone through early developmental stages, when tested as adults, were reported as swimming within thirty minutes after recovery, and in a manner indistinguishable from that of undrugged controls. These findings have been interpreted as excluding peripheral factors and effects of action in favor of an intraneural determination of development in amphibians. But Fromme (1941), replicating the experiment with tadpoles and with emphasis on analysis of the behavior, reported that drugged experimental subjects on recovery swam not only more slowly than the normals but with indications of deficient coordination. The effects of action, therefore, along with other possible feedback effects of function which have not yet been clearly identified, may play a role in the behavioral development of these amphibians.

Frequently cited also as evidence for an intraneural determination of patterns in lower vertebrates are the operative experiments of von Holst (1935) with teleosts and other fishes, from which rhythmic movements were reported in sections of trunk ostensibly deprived of their sensory connections. But Lissman (1946) in similar research found that carrying de-afferentation beyond a certain point abolished the locomotor rhythms completely; hence these activities evidently required some degree of afferent supply to function.

The fact deserves emphasis that these experiments were carried out with advanced or *adult* stages of lower vertebrates, hence can furnish no clear picture of how the respective patterns develop. Furthermore, their applicability to behavioral development in higher vertebrates remains unstudied, notwithstanding the frequency of textbook use of the results. How can such evidence be validly applied to higher levels while the extent of its relevance to the ontogeny of *lower* vertebrates remains unclear?

Rather than a strict intraneural rise of patterns, a view more in keeping with the evidence is that the nervous system develops in functional relationships to its setting, and that systems of intervening functional variables link it to the genes on the one hand and developing behavioral processes on the other. This view is strongly favored by a variety of evidence (Schneirla, 1956), as, for example, from experiments on the micturition pattern of the dog. Although the concept of a direct hormonal priming of an intraneural mechanism appeared to find support in earlier studies, subsequent research has shown that without its afferent components the adult pattern cannot function. Such mechanisms, also related deviously to hormonal processes, may well be essential to the development of neural properties underlying species-typical action patterns.

In view of the little-explored aspect of functional relationships in ontogeny, Coghill's conclusion that in behavioral development, generalization is primary, individuation secondary, may need heavy qualification. Serious questions pointing against the generalization-to-individuation view are raised by the finding of Carmichael and Smith (1939) with early guinea-pig embryos. They showed that stimulation of one limb produces discrete movement of that member or mass action according to whether stimulus intensity is weak or strong, respectively. In birds and mammals, particularly, early activities tend to be variable and indistinctly patterned. The pioneer investigations of Kuo (1932) with chick embryos, which disclosed varying successions of local and inclusive action as routine in that animal at early stages, also indicated significant relationships between movements and local actions as of leg and thorax. A swinging head movement was observed in those stages which might then in one sense have been considered specific, but which would have to be regarded as generalized in relation to different head responses appearing in later stages as in feeding, preening, and attacking. Moreover, the latter types are all in a developmental sense successors of the former.

Development must be examined from a functional and behavioral approach, going beyond the limits of embryology and genetics, for an adequate theory of the interrelationship of maturation and experience variables as differently related at successive stages. The mature chick embryo could not chip the egg, nor after hatching initiate and improve its discriminative pecking, unless increasing muscular strength and sensitivity facilitated the head lunge together with adequate postural support. Certainly, in different types of organisms, typical developmental systems may involve many close and subtle interrelationships of these variables characteristic of phyletic level and species, not readily understandable in terms of sharp differentiation of intrinsic and extrinsic factors. Such processes should become better understood when developmental research graduates from the descriptive to the systematic analytical stage.

To illustrate how involved these relationships can be even on an elementary level in vertebrates, we may consider the results of Tracy (1926) on toadfish ontogeny. At an early stage the larva of this fish, lying quiescent on the bottom for regular intervals, intermittently exhibits quick jerking movements, which the investigator termed "spontaneous" because they have no apparent extrinsic cause. But these movements, from Tracy's evidence, occur during each quiescent period through accretion in the blood of metabolic products as growth processes reduce available oxygen. Then, at a critical point in the build-up, an excitation somehow arises. Perhaps the existing biochemical condition (anoxia) directly causes motor neurones to fire, perhaps it acts indirectly as by altering thresholds—at any rate, sensory impulses from somewhere arouse a motor discharge to muscles. The rhythmicity of the process seems to rest on the fact that each movement phase serves to alter the embryo's condition, restoring the oxygen supply so that a further resting phase ensues—then a new accumulation of metabolites sets off another action phase, and so on. It is difficult to understand these processes from the facts of growth alone (Tracy, 1959).

The developmental role of the medium is emphasized strikingly by Waddington's (1959) experiments in which inbred strains of fruit-flies, after being subjected for many successive generations to an extrinsic agent, ether vapor, exhibit a distinct new structural character evidently due to the existence of genotypic changes. The point is that not only extreme extrinsic stresses (perhaps accounting for exceptional phenotypes) but also normal effects of the "standard" developmental situation are to be reckoned with at all stages in ontogeny. Clearly, the idea of an "original nature" in each species, innately determined through direct genic effects and modified only at "later stages" as through learning, must be held invalid. Who has determined, for any organism, the stage at which the "original nature" is established and ready for modification? A growing file of evidence shows that developing organisms enter into constant and progressively changing interactions with their environs, and that the developmental medium therefore participates crucially in ontogeny at all stages. It is these effects we are representing by the term "experience," which is not at all reducible to *learning*, although nativists often indulge in this reductionistic misinterpretation. "Learning" is to be considered one prominent outcome of experience at later stages in organisms capable of it.

Through a different kind of reductionist practice, characteristic rhythmic processes are conventionally considered "endogenous" because "no apparent external cause" is involved. But, as in the example already considered, the "cause" may be a set of external effects readily overlooked because they enter into subtle reactions with intraorganic processes. Sometimes the key external agent is not highly evasive once adequate investigative methods are used. For example, Harker (1953) found that the normal day-night activity rhythm of adult mayflies does not appear unless the individual has been subject to at least one twenty-four-hour light-dark cycle in the egg or larval stage. This rhythm accordingly does *not* appear in mayflies raised as eggs and larvae in continuous light or in continuous darkness. Somehow, specific physical changes acting for only a limited time early in ontogeny set up trace effects so implicated in growth that adult behavior is influenced. Despite the popularity of the "innate clock" concept, a prominent investigator in this field (Brown, 1959) has been led by a variety of evidence to conclude that the species-typical rhythmic functions of many animals may owe their timing to periodic external energy effects so obscure as to escape detection save by the most delicate techniques.

A useful generalization therefore is that, in any animal, extrinsic agencies common or exceptional to the developmental medium may contribute to ontogeny in ways characteristic of level and species, entering into developmental functions by merging with maturative processes appropriate to the stage. Knowledge of such phenomena often is arrived at only indirectly and by degrees. Now, however, the evidence seems strong enough to justify the deductive term "trace effects" for organic changes introduced through experience in early stages, even though they may not demonstrably influence action until much later in the individual's life. One example has been given from Harker's work; another one of importance comes from the research of Hasler and Wisby (1951) with Pacific salmon.

These fish, initially spawned and raised to the downstream-migration phase in specific headwater streams, leave the ocean and ascend the river system prior to spawning. And in a reliable frequency, as field tests and related laboratory experiments show, individual salmon tend to turn at branches toward the tributary carrying the chemical essence of headwaters in which the individual developed from the egg. What the organic trace effect may be is not yet known; that one exists to affect adult behavior seems clear.

Factors of *experience* influence development constantly and vitally. This concept covers a wide range of effects from biochemical changes set up variously (as illustrated) to more specialized types of afferent input including patterns having trace effects usually classed as "learning." Although, in studies of conditioning in embryonic chicks, Gos (1935) found no evidence of specific training to a touch-shock pairing before about sixteen days of age, simpler forms of change were indicated before that. In subjects given the test pairings after about the tenth day, there was observed a change in general responsiveness to the extrinsic stimuli used. Some trace effect, although a diffuse one, was thereby indicated. Accordingly, it is possible that, normally, experiences common in the species environment, such as a periodic turning of the eggs or knocking about in the incubative routine, might set up trace effects significant for later behavior, as perhaps in the form of a relative susceptibility to emotional excitement under certain general conditions. The response in such cases may be a different one in the animal's repertoire at the later stage. For example, it is possible that exposure of a larval solitary wasp during feeding to the odor of its typical species prey may induce a susceptibility of the sexually mature adult to this odor when that prey is encountered, eliciting the responses of stinging and ovipositing (then in the individual's repertoire).

The time at which conditioning and learning enter as influences in development, in animals having these abilities, is therefore likely to prove a relative matter, subject to how strict the definition is. Although evanescent conditioning results have been demonstrated in a few cases in birds and mammals prior to hatching or birth, the question remains as to when specific and more lasting patterns become possible. In tests with young puppies at different ages, Fuller *et al.* (1950) found no evidence before the twentieth day of a definite conditioning of the flexion response by electrical shock to exteroceptive stimuli such as a buzzer sound.[2] Since then, however, other investi-

[2] The growth process prerequisite to such attainments may be a sufficient myelinization of fibers in the corticospinal motor tracts known to occur at about that age in the dog.

gators have demonstrated general conditioned responses to gustatory and tactual stimuli in neonate mammals only a few hours or a few days old. The latter finding is consistent with results of studies in our laboratory (Rosenblatt *et al.*, 1962) showing that significant changes appear in kittens soon after birth in their orientation to the mother, with individually differentiated responses appearing within the first few hours. These adaptations concern both a gradually more efficient approach to the female's abdominal surface and processes in local orientation leading to attachment at a given zone in the nipple series (e.g., fore or aft). These adjustments clearly form an important basis for wider social adaptation in the litter situation. Learning processes are no doubt implicated, although (as preliminary analysis indicates) rudimentary and general in nature.

Limitations of the Distinction "Native versus Acquired" in Perceptual Development

Recently there has been a great increase in studies directed at demonstrating the effects of early experience in mammals, with results indicating that for young rats and dogs experiences in the later part of the litter period, from trauma to varying environmental heterogeneity, may identifiably affect later behavior. Although such research is still generally preliminary in nature, the results suggest that early experience may have a variety of effects, from a differential reaction to stimuli (Gibson and Walk, 1956) to general effects as of emotional excitability level and susceptibility to stress. To understand how varied such effects may be, particularly in their significance for the typical species behavior pattern, we must again widen our theoretical perspective to include very early stages and lower vertebrates.

We must first discard the traditional notion, still prevalent among ethologists, that specific behavior patterns arising in an organism raised in isolation from the "natural" environment and from species mates must *ipso facto* be "innate" or "inherited" in the sense of being exclusively gene-directed. Although the method of isolation is clearly useful, a compulsive nativistic interpretation of results is doubtful, as it involves a serious *non sequitur*. Perhaps the worst of the unexamined assumptions is that the experimenter knows *a priori* that the environment of isolation is not equivalent in any important respect to the typical species environment and the influence of species mates. This is a difficult matter to test, but let us consider the possibility that the isolated individual may influence its own development, through its own properties, in ways somewhat equivalent to the effects of species mates. Some time ago I suggested (1956) this might be possible through odor effects in social insects, and through visual effects (e.g., self-reflection) in fish. The latter possibility has been investigated in our laboratory (Shaw, 1962). Platyfish fry were removed from the female by Caesarian operation and raised in individual aquaria, isolated from stimuli from species mates and other animals until they were sexually mature. Individuals raised alone, surrounded by plates of ground glass, scored much lower in mating tests at maturity than subjects raised in community tanks during the first month with species mates and then placed in isolation. Early social experience thus proves essential to the normal development of the mating pattern in this fish. Experiments testing the hypothesis that self-stimulation may account for the lower scores of these isolated subjects are now under way. Scattered evidence suggests that in lower mammals, the female's experience in touching and licking her genital zones and posterior body contributes a perceptual orientation to those areas as familiar, and also facilitates a perceptual adjustment, by licking and not biting, to objects such as neonate young bearing equivalent olfactory properties.

Results from our investigations (Schneirla *et al.*, 1963) of maternal behavior in the domestic cat militate against assumptions of an inborn pattern, and speak in favor of progressive self-stimulative processes. Earlier development contributes certain specific organic processes; evidently also an orientation to her own body and perceptual differentiation of the genital area in particular. Analysis of results points to the crucial importance of organic processes such as uterine contractions and of their by-products: birth fluids, afterbirth, the neonate, and the membranes—each in its turn with potent stimulative effects. The processes of licking in particular, occurring in response to the attractive birth fluids, account for a progressive transfer of the female's attention from her own body to fluids on the substratum and through responses to these (as an incidental outcome) to the still wet neonate kitten. Abdominal contractions arouse attention and direct it posteriorly; fluids and further products promote and channelize the adjustment in ways basic for the establishment of a reciprocal stimulative relationship between female and kitten. Thus the pattern of parturitive behavior seems not to be organized from within, but rather to be assembled through the effect of stimuli furnished incidentally via the organic changes of parturition, literally a series of feedback effects of an essentially self-stimulative nature, merging into a perceived environment.

Self-stimulative processes appropriate to the stage evidently contribute to development in many and varied ways typical of the level.

A notable example offered for an early vertebrate stage is Kuo's (1932) evidence that from early organic functions in the chick embryo, such as beating of the heart and rhythmical pulsations of the amniotic membrane, stimulative effects may arise contributing to the ontogeny of the species-characteristic head-lunge response. His evidence indicates that in these early stages, self-stimulative processes (e.g., an incidental pushing up of the head by the thorax with dilatation of the heart) give rise to extrinsic stimuli such as tactual effects which (through their combination with proprioceptive stimuli and other causes) later can produce the head action alone. It is not that the heart "teaches the chick to peck," as a facetious nativist once remarked, but that incidentally its beating contributes to the development of a significant pattern. Actually, the process effected here resembles *sensory integration* (Maier and Schneirla, 1942) more than it does the typical conditioning pattern.

The isolation method is often as significant for what *does not* appear in the absence of the typical developmental situation or of species mates as for what does appear. Many lepidopteran larvae, when transferred away from their food plant, cannot survive; emphasizing the fact that, by laying her eggs in that situation, the female normally insures the presence of a set of experiences to which the caterpillar can adapt. At an early age, marsupial embryos shift from the mother's uterus to the pouch, a different situation still intimately associated with the mother and also involving a standard set of experiences to which the normally developed embryo can adapt. In ungulates the mother is a prominent feature of the environment to which the neonate soon adjusts by means of olfactory and visual stimulation she furnishes (Blauvelt, 1955). Oriented also by her directionalized licking, the neonate soon establishes a nursing relationship. Developmental environments in general tend to be species-standardized, so that experiences are more or less prestructured at all stages. This state of affairs indicates that developmental processes must have been significant in evolution (Orton, 1955), also that behavior must have had a prominent role in processes of natural selection prerequisite to species-typical adjustments and the environmental conditions of development.

In mammals generally, important aspects of the environment for embryo and neonate are prestructured by the mother. From our research, her role in patterning the situation of parturition has already been described for the domestic cat. In this animal the young are altricial, or blind-born, and the nest, localized and odor-saturated at the time of parturition, furnishes an orientative base for the young in their early perceptual development (Schneirla and Rosenblatt,

1961; Rosenblatt *et al.*, 1962). Soon after delivery, the female insures the presence of the neonates in this place by retrieving the scattered ones; she also introduces a directionalization to their variable movements by lying down and arching her body around them. The kittens, aroused to action and further directionalized by the female's licking of their bodies, are attracted to the female's abdomen by tactual and thermal stimulation from her. Thus, through a variety of experiences centering around the parent, the neonates arrive at the abdomen, nuzzle and attach to a nipple. Our studies show that from the first hour the neonate kittens not only improve in their ability to reach the female and re-attach but also to arrive at a given part in the nipple series.

Early experiences with a prestructured environment thereby furnish the basis for a social bond between parent and kittens, as well as a home base for the young. As our research suggests, attainment both of parent and of the nest reduces tension and introduces gratifications facilitating learning. During the first neonate week the adaptations of the kittens to female and nest improve steadily, paralleling sensory and locomotor development but also being clearly dependent upon processes of progressive learning and perceptual development. Sound production, "crying," indicates disturbance in disoriented kittens, and reductions in the intensity and pitch of this reaction in orientation tests through the first week offer clues to an orientative adjustment of increasing scope.

These early perceptual advances occur through integrations of tactual and olfactory cues progressively expanding the kitten's space from the home site. Navigation about the home cage improves slowly on this basis, centered on proximal cues from walls and floor. Although the eyes open at about one week, the proximal pattern of organization continues in force for a few days thereafter; then in a few days the kitten begins to make its way more freely on a visual basis, now reaching the home corner on the diagonal in tests instead of near the walls as before. The role of direct experience in this process is revealed particularly by the fact that kittens given the daily tests are definitely superior to untested kittens in attainment of the stage of visual perceptual control.

As analyses of results show, the gains of this stage are not altogether new but represent a reorganization in visual terms of earlier and more restricted patterns of proximal control on which visual perception is based. These results impressively illustrate the principle of qualitative advancement in perceptual development through an overlapping of stages in which gains from the maturational and experiential integrations of earlier stages provide a basis for new patterns under new conditions. On a lower vertebrate level, it is possible that tactual and proprioceptive integrations controlling head movements in the embryo provide a basis for the prompt ("spontaneous") responsiveness of early terrestrial stages in amphibians and reptiles with head lunging to visual movement (Maier and Schneirla, 1935, Chap. IX).

Operation of the same principle at a higher vertebrate level is represented by the findings of Nissen *et al.* (1951), who reduced and modified the normal opportunities of young chimpanzees for tactual and manipulative experience, first by binding the arms from birth, then after the fifth week by encasing all four extremities in tubes. When the subjects were freed of these impediments at thirty-one months, they were markedly deficient, compared to normal subjects, not only in discriminating touched points on the body but also in their perceptual adjustments to seen objects. In the same sense, not only growth but also action through experience impels normal processes of perceptual development in the human infant (Werner, 1940).

I have suggested (1959*b*) that "The socialization of early physiologically given biphasic excitatory states and the specialization of motivation and emotion advance hand in hand in the education of the infant mammal in perceptual processes and action." This principle, found applicable to more limited developmental attainments at lower phyletic levels, is illustrated in the ontogeny of the smile and of reaching, two adjustive processes crucial to man's perceptual and motivational development. Although actions such as head turning and arm extension toward the stimulated side, and a fleeting oral grimace, have been observed in neonate infants and even in the embryo, it is unlikely that these actions then represent "seeking" tendencies as Schachtel (1959) has suggested for the former, or innate social responses as Koehler (1954) has suggested for the latter. Rather, it is significant that both of these actions first arise as crude and variable responses to a variety of stimuli, but evidently always as incidental "weak" responses to low-energy stimulation. Neither can be called "purposive" in early stages; rather, analysis shows that reaching, through "progressive integrations of vision with preexisting tactuoproximal adjustments, becomes first a specific approach response, then a perceptually directed seeking response," and that the grimace becomes a "smile" or social approach response only through comparable processes in perceptual development (Schneirla, 1959*b*).

The purposive adjustments of the older human infant are therefore not themselves given as "instincts," but rather are the products of an involved perceptual development through which maturative-experiential processes of earlier stages are radically modified into externally meaningful patterns. This principle, concerning the appearance of adaptive behavioral adjustments through ontogeny, has been illustrated before in this essay in terms of ontogeny at lower phyletic levels—but always intimately in relation to the developmental content appropriate to the level.

Each type of animal grows into its own world with different properties and acquires its adaptive patterns, whatever their nature or scope, through the functioning of successive sets of intervening variables operating in the global situations of respective developmental stages. The emergence of psychological properties peculiar to the developmental circumstances will become increasingly understandable in the future, as investigators correct the limitations of concepts such as "growth" and "learning" by devising improved explanatory Gestalts in better tune with reality.

BIBLIOGRAPHY

ALLEE, W. C., EMERSON, A. E., PARK, O., PARK, T., and SCHMIDT, K. P. 1950. *Principles of animal ecology.* Philadelphia: Saunders.

BEACH, F. A. 1947. Evolutionary changes in the physiological control of mating behavior in mammals. *Psychol. Rev.*, 54, 297–315.

BLAUVELT, HELEN. 1955. Dynamics of the mother-newborn relationship in goats. *Group Processes, First Conference.* New York: Josiah Macy, Jr., Foundation, pp. 221–258.

BROWN, F. A. 1959. The rhythmic nature of animals and plants. *Amer. Scientist*, 47, 147–168.

CARMICHAEL, L. 1928. A further experimental study of the development of behavior. *Psychol. Rev.*, 34, 253–260.

———, and SMITH, M. F. 1939. Quantified pressure stimulation and the specificity and generality of response in fetal life. *Jour. Genet. Psychol.*, 54, 425–434.

CLARK, EUGENIE; ARONSON, L. R., and GORDON, M. 1954. Mating behavior patterns in two sympatric species of xiphophorin fishes: their inheritance and significance in sexual isolation. *Bull. Amer. Mus. Nat. Hist.*, 103, 139–335.

COGHILL, G. E. 1929. *Anatomy and the problem of behavior.* New York: Macmillan.

DOBZHANSKY, T. 1950. Heredity, environment and evolution. *Science*, 111, 161–166.

FROMME, A. 1941. An experimental study of the factors of maturation and practice in the behavioral development of the embryo of the frog, *Rana pipiens. Genet. Psychol. Monogr.*, 24, 219–256.

Fuller, J., Easler, C. A., and Banks, E. M. 1950. Formation of conditioned avoidance responses in young puppies. *Amer. Jour. Physiol.*, **160**, 462–466.

Gesell, A. 1950. Human infancy and the ontogenesis of behavior. *Amer. Scientist*, **37**, 529–553.

Gibson, Eleanor, and Walk, R. D. 1956. The effect of prolonged exposures to visually presented patterns on learning to discriminate them. *Jour. Comp. Physiol. Psychol.*, **49**, 239–242.

Gos, M. 1935. Les reflexes conditionnels chez l'embryon d'oiseau. *Bull. Soc. Sci. Liège*, **4–5**, 194–199; **6–7**, 246–250.

Hall, C. S. 1951. The genetics of behavior. *Handbook of experimental psychology*, Chap. 9, 304–329. New York: Wiley.

Harker, Janet E. 1953. The diurnal rhythm of activity of mayfly nymphs. *Jour. Exper. Biol.*, **30**, 525–533.

Harnly, M. H. 1941. Flight capacity in relation to phenotypic and genotypic variations in the wings of *Drosophila melanogaster*. *Jour. Exper. Zool.*, **88**, 263–274.

Hasler, A. D., and Wisby, W. J. 1951. Discrimination of stream odors by fishes and its relation to parent stream behavior. *Amer. Nat.*, **85**, 223–238.

Hebb, D. O. 1953. Heredity and environment in mammalian behavior. *Brit. Jour. Anim. Behav.*, **1**, 43–47.

Holst, E. von. 1935. Über den Prozess der zentralnervösen Koordination. *Pflüg. Arch.*, **236**, 149–158.

Hunter, W. S. 1947. Summary comments on the heredity-environment symposium. *Psychol. Rev.*, **54**, 348–352.

Koehler, O. 1954. Das Lächeln als angeborene Ausdrucksbewegung. *Zeitschr. menschl. Vererb.-u. Konstitutionslehre*, **32**, 390–398.

Kuo, Z. Y. 1930. The genesis of the cat's response to the rat. *Jour. Comp. Psychol.*, **11**, 1–30.

———. 1932. Ontogeny of embryonic behavior in Aves. IV. The influence of embryonic movements upon the behavior after hatching. *Jour. Comp. Psychol.*, **14**, 109-122.

Lissman, H. W. 1946. The neurological basis of the locomotory rhythm in the spinal dogfish (*Scyllium canicula, Acanthias vulgaris*). II. The effect of de-afferentation. *Jour. Exper. Biol.*, **23**, 162–176.

Maier, N. R. F., and Schneirla, T. C. 1935. *Principles of animal psychology*. New York: McGraw-Hill.

———, ———. 1942. Mechanisms in conditioning. *Psychol. Rev.*, **49**, 117–134.

McGraw, Myrtle. 1946. Maturation of behavior. *Manual of child psychology*, Chap. 7, 332–369. New York: Wiley.

Moore, K. 1944. The effect of controlled temperature changes on the behavior of the white rat. *Jour. Exper. Psychol.*, **34**, 70–79.

Needham, J. 1929. *The skeptical biologist.* London: Chatto.

Nissen, H., Chow, K. L., and Semmes, Josephine. 1951. Effects of restricted opportunity for tactual, kinesthetic, and manipulative experience on the behavior of a chimpanzee. *Amer. Jour. Psychol.*, **64**, 485–507.

ORTON, GRACE. 1955. The role of ontogeny in systematics and evolution. *Evolution*, **9**, 75–83.

ROE, ANNE, and SIMPSON, G. G. 1958. *Behavior and evolution.* New Haven: Yale Univ. Press.

ROSENBLATT, J. S., TURKEWITZ, G., and SCHNEIRLA, T. C. 1962. Development of suckling and related behavior in neonate kittens. Chap. 14, pp. 198–210, in *Roots of behavior*, E. L. Bliss (ed.). New York: Harper (Hoeber).

SCHACHTEL, E. G. 1959. *Metamorphosis—on the development of affect, perception, attention, and memory.* New York: Basic Books.

SCHNEIRLA, T. C. 1941. Social organization in insects, as related to individual function. *Psychol. Rev.*, **48**, 465–486.

———. 1947. Levels in the psychological capacities of animals. *Philosophy for the future.* New York: Macmillan, pp. 243–286.

———. 1962. Psychology, comparative. *Encyclopaedia Britannica*, **18**, 690Q–703. Chicago: Encyclopaedia Britannica, Inc.

———. 1956. Interrelationships of the "innate" and the "acquired" in instinctive behavior. *L'instinct dans le comportement des animaux et de l'homme.* Paris: Masson et Cie, pp. 387–452.

———. 1957. Theoretical consideration of cyclic processes in doryline ants. *Proc. Amer. Phil. Soc.*, **101**, 106–133.

———. 1959*a*. L'apprentissage et la question du conflit chez la fourmi. Comparaison avec le rat. *Journal de Psychologie*, **57** (1), 11–44.

———. 1959*b*. An evolutionary and developmental theory of biphasic processes underlying approach and withdrawal. *Nebraska symposium on motivation, 1959.* Univ. of Nebraska Press, pp. 1–42.

———, and ROSENBLATT, J. S. 1961. Behavioral organization and genesis of the social bond in insects and mammals. *Amer. Jour. Orthopsychiatry*, **31**, 223–253.

———, ———, and TOBACH, ETHEL. 1963. Maternal behavior in the cat. Chap. 4, pp. 122–168, in *Maternal behavior in mammals*, Harriet L. Rheingold (ed.). New York, London: Wiley.

SHAW, EVELYN. 1962. Environmental conditions and the appearance of sexual behavior in the platyfish. Chapter in *Roots of behavior*, E. L. Bliss (ed.). New York: Harper (Hoeber).

SPIETH, H. T. 1952. Mating behavior within the genus *Drosophila* (Diptera). *Bull. Amer. Mus. Nat. Hist.*, **99**, 401–474.

STERN, C. 1954. Two or three bristles? *Amer. Scientist*, **42**, 212–247.

TRACY, H. C. 1926. The development of motility and behavior reactions in the toadfish (*Opsanus tau*). *Jour. Comp. Neurol.*, **40**, 253–369.

———. 1959. Stages in the development of the anatomy of motility of the toadfish (*Opsanus tau*). *Jour. Comp. Neurol.*, **111**, 27–81.

VAN DER KLOOT, W. G., and WILLIAMS, C. M. 1953*a*. Cocoon construction by the Cecropia silkworm. I. The role of the external environment. *Behaviour*, **5**, 141–156.

Van der Kloot, W. G., and Williams, C. M. 1953*b*. Cocoon construction by the Cecropia silkworm. II. The role of the internal environment. *Behaviour*, **5**, 157–174.

Waddington, C. H. 1959. Evolutionary systems—animal and human. *Nature*, **183**, 1634–1638.

Watson, J. B. 1914. *Behavior: an introduction to comparative psychology.* New York: Holt.

Weiss, P. 1954. Some introductory remarks on the cellular basis of differentiation. *Jour. Embryol. Comp. Morph.*, **1**, 181–211.

Werner, H. 1940. *Comparative psychology of mental development*, 3d edn. New York: International Universities Press, 1957.

ARTICLE III

MECHANISMS IN CONDITIONING[1]

N. R. F. MAIER AND T. C. SCHNEIRLA

I. Conditioning *vs.* Trial-and-Error Learning

In recent years there has developed a tendency to break down the theoretical distinction between the classical notion of conditioning and that of selective or "trial-and-error" learning. This tendency has led investigators to carry out experiments closely resembling "problem-box" tests, treating the results so firmly under the dominance of conditioning concepts that possible relationships to the findings of problem-box studies proper may be overlooked. To be sure, Skinner (24) and Hilgard and Marquis (13) have differentiated conditioning and problem-box procedures, but they do not carry this distinction through to the basic mechanisms. We believe that closer analysis will disclose some fundamental differences of great importance.

For some time it has been recognized that contiguity in experience is the essential factor in the process of conditioning. On the other hand, in trial-and-error learning the selection process stands out clearly, and to deal adequately with this instance of learning it has been necessary to recognize not only the factor of contiguity but also the factor of motivation. Hence "reward" and "punishment" have figured prominently in discussions of trial-and-error learning.

It is an observed fact that when responses are accompanied by punishment they are in time less likely to come to expression, whereas those accompanied by reward are more likely to come to expression. The mechanisms responsible for these effects on behavior are, however, not agreed upon. Thorndike's earlier notion that punishment "stamps out" the learning and that reward "stamps in" the learning was an attempt to explain the facts in terms of associative bonds formed only in rewarded instances. Perhaps the problem can be dealt with more adequately if it is assumed that associations

[1] Reprinted from *Psychological Review,* vol. 49, No. 2, 1942, pp. 117–134, by permission of the Editor and the publishers.

are formed in both cases, a position Thorndike (25) now accepts. The view is that in the one case the animal learns what not to do and in the other it learns what to do—i.e., reward determines which responses will come to expression and punishment determines which will be inhibited. This point of view introduces a selection process in addition to an association process, and performance rather than the formation of associations is affected by the kind of incentive involved.

These considerations are sufficient to illustrate the fact that the particular roles of reward and punishment have for some years constituted a most troublesome theoretical problem for investigators working with maze, problem-box, and discrimination learning.

Recent experiments (1, 2, 3, 9, 18, 26) have shown that motivation may be a factor of importance in conditioning. In order to account for the facts, Culler (6) believes it is necessary to assume that the unconditioned stimulus has a dual function: that of giving rise to a response and that of serving as an incentive. Perhaps the recognition of the role of motivation in what are regarded as conditioning experiments is largely responsible for the current tendency to believe that the gap between the two forms of learning is being narrowed.

Before accepting a theoretical fusion of what may well be two distinctive forms of learning, it seems desirable to examine the evidence carefully to see whether certain of the differences between these processes may not be of basic importance. This precaution is suggested by known cases in which behavior mechanisms of qualitatively different natures present external similarities of relatively secondary importance. For instance, learning and maturation both exhibit a trend from initially generalized response to more specific behavior, yet much ground would be lost if basic differences in mechanism were to be subordinated to this feature of similarity. Only a careful examination will show whether similarities or differences more adequately represent essential behavior mechanisms. Thus emphasis upon certain similarities may well have led investigators to overlook or to minimize some differences of primary importance between the processes of conditioning and of trial-and-error learning.

In a general treatment (21) we have analyzed the learning process in terms of concepts primarily bearing upon what occurs in the animal rather than in terms of the situation. This analysis led us to distinguish between association formation which depends upon contiguity, and selection which depends upon reward or punishment. It seems desirable to apply that distinction here, beginning with a rudimentary diagram showing how the association and selection processes operate in conditioning and problem-box situations.

In Fig. 1 the typical problem-box setting is diagrammatically represented as a situation (*sit.*) which elicits a series of responses (R_1, R_2, $\cdots$ R_x, R_y). The order in which these reactions appear may depend upon the excitatory threshold of each manner of responding, or it is possible that the responses are an expression of a tendency to vary behavior (15, 19, 20). In any case the responses arise as parts of the repertoire that the animal brings to the experiment. Now because of the experimental set-up only one of these reactions to some aspect of *sit.* can occur in contiguity with food and so become associated with it. It is this response which gains in preference and finally dominates over other activity. When it dominates we say that the animal has mastered the problem. The animal, however, has not learned a *new* response—rather a response with a low excitatory threshold has been made dominant, or we may say it has become selected or preferred.

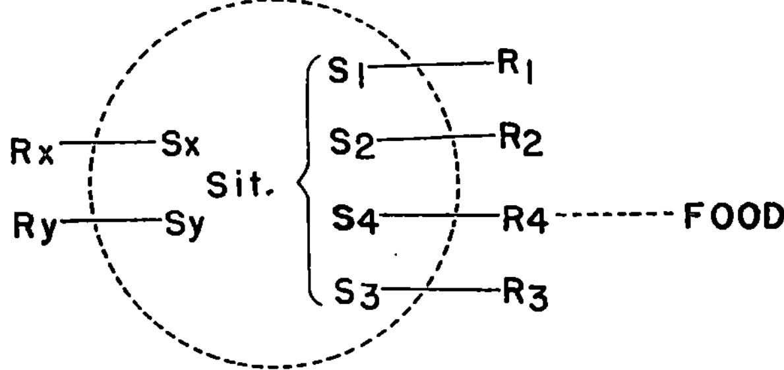

FIG. 1. Diagrammatic representation of selective learning.

In conditioning the situation is quite different. Here it is the unconditioned stimulus which determines the nature of the response. Since this stimulus initiates the response it cannot be a reward or a punishment for the response. The conditioned stimulus occurs repeatedly in contiguity with the unconditioned stimulus and soon it alone elicits the response. Initially the conditioned stimulus showed no signs of producing the response in question. We cannot therefore say there has been a shift in dominance. Rather, we may say that a previously ineffective stimulus has been transformed into an effective one, and that contiguity of the ineffective (conditioned) and the effective (unconditioned) stimuli has been the necessary condition.

Once the conditioned stimulus adequately elicits the response, however, the experimental conditions may become changed. If one then presents the unconditioned stimulus after the response occurs (the method of reinforcement often employed to prevent experimental extinction), the important aspect of the problem-box procedure is duplicated. Likewise, if the conditioned stimulus calls out the critical response promptly enough, the unconditioned stimulus (e.g., food; shock) will *follow* rather than precede the response. Here also the unconditioned stimulus can function as incentive even though no change in the external procedure has been introduced. In further trials the excitatory value of the conditioned response, or part of it, can thus be increased by selective learning. Since reinforcement occurs only when certain elements are present, the response may become progressively more specific, depending upon the nature of the unconditioned stimulus (e.g., food or shock used as unconditioned stimuli can serve as incentives more readily than light). Thus an animal conditioned to give foot-flexion in response to a bell, through the contribution of selective learning may gradually decrease the amplitude of the flexion to the minimum necessary to clear the grill. The animal's behavior now appears to be anticipatory in nature as is the case in all selective learning. From the above we can also see why conditioned responses frequently differ from the original unconditioned response.

That the *initial* phase of the conditioning procedure is an important one becomes apparent if we try to use the method of reinforcement (or reward) from the outset. Suppose we wish to train a dog to salivate to a bell. We place the dog in the bell situation and wait until he salivates before food is given. Under such conditions it would be extremely difficult, if at all possible, to train the animal. Since salivation to a bell is not a part of the dog's regular repertoire of behavior, one could not reward the act. Once, however, the bell has developed some excitatory value for salivation (by conditioning) the method of reinforcement can proceed.

Thus we may distinguish two phases in the conditioning *procedure*, a first phase in which a neutral stimulus is given excitatory value and a second in which a selective process operates. In Pavlovian experiments attention has been concentrated upon the characteristics of the first phase although the second phase may sometimes have been present. In the first phase contiguity alone determines the nature of events. It is in the second phase that the conditioning procedure can become comparable with trial-and-error learning.

Our analysis thus leads to agreement with Culler (**6**) in that the unconditioned stimulus may potentially have two functions, one for determining the response, the other for motivating it. Culler points out that these different functions of the unconditioned stimulus account for two stages in the development of the conditioned response, but he does not show why the stages appear as they do. The foregoing analysis emphasizes the fact that the successive stages grow directly out of the procedure, and makes it clear that it is the procedure rather than any intrinsic property of the unconditioned stimulus which leads to a shift in its function as training progresses.

The postulation of these two different phases in the procedure of conditioning permits us readily to integrate certain experimental findings which have seemed inconsistent with previous results, some of them among the reasons Culler has advanced for attributing a dual function to the unconditioned stimulus.

1. Culler and Mettler (7), also Girden, Mettler, Finch, and Culler (12), have found that conditioned responses can be established in decorticate animals virtually as readily as in normal animals. Shurrager and Culler (23) obtained conditioning with similar ease in dogs when only a small part of the spinal cord remained in function. However, the conditioned responses formed in neurally reduced dogs of these types are very generalized, and never reach the stage of a specific response. This seems to mean that association by contiguity is more primitive than selective learning, since it can proceed in the absence of cerebral cortex although selective learning is then excluded. The assumption that learning may take place on different neural levels is supported by experiments on conditioning under curare (10, 11). The postulation of two qualitatively different learning mechanisms makes understandable the fact that the first stage in conditioning can occur unimpaired although the second is excluded in spinal and decorticate animals.

2. Brogden (1) has found that after a conditioned response has been established a reward (food) may be substituted for the unconditioned stimulus (escape from shock) without leading to extinction of the response. If the unconditioned stimulus may acquire the role of incentive during the second phase of the conditioning procedure, it is to be expected that other incentives will function equivalently. That one incentive may be safely replaced by another in selective learning is a well-known fact.

3. Finch and Culler (9) and Brogden (2) found that they were able easily to develop a fifth order of conditioning although Pavlov had set the third order as the limit.

In order to accomplish this Finch and Culler first conditioned leg flexion to a tone. Since the flexion was produced by a shock grill on which the foot was placed, the animal soon escaped shock by a conditioned flexion. When this stage was reached, the selective phase could enter into the learning and prevent extinction. To establish the second order, a light-flash was presented with the tone. When the animal failed to flex the leg, shock to the thorax was administered and soon the light alone produced leg flexion. Since the tone produced leg flexion, a secondary conditioned response to light could be established and selection could then proceed as in the first order. By the time the third order of conditioning was reached it is reasonable to suppose that the leg flexion as a response to the general experimental situation was well within the animal's behavior repertoire. Under these conditions any new stimulus might call out leg flexion; hence the response was readily subject to selection and the first (or associative) stage of the conditioning procedure was probably not required. In establishing the third order, a stream of water was given with the light. Did the light in this case function as did the original unconditioned stimulus? We believe the light could have been omitted without changing the outcome. The point is that the stream of water followed by thoracic shock soon would have brought leg flexion to expression, since this response had been selected as a means for avoiding shock. If this was the case then it follows that trial-and-error learning rather than higher-order conditioning occurred in the above experiments.

The fourth and fifth orders of conditioning were then obtained with such ease that the experimenters believed they could have advanced readily to an indefinite number of higher orders. Since "higher-order conditioning" should introduce added difficulties with each further step, the absence of such difficulties here offers further reason for believing that our analysis is a plausible one.

The same analysis would apply to Brogden's experiment (2) in which a food reward replaced the "escape from thoracic shock" in the selective function.

There are no indications that the postulated second (selective) stage was involved in the Pavlovian experiments. This may be attributed to the use of a nonvoluntary response (salivation) and to the fact that Pavlov did not present reinforcement in a selective manner. We have thus implied that the exclusion of the selective stage would account for Pavlov's inability to carry conditioning beyond the third order. (Furthermore, the Russian experimenters had to contend with experimental extinction, since their critical response was not maintained by a selective learning.)

4. Brogden, Lipman, and Culler (3) conditioned guinea pigs to run in a rotator in response to a 1000-cycle tone. Shock was used as the unconditioned stimulus. In one group the running shut off the shock and so afforded relief, in another group the response was not thus rewarded.

During the first 75 to 100 trials the two groups showed essentially identical progress, the running in response to the buzzer being part of the pattern in about twenty per cent of the trials. After 100 trials, however, the two groups progressed quite differently. The group that was rewarded for running responded by running one hundred per cent of the time, after an average of 206 trials; the group not rewarded in this manner showed questionable increase in the tendency to respond by running. Instead, the animals of the latter group held their breath and tensed their body musculature. The experimenters believed that the group which responded to the tone by running duplicated the unconditioned response, whereas the other group showed anticipatory behavior.

It seems reasonable, however, to regard holding the breath, tensing the muscles, and leaping forward all as parts of the unconditioned response to shock. Since tone was coupled with shock, it could call out any part or all of this pattern. The experimenters measured only the running and this aspect was equally shown by both groups during what we regard as the associative stage.

The one group, however, was rewarded for running and, as a consequence, it is to be expected that this aspect of the response would become more and more selected. The other group may have minimized the shock effects by tensing the musculature (the experimenters considered this to be the case), and if this is true this aspect of the pattern should become selected. If the tensing did not relieve the shock effects then we should expect the response to tone to remain generalized. (The data are not sufficiently detailed to determine which of these alternatives occurred.)

In any case the final response was distinctly different in the two groups and we believe the second or selective stage to be responsible for the difference. The experimenters believe the form of the incentive resulted in two contrary forms of conditioning. Since selective learning is an established fact, it seems unnecessary to postulate two forms of conditioning to account for the findings.

5. Kellogg (14) emphasizes the presence of individual differences in the form of the conditioned leg flexion. From the types of this response which occur in different dogs he believes the notions of "stimulus substitution" and "anticipatory responses" in conditioning can both be shown. From our analysis Kellogg's results are to be expected. Individual differences in selective learning are the rule, since each animal selectively learns from its own repertoire. Because Kellogg's dogs were able to avoid shock by flexing the leg we believe the marked individual differences observed were due to the *selective learning* occurring after the first stage.

6. Loucks (18) found that the dog could not be conditioned to a buzzer when shock to the motor area of the cortex was used as the unconditioned stimulus for leg flexion. This indicates that the un-

conditioned stimulus must have effective sensory relations as well as a purely motor function, since only the sensory aspect of the unconditioned stimulus was lacking in this experiment. However, a change in procedure entirely altered the outcome of the experiment. When the animal was given a food reward each time leg flexion occurred, positive results were obtained.

Loucks contends that motivation made possible the establishment of a conditioned response otherwise unobtainable. According to our analysis conditioning did not occur in this experiment. Rather, cortical shock in the buzzer situation elicited a response, and reward was able to function selectively. Under these conditions foreleg flexion became associated with reward just as string-pulling becomes associated with reward in the problem-box situation. The first stage, i.e., the association between a neutral stimulus and the experience of the unconditioned stimulus, was thus omitted. Since selective learning is facilitated when an animal is "put through" a given act, it is not surprising that Loucks was able to obtain selective learning under conditions in which cortical shock offered a very effective way of manipulating the desired response.

That this distinction is real and not merely verbal in nature can be demonstrated through the fact that the two interpretations lead to opposite conclusions in certain critical instances.

Crisler (5) and Light and Gantt (17) have shown that conditioning can occur without the animal making the critical response during training. It is merely necessary to present the conditioned and unconditioned stimuli in contiguity. Thus, Light and Gantt prevented the unconditioned stimulus from producing a response by crushing the motor roots of the essential spinal nerves. On regeneration of the destroyed motor pathways it was found that the buzzer, which had been paired with the shock to leg (unconditioned stimulus), now produced leg flexion. This result could not have occurred had not the training somehow established an association between the buzzer and shock experiences. It thus follows that making the critical response is not essential for the development of true conditioning.

In trial-and-error learning, however, actively making the response is very important, since the critical action must come to expression if it is to be selected. Thus, if our contention that Loucks' experiment involved trial-and-error learning is correct, it follows that negative results would be obtained if the Loucks' experiment were repeated with the motor nerves crushed but using buzzer and cortical shock as he did. If, however, it is claimed that ordinary conditioning occurred, this modification should not affect the findings, since making the response is not essential to conditioning.

II. Sensory-Sensory *vs.* Sensory-Motor Connections

It appears to us that much of the confusion in the interpretation of conditioning data arises from a failure to analyze what has happened to the animal. As a rule, interpretations are based on what the animal *does*. Behind the animal's behavior are the mechanisms responsible for the behavior, and an improved understanding of the mechanisms should throw needed light on the overt occurrences. The point can be nicely demonstrated in connection with an extension of our preceding discussion.

In conditioning, we are told, the conditioned stimulus becomes a *substitute* for the unconditioned stimulus, which apparently means

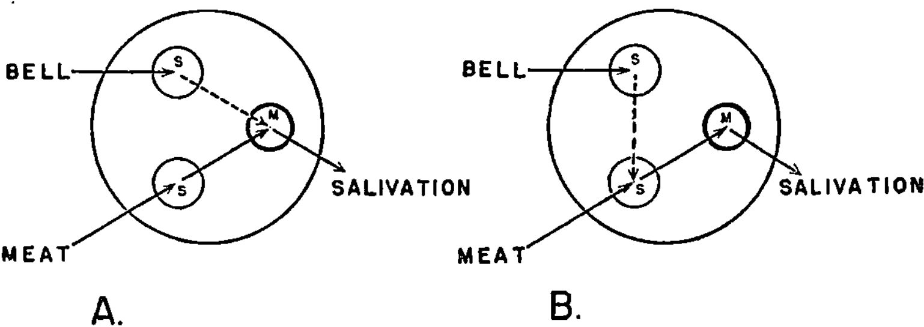

FIG. 2. Diagrammatic representation of two interpretations of the conditioning process. The large circles indicate the field of cortical function; the small circles indicate sensory and motor regions. A. Conditioning viewed as a *substitution* process. B. Conditioning viewed as an *extension* process.

that both occupy similar functional positions once training is effective. This generally accepted view is graphically presented in Fig. 2A. The dotted line indicates the functional connection which is assumed to arise through training.

An obvious alternative suggested by the traditional doctrine of the association of ideas is indicated in Fig. 2B. Here the dotted line suggests a very different functional connection. The first diagram indicates that a *sensory-motor* connection is established, whereas the second offers a *sensory-sensory* connection as the essential dynamic change. It is the latter which Maier and Schneirla (21, Chap. XV) have regarded as most probable.[2]

In the first case the term "substitution" is a pertinent description

[2] The implications of this question extend far beyond the field of mammalian conditioning. Thus we have based our theoretical comparison of psychological capacities among the lower vertebrate classes—and in fact among the higher invertebrate groups as well—(21, Chaps. V–X) upon *sensory integration* as the process critical for the qualitative nature of behavior.

of the postulated change, but in the second this term is misleading. Rather the term *extension* is more appropriate, as indicating the enlargement of a sensory pattern controlling the critical response. Although the second alternative is seldom considered it seems to best fit the facts. Let us consider some experimental findings which bear significantly on this question.

Loucks (18) found that cortical shock to the motor area could function as a conditioned stimulus, but *not* as an unconditioned stimulus. If conditioning depends upon an association between two sensory cortical patterns, then it is clear why cortical shock is thus limited in its functions. In Fig. 3A the dotted line represents our conception of the new dynamic relationship which is established when cortical shock is the conditioned stimulus. Shock to the leg

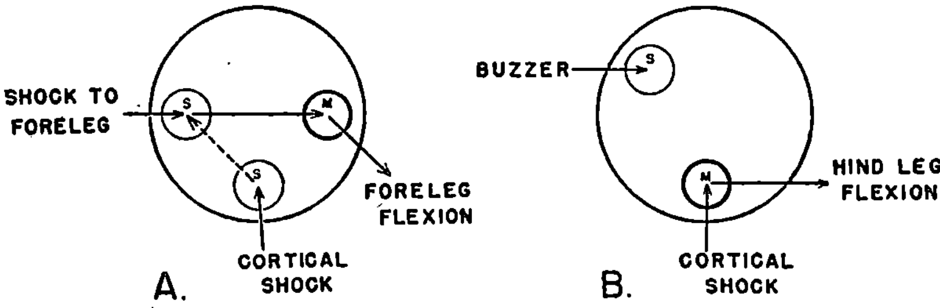

FIG. 3. Diagram showing the role of cortical shock in conditioning. A. Field of cortical function when cortical shock serves as the *conditioned* stimulus. B. Field of cortical function when cortical shock serves as the *unconditioned* stimulus.

furnishes the sensory component arousing motor cortex, which in turn elicits the unconditioned defense reaction. When cortical shock is introduced as unconditioned stimulus, however, the activated cortex directly gives rise to leg flexion and its function is then purely motor. Fig. 3B shows that a sensory component is absent in this experiment, and since negative results were obtained when a buzzer was used as the conditioned stimulus, we are led to regard the sensory aspect of the *unconditioned* stimulus as essential for positive results. If conditioned responses depend upon a sensory-motor connection, it is difficult to account for these negative findings. (We have already shown how the addition of reward can transform the negative to positive results.)

The outcome of an experiment by Shipley (22) also favors the sensory-sensory interpretation. He first conditioned a wink response to a light flash by presenting the light with a strike below the eye (*US*). Next he conditioned finger withdrawal to the same strike below the eye by presenting the strike (*CS*) with shock to finger (*US*).

The crucial test was now to reintroduce the light flash. In nine of his fifteen human subjects finger withdrawal was obtained with the flash, which had never been paired with shock to finger. If we postulate a sensory-motor connection as the essential relationship, the system is like that sketched in Fig. 4A. But the acquired connections involved here, x and y, provide no direct route between the cortical components of flash and finger withdrawal, and it is necessary to postulate highly questionable backward-running associations or kinesthetic components to account for the findings. If, however, we assume that conditioning develops sensory-sensory functional relations, the system is like that diagrammed in Fig. 4B. Here the

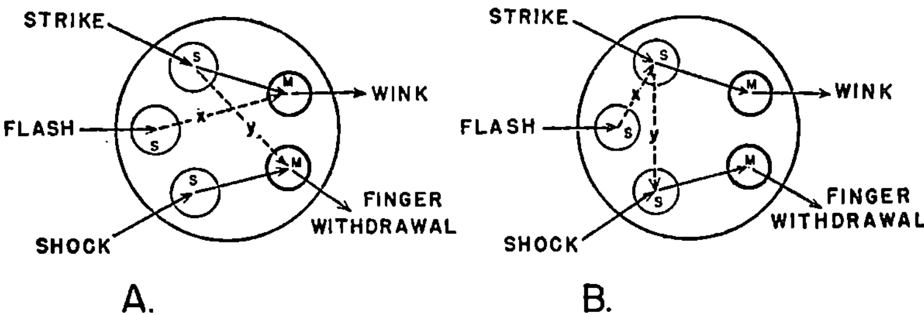

Fig. 4. Comparison of *sensory-motor* and *sensory-sensory* views as applied to Shipley's experiment. In A, the acquired connections, x and y, do not account for the association of "flash" and "finger withdrawal" functions. In B, the acquired connections, x and y, effectively account for the association of these cortical functions.

acquired associative relationships x and y furnish a direct functional connection between the cortical pattern set up by the flash and that controlling the finger withdrawal. Apparently Shipley did not consider this alternative.

In an experiment by Cason (4) on the conditioned wink response, a sound was paired with a shock below the eye. After this training the subjects reported that the shock when presented alone seemed weaker ("and lacking in something") than when it was accompanied by the buzzer. At first this difference had not been present. In order to account for this summation effect, it appears necessary to postulate the development of a sensory-sensory connection.

Other human findings are significant on this point. Recently Leuba (16) reported the conditioning of imagery processes. For example, a hypnotized subject was presented with the snapping of a cricket and the odor of creosote in combination. When awakened and presented with the snapping of the cricket, the subject was caused to sniff and report a peculiar odor. In this case also, the assumption of a controlling cortical pattern centering around a sensory-sensory connection serves to account most adequately for the results.[3]

III. Summary and Conclusions

For some time the problem of the relation between conditioning and selective (trial-and-error) learning has been a source of controversy in the study of learning. Because the conditioning phenomenon can be studied conveniently and described in relatively simple terms the possibility of reducing all learning to conditioning has been an attractive one. Recent evidence on the apparent function of motivation in conditioning has tended to broaden the conditioning concept so that the differences between selective learning and conditioning seem to have been broken down or at least substantially reduced. In this paper we have offered some cogent reasons for retaining the distinction.

Our analysis has served to point out that in the *first stage* of the conditioning *procedure* a neutral stimulus develops excitatory value for a response it previously did not control. The essential condition for this change in the animal is contiguity between the experience of the neutral stimulus and the experience of the unconditioned stimulus. Then a *second stage* may develop. Once the neutral stimulus has become a conditioned stimulus, the response it elicits may be rewarded or punished. Unconditioned stimuli such as food and shock can function in this stage in the same way that "reward" and "punishment" function in problem-box learning. Thus during this stage a selective learning is involved; as a result the conditioning procedure ceases to be unique and assumes the psychological characteristics of trial-and-error learning. Then, as in all selective learning, the critical response becomes more and more specific the more effectively it leads to reward or to escape from punishment.

Before a response can become selected, it must be in the animal's repertoire. It is characteristic of Pavlovian conditioning procedure to utilize a response which is not likely to be available in a given situation unless its unconditioned stimulus is present. However, in

[3] It may be asked why it was necessary to use hypnotized subjects to produce this effect. The answer seems to be that since implicitly aroused images are readily distinguished by the waking subject from direct perceptions, the imagery process is readily suppressed and not reported. Hence the *response* aspect of conditioning affords much better evidence of training and accordingly is commonly emphasized in the literature. However, the significant fact that normal subjects may anticipate the unconditioned stimulus supports our view that the counterpart of Leuba's sensory pattern also exists in the conditioning of waking individuals.

problem-box and other trial-and-error situations, the critical response is available in the situation and a specific unconditioned stimulus is not required to bring it to expression.

Thus selective learning depends upon reward and punishment as well as on contiguity, whereas associative learning requires only contiguity in experience. Failure to distinguish these as qualitatively different forms of learning leads to inconsistencies, since it results in overlooking the possibility that different learning situations may involve the two mechanisms in different ways. We have pointed out that the conditioning procedure first introduces purely associative learning, but then may admit selective learning depending upon certain special conditions. The principal conditions determining this change concern a possible function of the unconditioned stimulus as incentive and probably also the degree to which the critical response is voluntary and thus subject to selection. If these conditions are fulfilled, the conditioning procedure frequently used may result in a pattern which has the characteristics of both associative and selective learning.

Since the first stage of the conditioning procedure is unique in requiring an unconditioned stimulus, we believe that the term "conditioning" should be restricted to the changes occurring during this stage. It is the resulting limited routine type of response which Pavlov originally designated as "conditioned." Performing tests which are in reality problem-box experiments and terming the process "conditioning" does not reduce problem-solving to conditioning, but merely confuses the issue.

Restricting the term "conditioning" in the above sense serves to focus attention upon the nature of the associative process. We have offered evidence and considerations favoring the view that the essential change, as studied in conditioning, involves a new dynamic relationship between the sensory cortical pattern of the conditioned stimulus and that of the unconditioned stimulus. Thus we maintain that the key process is a modified sensory integration. The commonly accepted view of conditioning as depending upon a new sensory-motor connection seems inconsistent with experimental findings. Rather than speak of conditioning as "stimulus substitution" it is more advisable to regard it as the extension or qualitative enlargement of the pattern of stimulation which will elicit the critical response.[4]

[4] Comment by Dr. Maier on Dr. Stephens' paper, "Expectancy *vs.* effect—Substitution as a General Principle of Reinforcement," appearing in this issue:

The latter portion of Dr. Stephens' paper shows that he has been confronted

References

1. Brogden, W. J. Unconditioned stimulus-substitution in the conditioning process. *Amer. Jour. Psychol.*, 1939, **52**, 46–55.
2. ———. Higher order conditioning. *Amer. Jour. Psychol.*, 1939, **52**, 579–591.
3. ———, Lipman, E. A., and Culler, E. The role of incentive in conditioning and extinction. *Amer. Jour. Psychol.*, 1938, **51**, 109–117.
4. Cason, H. The conditioned eyelid reaction. *Jour. Exper. Psychol.*, 1923, **5**, 153–196.
5. Crisler, G. Salivation is unnecessary for the establishment of the salivary conditioned reflex induced by morphine. *Amer. Jour. Physiol.*, 1930, **94**, 553–556.
6. Culler, E. A. Recent advances in some concepts of conditioning. *Psychol. Rev.*, 1938, **45**, 134–153.
7. ———, and Mettler, E. A. Conditioned behavior in a decorticate dog. *Jour. Comp. Psychol.*, 1934, **18**, 291–303.
8. ———, Coakley, J. D., and Shurrager, P. S. Differential effects of curare upon higher and lower levels of the central nervous system. *Amer. Jour. Psychol.*, 1939, **52**, 266–273.

with the same problem with which we have been primarily concerned. His solution also shows some similarities to ours. The basic difference in my estimation is that Stephens' paper represents an attempt to reduce (or elevate) conditioning to trial-and-error learning, whereas our paper attempts to reduce only the latter portion of conditioning (the stage at which reinforcement occurs) to trial-and-error learning. Since he does not break the conditioning procedure into two phases, his interests may turn to a quantitative aspect of the problem. We emphasize a qualitative stage and cannot deal with quantitative aspects until the qualitative stages have been distinctly separated. The stages being qualitatively different demand quite different units of measurement.

That our first stage (association by contiguity) is a real one is supported by Stephens' attempt to explain the conditioned leg flexion in which the leg flexion does not achieve avoidance of shock. According to his analysis such conditioning procedure would lead to a tendency other than leg flexion.

The very fact that this procedure produces responses *simulating* the unconditioned response leads to the development of the notion of conditioning. Only recently has the emphasis been placed upon certain differences between the conditioned and the unconditioned responses. The differences have been explained by regarding the conditioned responses to be anticipatory in nature. Stephens goes a step further in differentiating the two responses and in so doing fails to account for the similarities.

Stephens' criticism of expectancy when used as an explanatory principle is very convincing. It seems that our sensory-sensory integration accounts for instances in which anticipatory behavior is apparent and at the same time does not require the concept of expectancy to explain it.

It also seems that Stephens' analysis of selection would be more convincing if he clearly distinguished between conditioning and problem-box situations. To regard food as an unconditioned stimulus in a problem-box situation in which a string must be pulled so broadens the meaning of the term that its basic connotation is lost.

9. FINCH, G., and CULLER, E. Higher-order conditioning with constant motivation. *Amer. Jour. Psychol.*, 1934, **46**, 596–602.

10. GIRDEN, E. Cerebral mechanisms in conditioning under curare. *Amer. Jour. Psychol.*, 1940, **53**, 397–406.

11. ———, and CULLER, E. Conditioned responses in curarized striate muscle in dogs. *Jour. Comp. Psychol.*, 1937, **23**, 261–274.

12. ———, METTLER, F. A., FINCH, G., and CULLER, E. Conditioned responses in a decorticate dog to acoustic, thermal, and tactile stimulation. *Jour. Comp. Psychol.*, 1936, **21**, 367–385.

13. HILGARD, E. R., and MARQUIS, D. G. *Conditioning and learning.* New York: D. Appleton-Century Co., 1940. Pp. 429.

14. KELLOGG, W. N. Evidence for both stimulus-substitution and original anticipatory responses in the conditioning of dogs. *Jour. Exper. Psychol.*, 1938, **22**, 186–192.

15. KRECHEVSKY, I. Brain mechanisms and variability: I. Variability within a means-end-readiness. *Jour. Comp. Psychol.*, 1937, **23**, 121–138.

16. LEUBA, C. The use of hypnosis for controlling variables in psychological experiments. *Jour. Abn. (Soc.) Psychol.*, 1941, **36**, 271–274.

17. LIGHT, J. S., and GANTT, W. H. Essential part of reflex arc for establishment of conditioned reflex. Formation of conditioned reflex after exclusion of motor peripheral end. *Jour. Comp. Psychol.*, 1936, **21**, 19–36.

18. LOUCKS, R. B. The experimental delimitation of neural structures essential for learning: The attempt to condition striped muscle responses with faradization of the sigmoid gyri. *Jour. Psychol.*, 1935, **1**, 5–44.

19. MAIER, N. R. F. Reasoning in white rats. *Comp. Psychol. Monogr.*, 1929, **6**, pp. 93.

20. ———. The specific processes constituting the learning function. *Psychol. Rev.*, 1939, **46**, 241–252.

21. ———, and SCHNEIRLA, T. C. *Principles of animal psychology.* New York: McGraw-Hill, 1935. Pp. 529.

22. SHIPLEY, W. C. An apparent transfer of conditioning. *Jour. Gen. Psychol.*, 1933, **8**, 382–391.

23. SHURRAGER, P. S., and CULLER, E. Conditioning in the spinal dog. *Jour. Exper. Psychol.*, 1940, **26**, 133–159.

24. SKINNER, B. F. *The behavior of organisms.* New York: D. Appleton-Century Co., 1938. Pp. 457.

25. THORNDIKE, E. L. *Human learning.* New York: Century, 1931. Pp. 206.

26. ZENER, K., and MCCURDY, H. G. Analysis of motivational factors in conditioned behavior: I. The differential effects of changes in hunger upon conditioned, unconditioned, and spontaneous salivary secretion. *Jour. Psychol.*, 1939, **8**, 321–351.

ARTICLE IV

FRUSTRATION THEORY: RESTATEMENT AND EXTENSION[1,2]

N. R. F. MAIER

In presenting a new point of view one must of necessity differ with the thinking of others. The perception of these differences may also vary considerably, so that a communication problem is introduced. The first step in resolving the disagreements resulting from an initial presentation of a viewpoint seems to be to understand the sources of the disagreements, assuming that emotional factors can be regarded as beside the point of a scientific discussion. Some of these differences are purely semantic; others are due to the emphasis the writer puts on the aspect of his views that are new, so that readers may assume that he has neglected or overlooked other important relationships; while the remainder are due to alternate ways of interpreting the same phenomena. The first two sources of disagreement can conceivably be settled by further writing and discussion, but the third needs to be resolved by research. Before new research can adequately settle the relative adequacies of differing points of view, however, the basic differences must be separated from differences resulting from faulty communication.

For the sake of progress in science it is also desirable to consider a fourth source of difference. No point of view remains static, and it is questionable whether an author can reinstate the meaning he originally intended when he later elaborates his position. His views must grow as a result of additional years of speculation, and pointed criticisms help him clarify his distinctions and stimulate him to come to deal with relationships that he may have neglected. Failure to communicate these later developments may unwittingly lead to misunderstandings and disagreements.

The frustration theory of the writer (16) seems to be a case in point.

[1] Reprinted from *Psychological Review*, vol. 63, No. 6, pp. 370–388, by permission of the Editor and the publishers.

[2] I wish to acknowledge my indebtedness to my colleagues Professors R. W. Heyns and E. B. McNeil, whose critical comments and constructive suggestions have greatly improved this paper.

Since its presentation in book form, there have been many descriptions, verbal statements, and critical research reports that make assumptions about the theory that seem unjustified to the author of the theory. Without attempting to determine which of the causes mentioned above makes for the discrepancy between the thinking of the author and that contained in restatements of the theory, it seems that some clarification and sharpening of distinctions is in order. It is hoped that such an analysis will clarify the thinking of proponents and opponents, obviate controversy due to faulty communication, and point up crucial experiments that can be formulated. Experimental findings will be used either to separate alternate points of view or to indicate the points at which alternate views are tenable. Since the theory of frustration cuts across learning, motivation, therapeutic processes, and personality theory, the concepts involved need to find a proper place in a general behavior theory.

Clarifying Questions

Is a Frustrated Person Without a Goal?—The writer's book title, *Frustration: The Study of Behavior Without a Goal,* may be taken to mean that a frustrated person has no goal. This was not the intent in selecting the title. The thought that the title was meant to convey was that a frustrated person's *behavior* is without a goal, i.e., the behavior sample under discussion lacks goal orientation. Suppose someone steps ahead of another who is in line waiting to purchase a railroad ticket. An argument ensues, followed by a fight. Both men are arrested and neither makes his train. However, the need for a railroad ticket remains for each, and the more intense the need is, the more frustrating is the interruption. The fighting, however, is not ticket-getting behavior. According to the theory, the fighting is frustration-instigated behavior. This position differs conceptually from a viewpoint that states that a new goal replaces the one of obtaining a ticket. Examples of possible substitute goals in this example are (*a*) obtaining relief from tension, (*b*) demonstrating superiority, (*c*) teaching a bully his place, and (*d*) doing injury to an enemy. Even though relief or satisfaction may sometimes result from such aggression, this fact should not be accepted as proof that the desire either for relief or for these satisfactions was an instigator. If such causation exists, it must be shown that it is an essential requisite to such behavior. According to frustration theory, adaptive or need-satisfying behavior is not assumed. Motivation therefore is separated from causation as an explanatory concept. Research is needed to support or reject these basically differing assumptions.

Is the Frustration-Fixation Relationship Circular?—Circular thinking in this instance would seem to run as follows: frustration produces fixation, and the appearance of a fixation in an animal indicates that it has been frustrated. To the extent that this relationship has been communicated, the author has been lax in that he has failed to clarify the process by which he reached certain conclusions. An attempt to reproduce the logical steps which led to the qualitative distinction between behaviors produced by frustration and behaviors produced by motivation learning is given below.

1. When rats were placed in an insoluble discrimination problem situation, all of them adopted a specific response of some sort (usually a position response), and they were persistent in executing this response when forced to make a choice. The consistency with which animals executed the response selected was a new experience to the observers, but this phenomenon did not suggest fixation. What was important was the fact that when a soluble problem subsequently was presented to the animals, some abandoned their position responses and adopted the discrimination response within standard learning limits (less than 100 trials), whereas the majority never learned (600 trials) and instead persisted, without a single variation, in their initial responses. The bimodal character of these learning scores was striking, particularly the fact that the majority of animals could not learn the problem at all.

2. In order to determine whether the failure of these animals to learn was due to the basic strength of the first response, control groups were added. These groups were required to learn an initial response that matched the responses spontaneously adopted by the experimental groups in the insoluble problem situation. When later tested on the discrimination problem, it was found that the majority of them learned the new problem in standard time, but a minority failed entirely. (It was the practice of earlier experimenters to discard the few animals that failed to learn problems of this kind.)

The following facts needed to be explained: (*a*) unexpected and apparently unreinforced responses of unusual persistence occur spontaneously; (*b*) the number of animals having such responses varies with the situation, there being few such animals in simple learning problems and many in insoluble problems; and (*c*) animals that learn and those that fail to learn form two discontinuous populations.

The following hypotheses were developed to explain the findings.

1. Responses that persist when adaptive alternatives are clearly available are different in kind from habits, since habits are subject to change. The new type of response was given the name *abnormal*

fixation, so as not to confuse it with the term *fixation* in habit formation. (For experimental purposes the criterion adopted for abnormal fixation was a specific unadaptive response that persisted beyond 200 trials, or more than twice the number required for substituting one habit for another. Tests had shown that animals that fail to learn in 200 trials also fail with more trials.)

2. Frustration causes abnormal fixations. (This hypothesis was suggested by the fact that an insoluble problem should be frustrating to more animals than a learning problem.)

3. Frustration is a physiological mechanism that determines the nature of behavior when the frustration threshold has been passed over. (The possibility was implied that the autonomic nervous system dominated behavior.)

4. Individuals differ in their frustration thresholds.

Thus the concept of abnormal fixation was developed to account for certain experimental findings, particularly the bimodal nature of certain distributions in scores. A comparable term is the concept of a habit. Reinforcement produces habits, and the appearance of a habit in an animal indicates that it has been rewarded after making the response. However, one does not regard this relationship as circular.

Once, however, the causal relationship between frustration and fixation has been established, one can use differences in the number of animals showing fixated behavior as a measure of the number of frustrated rats. This was done in subsequent studies, in which the concept of the frustration-fixation relationship was used as a working hypothesis.

For example, in later studies these conclusions were drawn.

(*a*) Increases in the length of the insoluble problem period increase the number of animals whose frustration thresholds will be exceeded (24).

(*b*) An increase in the difficulty of a learning problem increases the number of animals that will be frustrated (21).

(*c*) Frustration produces fixated behavior, which has a degree of rigidity (simulating compulsive behavior) unlike that found in habits (9, 15, 16, 26, 27, 28).

(*d*) The presence of bimodal distributions in stressful situations should be examined to determine whether or not frustration has occurred (9, 21, 27, 28). (Bimodal distributions may have a number of causes and frustration is one of them. The appearance of bimodal distributions requires an explanation in any case, and failing to test for it amounts to making an unproved assumption.)

Each of these conclusions is based upon the number of rats showing abnormal fixations, but one can argue that reasoning has been circular only if one questions the original conclusion, which was that the fixations in the initial experimental condition (25) were produced by frustration and not by learning. To question this conclusion, one must go to the facts upon which it rests, not to subsequent studies. The purpose of the follow-up studies was one of extending our knowledge of the frustration mechanism, rather than one of repeating or re-establishing the case for the qualitative distinction between the behavior of frustrated and nonfrustrated animals.

Is Escape from Punishment a Goal?—Can we assume that fixations are caused by an animal's attempt to escape punishment on the jumping stand? This represents an alternative explanation of why an animal may persist in an apparently unadaptive response and fails to learn the discrimination problem. According to this view, the animal is rewarded by its response because it escapes punishment, and hence it fails to learn the better way to escape that is offered. What must be explained, however, is the bimodal distribution of scores and why this kind of response differs from reinforced responses. There are many explanations for persistent behavior, and Dollard and Miller (3) make this their task when they reduce abnormal fixations to escape learning. Researches on avoidance learning present evidence in support of the persistence of behavior produced by punishment, and the question may be raised as to whether the frustration-fixation hypothesis is an alternative concept for interpreting the data in their experiments. It is also possible that the interpretations may be supplementary rather than in conflict, particularly if one entertains the notion of two behavior mechanisms: motivation and frustration. Thus punishment may serve either as a negative incentive or as a frustrating agent, depending on its intensity.

Is Anxiety Reduction a Goal?—Anxiety reduction has been previously discussed (19) as an alternative explanation for abnormal fixation phenonema, and it, too, assumes the task of explaining persistent behavior. Anxiety reduction is described as a kind of reward for the fixated response, which means that the response is actually reinforced by the tension reduction that follows the jump. All that is required is the postulation that animals have a *need* to reduce tensions. Why an even more rewarding response (one that includes tension reduction and an unpunished jump to food) is not adopted, as occurs in other learning problems, also needs to be explained.

However the bimodal nature of the distribution of scores seems to be a difficulty that prevents the anxiety reduction theory from being an alternative viewpoint. Maier and Ellen previously pointed out that Farber's data (7) contained a bimodal distribution, and Farber has attempted to explain his exceptional cases by genetic strain differences and the probability of skewed distributions (7, 8). Whether or not the bimodal distributions in other sets of data (25, 27, 28, 31, 32, 37) can be reduced to artifacts of a similar nature remains a problem.

Can Partial Reinforcement Produce "Fixated" Responses?—Partial reinforcement has been used to explain the fact that extinction of conditioned responses formed on the basis of 50 per cent reinforcement is slower than on the basis of 100 per cent reinforcement. If the strength of fixated responses could be reduced to partial reinforcement, it might be unnecessary to postulate a separate frustration mechanism. Since an insoluble problem also involves 50 per cent reinforcement (and 50 per cent punishment as well), and since fixations are produced more frequently in the insoluble problem test than in the soluble problems, it seems plausible to link fixations either with the frustrating effects of the insoluble problem or the partial reinforcement of the insoluble problem.

Wilcoxon (37) performed an experiment to test the applicability of the partial reinforcement concept to persistent behavior in the jumping apparatus. He used three groups of rats, two of which were duplicates of groups used in our initial experiment (25). His Group I was like our Groups I and II which learned under standard reinforcement conditions (25). The animals were required to learn a position response in a Lashley-type jumping apparatus, and they were rewarded for a jump to the correct side and punished for a jump to the incorrect side. This situation gave the animal punishment on 50 per cent of the initial responses, and this percentage declines to 0 per cent as the animal learned the problem.

His Group III was like our Group III, since the animals were exposed to the insoluble problem situation. These animals were confronted with a problem situation in which neither card nor position was consistently rewarded. As a result animals received punishment on 50 per cent of their trials throughout the training period.

His Group II was new, and was required to learn under conditions of partial reinforcement. The rats were treated like Group I animals except that a jump to a particular side was rewarded on every *other* trial. Thus the animal was punished *every* time it jumped to the negative side and *every other* time it jumped to the positive side. The pattern of reward and punishment resulted in animals' learning to jump to the positive side. This procedure results in animals' being punished on three out of four trials when they jump at random during the early stages of learning, and on two out of four trials when they have mastered the problem.

Table 1 shows the three kinds of groups used by Wilcoxon, the initial and final punishment-reward ratios for each, the percentage of fixated position responses obtained by Wilcoxon, and the percentage of fixated position responses obtained by Maier, Glaser, and Klee (25). For comparable groups the results are in agreement, and in

TABLE 1.—FREQUENCY OF ANIMALS WITH FIXATED BEHAVIOR IN WILCOXON'S STUDY

Initial Training Situation	Initial Punishment Frequency (Random Trials)	Lowest Punishment Frequency After Learning	Wilcoxon Per Cent Rats Fixating	Maier *et al.* Per Cent Rats Fixating
Position reward				
Wilcoxon's Group I	50	0	38	30
Insoluble problem				
Wilcoxon's Group III	50	50	58	64
"Partial" reinforcement				
Wilcoxon's Group II	75	50	92	—

neither experiment was there a problem of classifying fixated and nonfixated behaviors. The point of difference arises over the interpretation of the fact that the partial reinforcement group had 92 per cent fixated position responses whereas the insoluble problem had 58 per cent. Wilcoxon argues that partial reinforcement makes for fixations, and that the insoluble problem is a variation of a partial reinforcement situation. The experiment is not a crucial one, because according to frustration theory the greatest number of fixations should occur in the group that received the most punishment, and this is the so-called partial reinforcement group. That the frequency of fixations increases with the frequency of punishment has been reported in a number of studies (24, 27, 31, 32). Unpublished data from Takeshi Shimoyama of Tokyo show a consistent increase in the number of fixated rats with increased frequency of punishment. He used four groups (10 rats each) with punishment frequencies of 0, 30, 50, and 70 per cent, and obtained position fixations in 0, 2, 3, and 6 rats, respectively.

The partial reinforcement explanation has an advantage in that it does not make it necessary to postulate a new behavioral mechanism, whereas frustration theory does not need the concept of partial reinforcement to explain the effects of punishment and inconsistencies with reinforcement results. What is most important, however, is that the frustration mechanism is needed to account for the bimodal distribution of scores obtained in all of the groups; since partial reinforcement applies to all animals in a group, it should not cause a split.

What Factors Influence the Number of Animals That Will Fixate a Response?—Bimodal distributions are obtained in experiments of discrimination learning, but the proportion of animals that are unable to learn, *vs.* those that learn readily, varies. Chance variation and strain differences are to be expected, but a number of other causes have been experimentally demonstrated.

The following results indicate some of the relevant factors.

1. Punishment on every trial produces more fixations in rats (also more rapid learning for those that learn) than punishment on every other trial (27).
2. Sixteen days in an insoluble problem produces more fixations than eight days (24).
3. Mild electric shock given to human subjects on 75 per cent of the trials in an insoluble problem for 50 trials produces more fixations than the same absolute number of shocks distributed over 150 trials (31).
4. Feeding after electric shock results in a smaller number of fixations than shock alone (7).
5. Guidance during the early stages of learning prevents fixations from forming later on when the rats are placed in an insoluble problem (20).
6. Difficult learning problems result in more fixations than easy learning problems (21, 25, 27).
7. Learned position responses that correspond to the natural preferences of the rat are less likely to become fixated than position responses that are on the side opposite the natural preference (25).
8. The use of the word "wrong" instead of electric shock, to indicate an error, produced fixations in human subjects about as frequently as electric shock to the fingers (31, 32).
9. Failure in fulfillment of expectancies seemed to be an important influence in the formation of fixations in a discrimination learning problem in rats (22).

Since punishment and experience of failure are associated with situations producing high rates of fixation, the frustration-fixation hypothesis is supported. It also seems necessary to regard the transition from problem-solving or trial-and-error behavior to stereotyped behavior to be rather sharp, or even of an all-or-nothing character. This suggests that there is a frustration threshold.

A further point of interest and possible importance is mentioned here in the hope that it may encourage other experimenters to report similar observations. This is the role of the experimenter in influencing the behavior of animals, particularly under stress. Some years ago two research assistants were working in adjacent rooms on related problems, each with three groups of twelve or more rats from the same colony, over a period of a semester. One of them obtained the usual number of fixated position responses (over 50 per cent) in each of the successive groups with which he worked; the other was unable to obtain a single fixation. Although they compared procedures on preliminary training, methods of testing, and other general routines, they were unable to determine the reason for the differences. Motivational considerations also failed to throw light on the matter. The researcher who was unable to obtain fixations required them for his doctoral dissertation, so that his results did not correspond with his motives. However, it was discovered that he felt sorry for the rats, and this may have caused him to pet the rats between trials somewhat more than other researchers. This possible influence might be analogous to feeding after shock, which reduced the number of fixated rats in Farber's experiment (7).

Can Abortive Behavior Prevent Punishment?—The fact that animals learn to jump abortively when placed in an insoluble problem serves to reduce any punishment that a locked window produces. This method of jumping may create a condition that prevents rats from discovering that the problem is soluble, and hence interferes with learning the discrimination problem. To test this possibility Wilcoxon (37) separated the rats that learned from those that failed to learn (fixated). He found that the latter group had more abortive responses than the former, and concluded that this group difference supported the above hypothesis. This explanation could conceivably account for the bimodal phenomenon (animals that learn *vs.* animals that fixate). However, Wilcoxon's conclusions must be questioned because his learners came primarily from Group I and his nonlearners from his Groups II and III. The fact that his learners and nonlearners came from different groups makes it impossible to evaluate the function of the basic variable which he purposely introduced into the experiment to test the influence of partial reinforcement. But even if this difference in abortive jumping between learners and nonlearners is obtained under comparable conditions, it is not a crucial comparison because it does not exclude the equally plausible but opposite interpretation, which is that failure to learn causes abortive jumping. In order to distinguish cause from effect, it is necessary to study how abortive jumping develops in relation to the progress made in learning.

Maier, Glaser, and Klee (25) showed that animals placed in an insoluble problem soon develop abortive jumping, which increases as the animal is retained on the insoluble problem. However, when the problem is made soluble, the abortive jumping is expressed differentially—that is to say, it declines in frequency for the trials on which the positive card is on the side to which the animal has been jumping, but increases for the trials on which the negative card is on the side to which the animal has been jumping. This, then, is the first evidence that the rat has learned to discriminate between the stimulus cards. For animals that do not show fixations, this *differential* abortive jumping begins to appear about the same time that the position response is replaced by the discrimination response; in other words, the animal now follows the positive card from side to side, and abortive jumping disappears because the animal no longer jumps to the negative card. For animals that fixate, the differential abortive jumping persists and becomes even more clear cut, since this discrimination between the stimulus cards is not followed by abandonment of the position response. The rat with the fixated position response fails to follow the positive card from side to side, and thus jumps abortively whenever the negative card is on the side of the fixation. The greater total of abortive jumps in rats with fixations than in rats that learn, therefore, may be an artifact, i.e., they jump abortively more often because they jump to the negative card over a longer period of time.

In a recent study (23) it was possible to analyze the data to test the hypothesis suggested by Wilcoxon. Table 2 shows that there are 73.3 per cent abortive jumpers among rats with fixations, and 58.6 per cent abortive jumpers among rats that give up the position response. Although the difference in percentages is not significant, the trend might suggest that abortive jumping hindered learning of

TABLE 2.—COMPARISON OF ABORTIVE JUMPING IN RATS WITH FIXATED AND NONFIXATED POSITION RESPONSES

	N	Per Cent Jumping Abortively	Per Cent Showing Differential Abortive Jumping	Average Days to Learn Differential Abortive Jumping
Rats with Fixated position responses (nonlearners)	75	73.3	96.4	6.6
Rats with Nonfixated position responses (learners)	29	58.6	94.1	6.1

the discrimination response. However, one may also measure learning by the *differential* abortive jumping. If jumping abortively makes the rat less prone to discriminate between the positive and negative cards, then initial abortive jumping should not lead to differential abortive jumping. However, Column 3 shows that 96.4 per cent of the animals that persist in the position response, and 94.1 per cent of the animals that adopt the discrimination response, show differential abortive jumping. These percentages are too high and too much alike to suggest that abortive behavior can be a causal factor in the phenomenon of fixation. The rates of learning to discriminate, as measured by the number of days required to show differential abortive jumping, are shown in Column 4. These measures also are too similar (6.6 days *vs.* 6.1 days) to support Wilcoxon's hypothesis.

Can Several Learning Concepts Be Combined to Reinterpret the Michigan Findings?—Wolpe (38) has made a systematic attempt to reduce the phenomena of abnormal fixation to learning principles by dealing with the experimental evidence on which the concept was based. He draws freely from various learning principles, and these, if not integrated, may create the dilemmas that Eglash (5, 6) has discussed in his attempts to test the adequacy of various learning concepts for interpreting the experimental findings upon which frustration theory rests. Wolpe's case will be made stronger if he can reply to Eglash and also meet the objections and questions raised below.

Wolpe accepts the anxiety reduction theory in dealing with Farber, Mowrer, and others, but at the same time he recognizes the inadequacy of the theory for dealing with the Michigan data. In order to interpret these data, he postulates that the air blast is the primary drive and that jumping reduces the air-blast drive and hence reinforces jumping. "When jumping in a particular way has thus been repeatedly reinforced it becomes firmly established as the habitual response to the air-blast stimulus, and the more firmly it is established the weaker does the competing alternative response tendency become" (38, page 114). Thus the animal does not use its knowledge of the learning discrimination response because the air blast, not the stimulus card, is the stimulus. (It is assumed that Wolpe would deal similarly with the experiments in which electric shock was used to force a response.) This escape type of approach to discrimination learning postulates a primitive kind of S-R behavior in which one stimulus dominates the behavior to the exclusion of all others. Such stimulus dominance is not borne out by the behavior of rats in other learning situations where almost every change elicits curiosity. Five experiments on fixations seem to be in conflict with this interpretation, but some of these were too recent to be accessible to Wolpe.

Ellen (4, 17) confronted rats having fixated position responses with three stimulus cards, two of them being on the side of the animals' position fixation, the other on the side opposite the position fixation. When one of the former two was positive and the other negative, the animals tended to jump to the positive card. However, they would never select the positive card that was placed on the side opposite the fixation. Thus the animals *will* jump to the positive stimulus card and avoid the negative, but only if this discriminatory behavior does not interfere with the fixated position response. Giving the animal a chance to succeed within the scope of its fixation was a type of therapy Ellen was evaluating in these experiments, and he obtained positive results of a significant magnitude. For the purposes of this discussion, his findings demonstrate that the animals' behavior is not entirely dominated by escape in experiments with the Lashley apparatus.

Klee (14) obtained fixations in rats by using only the hunger drive to force a response. Although this was an extremely time-consuming process, and retraining had to be instituted to prevent actual starvation, fixations were produced. In this situation, a jump from the stand could not have served a reinforcement value since hunger tension was not reduced by a jump. Perhaps one might make a case that the animal was rewarded by being returned to its cage after being on the stand for many hours, but this supposition seems somewhat tenuous.

A third experiment that bears on the issue raised by Wolpe is one by Feldman (9). He has shown that if rats are required to jump to stimulus cards on some trials, and on others are presented with short pathways leading to the same cards, they will respond differently to the cards under these two conditions. Some rats had one fixation for jumping and developed another for walking when a pathway was introduced, but the same primary stimulus (electric shock to tail) was used for forcing both the walking and jumping methods for leaving the jumping stand. Interesting, too, is the fact that rats with fixated jumping responses (which Wolpe would describe as the learned responses to the primary stimulus that was applied to the jumping stand) never attempted to jump when the walking pathways were in place. Secondary stimuli influence the primary response, and this type of adjustment to a larger situation is denied by Wolpe when he speaks of secondary drives being unable to compete with the primary drive.

The fourth obstacle confronting Wolpe's basic assumption is the fact that the number of fixations increases as the second response to be learned is increased in difficulty (**21, 25**). If animals merely responded to the primary drive and were not influenced by the discrimination problem with which they are confronted, the number of fixations should not be so directly influenced by the cards used.

Finally, it would seem that rats that had previously acquired fixations should develop them more quickly than others. It has been found (**20, 28**), however, that rats cured of fixations are less prone to develop them a second time. Although a clearly formulated explanation for this finding has not as yet been supplied, it is in conflict with Wolpe's position. It is the writer's opinion that guidance therapy increased the frustration threshold, and hence reduced the occurrence of fixations.

Experimental findings that Maier and Ellen (**19**) described as crucial in invalidating the anxiety-reduction theory are dealt with by Wolpe in an interesting manner. Although the basic assumption on which his viewpoint rests has been questioned above, his interpretations of the data are sufficiently ingenious to warrant consideration and comment.

(*a*) The fact that more animals fixate the response in progress when 100 per cent punishment is used for breaking it than when 50 per cent punishment is used, is explained by saying that the 100 per cent punishment condition causes the rat to require more of an air blast to elicit a jump and hence to experience greater reinforcement of the ongoing jumping response[3] than does the 50 per cent punishment condition. It is not known whether 100 per cent punishment produces greater resistance than does 50 per cent punishment, but even if this were true, the argument seems to hang on tenterhooks. Would it not also follow that rats in the 100 per cent punishment condition would more readily learn to jump, having had more reinforcement? As a result they should learn sooner and therefore reduce the amount of air blast. Reducing the air-blast exposure would decrease the reinforcement of the 100 per cent group, thus allowing the 50 per cent punishment group to catch up.

(*b*) Rats in the 100 per cent punishment group that give up their position responses do so on fewer trials than rats that are in the 50

[3] Wolpe describes the ongoing response as a fixation, but he is actually referring to position stereotypes and position habits. This error is beside the point for the purpose of the present discussion, but is mentioned in order to correct a frequent misconception. According to frustration theory, not every persistent response is an abnormal fixation, but only those that have been shown to fall outside the normal range. The most frequent criterion is 200 trials of no change when a correct response is clearly available.

per cent punishment group. This is explained by saying that only rats that learn quickly avoid developing the fixations described above. This is in general agreement with the explanation Maier and Klee used (27), except that they indicated that pressure from behind (shock) increases motivation until the pressure reaches the point where it is frustrating to animals. Animals would differ at which point pressure ceases to be motivating and becomes frustrating.

(*c*) To explain the bimodal distribution of scores, Wolpe uses the principle of response oscillation. When the response strength of an alternative response is strong enough, it may occasionally be expressed, and the fixation may be broken. In other animals the alternative response does not happen to occur while the relative habit strengths are right. This apparently means that if an animal tries out another response, it can discover the virtues of this response and adopt it if the strength of the primary response is not too great. However, this concept seems to overlook the compulsive character of the fixated response.

(*d*) Wolpe deals with some of the evidence that has been used to make a case for the compulsive behavior. He points out that an animal does not jump to an open window when the opportunity is given because the response to the air blast is so strong that it completely dominates the animal's behavior. This is the kind of attention one may expect in panic, but it seems to be a rather extreme assumption to state that a rat in a problem-solving situation can be so completely controlled by a stimulus.

(*e*) Failure of trial-and-error behavior to occur when a fixated response is punished is given as evidence that a fixated response is a strong habit. He points out that the rigidity of the position response depends upon the fact that reinforcement occurs on all occasions. Thus the compulsive character of behavior is on a continuum with strong habits.

(*f*) The fact that guidance may effectively break a fixation in a few trials is not regarded as an unusual phenomenon by Wolpe. Guidance, he points out, makes an alternative response possible and so the very strong habit is dropped for another. Why do relative habit strengths sometimes change so quickly, while at other times the outstanding characteristic of competing responses is the complete domination of a single habit? Are the explanations for points (*e*) and (*f*) parts of one learning theory, or do they represent principles from competing learning theories? Detailed analysis of the data on the effects of guidance demonstrate that an animal does not adopt an alternative response immediately; rather, the effect of guidance is that it causes the rat to give up the fixated response. Only later is the new response learned. Maier and Klee (28) have shown that guided animals show a gap between giving up the fixated response and adopting the new response, whereas nonguided animals do not give up the response in progress until an alternative is practically learned.

Wolpe does not claim that frustration theory has been disproven by his discussion. Rather he feels the theory is unnecessary because an explanation in terms of learning is possible.

Can the Difference Between Fixations and Habits Be Reduced to Two Kinds of Learning?—In 1935 Maier and Schneirla treated conditioning and trial-and-error learning as qualitatively different forms of learning, and discussed the topics in separate chapters of their book (29). Later (30) they pointed out how conflicting evidence in learning data could be reconciled by making a qualitative distinction between trial-and-error learning, which always requires reinforcement in the form of reward (or punishment), and Pavlovian conditioning, which requires reinforcement in the form of the presentation of the unconditioned stimulus. Confusion arose because in some instances the same object (food) could serve either as a stimulus for salivation or as a reward for the same response. Skinner (35) later found a similar distinction (respondent *vs.* operant conditioning) essential to his organization of data, while Mowrer (33) raised the same issue when he postulated the dual nature of learning: conditioning of autonomic responses (such as fear reactions) and instrumental (trial-and-error) learning. Recently Solomon and Wynne (36) have used this dual nature of learning as a point of departure for explaining the rapid acquisition and unusual persistence of avoidance responses. They believe that different sets of learning concepts must be used when pain-fear stimulation is intense enough to cause trauma, and when dealing with mild avoidance-producing stimuli.

In introducing stimulus intensity (classified as trauma-producing) as a learning variable, Solomon and Wynne must assume that there is a threshold that has to be exceeded to obtain the change, and they indicate that a state of anxiety, with the resultant autonomic responses and feedback, is produced with such stimuli. This viewpoint is in the direction of the frustration theory, but it fails to take the complete step of postulating a dual mechanism. It can explain why mild and intense punishment produce different reactions, and, if the threshold concept is clarified, it can explain bimodal distributions. The theory is too limited, however, to incorporate regressive and aggressive behaviors, and it does not escape the onus of postulating a qualitative distinction, since behavior developed in a state of trauma differs from that learned under nonanxiety-producing conditions.

There is considerable evidence which indicates that all learning data cannot be incorporated under one set of principles, but the nature of the essential difference is controversial. The fact that the Pavlovian type of conditioning experiments often utilized autonomic responses may have caused Mowrer, as well as Solomon and Wynne, to make the separation along physiological lines. However, a conditioned eyewink is an example of strict Pavlovian conditioning, but it is not associated with anxiety. It would also be inappropriate to classify eyelid conditioning with instrumental learning.

Extensions of Frustration Theory

Physiological Basis for Frustration Mechanism.—In describing the frustration mechanism, emphasis has been given to differentiating it from the motivation process at the behavior level. Is there a physiological justification for postulating a second type of intervening variable? The one that naturally comes to mind is to associate frustration with the autonomic nervous system. The concept of threshold is associated with emotional behavior and the arousal of the autonomic nervous system. Emotional behavior is often described as irrational and uncontrollable, and placed in contrast with rational behavior. Although frustration and emotion in many respects are overlapping concepts, they are by no means equivalent terms. For example, the emotions include love and pleasant feeling tones as well as fear, hate, and unpleasant feeling tones; whereas frustration is usually used as a more specific and less inclusive term. For the present it seems appropriate to link rage and terror with frustration, but to exclude the pleasant emotions and mild emotional feeling tones. Since frustration theory clearly separates hate and intense fear states from emotions of love, it becomes imperative to find differing autonomic reactions during states of love and hate if support for this physiological mechanism is to be obtained.

Jost has analyzed a variable in polygraphic measurements that may throw new light on this problem (11). He emphasizes the *patterning* of the various polygraph indices rather than the individual measures of heart rate, blood pressure, galvanic skin resistance, and muscle tension, and describes some interesting individual differences in the manner of reacting to conflict. Using stressful and frustrating situations, Jost, Ruilmann, Hill, and Gulo (12) were able to differentiate between the autonomic reactions of hypertensive patients and a normal control group. Jurko, Jost, and Hill (13) also obtained significant differences in physiological measures between nonpatient, psychoneurotic, and early paranoid schizophrenic groups. The authors present evidence favoring differences in patterns of discharge between the groups, and suggest the concept of levels of energy discharge. Psychoneurotics tend to show hyperadaptation, whereas schizophrenics show disadaptation. They used the Rosenzweig Picture Frustration Test to produce the emotional changes. These studies are suggestive in that they reveal that persons suffering from emotional problems show physiologically different reactions to frustration than normals. In an earlier study, Sherman and Jost (34) had found physiological measures to be sufficiently unique to serve a diagnostic purpose.

Jost (11), as well as Solomon and Wynne (36), stresses the notion that the autonomic functions may influence higher neural centers by some feedback mechanism. This type of control over higher centers would seem to be consistent with frustration theory, and studies along this line should lead to further development and refinement of the theory.

In relating the autonomic nervous system and the frustration process one cannot assume that autonomic functions are absent when a state of frustration does not exist. Rather, the experimental question becomes one of determining whether a level, or unique patterning, of autonomic functioning can be associated with frustration. Certainly pleasant and unpleasant feeling states can occur without assuming frustration. At what level does anger change to rage, fear to terror or panic, and choice behavior to compulsion? Frustration theory would demand some kind of sharp transition, either in the form of a dropping out of voluntary control mechanisms or a sheer dominance of autonomic processes as provocation exceeded a certain point. The possibility of an intermediate condition, such as conflict between the motivation and frustration processes, needs exploration, and it may be the kind of mechanism that underlies the state of anxiety.

Another physiological mechanism that conceivably may be associated with the frustration process is suggested by the researches of Coghill (2). It will be recalled that he found behavior changes in the developing salamander to run parallel with neural growth; new behaviors appeared rather suddenly as essential neural links were formed. According to Coghill's work, the gross bodily functions develop first, and require only a limited number of cells in a chain of neurons. For example, the limb responses of the salamander occur in conjunction with waves of contraction that move down the side of the body. A single stimulus thus releases this total behavior.

Later, as the limbs develop further, additional neurons are grown so that each muscle group has its own specific neurons and therefore controls a smaller amount of behavior. Eventually each movement has its own controls. The process of development therefore is from gross or total behavior to specific or small units of behavior. This view of the developmental process, known as *individuation,* is in contrast to the view that treats the reflex as the unit of behavior.

According to Coghill's analysis, it follows that each muscle group may be activated either as part of the larger pattern or as a separate response. Specific or individuated behavior depends upon the acquisition of additional neural mechanisms which are superimposed upon the more primitive mechanisms controlling the grosser patterns. However, the newer responses can make their appearance only if there is a successful inhibition of the older patterns. Thus mature behavior is both a process of excitation of the more recently developed pathways and a process of inhibition of the older pathways.

Up to this point the anatomical evidence is fairly rich, and the speculation is supported by the evidence. The next step requires merely that we assume that frustration removes inhibition and permits the grosser responses to take over in much the same ways that removal of the cortex removes spinal inhibitions. Since frustrated responses are more global and less subject to cortical control, the required assumptions are not unreasonable. Certainly the general notion seems worthy of further refinement.

If the evidence for qualitative difference in behavior is sound, it becomes important to locate the physiological basis on which it rests. Research along this line could do much to increase our knowledge of the control of behavior.

In attempting to relate frustration-instigated behavior both to the autonomic nervous system and to grosser segments of unindividuated behavior, a relationship between emotion and primitive mass behavior is implied. This connection is not out of line with basic facts, because emotions are integrated at the thalamic level, they are general or even total responses, and they are pretty much beyond voluntary control. Since the cortical mechanisms are new and superimposed upon the phylogenetically older subcortical structures, it is understandable if their influence is confined to ideation and motivated or rational behavior.

The Function of Catharsis.—The benefits attributed to catharsis vary and they perhaps are accepted more readily by the clinician than by the theoretician. From the point of view of frustration theory, catharsis should have a therapeutic value of two sorts.

When in a state of frustration, during which the autonomic nervous system sends sensory impulses (feedback) to the central nervous system, an individual's behavior becomes the end of a sequence. The behavior is the organism's response to the frustrating condition as it exists at that moment. Since the frustrated behavior is the final link in the causal chain, anything that facilitates its expression tends to end the cycle. Behaviors released along harmless channels (where there is no retaliation) most effectively remove the stimulus conditions setting off the state of frustration, whereas forms of physical violence that are met with superior force are least beneficial because they set up a new cycle of frustration. The unit of behavior controlled by the frustration process is in contrast with that controlled by the motivation process, because the latter requires a goal or need reduction to terminate its cycle, and behavior expressed serves as a means to this end.

The second way in which catharsis serves a therapeutic function according to frustration theory is through a clarification of the problem. As long as a state of frustration exists, the autonomic nervous system sends distracting sensory impulses to the brain. These are experienced as feelings (in accordance with the James-Lange theory) and tend to confuse the cognitive picture of the problem situation. Clarification of the perception of the actual problem should occur, both because verbal expression would tend to unscramble the two conflicting sources of data (externally *vs.* internally aroused sensations) and because general expression should reduce the extent of the confusing feeling data.

The Unconscious and the Logic of Feeling.—Freudian theory describes the need to bring the unconscious mind to consciousness in overcoming compulsions. Many forms of conflict are therapeutically treated as though they were the result of contradictory influences of the conscious and the unconscious minds. In brief, two different kinds of influences on behavior are postulated. Frustration theory sets up a similar dichotomy: the logic of ideas (i.e., of the mind) and the logic of feeling (i.e., of the viscera). Since these logics are qualitatively different, an aspect of therapy in frustration theory depends upon removing the conflicts between thinking and feeling.

A little further speculation brings these two sets of dichotomies close together. Feelings are vague and difficult to describe, but most compelling. Was it not natural for a theorist, writing at the time when psychology was dominated by philosophy, to describe these confusing and yet compelling forces as an unconscious mind? In the writer's thinking this simple step of substituting feeling for the unconscious mind has given him a greater appreciation of the original psychoanalytic concepts, and at the same time it clarifies the place of emotion in psychological conflicts.

Preoccupation with the Self and Therapy.—If the state of frustration is characterized by autonomic feedback, it follows that feelings will dominate consciousness at such times; as a result, perceptions and thoughts will have a subjective rather than objective reference. The frustrated person, therefore, must be unusually aware of a mass of feelings or sensations of internal origin. When these sensations are excessive, they would occupy his attention and consequently make him less aware of the external world as well as insensitive to the feelings of other persons. His social behavior would appear selfish and inconsiderate because he would be preoccupied with his own feelings and his relationship to other persons, rather than concerned with problems involving relationships to external objects. He would be a good observer of how another person's actions affect him, but quite unaware of the way his own behaviors affect the feelings of others. Such a person would be preoccupied with himself and the nature of the self.

Viewed in this manner, excessive preoccupation with the self would tend to make a person ineffective in group endeavors. A person who contributed to a discussion only for the purpose of gaining something for himself or even for the purpose of making an impression on others would not be functioning at his best in assisting the solving of external problems. Behaviors such as self-aggrandizement and self-effacement also seem to be by-products of an excessive awareness of the self, and serve only incidental purposes.

Even the solving of human-relations problems would not be promoted by a person whose perceptual organization was dominated by sensations of internal origin. Any sensations that tended to increase the awareness of the self would be inclined to cause a person to see himself as a central figure in a social situation. This kind of social outlook would readily lead to misunderstandings and conflicts. Good human relations would seem to require sensitivity for the feelings of others.

The condition of paranoia illustrates how the external world may be distorted by being viewed against a background of feeling. The self seems to be the center of the universe, and all events and actions of others become related to the self. Thus the meanings and interpretations of events and actions are influenced by the observer's role in the universe and by the way he feels. Although the paranoid personality is aware of his external world, he is not able to separate the sensations of external origin from the mass of feelings created by his anxieties.

The distracting role of visceral sensations becomes even more apparent in the condition of schizophrenia. In describing the nature of the schizophrenic process, Jenkins (10) has emphasized both the early frustration and the lack of interest in external things. He describes the self-consciousness of patients as painful and maladaptive. The schizophrenic becomes insensitive to minor social cues (intonations, gestures, and so forth), and more and more his consciousness centers "in processes which are the very essence of the self" (10, p. 9). Thus the schizophrenic does not seem to retreat from the external world; rather, his inner world takes over. It appears that the withdrawal behavior of schizophrenics may not be a form of escape from the external world, but one of being engulfed by his internal world. This distinction would seem to have basic therapeutic implications.

The fact that the literature on clinical psychology deals considerably with the problem of the self, while the literature on normal personality shows little concern with the problem, supports the hypothesis described above. If this idea is pursued a step farther, it suggests the notion that good adjustment is a process of being aware primarily of external stimulation. Therapy, accordingly, would become the process of reducing the awareness of the self as well as the amount of subjective sensations, and increasing the awareness of external stimulation, including the behavior of other persons. Human relations programs, utilizing discussion and role-playing procedures, are examples of training methods that have been used to increase the sensitivity of an individual to the feelings of others. It is possible that too much concern with the self may be detrimental, and that the type of interest a poorly adjusted person portrays may delay his adjustment.

Both therapy and human-relations training have as their objective the building of constructive attitudes and improved communication between client or trainee and other persons. Training in human relations has profited much from clinical psychology and it is now possible that clinical psychology may gain something from the experiences in human relations training where group methods and concern for the other person are emphasized in the methodology (18).

Sensory Deprivation and the Self.—There are two ways in which a person might conceivably become preoccupied with his internal sensations, if we assume that domination is a relative matter. The examples cited above are taken from abnormal psychology and illustrate the condition of being trapped or overwhelmed by an overabundance of sensations of internal origin. The second possible type of preoccupation with internal stimulation could result from a sharp reduction in external stimulation.

The researches associated with extreme sensory deprivation indicate that when external stimulation is greatly reduced, the content of consciousness becomes startingly different and strange to the person so treated. Much of this research is, as yet, unpublished and it is difficult to anticipate how the data will add up, but there is a strong hint that abnormal emotional experiences and anxiety states are produced in this manner. For example, Bexton, Heron, and Scott (1) reported various kinds of hallucinations and dream-like imagery when their subjects were placed on a comfortable bed in a lighted, soundproof cubicle for an extended period. To reduce sensations further, subjects wore translucent glasses, gloves, cardboard cuffs that extended beyond the finger tips, and a U-shaped rubber pillow which restricted movement.

Subjects found the experience unpleasant; they showed emotional lability, were easily upset, and irritability increased as the experimental period continued. In addition to visual, auditory, tactual, and kinaesthetic hallucinations, some subjects described strange experiences. Two subjects reported that they felt as if they had two bodies; in one case the bodies were side by side and in the other, they overlapped. Other subjects reported feelings of "otherness" and bodily "strangeness." Others reported their heads or minds as being detached from the body.

The authors conclude that their findings represent evidence of a person's dependence upon aspects of the sensory input that previously had not been recognized. It would seem to require but little stretch of the imagination to interpret the experiences reported as containing a description of an exaggerated awareness of the self, and to regard the hallucinations produced under sensory deprivation to be similar to those reported by mental patients.

The fact that sensory deprivation tends to produce a wide variety of new and strange experiences for subjects, which also seem to be highly unpleasant (1), has led to a good deal of speculation. Since the findings came as a distinct surprise to psychologists generally, it is apparent that our present knowledge of human experience does not permit us to predict the effects of extreme changes. In other words, accepted theory does not permit realistic speculation that wanders too far from previously explored territory.

If, however, the phenomena are viewed from the frame of reference of frustration theory, no new assumptions are needed. Rather the findings supplement the interpretation that this theory places upon mental illness, and even lead one to entertain seriously the consideration of a form of therapy for certain conditions of schizophrenia that is either new or that previously lacked conceptual support. This is the use of intense external stimulation to offset the excessive internal stimulation. It is possible that music therapy served this distracting purpose for some patients, but the general idea suggested by these experiments goes beyond this point. Among the research approaches that come to mind are the use of amplification devices in speaking to patients; making radical changes in the environments of patients, such as the use of bright lights and sharp contrasts; and varied methods designed to increase activity. It is conceivable that the intensely stimulating environment would serve to release withdrawn patients from the domination of stimulation of internal origin and thereby make them more accessible to psychological therapy. Drugs that depress visceral functions should serve a similar purpose.

Summary and Conclusions

This paper has two major objectives: (*a*) to clarify vague or misunderstood conceptions of the author's theory of frustration, and (*b*) to extend the theory and its implications.

Nine questions are discussed in an attempt to achieve the first objective. The greatest source of confusion and difference in viewpoint is caused by the postulation of a qualitative distinction between frustration-instigated and motivation-controlled behavior. This distinction requires a re-examination of psychological causation and the part played by learning in behavior modification, but each alternative approach seems to satisfy one portion of the evidence at the expense of another portion. The systematic evaluation of the facts regarding the varied issues raised by the questions should clarify problems for future research.

Other issues, around which questions conceivably could be raised, have to do with the *threshold* concept implied in frustration theory and the concept of *availability* as the selective mechanism in frustration. However, these seem to be secondary issues at the present time because they depend upon the initial acceptance of the qualitative distinction. It is hoped that the primary assumption of a qualitative distinction will not continue to prevent the investigation of the many new problems that are suggested by the assumption.

The new developments in frustration theory deal with the implications of two sources of sensory data: those of internal origin and those of external origin. The discussion includes (*a*) suggestions for a physiological basis for two qualitatively different types of behavior, (*b*) an analysis of the therapeutic value of catharsis, (*c*) a comparison between the logic of feeling and the functions of the unconscious, (*d*) the concept of the self in the light of frustration theory and its implications for therapy, and (*e*) an interpretation of the results of sensory deprivation in the light of the two sources of sensory data.

References

1. Bexton, W. H., Heron, W., and Scott, T. H. Effects of decreased variation in the sensory environment. *Canad. Jour. Psychol.*, 1954, **8**, 70–76.
2. Coghill, G. E. *Anatomy and the problem of behavior.* London: Cambridge Univ. Press, 1929.
3. Dollard, J., and Miller, N. E. *Personality and psychotherapy.* New York: McGraw-Hill, 1950.
4. Ellen, P. The compulsive nature of abnormal fixations. *Jour. Comp. Physiol. Psychol.*, 1956, **49**, 309–317.
5. Eglash, A. Perception, association, and reasoning in animal fixations. *Psychol. Rev.*, 1951, **58**, 424–434.
6. ———. The dilemma of fear as a motivating force. *Psychol. Rev.*, 1952, **59**, 376–379.
7. Farber, I. E. Response fixation under anxiety and non-anxiety conditions. *Jour. Exp. Psychol.*, 1948, **38**, 111–131.
8. ———. Anxiety as a drive state. In M. R. Jones (ed.), *Nebraska symposium on motivation.* Lincoln, Nebraska: Nebraska Univ. Press, 1954. Pp. 1–55.
9. Feldman, R. S. The specificity of the fixated response in the rat. *Jour. Comp. Physiol. Psychol.*, 1953, **46**, 487–492.
10. Jenkins, R. L. Nature of the schizophrenic process. *Arch. Neurol. Psychiat.*, 1950, **64**, 243–262.
11. Jost, H. *The use of polygraphic techniques in psychophysiological research and clinical psychology.* Chicago: Associated Research, Inc., 1953. Pp. 11.
12. ———, Ruilmann, C. J., Hill, T. S., and Gulo, M. J. Studies in hypertension: I. techniques and control data; II. central and autonomic nervous system reactions of hypersensitive individuals to simple physical and psychological stress. *Jour. Nerv. Ment. Dis.*, 1952, **115**, 35–48; 152–162.
13. Jurko, M., Jost, H., and Hill, T. S. Pathology of the energy system: an experimental-clinical study of physiological adaptive capacities in a non-patient, a psychoneurotic, and an early paranoid schizophrenic group. *Jour. Psychol.*, 1952, **33**, 183–198.
14. Klee, J. B. The relation of frustration and motivation to the production of abnormal fixations in the rat. *Psychol. Monogr.*, 1944, **56**, No. 4 (Whole No. 257).
15. Kleemeier, R. W. Fixation and regression in the rat. *Psychol. Monogr.*, 1942, **54**, No. 4 (Whole No. 246).

16. MAIER, N. R. F. *Frustration: the study of behavior without a goal.* New York: McGraw-Hill, 1949.

17. ———. The premature crystallization of learning theory. In *The Kentucky symposium: Learning theory, personality theory, and clinical research.* New York: Wiley, 1954. Pp. 54–65.

18. ———. *Principles of human relations: applications to management.* New York: Wiley, 1952.

19. ———, and ELLEN, P. Can the anxiety-reduction theory explain abnormal fixations? *Psychol. Rev.*, 1951, **58**, 435–445.

20. ———, ———. Studies of abnormal behavior in the rat: the prophylactic effects of "guidance" in reducing rigid behavior. *Jour. Abnorm. Soc. Psychol.*, 1952, **47**, 109–116.

21. ———, ———. Reinforcement *vs.* consistency of effect in habit modification. *Jour. Comp. Physiol. Psychol.*, 1954, **47**, 364–369.

22. ———, ———. The effect of three reinforcement patterns on position stereotypes. *Amer. Jour. Psychol.*, 1955, **68**, 83–95.

23. ———, ———. Studies of abnormal behavior in the rat: XXIV. Position habits, position stereotypes, and abortive behavior. *Jour. Genet. Psychol.*, 1956, **89**, 35–49.

24. ———, and FELDMAN, R. S. Studies of abnormal behavior in the rat: XXII. Relationship between strength of fixation and duration of frustration. *Jour. Comp. Physiol. Psychol.*, 1948, **41**, 348–363.

25. ———, GLASER, N. M., and KLEE, J. B. Studies of abnormal behavior in the rat: III. The development of behavior fixations through frustration. *Jour. Exp. Psychol.*, 1940, **26**, 521–546.

26. ———, and KLEE, J. B. Studies of abnormal behavior in the rat: VII. The permanent nature of abnormal fixations and their relation to convulsive tendencies. *Jour. Exp. Psychol.*, 1941, **29**, 380–389.

27. ———, ———. Studies of abnormal behavior in the rat: XII. The pattern of punishment and its relation to abnormal fixations. *Jour. Exp. Psychol.*, 1943, **32**, 377–398.

28. ———, ———. Studies of abnormal behavior in the rat: XVII. Guidance *vs.* trial and error in the alteration of habits and fixations. *Jour. Psychol.*, 1945, **19**, 133–163.

29. ———, and SCHNEIRLA, T. C. *Principles of animal psychology.* New York: McGraw-Hill, 1935.

30. ———, ———. Mechanisms in conditioning. *Psychol. Rev.*, 1942, **49**, 117–134.

31. MARQUART, D. I. The pattern of punishment and its relation to abnormal fixation in adult human subjects. *Jour. Gen. Psychol.*, 1948, **39**, 107–144.

32. ———, and ARNOLD, L. P. A study of the frustration of human adults. *Jour. Gen. Psychol.*, 1952, **47**, 43–63.

33. MOWRER, O. H. On the dual nature of learning—a reinterpretation of "conditioning" and "problem-solving." *Harv. Educ. Rev.*, Spring, 1947, 102–148.

34. SHERMAN, M., and JOST, H. Diagnosis of juvenile psychoses. *Amer. Jour. Dis. Child.*, 1943, **65**, 586–592.

35. SKINNER, B. F. *The behavior of organisms.* New York: Appleton-Century, 1938.

36. SOLOMON, R. L., and WYNNE, L. C. Traumatic avoidance learning: the principles of anxiety conservation and partial irreversibility. *Psychol. Rev.*, 1954, **61**, 353–385.

37. WILCOXON, H. C. "Abnormal fixation" and learning. *Jour. Exp. Psychol.*, 1952, **44**, 324–333.

38. WOLPE, J. Learning theory and "abnormal fixations." *Psychol. Rev.*, 1953, **60**, 111–116.

ARTICLE V

SELECTOR-INTEGRATOR MECHANISMS IN BEHAVIOR[1,2]

N. R. F. MAIER

Psychological theorizing would be essentially simplified if the measurement of the response (or output) and the measurement of the stimulus (or input) told the whole story. However, the organism itself makes its own unique contribution to both the stimulus and the response, and continues to be a source of confusion. This confusion becomes apparent when one measures ability by means of performance. Since the only measure of ability is through this approach, indirect methods must be used to separate ability from performance. These indirect approaches often lack refinement and may be questioned, yet it would be unfortunate to overlook a distinction merely because it could not be proved to a critical audience. Nevertheless, the strict scientist could conceivably discourage a researcher from making qualitative distinctions by requiring him to furnish conclusive evidence for them. The burden of proof, according to the law of parsimony, seems to rest with the person who makes the qualitative distinctions.

The number of basic mechanisms or qualitative distinctions that psychology must make or the number of evolutionary steps that it must postulate has been determined by nature. It is the scientist's job to make qualitative distinctions where they do indeed occur and to make quantitative measures where these are relevant. Progress in science depends not upon the simplicity or the complexity of its theories but rather upon whether the theories make distinctions of the proper kind.

Since performance is measurable, what are some of the problems regarding the inference of ability? Certainly the ability to perform may differ from the performance. To guard against this source of

[1] Reprinted from *Perspectives in Psychological Theory—Essays in Honor of Heinz Werner*, B. Kaplan and S. Wapner (eds.), 1961, New York: International Universities Press, by permission of the Editors and the publishers.

[2] The writer is indebted to Dr. L. R. Hoffman and Miss Melba Colgrove for their critical comments and suggestions.

error one can state that ability to perform on a given occasion depends upon the performance on that occasion, but does this statement advance our knowledge? An ability can conceivably remain constant, but the ability to express it may vary. When a violinist's performances vary, does he lose and gain ability or does only his performance vary?

It seems that there are many factors that influence performance, and ability is but one among others. Motivation, morale, fatigue, and adjustment are examples of a few of them. It is recognized that performance cannot exceed ability (except in so far as chance plays a part), but that ability may exceed performance. This means that the loss of an ability cannot be assumed merely because it is inaccessible under certain circumstances. However, an ability may be inferred if it can be brought to expression. This suggests that progress may more readily be made if many ways for producing behavior are studied. Researches that limit behavior are most likely to overlook latent abilities. It also suggests that performance be studied under a variety of conditions so as to separate variations in ability from variations in performance.

Let us examine some samples of specific data to clarify problem areas in which present psychological concepts fail to be helpful in understanding variations in behavior. Research by Zucker (1943) raises a problem in point. He found that delinquent and nondelinquent populations of children completed the items in a story-completion test differently. Seventy-five per cent of the solutions of delinquents were characterized by having nonconstructive endings, whereas only 24 per cent of the nondelinquents completed their stories in this manner.

What are the implications of these findings? It might be argued that the learning or biological make-up of these two populations of children were different or it might be assumed that they performed differently and were of like make-up.

Further light is thrown on this question by an experiment by Edwards (1954). He used the same story-completion tests as Zucker, but instead of using delinquent and nondelinquent populations, he used sixth-grade school children in a role-playing situation. Three groups were compared. One group was given a role which described the home picture as one of being rejected by the parents; a second group was given the role which described the home as being warm and accepting; and the third group was uninstructed, and therefore the children were left to assume their own real life roles. However, the school was located in a depressed area, which was a source of an actual delinquent population.

Story completions of a nonconstructive or delinquent variety showed different percentages for these three groups. The results were as follows: 76 per cent for the rejected group; 32 per cent for the accepted group; and 67 per cent for the control group.

It may be assumed in this instance that the three populations had similar behavior make-ups since the division of the groups was random. It also follows, since the roles supplied were unrelated to the stories, that no specific learnings were given. The difference in behavior, therefore, must be attributed to some performance factor other than the difference in ability. The role instructions, which may have set up a kind of "feeling," evidently determined the types of behaviors that would be expressed and organized in a given situation. This suggests that delinquent and constructive behaviors need not be specifically learned. Each individual has adequate constructive and nonconstructive behaviors available for expression, so that the actual behavior expressed is a sample revealed on a given occasion. Since Edward's findings so accurately duplicate those of Zucker, the importance of some selection process becomes very apparent. This statement does not preclude the possibility that learning may supplement the degree of ability and create some basic differences in children reared under different conditions.

From the above analysis it follows that individual differences in behavior must be viewed in two ways: (*a*) differences due to the variation in ability, and (*b*) differences in behavior samples shown on a given occasion.

The value of this distinction also becomes apparent from an analysis of three experiments on variability, conducted by Krechevsky (1937). He compared normal and brain-injured rats on three different measures of variability, and although the results were clear-cut in each instance, each led to a different conclusion.

In the first experiment he used a checkerboard maze, which offered 20 routes to food, all of equal length and all rewarded. In this free type of situation the rat could confine its runs to one or several routes. He found that normal rats averaged 5.46 routes in 15 runs, whereas the brain-injured rats averaged 3.00 routes. This raised the question of the nature of the effect of the brain injury. Has the ability to show variable behavior been reduced in the animal or has the operation affected the performance process? Suppose the operation alters the need to show variable behavior. In such event, operated rats would show less variable behavior in a free situation than normals since such behavior would be less need satisfying for them.

Krechevsky's second study throws some light on this matter. He taught normal and operated rats two routes to food. One route was varied from time to time; the other was kept constant and was somewhat shorter. No errors could be made on either route since there were no blind alleys. In the test situation the animal was confronted with a choice between the two routes. It was found that normal rats took the variable route 57.1 per cent of the time, whereas operated rats took it 34.3 per cent of the time.

In this instance the operated rats chose the efficient (shorter) route, and one might conclude that they were of superior ability. But was this the reason for the choice? They may have avoided the variable route because it offered difficulty or they may have selected the constant path because they disliked variability.

The interpretation that the operated rats were wise and chose the shorter path was eliminated by Krechevsky's control experiment in which he used both a long and a short path, but eliminated the variability. With variability removed, normal rats lost their preference for the long path since their choices of it went from 57.1 to 47.0 per cent. Operated rats, however, increased their choices for the long path from 34.3 to 41.3 per cent. In this instance the normal rats seem the wiser, but the scores approach similarity because variability influenced normal and operated rats in the opposite ways. It seems that operated rats avoid variable situations, but is this due to lack of ability or is it a difference in preference?

To determine whether the operation affected the basic ability, we must have a situation that makes a variable path desirable. Can operated rats cope with variable situations if the matter of selection due to preference is eliminated?

Krechevsky's third experiment supplies the answer to this question. In this experiment he made the variable path shorter than the constant path. In Table 1 the results of the different test conditions are summarized. The last line of the table shows that this change in the length of paths had little effect on the normal rats. They selected the variable path 55.0 per cent of the time, which is about the same as the previous 57.1 per cent obtained when it was the longer path. In choosing between the two paths about an equal number of times, normal rats found a further way to express a desire or a need to vary behavior: they increased their variability by frequently alternating between the two routes on successive trips.

Making the variable path the shorter of the two, however, markedly influenced the performance of operated rats. They now chose the variable path 51.3 per cent of the time instead of the former 34.3 per cent. The shorter length of the path, therefore, offset the disadvantage of variability.

TABLE 1.—CHOICE BETWEEN A VARIABLE AND A CONSTANT PATH

Test	Per Cent Variable Path is Chosen by	
	Normal Rats	Operated Rats
Experiment 2A Longer Variable Path *vs.* shorter Constant Path	57.1	34.3
Experiment 2B Longer Path Not Varied *vs.* shorter Constant Path	47.0	41.3
Experiment 3 Shorter Variable Path *vs.* longer Constant Path	55.0	51.3

These data clearly indicate that the ability to show variable behavior can be brought to expression in operated rats. It must be concluded, therefore, that this ability is not destroyed by operation. If only adaptive ability had been considered, the first experiment would have suggested that operated rats were inferior to normals, the second experiment would have shown them to be superior, and the third would have shown them to be equal.

The results are consistent with each other, however, if we interpret them to mean that brain operations changed the animals with respect to the mechanisms that determine the expression of variable behavior. Apparently normal rats prefer variable situations, whereas operated rats prefer constant situations, and this difference in preference causes them to behave differently under certain conditions.

Possibly an even more dramatic example of the pitfalls inherent in the process of inferring ability from performance is provided by the recent work of McCleary (1959). He found that cats with bilateral lesions in the area of the subcallosal cortex returned promptly to a food trough at which they had just received severe electric shocks to the mouth. Normal cats under similar circumstances did not return for a matter of days even though it was their sole source of food. It would be tempting to conclude that the lesion had interfered with the animals' ability to become frightened of the food trough. A control experiment, however, demonstrated that the ability to acquire a fear was perfectly intact. In a standard shock-avoidance task the operated animals learned an avoidance response as promptly as did the normals. The author concluded that the brain lesion interfered with normal emotional behavior or did not, depending on the performance required of the animal.

ABILITY *vs.* PERFORMANCE

In order to cope with some of the problems raised when the differences between the ability to behave and performance are not clearly differentiated, it may be well to make this basic distinction at the outset rather than cope with it late in the theorizing process. Thus the ability to behave may be thought of as an organism's behavior potential. This represents the behavior repertoire. Every animal has the ability to express a variety of behaviors, but on a given occasion only certain of these many latent behaviors are expressed. Instead of assuming that the stimulus is the only factor that initiates behavior, and thereby making the stimulus a part of the behavior description, it will be assumed here that other possibilities for initiating behaviors exist. These initiators may be either in the external world or in the organism itself. The processes that determine the behaviors that are brought to expression will be regarded as selector-integrator mechanisms. The name implies not only that certain behaviors are selected from alternative possibilities, but also that the parts of behavior are integrated or organized into some pattern.

Viewed in this manner the behavior repertoire may be thought of as analogous to the cells of a memory drum of a computer. The selector mechanism would be the particular program which would make a specific selection from among stored information. The information may be stored in the form of simple behavior segments or memories, or as complex built-in behavior sequences or systems of meanings. The latter would be analogous to subroutines stored in a computer. The selection obtained from the program would represent the elements or sequences of behavior expressed on a given occasion. Furthermore, the behaviors selected could be organized by the program into a specified sequence or configuration to provide a desired output. This aspect of the programming process would be analogous to the integrative function of the selector-integrator mechanism in behavior. Thus the organization of the behavior pattern would be determined by the way the parts were stored as well as by the way they were integrated at the time of expression. It can be seen that stored learned sequences might be helpful on some occasions, but might be a disadvantage and need breaking up on other occasions.

Whether or not the analogy is fully accurate, it does make the essential separation: it distinguishes between (*a*) the behavior repertoire or ability; and (*b*) a selection-integration process, which selects and integrates a sample from the repertoire to produce performance. Although some theories of motivation make a similar distinction, it will be seen that this similarity ceases when a variety of selector-integrator mechanisms are introduced.

Returning to the facts about delinquency already mentioned, it becomes apparent that delinquent and constructive behavior need not assume different behavior repertoires. As a matter of fact, delinquency seems to be a phenomenon of the selector-integrator mechanism. The same is true of variable behavior in rats. It is incorrect to interpret Krechevsky's results as due to a change in the behavior repertoire. Rather the results readily fall into place if it is assumed that the brain injury has altered the selector-integrator mechanism.

Having made the basic distinction between two aspects of behavior, the next step is to examine each of them more carefully and make distinctions within these aspects. These qualitative distinctions should be made at the initial stages of theorizing since the developmental process may be different for each process. If certain processes fail to be basically different, they can be combined with others. The assumption of too many qualitative differences, therefore, can always be corrected, but the assumption that qualitative differences are not present, when they do in fact exist, makes all research based upon this false assumption valueless.

The Behavior Repertoire

The arousal of instinctive patterns or the selection of a habit from a family hierarchy (Hull, 1935) would be the function of the selector-integrator mechanisms. An organism's behavior repertoire may be assumed to consist of the abilities obtained from maturation and through learning. Thus bodily structures, sensory capacity, and neural connections (influenced by heredity and growing conditions) plus the changes produced by learning would establish the functional linkages or patterns that would be available for expression. This much is not unique and is consistent with the conceptualization in most theories.

In addition to the functional linkages, which tie segments of behavior together (association formation), it is also necessary to include a closely related process—that of retention or memory. Learning is a function of the ease with which associative connections are formed, whereas retention depends upon the durability of these linkages. Lashley (1929) has shown retention (as measured by relearning) to be differently related to brain injury than original learning. Since the measurement of retention utilizes relearning scores, one would expect the element of learning to be present in both measures. Despite this fact, the ability to retain and the ability to learn have been found to be only moderately correlated.

Unlearning also is different from forgetting. Buytendijk (1931) found brain-operated rats equal to normals in learning a position response, but when he required them to reverse their habits, the operated rats were clearly handicapped. Maier (1932*a*, 1932*b*) found that lesions which had marked effects on a test measuring the ability to reorganize experience had no effect on the formation of a simple association. It is permissible to postulate basic differences, since there is no evidence showing that learning, retention, and unlearning merely are different aspects of the same process.

It would appear that the ability to form behavior units would influence the behavior repertoire in an additive way, but that the ability to disassociate or to break up learned patterns would be a different process and its function would be to make for greater flexibility and ease in the reorganizations of experience. If we assume that creativity is more than learning, and that solutions to problems sometimes may be new integrations rather than whole units transferred from other situations, the concept of fragmentation becomes as essential to the behavior repertoire as does the concept of learning.

In general, learning has been treated as if it were an asset to adjustment. It seems, however, that its possible disadvantages should also be recognized. Frequently, parts of learned patterns are desired, while the whole pattern would be a detriment. This disadvantage was found to be apparent in problem-solving situations, in that subjects persist in following habitual lines of thinking (Maier, 1930, 1931*b*, 1945). However, once this condition is recognized, it becomes evident in other situations. For example, speech sounds in a particular language are specific combinations. Since a different language requires other basic combinations, it is difficult for an adult to lose his accent; a child, however, readily learns to speak a new language without an accent. On the other hand, adults who speak many languages have little accent, perhaps because their basic speech sounds are fragmented through the learning of several languages.

Basic research supporting this hypothesis is not lacking. Padilla (1930) found chicks unable to acquire the normal pecking movements when kept in darkness for two weeks. It appeared that the reflexes associated with pecking had become combined with feeding from a medicine dropper, and these combinations interfered with the learning of normal pecking because the initial elements were no longer available. However, shorter periods in darkness had been found to be an asset to the acquisition of pecking because of the increased maturity (Shepard and Breed, 1913; Bird, 1925; and Moseley, 1925).

Since individuals differ in their aptitudes for acquiring learning, it is to be expected that the ability to disassociate or fragment behavior segments will also be an aptitude that will show variations between individuals. For this reason flexibility is an appropriate part of the behavior repertoire, even though it may not reveal itself as clearly as does habit formation.

Types of Selector-Integrator Mechanisms

In considering the methods by which behavior is selected from a given behavior repertoire, psychological processes formerly regarded as unrelated become grouped together. Most of these processes, being well known, need not be discussed in detail. However, an attempt will be made to describe them with reference to the functional meanings that emerge when they are treated as selector-integrator mechanisms.

Locus, Intensity, and Form of the Stimulus.—The most primitive type of behavior selection is determined by the *location* of the stimulus. The response of an amoeba, for example, depends upon where it is stimulated; i.e., different behaviors can be elicited from its repertoire merely by stimulating its various parts. The *intensity* of the stimulus also acts selectively. Generally speaking, mild stimulation activates approach behavior, whereas strong stimulation activates withdrawal.

This relationship between approach and withdrawal, as a function of stimulus intensity, seems to be preserved throughout the animal kingdom. Not only does the direction of movement of the amoeba depend on stimulus intensity, but whether or not a dog will show fear of or approach the hand depends upon how quickly the hand is moved. The fear behavior shown by goslings when the hawk flies overhead, as contrasted with their preparing for feeding when the mother approaches, seems to depend upon rate of change in stimulation. The build of the hawk is such as to produce rapid change in the visual field, whereas the long neck and slim body of the goose yields a more gradual change.

As organisms increase in complexity of bodily make-up, stimulation effects spread, first to adjacent regions; later, with the development of conductive tissues, more remote parts of the body are included in the behavior. The *location* of the stimulus thus becomes less decisive and the *form* of stimulation becomes a critical selector as one proceeds up the evolutionary ladder. The highest development of this type is associated with the appearance and refinement of sense organs. Through the development of sense organs, the environment becomes capable of selecting different behaviors from an organism merely by stimulating it with different forms of energy.

Although the size of an animal's repertoire of behavior also increases as the selector mechanisms evolve and improve, it is clear that the behavior expressed will be selective, depending upon the stimulus conditions. The development of this type of control over behavior is unique in that it is entirely external to the animal.

Stimulus Set and Attention.—Even in the simplest animals the behavior expressed includes more than the influence of the stimulus. In the amoeba the same stimulus has different effects, depending upon the movement pattern in progress at the time. Thus, if the amoeba is moving in a particular direction, this temporary "anterior end" is dominant over other portions of the body (Maier and Schneirla, 1935). As soon as an activity pattern is established a physiological gradient is in evidence, and this condition in the organism influences the behavior that a particular stimulus can produce at a given time. This physiological gradient is perhaps the most primitive kind of set.

Such terms as stimulus set, mental set, and expectations refer to conditions in the organism that alter the way an organism will respond to stimulation. In general, this type of influence over behavior has been neglected in research, particularly in that on lower animals, so that much of the comparative evidence depends on observational data.

While studying delayed reactions Tinklepaugh (1928) described some unusual behavior in a monkey. Although a monkey normally will eat lettuce, he refused it when the experimenter fooled him by causing him to expect a banana. In this experiment Tinklepaugh pretended to place a slice of banana under a cup, but palmed the banana and substituted the lettuce instead. When the monkey turned up the cup and found lettuce he became excited and refused to eat the lettuce. Apparently the behavior elicited was influenced by an expectation or stimulus set.

Hull (1935) introduced a provision for the influence of expectation in order to show how behavior described by Maier (1929) as reasoning in rats might be reduced to S-R theory. Hull thus attempted to account for the appearance of a new adaptive response by making it a selection from the "family of habits." This approach tends to minimize the importance of the set by making it a part of the habit strength and thus a part of the behavior repertoire.

The phenomenon of attention is similar to a set or expectation in that it represents a condition of the organism that influences receptivity or responsiveness to stimulation of a particular kind. The influence of attention in reducing reaction time, increasing sensitivity for one stimulus and excluding others, increasing discrimination, etc., are well-known phenomena. Although the phenomenon of attention played a prominent part in the history of psychology, modern theory has neglected it. However, recent research by Hernández-Péon, Scherrer, and Jouvet (1956) has provided empirical evidence that the nervous system is organized in such a way that stimuli irrelevant to the objects being attended to are not permitted to stimulate the cortex. When the cat in this instance was attending visually (and presumably olfactorily) to the mice in the bottle, the clicks stimulating its ear were prevented from intruding on the cortical representation of the stimulus. In attention, then, the nervous system is organized to be sensitive to certain stimuli and to suppress irrelevant ones.

Usually the term attention has been applied to human beings only, it being assumed that attention is a condition of awareness and that introspection was needed for its study. However, Maier (1930) demonstrated that the performance of rats in a difficult problem situation was best (*a*) in new test situations; (*b*) on the first four as against the last four trials of the day; (*c*) when changes in the situation were made; and (*d*) after several days' rest. The results of each of these conditions are in conflict with what one would predict from learning, but each involves a factor (novelty or change) that leads to obtaining attention. Thus an animal frequently violates a principle of learning when the behavior selected is influenced by attention. Frank (1932) found herself unable to duplicate her own data, in a study of normal, underfed, and rachitic rats, when she repeated the experiment in another laboratory. The various inconsistencies in her data, however, became reconciled when the findings were analyzed in terms of different distractors offered in the two laboratories. Consideration of both motivation and distraction led to the correct predictions regarding the subsequent analyses of her data. Depending upon the way distraction and motivation interacted in selecting behavior from the three groups of rats, the same experimental conditions caused the three groups of rats to behave differently. Analyzed in terms of learning ability alone, these findings are entirely ambiguous.

The specific natures of stimulus set and of the attention mechanism are unknown and it is quite possible that their evolution has a varied history. However, both set and attention seem to select the stimuli and put the organism in readiness for its response. In this sense attention is closely related to perception, but the laws of attention

(Pillsbury, 1908) and the laws of perception (Koffka, 1935) are quite different.

Perception.—The way an individual organizes sensory data determines not only what is learned (Maier, 1939*b*) but also which of several possible behaviors will be expressed. Kuo (1930) clearly demonstrated that cats when shocked in the presence of rats may learn to fear (*a*) rats, (*b*) the shock-box, or (*c*) a rat in the shock-box. Thus the learning is not a function of the relationship in the environment, but of the relationship that a given organism perceives. In a similar manner the behavior selected for expression is influenced not merely by the stimulus situation, but by the way the sensory data are organized by the individual. For example, if a person hears a noise which he interprets as that produced by someone clicking his false teeth, he may show irritation; whereas if he perceives the same noise as caused by a faulty radiator, he may be able to dismiss it entirely. Situations perceived as threatening release one kind of behavior, while those perceived as supportive release another kind.

The concept of equivalent stimuli developed by Klüver (1933) holds that a given response will be elicited by a variety of situations. Stimulus conditions that elicit the same response are said to be equivalent with respect to that response, whereas stimulus conditions that fail to elicit a given response are said to be nonequivalent. The concept of equivalence in stimuli is not merely a matter of stimulus generalization, but is one of perceptual organization. Klüver pointed out that an animal might readily discriminate between the stimuli it found to be equivalent.

Because animals show wide individual differences in the stimuli they find to be equivalent, it follows that the influence of *transfer of training* and the *concepts* aroused by a given situation will be an individual matter. The researches of Lashley (1938), Maier (1939*b*, 1941), and Wapner (1944) demonstrated that, even in the rat, the stimuli essential for arousing a previously trained response varied greatly from one individual to another. These researches revealed that two rats responding similarly in a training situation might react quite differently to presentations of the same set of new stimuli. Maier (1941) and Wapner (1944) demonstrated further that brain injury increased the range of equivalent stimuli by causing the animal to be less responsive to differences than formerly. Thus operated rats expressed their learned response in a greater number of situations. Retesting the rats with the same battery of new stimulus combinations had an opposite effect and made rats more discriminating. Hence, extra exposure made the learned response less accessible. These findings support the view that perceptual organization determines the behaviors that will be selected from an animal's repertoire.

The work of Koffka (1935), Werner (1940), Werner and Wapner (1949, 1954), Wapner and Werner (1957), and Witkin, Lewis, Hertzman, Machover, Meissner, and Wapner (1954) demonstrates that the laws governing perceptual organization not only differ from those of learning but are less likely to include concepts related to reinforcement and more likely to include concepts related to the individual and the total stimulus situation. Thus it is important to make a basic distinction between the functions of perception and functions of learning. When learning is considered a part of the behavior repertoire and perception is regarded as a selector-integrator mechanism, an exceptionally sharp separation between them results. This separation, however, does *not* preclude the possibility that perceptions are learned and that responses associated with them will be aroused by a particular perception. However, it does mean that a well-learned response will cease to be available if something occurs to change the perception in a recall situation. Gottschaldt (1926) caused recognition failures to occur in well-learned responses by embedding stimuli in such a fashion that different perceptual organizations were formed during the recall period. Methods for altering perception are more varied and extensive than those for altering associations, and the part played by each must be differentiated when a distinction between the behavior repertoire and the behavior sample selected for expression is made.

Needs and the Motivation Process.—Needs, like attention, refer to a condition of the organism, but the condition is a physiological one, especially with respect to such basic needs as hunger, thirst, sex, and the maternal urge. A hungry animal is sensitive to food in its environment and will locate it when a well-fed animal will fail to react to food cues. I have repeatedly observed that rats who are on their way to a known food position will hop over food placed on an elevated path. However, rats that are exploring a series of paths will never miss such a piece of food. If hurried, an exploring animal will stop suddenly and back up to the food it has passed over. Thus hunger and stimulus-set seem to combine in that hunger increases sensitivity to food, while set limits the sensitivity to a specific expectation.

Hungry men think and dream of food as Guetzkow and Bowman (1946) have shown. Hunger also initiates organized behavior sequences, even though an activity such as restlessness may be random. If an animal has learned the location of food, certain learned responses will be selected from its behavior repertoire. In such instances the animal will not only go to the locations where food has formerly been found (i.e., express reinforced habits) but will systematically explore a general region. This "hunting pattern" is both typical and clearly distinctive in that such behavior is recognized by other animals. Thus the prey flees when the lion is hungry and is on the hunt, but grazes undisturbed when the lion takes a leisurely stroll.

Hunger, therefore, not only selects stimuli from the organism's environment, but selectively releases and integrates a number of different responses so that the behavior shows variety, though variety of a select nature.

Thirst, maternal, and sex needs serve a similar function, although each is unique in its selectivity of stimuli. In this respect each need becomes a separate selector mechanism. However, for the present, it seems appropriate to group physiological needs together in accordance with prevalent theory.

Social needs function in a manner similar to the physiological ones, but in this instance it is difficult to locate the stimulating condition or organ. Many of these needs are of an acquired (learned) nature, such as a boy's need for a bicycle. In such cases it would be reasonable to assume that the location of the need was in the brain. Nevertheless, a boy with a need for a bicycle is selective in his observations, and his behaviors are clearly oriented toward obtaining the bicycle. Like all need conditions, social needs select goal-oriented behaviors.

It will be apparent that acquiring a need involves a different kind of learning from acquiring a conditioned response or gaining an education. If responses are learned through reinforcement, it is obvious that the boy who has no bicycle has the strongest need, and yet it is the neighbor boy with a bicycle who experiences the reinforcement. Social pressure, the mere observation of what others have, is one of the important ways in which needs are acquired. A boy who acquires a need for a bicycle may undergo a great behavior change, not because his behavior repertoire has been altered, but because this acquired need selects from his repertoire specific behaviors previously present but unexpressed.

Attitudes.—Attitudes represent relatively permanent sets which selectively influence the responses elicited—by determining the facts that are observed, as well as by influencing the interpretations with which these observed facts are invested. When defined as selectors of stimuli, attitudes are similar to attention in that they determine what is observed. Thus an unfavorable attitude toward a racial group causes its faults to be observed and hence reacted to, whereas a favorable attitude causes its best behaviors to be seen. Maier and Lansky (1957) have shown that even in a role-playing situation, persons exposed to the same facts will select and react, primarily, to those facts which are consistent with the attitude supplied. Likewise, interpretations of facts and organizations of behavior are guided by an attitude.

A person holding an unfavorable attitude toward Negroes would be similar to a person who had been instructed to pay attention to all of the undesirable things Negroes did. The difference would be that, generally speaking, the acts of attention are constantly changing and are subject to shifting interests as well as to changes in the intensity of stimulation; whereas attitudes are relatively constant and less dependent upon stimulus intensity.

A second difference between attitude and attention is the emotional involvement or feeling that accompanies attitudes. Liking or disliking are characteristic emotional accompaniments of attitudes, while attention is primarily a cognitive function; although it must be pointed out that autonomic responses are associated with stages of attention. It is generally recognized that attitudes, particularly unfavorable ones, represent emotional sets, whereas attention is more a function of intelligence. Often the two selective conditions accompany one another—in that the attitude determines the kinds of stimuli on which attention will be focused.

Direction in Thinking.—The controversy over whether the ideas recalled during productive thinking are more than random and more than a function of the strength of associative bonds is an old one. Psychologists who held that there were thought processes that could not be reduced to the laws of recall suggested various kinds of selective factors. For Selz (1913) the "determining tendency" served a selective function, for Wertheimer (1925) the reorganization function was a kind of closure, and for Köhler (1929) the sensefulness of the behavior was the product of insight. My concept "direction in thinking" (Maier, 1930, 1931*a*) performs a similar function, but combines the selective function of the determining tendency with the integrative function of Wertheimer's closure.

The string problem (Maier, 1931*b*), which requires the tying together of the ends of two strings hanging from the ceiling, illustrates the selective function of a "direction" in problem solving. When the two strings are far enough apart so that the subject cannot reach one while holding onto the other, he may envisage his difficulty to be either of the following:

(1) My arm is too short.
(2) The strings are too short.
(3) The end of the first string won't stay in the middle while I go to get the second one.
(4) The second string won't come to me while I hold on to the first one.

Each of these difficulties is associated with a general selective function or a direction in thinking. This direction greatly determines both (*a*) what is perceived in the environment, and (*b*) the ideas or thoughts that are recalled. The first difficulty mentioned above causes sticks, poles, and objects with handles to be readily perceived; while mental images of such objects and experiences having to do with extending the reach are recalled. The second difficulty causes strings, window-shade cords, and belts to be noticed; while various ideas having to do with lengthening a string are recalled. The third difficulty inclines the person to think of chairs that might be moved between the strings so that one of the strings could be tied to it; while impractical methods of anchoring one cord in the center are tried out. The fourth difficulty causes electric fans to be noticed or requested; at the same time, such activities as opening a window to permit a breeze to blow the one cord toward the center while the other is carried there may be performed.

These selective perceptions and recalls indicate that the activities expressed are neither random nor based upon strength of associative connections because the direction not only selects relevant objects and memories but also violates the principles of recall based upon association. The research of Lewin (1926) on nonsense syllables clearly showed that a strong association between a pair of nonsense syllables could be inhibited readily by merely instructing the subject to rhyme each syllable as it was exposed.

Very close to the concept of direction is that of "hypothesis" in behavior developed by Krechevsky (1932). He found that rats in a discrimination-problem situation are following a system rather than making random choices. Thus they may make some reactions to brightness, then a series of reactions to position, followed by a sequence of alternations, etc. Even in the rat, it appears that a selective mechanism was needed to account for the lack of randomness in trial-and-error behavior.

Although an attitude and a direction in thinking are similar in that they both influence perceptions and recall, there are basic differences. Attitudes, as already suggested, involve emotions, whereas directions in thinking exert primarily a cognitive influence. Furthermore, directions are more subject to change and can be altered by instruction (Maier, 1933); whereas attitudes, by comparison, are rigid when viewed rationally. Then too, directions in thinking are largely influenced by the perception of the difficulty or immediate situation, whereas attitudes usually have a longer history and hence are brought to the situation rather than derived from it.

An Obstacle as a Problem Situation.—A problem exists when a response to a given situation is blocked. Ordinarily a hungry animal responds to food by approaching it and eating. However, when an obstacle blocks the approach, variable behavior comes to expression. Thorndike (1898) described such behavior as "random trial and error," whereas Adams (1929) found an element of selectivity and insightfulness in the behavior expressed; but both recognized the characteristic of variability in the behavior. A somewhat higher status was accorded to "trial-and-error" behavior when Dewey (1910) described problem solving as "mental trial and error." In this manner Thorndike's simple mechanism was generalized to the thinking process of man. The fact that Dewey's model of thinking also emphasized the point that the person, when confronted with a problem, goes from one idea to another is, however, of importance since variability still emerges as the dominant characteristic of problem-solving behavior.

Variability is a biologically sound mechanism for problem solving, since it is apparent that when a given sample of behavior is inadequate another sample may be effective. In brief, a problem stimulates an animal to run through his repertoire of behavior. If the solution is included in this repertoire, the problem is solved.

Direction in thinking may shorten this process by confining the variability to certain areas in the behavior repertoire. Thus, in the case of the string problem, variability may be restricted to trying out different ways in which to increase the reach of the arm. However, variability also may operate so that the directions in thinking are varied. Sherburne (1940) found that superior problem solvers varied their directions in thinking, whereas less capable individuals tended to confine their variability within a single direction.

Variability, it appears, may be restricted or range widely but still serve no other purpose than that of selecting latent solutions from a repertoire stored in the animal. However, what Wertheimer (1945) calls reorganization, and what I have called the combination of parts of isolated experiences, further increases the scope of the variability process. One may think of *reproductive* thinking as the application of an old solution to a new problem situation, and of *productive* thinking as yielding a solution made up of parts of several old solutions or ideas.

In productive thinking the combination of elements results in a new integration or product. For this to be accomplished, old combinations must be broken down or fragmented and new integrations formulated. For example, when the materials in the string problem are so limited that none of them can be used for extending the reach, when ways to lengthen the cord are impossible, and when things for anchoring the string do not exist, the problem remains insoluble as long as only these directions are explored. If only a pair of pliers is available, none of the above approaches is feasible. However, one of the strings can be transformed into a pendulum, so that the swinging string will then come to the problem solver while he is holding the other string. In this instance the pliers change their function and become a weight; the string, which previously dragged on the floor and was too long, must be shortened; and pendulums, which formerly served as timing devices, become methods for getting objects to come to the problem solver. That such experiences are already in an individual's repertoire can be argued. The controversy as to whether there is such a thing as creative problem solving is characterized by this kind of argument. Although I have reported experimental evidence elsewhere (Maier, 1929, 1938, 1940, 1945) in support of this developing of new integrations, for our present purposes this point is irrelevant. If new combinations do in fact occur in problem situations, the principle of variability is operating. The creative person continually varies the way he fragments experiences and forms recombinations, so that variability is present in the thinking of an inventor as well as in the problem solving of a chicken. The inventor and the chicken differ, however, with respect to their repertoires, their directions of thinking, the plasticity of their nervous systems, their ability to fragment past learning, and in their behavioral equipment.

A problem situation does not indefinitely initiate variability, however. With repeated failure and a depletion of possible variable responses, substitute goals or escape may take the animal out of the problem. If these opportunities are not available, if the need for the incentive is great, or if pressure forces the animal to obtain a solution to a problem, the situation may become a frustrating condition.

Operationally, a frustrating situation and a problem situation are alike, although they may differ in the degree to which failures and pressures are involved. The behaviors, however, are qualitatively different, indicating that a change takes place in the organism. Thus a problem may frustrate one individual while initiating problem-solving behavior in another. This is why stressful situations divide behaviors into two basic classes: problem-solving responses and frustrated responses (Maier, 1946; Jenkins, 1957).

An Obstacle as a Frustrating Agent.—The transition from problem-solving behavior to frustrated behavior is rather sharp and apparently depends upon the tolerance or frustration threshold of the individual. Increasing either the degree or the period of stress noticeably increases the number of individuals exhibiting frustrated behavior.

(See Maier and Klee, 1943; Maier and Feldman, 1948; Marquart, 1948; and Shimoyama, 1957.)

Three types of behavior have been experimentally linked with frustration. Dollard, Doob, Miller, Mowrer, and Sears (1939) have demonstrated experimentally that frustration leads to *aggression.* Individuals who previously showed co-operative behavior will show hostile and destructive behavior if kept in the same situation for periods of time. The research of Barker, Dembo, and Lewin (1941) linked frustration with *regression.* Their studies revealed that the play of children will regress and become less mature when frustration is induced. For example, a five-year-old child may be able to play with a toy as a normal five-year-old or as a three-year-old, depending upon whether or not he is frustrated.

The third type of frustrated behavior was investigated by Maier and his students. My book (1949) brought together a series of studies showing that frustration leads to rigidity or *abnormal fixation.* The same animals that were capable of showing variable behavior on other occasions became rigid and inflexible under stress. Patrick (1934) and Marquart (1948) have shown similar results in their researches with human subjects.

The condition of frustration, therefore, is known to be responsible for the expression of behavior having the characteristics of aggression, regression, and fixation. Perhaps there are others, not yet experimentally isolated. The characteristic of *resignation* seems to be one of these. However, enough is known to make the state of frustration in the organism an important selector and integrator of behavior.

The characteristics of aggression, regression, and fixation may be combined in a single act or they may be revealed separately. Although the behavior expressed under frustration may show variations, as yet not fully understood, the characteristics are distinctive enough to contrast markedly with those expressed when the organism is in a problem-solving condition, where the goal to be achieved is central to the selection of behavior.

Further studies on the selective factors operating during frustration are needed so that predictions as to the specific acts to be expressed can be made. Thus, a delinquent child may show aggression, but whether the behavior takes the form of vandalism, fighting, or stealing is not differentiated by such a term. An additional concept, that of availability (Maier and Ellen, 1959), seems to determine the specific form that aggression will take. If a child who is a member of a gang feels hostile, the suggestion of a gang leader might cause the whole group to express their hostilities in a specific manner. Regression, which often occurs in conjunction with aggression, renders the individual suggestible, while the leader's suggestion makes the specific act available.

Although aggression, regression, and abnormal fixation need to be further explored, it is obvious that they are clearly associated with frustration. Much of frustrated behavior combines all of these characteristics, but they obviously vary greatly in proportion. To some extent the behavior repertoire may be a factor in determining the type of behavior expressed by a frustrated individual, and certainly the principle of availability would be a factor in the selection and integration of the specific acts selected (Maier and Ellen, 1959). Since the *behavior in progress* at the time of frustration is that which is most available, a gentle response may turn into a destructive act; it may become rigid or ritualistic, or it may become immature when the condition of frustration replaces that of problem solving. These changes can take place without there being any change in the behavior repertoire.

Research Needs

Of the selector-integrator mechanisms described in this paper some are located in the environment, while others represent a condition of the organism. The dimensions of the stimulus as selectors are the mechanisms most closely associated with the environment, while stimulus-set, attention, need, and direction in thinking are completely in the organism. Perception and the obstacle situations represent a combination of organismic and environmental influence. Perception is an individual matter in that the same stimuli may be organized differently, yet the environment also plays a part in that such external factors as proximity, closure, continuity of lines, etc., influence perceptual organization. Since an obstacle in the situation may initiate either problem-solving behavior or frustrated behavior, the resulting selection is an individual contribution. Nevertheless, in that the obstacle is in the environment, this aspect of the mechanism is external to the individual.

Whether or not the eight selector mechanisms can be reduced by combining some of them or whether further distinctions are needed is a question that can best be answered through research. Certainly, however, when different laws are needed to describe an influence on behavior expression, the distinctions should be maintained.

In the light of our present knowledge it seems unwise to arrange the selector-integrator mechanisms in any evolutionary order. Each mechanism, as well as each aspect of the behavior repertoire, seems to range from the simple to the complex. Perhaps our understanding of the developmental process would be increased if this process could also be viewed within the limited framework constituted by each selector-integrator mechanism and by each portion of the behavior repertoire. This approach would necessitate an increase in the number of qualitative distinctions, but might well eventuate in a greater consistency of quantitative findings.

The reader will undoubtedly have observed that the selecting aspect of the selector-integrator mechanism has been more adequately treated than the integrating aspect. This suggests a possible deficiency in our research coverage.

Prevalent theory treats the organization of behavior as if it were built into the organism. Reflex integration has been quite adequately explained by the way neural connections and thresholds are arranged (Fulton, 1926). Coghill (1929) correlated neural development and the behavior changes of the developing salamander from the head-flexion stage through the coil and "S" stages to swimming movements, thereby demonstrating that the integration of innate behavior is built into the organism.

Learning theorists followed the same lead and assumed that the behavior pattern expressed was a learned one. Learned responses were initially conceptualized as chains of associative bonds, but through the Gestalt influence behavior was likened to patterns of action. There appears to be a basic validity in assuming that the organization of behavior can be traced to the behavior repertoire, but is this the complete answer?

The feeding response of the amoeba varies when it engulfs a particle of food, depending upon the stimulus conditions. The amoeba does not run through a sequence of movements as does a squirrel burying a nut on a table; going through the full sequence of the movements from digging to covering, but accomplishing nothing. Such "blind" behaviors may occur, but are these typical?

Schneirla's researches (1941, 1952) on the army ant show how colony behavior is integrated around stimuli supplied while a behavior sequence is in progress. Comparable studies are needed on higher animals to show both the contributions made by the environmental and by the various internal conditions of the organism to the particular integration revealed on a given occasion. It is clear from Schneirla's studies that the behavior brought to expression by the selector-integrator mechanisms is organized in accordance with conditions as they exist at a given time.

It is well known that when a cat pulls a string to escape from a problem box, it does not always accomplish this with the same set of movements. Pulling the string may be done in a different way each time. Variations include the use of either the right or left paw, and the teeth. Guthrie and Horton (1946) contend that these variations represent different learnings and point to their own research in which the learned pattern is highly specific. Their problem box utilized a post that had to be tilted in order to free the confined animal. Since cats do not run into objects, the only way the correct response could be discovered by them was through accident. It was found that cats solved the problem by backing into the post and that these backing responses were highly specific. Their unique findings can be explained by assuming that the cat is unaware of what it is doing when it accidentally backs into the post, and so it can only learn a "blind" act. The Guthrie-Horton situation restricted the learning of the accomplishment and consequently forced the animal to learn a kinesthetic act which of necessity is specific.

When strings are to be pulled the cat does not avoid them as it does a post. Instead it manipulates them and so it can observe what it is doing. Thus the cat's behavior can be oriented toward the pulling of the string rather than guided by specific movements controlled by kinesthetic cues. Pulling strings can be done in a variety of ways and each way of pulling them seems to be organized at the time of expression. No disorganized mixtures of previous responses seem to appear and the behavior adapts itself to changing conditions. Successes and failures in getting hold of the string supply a "feedback" which influences the performance.

What is called "goal-oriented behavior" is activity that is directed toward some end, and various means are used to achieve this end. If the concept of selector-integrator mechanisms as an approach to behavior is a sound one, it also follows that each of the means to an end has some integration, and it would appear that the nature of the goal plays a prominent part in this integration. Thus I should like to postulate the existence of equivalent behaviors as well as equivalent stimuli. Behaviors that accomplish the same objective would therefore be regarded as equivalent, with a given animal expressing new samples of behavior in its attempts to achieve the same goal.

Evidence in support of this view is not entirely lacking. Animals that have run through a maze will do so after cerebellar injuries (Lashley and McCarthy, 1926). Such injuries destroy the equilibrium and prevent normal locomotion, but the rats still take the correct pathways although they roll and drag themselves along. These movements, expressed for the first time, are not random but are organized to yield a particular end result.

Both Brunswik (1943) and Tolman (1955) have recognized the fact that the same end may be accomplished by different behaviors. They suggest that performance be defined in terms of the accomplishment rather than the pattern of movements, since the latter cannot be predicted. Thus learning would be measured by *what* the behavior accomplished rather than by *how* something was accomplished. This change in definition, however, does not escape the necessity of explaining how the integrated behavior pattern that appears spontaneously comes to expression.

Mary Henle (1956) drew attention to a large area of activity that has apparently been generally overlooked. She points out that motivational theories fail to deal with the many activities that are carried on in the goal region. She distinguishes between "satisfaction and release of tension in the case of goal striving as contrasted in many cases with regret when it becomes necessary to terminate activity in the goal region." The continued enjoyment of friends or the pleasure of a ride in the country represent activities that are continued because they are enjoyed, but need not lead to some further goal. These activities not only are selected, but they are integrated and organized into functional patterns not previously laid down. Sets, attitudes, needs, problems or obstacles, and stimulation all enter into the selection and integration of such behavior.

If response integration is viewed as a special problem, it opens a field for investigation corresponding to that of sensory integration or perception. At one time it was thought that perceptual organization was largely a matter of the sense organs and of learning. Modern perception theory, as already indicated, includes personality variables and makes each perceptual experience an integrated unit. The study of response integration may lead to equally rich findings.

SUMMARY

The behavior theory suggested is characterized by the postulation of a clear-cut distinction between the *ability* to behave and the *actual* behavior that is expressed on a given occasion. It is assumed: (1) that an organism has a store of potential behaviors which constitute its behavior repertoire; and (2) that a number of selector-integrator mechanisms operate to determine which of these potential behaviors will be expressed, and thus constitute the organism's performance at a given time. The analogy of the computer is suggested —in that the behavior repertoire is analogous to the stored data, whereas the selector-integrator mechanisms are analogous to the programs which the computer follows. Behavior then becomes analogous to the output of the computer. Thus the behavior that constitutes performance is dependent upon both the repertoire and the selector mechanism in much the same way that the product of the computer is dependent upon both its stored data and its program.

The behavior repertoire includes (1) the influence of heredity and growth (so-called innate elements and sequences of behavior); (2) learning (consisting of patterns, memories, habit sequences, and skills); and (3) the ability to unlearn or to break up learned patterns to make elements of these patterns available (primarily essential to problem solving).

Eight selector-integrator mechanisms, which may operate at the same time to select and organize behavior, are differentiated. These are (1) the locus, intensity, and form of the stimulus; (2) stimulus set and attention; (3) perception; (4) attitudes; (5) needs; (6) direction in thinking; (7) the obstacle as a problem; and (8) the obstacle as a frustration instigator.

Although motivation is traditionally regarded as the mechanism that brings behavior to expression, it is apparent from the above-mentioned selector integrators that motivation is only one of eight possible mechanisms. The viewpoint that it takes a stimulus to trigger off a response is also represented, but only as one of the eight mechanisms determining the sample of behavior expressed. At this time, it seems that eight mechanisms are necessary to account for the performance problems raised. Perhaps this number can eventually be reduced, but it is also possible that additional ones will be needed. The number of different mechanisms postulated is of secondary concern. The important point is that a break has been made from the limitations imposed by the current use of a minimum number of mechanisms. The law of parsimony is valid only when it is invoked to choose between equally adequate theories; it does not favor the simple explanation when simplicity alone is its major virtue.

This reorganization of psychological concepts introduces some changes in functional meanings and also modifies the relative importance of the various contributors to behavior. Behavior now becomes a sample selected and integrated from the general repertoire of an animal at a given time. It is possible that this approach will expose a variety of ways in which growth and evolution proceed, if development is viewed from the standpoint of (*a*) the acquisition of the behavior repertoire, and (*b*) the refinements of the selector-integrator mechanisms. The developmental approach may be the best way to demonstrate the presence of qualitative distinctions among the various selector-integrator mechanisms and among the factors that contribute to the behavior repertoire.

Areas deficient in research naturally come to light when a field of study is viewed from a different perspective. The need to know more about the ability to escape the restrictions imposed by learning is one problem raised in considering the constitution of the behavior repertoire. Thus far, traditional approaches have made innate and learned patterns of behavior the subject of investigation, but they have not dealt with the way these interfere with the emergence of new behavior. The ability to disassociate and to fragment linkages may be as important as the ability to form associations.

Although the selector-integrator mechanisms mentioned have been the subject of numerous investigations, they have not been studied as parallel or interacting determiners of performance. The integrative aspect of these mechanisms seems to be the most neglected subject.

It is suggested that the concept of *equivalence* in behavior be introduced in order to group together the various behavior patterns that achieve the same results. This concept is the complement of the concept of equivalent stimuli, which groups together various stimuli that elicit the same response.

The traditional view, that the organization of behavior is controlled by the behavior repertoire, seems to be only partly correct. It appears that much of behavior is integrated at the time of expression, somewhat as sensory data are spontaneously organized in perceptions. One can perceive only one of several possible perceptual organizations at a given time (i.e., figure-ground reversals), and it seems that only one, of several possible patterns of movement, will make its appearance at a given time. The one behavior pattern which is expressed, however, makes its appearance as an organized whole, having a clearly apparent function, adapted to changing circumstances. Study of the spontaneous integration of behavior may conceivably lead to a set of principles of behavioral organization comparable to the principles of perception.

BIBLIOGRAPHY

ADAMS, D. K. 1929. Experimental studies of adaptive behavior in cats. *Comp. Psychol. Monogr.*, **6**, 1–168.

BARKER, R., DEMBO, T., and LEWIN, K. 1941. *Frustration and regression: an experiment with young children.* Iowa City: Univ. of Iowa Press.

BIRD, C. 1925. The relative importance of maturation and habit in the development of an instinct. *Pedagogical Seminary,* **32**, 68–91.

BRUNSWIK, E. 1943. Organismic achievement and environmental probability. *Psychol. Rev.*, **50**, 255–272.

BUYTENDIJK, F. J. 1931. Le cerveau et l'intelligence. *Journal de Psychologie*, **28**, 345–371.

COGHILL, G. E. 1929. *Anatomy and the problem of behavior*. New York: Macmillan.

DEWEY, J. 1910. *How we think*. Boston: Heath.

DOLLARD, J., DOOB, L. W., MILLER, N. E., MOWRER, O. H., and SEARS, R. R. 1939. *Frustration and aggression*. New Haven: Yale Univ. Press.

EDWARDS, W. 1954. Unpublished research reported by N. R. F. Maier in *The Kentucky symposium*. New York: Wiley, pp. 54–65.

FRANK, M. 1932. The effects of a rickets-producing diet on the learning ability of white rats. *Jour. Comp. Psychol.*, **13**, 87–105.

FULTON, J. F. 1926. *Muscular contraction and the reflex control of movement*. Baltimore: Williams & Wilkins.

GOTTSCHALDT, K. 1926. Über den Einfluss der Erfahrung auf die Wahrnehmung von Figuren. *Psychologische Forschung*, **8**, 261–317; **12**, 1–87.

GUETZKOW, H. S., and BOWMAN, P. H. 1946. *Men and hunger*. Elgin, Ill.: Brethren Publishing House.

GUTHRIE, E. R., and HORTON, G. P. 1946. *Cats in a puzzle box*. New York: Rinehart.

HENLE, M. 1956. On activity in the goal region. *Psychol. Rev.*, **63**, 299–302.

HERNÁNDEZ-PÉON, R., SCHERRER, H., and JOUVET, M. 1956. Modification of electric activity in cochlear nucleus during "attention" in unanesthetized cats. *Science*, **123**, 331–332.

HULL, C. L. 1935. The mechanism of the assembly of behavior segments in novel combinations suitable for problem solution. *Psychol. Rev.*, **42**, 219–245.

JENKINS, R. L. 1957. Motivation and frustration in delinquency. *Amer. Jour. Orthopsychiatry*, **27**, 528–537.

KLÜVER, H. 1933. *Behavior mechanisms in monkeys*. Chicago: Univ. of Chicago Press.

KOFFKA, K. 1935. *Principles of Gestalt psychology*. New York: Harcourt, Brace.

KÖHLER, W. 1929. *Gestalt psychology*. New York: Horace Liveright.

KRECHEVSKY, I. 1932. "Hypotheses" *vs.* "chance" in the presolution period in sensory discrimination. *Univ. Calif. Publ. Psychol.*, **6**, 27–44.

———. 1937. Brain mechanisms and variability: Part I. Variability within a means-end-readiness. Part II. Variability when no learning is involved. Part III. Limitations of the effect of cortical injury upon variability. *Jour. Comp. Psychol.*, **23**, 121–138; 139–163; 351–364.

KUO, Z. Y. 1930. The genesis of the cat's response to the rat. *Jour. Comp. Psychol.*, **11**, 1–35.

LASHLEY, K. S. 1929. *Brain mechanisms and intelligence*. Chicago: Univ. of Chicago Press. (New York: Dover Publications, Inc., 1963.)

———. 1938. The mechanism of vision: XV. Preliminary studies of the rat's capacity for detail vision. *Jour. Gen. Psychol.*, **38**, 123–193.

——, and MCCARTHY, D. A. 1926. The survival of the maze habit after cerebellar injuries. *Jour. Comp. Psychol.*, **6**, 423–433.

LEWIN, K. 1926. Vorsatz, Wille, und Bedürfnis. *Psychologische Forschung*, 7, 330–385.

MAIER, N. R. F. 1929. Reasoning in white rats. *Comp. Psychol. Monogr.*, 6, 1–93.

———. 1930. Reasoning in humans: I. On direction. *Jour. Comp. Psychol.*, 10, 115–143.

———. 1931*a*. Reasoning and learning. *Psychol. Rev.*, 38, 332–346.

———. 1931*b*. Reasoning in humans: II. The solution of a problem and its appearance in consciousness. *Jour. Comp. Psychol.*, 12, 181–194.

———. 1932*a*. The effect of cerebral destruction on reasoning and learning in rats. *Jour. Comp. Neurol.*, 54, 45–75.

———. 1932*b*. Cortical destruction of the posterior part of the brain and its effect on reasoning in rats. *Jour. Comp. Neurol.*, 56, 179–214.

———. 1933. An aspect of human reasoning. *Brit. Jour. Psychol.* (*Gen. Sect.*), 24, 144–155.

———. 1938. A further analysis of reasoning in rats: II. The integration of four separate experiences in problem solving. *Comp. Psychol. Monogr.*, 15, 1–43.

———. 1939*a*. The specific processes constituting the learning function. *Psychol. Rev.*, 46, 241–252.

———. 1939*b*. Qualitative differences in the learning of rats in a discrimination situation. *Jour. Comp. Psychol.*, 27, 289–328.

———. 1940. The behavior mechanisms concerned with problem solving. *Psychol. Rev.*, 47, 43–58.

———. 1941. The effect of cortical injuries on equivalence reactions in rats. *Jour. Comp. Psychol.*, 32, 165–189.

———. 1945. Reasoning in humans: III. The mechanisms of equivalent stimuli and of reasoning. *Jour. Exper. Psychol.*, 35, 349-360.

———. 1946. *Psychology in industry*. Boston: Houghton Mifflin.

———. 1949. *Frustration: the study of behavior without a goal.* New York: McGraw-Hill.

———. 1954. The premature crystallization of learning theory. In *The Kentucky symposium: Learning theory, personality theory, and clinical research.* New York: Wiley, pp. 54–65.

———, and ELLEN, P. 1959. The integrative value of concepts in frustration theory. *Jour. Cons. Psychol.*, 23, 195–206.

———, and FELDMAN, R. S. 1948. Studies of abnormal behavior in the rat: XXII. Strength of fixation and duration of frustration. *Jour. Comp. Physiol. Psychol.*, 41, 348–363.

———, and KLEE, J. B. 1943. Studies of abnormal behavior in the rat: XII. The pattern of punishment and its relation to abnormal fixations. *Jour. Exper. Psychol.*, 32, 377–398.

———, and LANSKY, L. M. 1957. Effect of attitude on selection of facts. *Personnel Psychol.*, 10, 293–303.

———, and SCHNEIRLA, T. C. 1935. *Principles of Animal Psychology*, New York: McGraw-Hill.

MARQUART, D. I. 1948. The pattern of punishment and its relation to abnormal fixation in adult human subjects. *Jour. Gen. Psychol.*, 39, 107–144.

MCCLEARY, R. A. 1959. The influence of sub-collosal lesions on fear-motivated behavior in cats. Unpublished manuscript.

MOSELEY, D. 1925. The accuracy of the pecking response in chicks. *Jour. Comp. Psychol.*, 5, 75–97.

PADILLA, S. G. 1930. Further studies on the delayed pecking of chickens. Doctorate dissertation, University of Michigan.

PATRICK, J. K. 1934. Studies in rational behavior and emotional excitement: II. The effect of emotional excitement on rational behavior of human subjects. *Jour. Comp. Psychol.*, 18, 153–195.

PILLSBURY, W. B. 1908. *Attention.* New York: Macmillan.

SCHNEIRLA, T. C. 1941. Social organization in insects as related to individual function. *Psychol. Rev.*, 48, 465–486.

———. 1952. Sexual broods and the production of young queens in two species of army ants. *Zoologica*, 37, 5–37.

SELZ, O. 1913. *Über die Gesetze des geordneten Denkverlaufs.* Stuttgart: W. Spemann.

SHEPARD, J. F., and BREED, F. S. 1913. Maturation and use in the development of an instinct. *Jour. Anim. Behav.*, 3, 274–285.

SHERBURNE, B. J. 1940. Qualitative differences in the solution of a problem involving reasoning. Doctorate dissertation, University of Michigan.

SHIMOYAMA, T. 1957. Studies of abnormal fixation in the rat: I. Effects of frequency of punishment in the insoluble situation. *Japanese Jour. Psychol.*, 28, 203–209.

THORNDIKE, E. L. 1898. Animal intelligence; an experimental study of the associative process in animals. *Psychol. Rev. Monogr.*, 2, 1–109.

TINKLEPAUGH, O. L. 1928. An experimental study of representative factors in monkeys. *Jour. Comp. Psychol.*, 8, 197–236.

TOLMAN, E. C. 1955. Principles of performance. *Psychol. Rev.*, 62, 315–326.

WAPNER, S. 1944. The differential effects of cortical injury and retesting on equivalence reactions in the rat. *Psychol. Monogr.*, 57, 1–59.

———, and WERNER, H. 1957. *Perceptual development: an investigation within the framework of sensory-tonic field theory.* Worcester: Clark Univ. Press.

WERNER, H. 1940. *Comparative psychology of mental development*, 3d ed. New York: International Universities Press, 1957.

———, and WAPNER, S. 1949. Sensory-tonic field theory of perception. *Jour. Personality*, 18, 88–107.

———, ———. 1952. Toward a general theory of perception. *Psychol. Rev.*, 59, 324–338.

———, ———. 1954. Studies in physiognomic perception: I. Effect of configurational dynamics and meaning-induced sets on the position of the apparent median plane. *Jour. Psychol.*, 38, 51–65.

WERTHEIMER, M. 1925. *Drei Abhandlungen zur Gestalttheorie*. Erlangen: Verlag der Philosophischen Akademie.

———. 1945. *Productive thinking*. New York: Harper.

WITKIN, H. A., LEWIS, H. B., HERTZMAN, M., MACHOVER, K., MEISSNER, P. B., and WAPNER, S. 1954. *Personality through perception*. New York: Harper.

ZUCKER, H. 1943. The emotional attachment of children to their parents as related to behavior and delinquency. *Jour. Psychol.*, 15, 31–40.

SUPPLEMENTARY BIBLIOGRAPHY

ALLEE, W. C. 1938. *The social life of animals.* New York: Norton (rev. ed.: Beacon Press, 1951).

———, EMERSON, A. E., PARK, O., PARK, T., and SCHMIDT, K. P. 1949. *Principles of animal ecology.* Philadelphia: W. B. Saunders Co.

BEACH, F. A. 1948. *Hormones and behavior.* New York: Paul Hoeber.

———, HEBB, D. O., MORGAN, C. T., and NISSEN, H. 1960. *The neuropsychology of Lashley.* New York: McGraw-Hill.

BLISS, E. L. (ed.). 1962. *Roots of behavior—genetics, instinct, and socialization in animal behavior.* New York: Harper (Hoeber).

BRAZIER, M. A. B. (ed.). 1959. *The central nervous system and behavior.* New York: Josiah Macy, Jr., Foundation.

CARTHY, J. D. 1958. *An introduction to the behavior of invertebrates.* New York: Macmillan.

DANIELLI, J. F., and BROWN, R. (eds.). 1950. *Physiological mechanisms in animal behavior.* New York: Academic Press. (*Sympos. Soc. Exper. Biol.*, No. 4—chapters by Lashley, Lorenz, Tinbergen, and others.)

DARLING, F. F. 1937. *A herd of red deer.* London: Oxford.

DEMBOWSKI, J. 1955. *Tierpsychologie.* Berlin: Akad.-Verlag.

FISCHEL, W. 1948. *Die höheren Leistungen der Wirbeltiergehirne.* Leipzig: Barth Verlag.

FRAENKEL, G., and GUNN, D. L. 1940. *The orientation of animals.* London: Oxford Univ. Press. (New York: Dover Publications, Inc., 1961.)

FRISCH, K. VON. 1950. *Bees, their vision, chemical senses, and language.* Ithaca, N.Y.: Cornell Univ. Press.

HARLOW, H. F., and WOOLSEY, C. N. (eds.). 1958. *Biological and biochemical bases of behavior.* Madison, Wisc.: Univ. of Wisconsin Press.

HARRIS, D. B. (ed.). 1957. *The concept of development.* Minneapolis: Univ. of Minnesota Press.

HEBB, D. O. 1949. *The organization of behavior.* New York: Wiley.

———, and THOMPSON, W. R. 1954. The social significance of animal studies. In G. Lindsey (ed.), *Handbook of social psychology,* Cambridge, Mass.: Addison-Wesley.

HEDIGER, H. 1950. *Wild animals in captivity.* London: Butterworths.

HILGARD, E. R., and MARQUIS, D. G. 1940. *Conditioning and learning.* New York: Appleton-Century.

HULL, C. L. 1943. *The principles of behavior.* New York: Appleton-Century.

JEFFRIES, L. A. (ed.). 1951. *Cerebral mechanisms in behavior.* New York: Wiley.

JONES, M. R. (ed.). 1953– . *Nebraska symposia on motivation.* Lincoln, Nebr.: Univ. of Nebraska Press. (Annually since 1953.)

KLOPFER, P. H. 1962. *Behavioral aspects of ecology.* New York: Prentice-Hall.

KRUSHINSKII, L. V. 1962. *Animal behavior—its normal and abnormal development.* New York: Consultants Bureau (tr.). Moscow Univ. Press, 1960.

LANYON, W. E., and TAVOLGA, W. N. (eds.). 1960. Animal sounds and communication. Washington, D.C.: *Amer. Inst. Biol. Sci. Publ.*, No. 7.

LORENZ, K. 1937. The companion in the bird's world. *Auk*, **54**, 245–273. (Tr. from: 1935, Der Kumpan in der Umwelt des Vögels, *Jour. f. Ornith.*, **83**, 137–214.)

MAIER, N. R. F. 1949. *Frustration.* New York: McGraw-Hill (Ann Arbor, Mich.: Univ. of Michigan Press, 1961).

MORGAN, C. T., and STELLAR, E. 1950. *Physiological psychology.* New York: McGraw-Hill.

MUNN, N. L. 1938. *Psychological development.* New York: Houghton Mifflin.

PAVLOV, I. 1927. *Conditioned reflexes.* London: Oxford Univ. Press (New York: Dover Publications, Inc., 1960).

ROE, ANNE, and SIMPSON, G. G. (eds.). 1958. *Behavior and evolution.* New Haven, Conn.: Yale Univ. Press.

ROEDER, K. (ed.). 1953. *Insect physiology.* New York: Wiley. (Chapters on sensitivity, ecology, and behavior.)

SCHILLER, CLAIRE H., and LASHLEY, K. S. 1957. *Instinctive behavior.* New York: International Universities Press.

SCHNEIRLA, T. C. 1950. The relationship between observation and experimentation in the field study of behavior. *Ann. N.Y. Acad. Sci.*, **51**, 1022–1044.

———. 1952. A consideration of some conceptual trends in comparative psychology. *Psychol. Bull.*, **49**, 559–597.

———. 1959. Psychology, comparative. *Encyclopaedia Britannica.* (1948.)

SCOTT, J. P. 1958. *Animal behavior.* Chicago: Univ. of Chicago Press (New York: Doubleday, 1963).

SKINNER, B. F. 1939. *The behavior of organisms.* New York: Appleton-Century.

SPENCE, K. W. 1956. *Behavior theory and conditioning.* New Haven, Conn.: Yale Univ. Press.

STEVENS, S. S. (ed.). 1951. *Handbook of experimental psychology.* New York: Wiley.

STONE, C. P. (ed.). 1951 (rev. ed.). *Comparative psychology.* New York: Prentice-Hall.

SYMPOSIUM SINGER-POLIGNAC (Paris, 1954). 1956. *L'instinct dans le comportement des animaux et de l'homme.* Paris: Masson et Cie. (Articles by von Frisch, Hediger, Lehrman, Lorenz, Piéron, Schneirla, and others.)

TEMBROCK, G. 1961. *Verhaltenforschung: eine Einführung in die Tier-Ethologie.* Jena: G. Fischer Verlag.

THORPE, W. H., and ZANGWILL, O. L. (eds.). 1961. *Current problems in animal behavior.* Cambridge: Cambridge Univ. Press.

TINBERGEN, N. 1951. *The study of instinct.* London: Oxford Univ. Press.

TOLMAN, E. C. 1958. *Behavior and psychological man.* Berkeley, Calif.: Univ. of California Press.

WALLS, G. L. 1942. The vertebrate eye and its adaptive radiation. Bloomfield Hills, Mich.: *Cranbrook Inst. of Science Bull.*, No. 19, 785 pp.

WARDEN, C. J., JENKINS, T. N., and WARNER, L. *Comparative psychology.* 1935, vol. 1, Principles and Methods; 1936, vol. 3, Vertebrates; 1940, vol. 2, Invertebrates. New York: Ronald Press.

WATERS, R. H., RETHLINGSHAFER, D. A., and CALDWELL, W. E. 1960. *Principles of comparative psychology.* New York: McGraw-Hill.

WERNER, H. 1950. *Comparative psychology of mental development.* New York: Harper.

WOLFSON, A. (ed.). 1955. *Recent studies in avian biology.* Urbana, Ill.: Univ. of Illinois Press. (Chapters on behavior by Emlen, Farner, Griffin, Davis, and others.)

YOUNG, P. T. 1961. *Motivation and emotion.* New York: Wiley.

YOUNG, W. C. (ed.). 1961 (3d ed.). *Sex and internal secretions.* Baltimore: Williams & Wilkins. (Chapters by Beach, Lehrman, and Young on behavior.)

AUTHOR INDEX

SUPPLEMENTARY AUTHOR INDEX

SUBJECT INDEX

F

G

H

N

O

P

R

T

SUPPLEMENTARY SUBJECT INDEX

Psychology

YOGA: A SCIENTIFIC EVALUATION, Kovoor T. Behanan. A complete reprinting of the book that for the first time gave Western readers a sane, scientific explanation and analysis of yoga. The author draws on controlled laboratory experiments and personal records of a year as a disciple of a yoga, to investigate yoga psychology, concepts of knowledge, physiology, "supernatural" phenomena, and the ability to tap the deepest human powers. In this study under the auspices of Yale University Institute of Human Relations, the strictest principles of physiological and psychological inquiry are followed throughout. Foreword by W. A. Miles, Yale University. 17 photographs. Glossary. Index. xx + 270pp. 5⅜ x 8. T505 Paperbound **$2.00**

CONDITIONED REFLEXES: AN INVESTIGATION OF THE PHYSIOLOGICAL ACTIVITIES OF THE CEREBRAL CORTEX, I. P. Pavlov. Full, authorized translation of Pavlov's own survey of his work in experimental psychology reviews entire course of experiments, summarizes conclusions, outlines psychological system based on famous "conditioned reflex" concept. Details of technical means used in experiments, observations on formation of conditioned reflexes, function of cerebral hemispheres, results of damage, nature of sleep, typology of nervous system, significance of experiments for human psychology. Trans. by Dr. G. V. Anrep, Cambridge Univ. 235-item bibliography. 18 figures. 445pp. 5⅜ x 8. S614 Paperbound **$2.35**

EXPLANATION OF HUMAN BEHAVIOUR, F. V. Smith. A major intermediate-level introduction to and criticism of 8 complete systems of the psychology of human behavior, with unusual emphasis on theory of investigation and methodology. Part I is an illuminating analysis of the problems involved in the explanation of observed phenomena, and the differing viewpoints on the nature of causality. Parts II and III are a closely detailed survey of the systems of McDougall, Gordon Allport, Lewin, the Gestalt group, Freud, Watson, Hull, and Tolman. Biographical notes. Bibliography of over 800 items. 2 Indexes. 38 figures. xii + 460pp. 5½ x 8¾.
T253 Clothbound **$6.00**

SEX IN PSYCHO-ANALYSIS (formerly CONTRIBUTIONS TO PSYCHO-ANALYSIS), S. Ferenczi. Written by an associate of Freud, this volume presents countless insights on such topics as impotence, transference, analysis and children, dreams, symbols, obscene words, masturbation and male homosexuality, paranoia and psycho-analysis, the sense of reality, hypnotism and therapy, and many others. Also includes full text of THE DEVELOPMENT OF PSYCHO-ANALYSIS by Ferenczi and Otto Rank. Two books bound as one. Total of 406pp. 5⅜ x 8.
T324 Paperbound **$1.85**

BEYOND PSYCHOLOGY, Otto Rank. One of Rank's most mature contributions, focussing on the irrational basis of human behavior as a basic fact of our lives. The psychoanalytic techniques of myth analysis trace to their source the ultimates of human existence: fear of death, personality, the social organization, the need for love and creativity, etc. Dr. Rank finds them stemming from a common irrational source, man's fear of final destruction. A seminal work in modern psychology, this work sheds light on areas ranging from the concept of immortal soul to the sources of state power. 291pp. 5⅜ x 8. T485 Paperbound **$2.00**

ILLUSIONS AND DELUSIONS OF THE SUPERNATURAL AND THE OCCULT, D. H. Rawcliffe. Holds up to rational examination hundreds of persistent delusions including crystal gazing, automatic writing, table turning, mediumistic trances, mental healing, stigmata, lycanthropy, live burial, the Indian Rope Trick, spiritualism, dowsing, telepathy, clairvoyance, ghosts, ESP, etc. The author explains and exposes the mental and physical deceptions involved, making this not only an exposé of supernatural phenomena, but a valuable exposition of characteristic types of abnormal psychology. Originally titled "The Psychology of the Occult." 14 illustrations. Index. 551pp. 5⅜ x 8. T503 Paperbound **$2.00**

THE PRINCIPLES OF PSYCHOLOGY, William James. The full long-course, unabridged, of one of the great classics of Western literature and science. Wonderfully lucid descriptions of human mental activity, the stream of thought, consciousness, time perception, memory, imagination, emotions, reason, abnormal phenomena, and similar topics. Original contributions are integrated with the work of such men as Berkeley, Binet, Mills, Darwin, Hume, Kant, Royce, Schopenhauer, Spinoza, Locke, Descartes, Galton, Wundt, Lotze, Herbart, Fechner, and scores of others. All contrasting interpretations of mental phenomena are examined in detail — introspective analysis, philosophical interpretation, and experimental research. "A classic," JOURNAL OF CONSULTING PSYCHOLOGY. "The main lines are as valid as ever," PSYCHOANALYTICAL QUARTERLY. "Standard reading . . . a classic of interpretation," PSYCHIATRIC QUARTERLY. 94 illustrations. 1408pp. 2 volumes. 5⅜ x 8. Vol. 1, T381 Paperbound **$2.50**
Vol. 2, T382 Paperbound **$2.50**

THE DYNAMICS OF THERAPY IN A CONTROLLED RELATIONSHIP, Jessie Taft. One of the most important works in literature of child psychology, out of print for 25 years. Outstanding disciple of Rank describes all aspects of relationship or Rankian therapy through concise, simple elucidation of theory underlying her actual contacts with two seven-year olds. Therapists, social caseworkers, psychologists, counselors, and laymen who work with children will all find this important work an invaluable summation of method, theory of child psychology. xix + 296pp. 5⅜ x 8. T325 Paperbound **$1.75**